Design of VLSI Gate Array ICs

Ernest E. Hollis

Project Manager, Custom ELSI
Sanders Associates, Inc.
Nashua, NH

Lecturer
State-of-the-Art Program
Northeastern University
Boston, MA

Prentice-Hall, Inc., Englewood Cliffs NJ 07632

Hollis, Ernest E., (date)
  Design of VLSI gate array ICs.

  Bibliography: p.
  Includes index.
  1. Integrated circuits—Very large scale
integration—Design and construction. I. Title.
TK7874.H64   1986   621.395   86-12189
ISBN 0-13-201930-2

Editorial/production supervision: Gretchen K. Chenenko
Cover design: 20/20 Services
Manufacturing buyer: Gordon Osbourne

Cover photo courtesy of Mentor Graphics

The author believes that the information contained in this book is correct. However, no liability of any kind is assumed by the author or publisher for errors or omissions. In order to make the presentation of material clearer, it has sometimes (although rarely) been altered. Before using the information contained herein for design or other purposes, the reader should check with the appropriate vendors for the latest technical and policy information. No warranty is expressed or implied that any of the information contained herein is not protected by patent or other laws.

© 1987 by Ernest E. Hollis

All rights reserved. No part of this book may be
reproduced, in any form or by any means,
without permission in writing from the publisher and author.

Printed in the United States of America

10 9 8 7 6 5 4 3 2 1

ISBN 0-13-201930-2   025

PRENTICE-HALL INTERNATIONAL (UK) LIMITED, *London*
PRENTICE-HALL OF AUSTRALIA PTY. LIMITED, *Sydney*
PRENTICE-HALL CANADA INC., *Toronto*
PRENTICE-HALL HISPANOAMERICANA, S.A., *Mexico*
PRENTICE-HALL OF INDIA PRIVATE LIMITED, *New Delhi*
PRENTICE-HALL OF JAPAN, INC., *Tokyo*
PRENTICE-HALL OF SOUTHEAST ASIA PTE. LTD., *Singapore*
EDITORA PRENTICE-HALL DO BRASIL, LTDA., *Rio de Janeiro*

This book is dedicated

to my late beloved Dad, Mr. Ray Chase Hollis, Sr.,
and Mother, Mrs. Lucinda Foley Hollis

and to my beloved family

Richard Ray, David Matthew, and Susan McGraw Hollis
with much love and affection.

Contents

PREFACE   xvii

1 INTRODUCTION   1

    1.0   Introduction   1
    1.1   What Are Gate Arrays?   6
            1.1.1 Macros, 11
    1.2   Purpose of Customization   12
    1.3   Other Functions   12
            1.3.1 Memories and Microprocessors on Gate Arrays, 12
            1.3.2 Mixed Analog and Digital Functions in Gate Arrays, 13
    1.4   Spectrum of Possibilities   14
            1.4.1 Computer-Aided Design, 14
    1.5   Generations of Gate Arrays   15
    1.6   Two Frequently Used Terms That Have Multiple Meanings   16
            1.6.1 Cell, 16
            1.6.2 Gate, 16
    1.7   Other Programmable Methodologies   16

1.8   Gate Arrays versus Pure Custom and Standard Cells; Silicon Compilation   18

       1.8.1 Standard Cell Approach, 18
       1.8.2 Relative Die Sizes between Standard Cell and Gate Arrays, 18
       1.8.3 Die Size versus Yield, 19
       1.8.4 Benefits of Standard Cells, 20
       1.8.5 Converting from Gate Array to Standard Cell, 20
       1.8.6 Gate Arrays versus Standard ICs (Not Standard Cell ICs) and Full Custom ICs, 21
       1.8.7 Silicon Compilers, 21

1.9   A Guess at the Future   22

1.10   Why Use Gate Arrays?   23

       1.10.1 Benefits of Gate Arrays, 24
       1.10.2 Technical Advantages of Gate Arrays, 25
       1.10.3 Economic Advantages of Gate Arrays, 28
       1.10.4 Other Benefits of Using Custom LSI in General and Gate Arrays in Particular, 30

1.11   Why Not Use Gate Arrays?   32

1.12   Examples of Gate Array Use   33

1.13   VHSIC Gate Arrays   35

1.14   Standards for Gate Arrays and for CAD Design Tools   36

1.15   Summary   36

       References   36

2 GATE ARRAY TECHNOLOGIES   38

2.0   Introduction   38

2.1   Overview of Silicon Technologies Used in Logic   39

       2.1.1 TTL De Facto Standard, 40

2.2   Overview of Gate Array Technologies   41

       2.2.1 Combinations of Technologies—BIMOS and GaAs/Si, 41
       2.2.2 The Term "Substrate," 43

2.3   Bipolar Technologies   43

       2.3.1 Bipolar Structures, 43
       2.3.2 ISL/STL Technology, 47
       2.3.3 ECL Technology, 50

2.4   Fundamentals of MOSFETs   51

       2.4.1 Terminal Connections, 52

- 2.4.2 Enhancement versus Depletion Mode, 52
- 2.4.3 Symbols Used, 53
- 2.4.4 Cross-Sectional View, 54
- 2.4.5 Transistor Action, 55
- 2.4.6 Logic Voltages Pertinent to CMOS Devices, 56
- 2.4.7 Sizes of MOS Transistors, 56
- 2.4.8 Punch-Through and Avalanche, 58
- 2.4.9 Scaling of MOS Transistors, 59
- 2.4.10 Self-Alignment and Silicon Gate, 60
- 2.4.11 Simple Logic Structures Using MOS Transistors, 67

2.5 CMOS and Its Permutations   72

- 2.5.1 Variants of CMOS, 73
- 2.5.2 Latch-Up, 76
- 2.5.3 Propagation Delays, 77
- 2.5.4 Protection of Inputs from Electrostatic Discharge and Other Voltages, 84
- 2.5.5 A Useful Feature of Battery-Operated CMOS, 85
- 2.5.6 Interfacing CMOS to TTL, 85
- 2.5.7 Transition to CMOS Logic, 85

2.6 Technology of Interconnects   87

- 2.6.1 Distribution of Interconnects, 89

2.7 General Comparison among Technologies   91

- 2.7.1 Speed–Power Product: ECL versus CMOS, 93
- 2.7.2 Power Dissipation: CMOS versus Bipolar, 94
- 2.7.3 Drive Capabilities of Gate Array Technologies, 95
- 2.7.4 Differences between Designing with CMOS and ECL Technologies and Gate Arrays in Particular, 95

2.8 Isolation Methods   97

- 2.8.1 Oxide Isolation versus Junction Isolation for Bipolar Structures, 98
- 2.8.2 CDI Technique, 99
- 2.8.3 Oxide Isolation versus Junction Isolation for CMOS Structures, 101
- 2.8.4 Use of Insulating Substrates, 101
- 2.8.5 Dielectric Isolation, 101

2.9 Ring Oscillators   102

2.10 New Developments   104

- 2.10.1 Silicon-on-Insulator Techniques, 104
- 2.10.2 Vertically Stacked CMOS, 105
- 2.10.3 GaAs, 106
- 2.10.4 GaAs on Silicon, 110

2.11 Summary   111

Appendix 2A: Rudiments of Semiconductors   111

3 GATE ARRAY GRID SYSTEMS 118

3.0 Introduction 118

3.1 Definition of Gate Array Grid Systems 118

3.2 Single-Layer Metal Silicon Gate CMOS Arrays 120

3.2.1 Detailed Example, 120
3.2.2 CDI's Grid Structure, 127
3.2.3 SPI's Grid Structure, 127
3.2.4 MITEL's Grid Structure, 129

3.3 Two-Layer Metal CMOS Arrays 133

3.3.1 STC's Grid Structure, 133

3.4 Value of the Break in the Middle of the Gate 134

3.5 ISL Grid Structure 134

3.6 ECL Gate Array Structure 135

3.7 Guided Tour of the Inside of a Gate Array Chip 140

3.7.1 Latch-Up Protection, 142

3.8 Structures of Gate Arrays Employing the Gate Isolation Technique 143

3.9 Study of Yields of Different Gate Array Cell Structures 146

3.10 Summary 146

4 OVERVIEW OF GATE ARRAY DESIGN 149

4.0 Introduction 149

4.1 Summary of the Flow 152

4.1.1 The Starting Point—A Good Specification, 152
4.1.2 Definition of Other Tasks, 152
4.1.3 Logic Verification, 154

4.2 Mask Design 155

4.2.1 The "Floor Plan," 156
4.2.2 Methods of Mask Design, 158
4.2.3 Checking the Mask Design, 165

4.3 Processing, Packaging, and Testing 167

4.4 Times Required for the Different Steps 167

4.5 Vendor Interaction 168

	4.5.1	User Involvement in the Design, 169
	4.5.2	The Vendor's Point of View, 169
	4.5.3	Sell-Off, 171
	4.5.4	Where Can Changes Be Made? 171

4.6   Configuration Control of the Design   172

4.7   Example to Be Used in This Book   173

4.8   Technology Conversion   173

      4.8.1   Conversion of TTL to CMOS, 174
      4.8.2   Conversion of TTL to ISL/STL, 178

4.9   If You Make a Mistake . . .   179

4.10 Summary   180

5 FEASIBILITY, CIRCUIT ANALYSIS, AND PARTITIONING   181

5.0   Introduction   181

5.1   Feasibility   181

5.2   Technical Feasibility   183

      5.2.1   Technology Choices, 184
      5.2.2   Technology Selection Based on Logic Function, 187
      5.2.3   Size of the Array—Cell or Transistor Count, 187
      5.2.4   Laying Out on a Larger Chip Initially, 189
      5.2.5   Pin-Outs, 191
      5.2.6   Gate Array Families, 193
      5.2.7   Excluding Circuit Elements, 194
      5.2.8   Power Dissipation of the Package, 194
      5.2.9   Temperature Effects, 197
      5.2.10 Analysis of Propagation Delays, 200
      5.2.11 Freedoms and Restrictions of Gate Arrays, 203
      5.2.12 Retrofitting Existing Parts, 206

5.3   Partitioning   207

      5.3.1   Incorporating Two Similar Designs into One Circuit on One Gate Array, 204
      5.3.2   Partitioning a Circuit into Multiple Gate Arrays and across Printed Circuit Board Boundaries, 210

5.4   Distribution of Clock and Other High-Fan-Out Signals   211

      5.4.1   Single-Driver Approach, 211
      5.4.2   Problems with the Single-Driver Approach, 211
      5.4.3   Tree-Driver Approach, 212
      5.4.4   Typical Mistakes, 212
      5.4.5   CMOS Drivers, 212

5.5 System Partitioning—Rent's Rule 213

    5.5.1 Applying Rent's Rule to System Partitioning into Gate Arrays, 215

5.6 Work Aimed at Minimizing the Number of Gate Arrays Required by a System 215

5.7 Cost and Time Drivers in Gate Arrays 216

    5.7.1 Time Drivers, 216
    5.7.2 Time to Study Computer Printouts, 217
    5.7.3 Regularity of the Circuit, 217
    5.7.4 Adequacy of the Macro Library, 217

5.8 Summary 218

6 CAD FOR GATE ARRAYS 220

6.0 Introduction 220

6.1 CAD Tools Most Needed and Used 221

6.2 Simulators 222

    6.2.1 Overview, 222
    6.2.2 Logic Simulators versus Circuit Simulators, 222
    6.2.3 Typical Logic Simulators, 223

6.3 Circuit Simulation 224

    6.3.1 Uses and Problems of Circuit Simulation, 224
    6.3.2 Coding and Running the SPICE Simulator, 225

6.4 Checking the Layout of the Mask versus the Schematic 237

6.5 Schematic Entry 237

    6.5.1 Overview, 237
    6.5.2 Naming of Signals and Nodes, 241
    6.5.3 Netlist Extraction and Simulator Interface, 241
    6.5.4 Creation of New Library Elements, 241

6.6 Tegas Logic Simulator 242

    6.6.1 Purpose of This Section, 242
    6.6.2 Uses of Tegas, 243
    6.6.3 Features of Tegas, 243
    6.6.4 Makeup of Tegas5, 245
    6.6.5 Important Interfaces to and from Tegas, 245
    6.6.6 Libraries in Tegas, 246
    6.6.7 Comparative Analysis Using Subdirectories, 247
    6.6.8 Examples to be Used, 248
    6.6.9 Nonuniform Nesting, 248

6.7 General Flow of Logic Simulation 248

    6.7.1 Network Encoding of the Preprocessor, 249
    6.7.2 Simulation, 257

6.8 Summary 267

7 DESIGN FOR TESTABILITY AND TEST PATTERN DEVELOPMENT   269

7.0 Introduction 269

7.1 Overview of the Flow 271

7.2 Design for Testability 272

    7.2.1 Ad hoc Techniques, 273
    7.2.2 Structural Techniques, 273
    7.2.3 Comments on Methods for Design of Testable Circuits, 277

7.3 Testability Analysis Using COPTR 277

7.4 Important Definitions 281

    7.4.1 Faults and the "Stuck At" Concept, 281
    7.4.2 Master Fault File, 281
    7.4.3 Test Vectors, 283
    7.4.4 Detectability, 283
    7.4.5 "Bad Machine," 283

7.5 Fault Analysis Tools 283

    7.5.1 DFS Program, 283
    7.5.2 Mode 5, 284
    7.5.3 Path Sensitization, 285
    7.5.4 Outputs from Mode 5, 285
    7.5.5 One Lack, 288

7.6 Subtleties 288

    7.6.1 Faults That Cannot Be Detected, 288
    7.6.2 Decreasing the CPU Time of Fault Grading, 291

7.7 Effects of Fault Coverage on Yield 292

7.8 Test Vector Generation and Use 292

    7.8.1 Automatic Test Vector (Pattern) Generator, 293

7.9 Iterative Exhaustive Test Pattern Generation 294

7.10 Tester Interface 294

7.11 Suitability of the "Stuck-At" Model 295

7.12 Summary 295

8 WORKSTATIONS AND OTHER TOOLS 297

- 8.0 Introduction 297
- 8.1 Overview of Interactive Graphics Systems 298
  - 8.1.1 Definition of an IGS, 298
  - 8.1.2 Properties of an IGS, 298
  - 8.1.3 Mask Layout, 300
  - 8.1.4 Functions of IGSs, 300
  - 8.1.5 Digitizing, 304
  - 8.1.6 Example of a Coordinate List, 307
- 8.2 Workstations 311
  - 8.2.1 Problem of Definition, 312
  - 8.2.2 Two Categories, 312
  - 8.2.3 Parts of a CAD Workstation, 313
  - 8.2.4 Interaction with Gate Array Vendors, 314
  - 8.2.5 Networking, 314
  - 8.2.6 Comparison with the Second Category of Workstations, 315
  - 8.2.7 Benefits of Large Amounts of RAM, 315
  - 8.2.8 Typical Operation, 316
  - 8.2.9 Cursor Movement Devices, 322
  - 8.2.10 Menu Items, 322
  - 8.2.11 Correct by Construction, 322
  - 8.2.12 Hierarchical Data Base, 324
  - 8.2.13 Electrical Trace Function, 324
  - 8.2.14 Comparisons among Workstations, 325
  - 8.2.15 PCs Used as Workstations, 325
  - 8.2.16 Simulation Using Physical Modeling, 327
- 8.3 Placement and Routing 327
  - 8.3.1 Manual Routing Exercises, 328
  - 8.3.2 Automatic Placement and Routing Programs, 328
  - 8.3.3 Three Non-gate Array Layout Techniques, 339
- 8.4 Designing Remotely 344
  - 8.4.1 Advantages, 344
  - 8.4.2 Problems That Can Occur, 345
- 8.5 Special-Purpose Simulation Processors 347
  - 8.5.1 Zycad LE Series, 348
  - 8.5.2 Daisy Megalogician, 351
  - 8.5.3 Economics, 352
- 8.6 Hard-Copy Output 353
  - 8.6.1 Alphanumeric versus Graphic, 353
  - 8.6.2 Printers, 354
  - 8.6.3 Graphics Plotters, 355

	8.6.4 Electro-Static Plotters, 355
	8.6.5 Photoplotters, 355
8.7	Summary 356

9 DESIGN OF CMOS MACROS   358

9.0   Introduction   358

       9.0.1   Steps Involved, 359
       9.0.2   CMOS Macros, 360

9.1   FET Diagrams   363

9.2   Layout Methods   363

9.3   Inverters   364

9.4   NAND and NOR Gates   365

       9.4.1   Benefit of Lower "On" Resistance of the NFET, 365
       9.4.2   Disadvantages, 367
       9.4.3   NOR Gates, 368

9.5   Student Exercises   369

9.6   Isolation of Elements within a Cell   370

9.7   Phantom Transistors   372

9.8   Four-Input Gates   373

9.9   AND-OR-INVERT and OR-AND-INVERT Structures   373

       9.9.1   Symmetrical versus Asymmetrical, 376
       9.9.2   An Important Property of CMOS, 377
       9.9.3   Layers and Strings of FETs, 377
       9.9.4   Importance of Tying Node Capacitance to Power Supply Rails, 379
       9.9.5   Degenerate versus Nondegenerate AOIs, 379
       9.9.6   AOIs in Complex Structures, 379
       9.9.7   Chip Layouts of AOIs, 380
       9.9.8   Propagation Delays of AOIs, 384

9.10   Uses of AOIs   386

       9.10.1   Scratch-Pad Memory, 386
       9.10.2   Detection Circuits, 386
       9.10.3   Multiplexing, 386
       9.10.4   Counter Circuits, 387
       9.10.5   An Important Task AOIs Cannot Do, 387

9.11   Transmission Gates   389

       9.11.1   Disadvantages of TGs, 389
       9.11.2   Operation, 390

	9.11.3 Layout, 391
	9.11.4 Advantages in Terms of Space and Loading to a TG Pair, 391
	9.11.5 Operation of EX-OR Made from a TG Pair, 394
	9.11.6 TGs in Latches and Flip-Flops, 395
	9.11.7 Multiplexers Made from TGs, 397
9.12	Floatable Drivers (Clocked Inverters) 399
9.13	Input Structures 400
9.14	Output Structures 403
9.15	"On" Resistance Calculations 405
9.16	Tristate Buffers 408
9.17	Driving the Output Devices 410
	9.17.1 General Guidelines for Driving Buffers, 410
9.18	Paralleling of Structures 410
	9.18.1 Paralleling AOIs, 414
	9.18.2 Paralleling Gates Other Than Inverters, 414
	9.18.3 Other Benefits and Problems with Paralleling, 414
9.19	Schmitt Triggers 414
	9.19.1 One Shots (Monostable Multivibrators), 416
9.20	Use of Inverters for Delay and Other Purposes 416
	9.20.1 Use of Inverters for Delay, 416
	9.20.2 Use of Inverters and Logic Gates for Pulse Generation, 416
9.21	Routing Subtleties 419
9.22	Input Protection Networks 421
9.23	Summary 422

10 BIPOLAR MACROS   423

10.0	Introduction 423
10.1	ISL and STL 424
	10.1.1 Wire ANDing, 427
	10.1.2 Use of Wire ANDing to Obtain Logic Functions, 427
	10.1.3 Conversion of a Circuit to ISL or STL, 429
	10.1.4 Current Hogging, 434
	10.1.5 Array Grid Structure, 434
	10.1.6 Examples of Macro Design Using Internal Array, 435

10.2 ECL 440

    10.2.1 Mixed ECL and TTL I/Os on the Same ECL Gate Array Chip, 440
    10.2.2 Speed–Power of ECL Compared to That of CMOS, 442
    10.2.3 Two Major Families of ECL Parts, 444
    10.2.4 User–Designed ECL Macros, 445
    10.2.5 Basic ECL Structure, 445
    10.2.6 Delay–Saving Methods in ECL, 447
    10.2.7 Examples of ECL Gate Array Cells and Common Functions Made from Those Cells, 449
    10.2.8 I/O Structures, 456
    10.2.9 Equalizing Power on the Chip, 457

10.3 Summary 458

11 PROCESSING, PACKAGING, AND TESTING 459

11.0 Introduction 459

11.1 Making the Mask 460

    11.1.1 Runout, 461
    11.1.2 Mask Repair, 461

11.2 Processing 462

    11.2.1 Clean Rooms, 462
    11.2.2 The Process Itself, 464

11.3 Wafer Test 470

    11.3.1 Wafer Probing, 470

11.4 Packaging 473

    11.4.1 Chip Carriers, 475
    11.4.2 Pin Grid Arrays, 477
    11.4.3 High-Wattage Circuits, 478
    11.4.4 Die Bonding, 478

11.5 Test Circuits on the Gate Array Die and Wafer 480

11.6 Tester Interface 480

    11.6.1 Tester Interface Programs, 481
    11.6.2 Time Sets, 481
    11.6.3 High-Speed Testers, 481

11.7 Summary 482

APPENDIX A   REPRESENTATIVE GATE ARRAY VENDORS AND PRODUCTS   483

APPENDIX B	MAJOR WORKSTATION VENDORS	486
APPENDIX C	SIMULATOR CPU TIMES	487
APPENDIX D	TRADEMARKS	488
SUPPLEMENTAL READING		489
INDEX		502

Preface

Gate arrays are an important tool in the system designer's toolbox. As such, people in a wide variety of disciplines need to know about them. This book is aimed primarily at two somewhat disparate groups of people. The first group consists of system designers. The second group is made up of the decision makers and holders of development dollars who must decide whether to proceed with the gate array or some other approach.

The author has personally designed and been closely involved with the design of a large number of gate array chips. This involvement began in the early days of gate arrays and has continued to the present. The tools used have ranged from the almost completely manual to the almost completely computer aided. In addition, the author has also done designs of pure custom chips.

The author's background also includes design, implementation, testing (including a number of "firsts" in both the experimental and theoretical areas), and program or project management of systems in such diverse fields as fiber and electrooptics, microwaves, and modems. As such, he is not wedded to gate arrays as the "be-all, end-all" methodology for all system designs. Moreover, he is highly aware of the real-world problems that are faced in implementing systems of which components, such as gate arrays, are a part. They may be an important part, but they are only a part; nevertheless, their costs, risks, and development times must be weighed against the needs of the overall system. Time to market an economically and functionally competitive product is crucial in today's arena.

This philosophy, reflected in the content of the book, is somewhat different from that of the purist in the field who has dealt only with components or who has the option of dealing only with hypothetical problems that have neat, nicely mathematically quantifiable answers.

This is not a book on theory. However, one of the author's intents is to give readers sufficient background to enable them to understand the literature on gate arrays. Therefore, a number of concepts, such as basics of processing, scaling of devices, and Rent's rule, are introduced with appropriate references to the literature.

Another intent is to make the reader aware of new developments that may impact the design and implementation of gate arrays. Thus, topics such as joint gate CMOS, silicon-on isolator, GaAs, and iterative exhaustive test pattern generation are mentioned, again with appropriate references to the literature.

This book is an outgrowth of two types of courses on gate arrays given by the author over the years plus a course on the Tegas logic simulator also developed by the author. Portions of the latter course have been incorporated into the gate array course.

The gate array courses are of two types. One is an overview for managers that is typically given in an intensive one-day seminar. The other is a 22-hour course given in the evening. Even the longer course covers only a small portion of the available material. These courses have been taught in the State-of-the-Art program at Northeastern University by the author since 1980 and at Sanders Associates, Inc., Nashua, NH, prior to that. A portion has also been taught at the Hellman Associates seminars.

Because of the rapidly changing nature of the gate array business, the author usually teaches the courses differently each time despite the fact that the courses are given in successive quarters of the school year.

As part of the long design course (22 hours of instruction plus appropriate homework), the "students" visit Sanders Associates, Inc. to see first hand what had been talked about. They receive demonstrations of schematic entry, auto place and route, macro design on an IGS, wafer probers, and the laboratory that the author designed for final processing of gate array wafers. Sanders also allows them to try out a very brief simulation example. A demonstration of a workstation is also given.

The nature of the student taking the course, especially the long design course (which is really too short), has changed. More and more frequently students enroll who not only know what a gate array is, but have already designed one at one of the local design centers. The author also more frequently finds students in fields other than design who are interested in the subject.

The most extreme case of this occurred during the planning for the first all-day seminar on gate arrays in the Boston area a number of years ago. The seminar announcements were intended to go out to technical managers. Instead, because of inadvertently selecting the wrong computer code for the Northeastern University mailing list, the announcements were sent to bank executives. Four of the latter actually signed up!

The book is organized into five sections. The first section deals with basic concepts. It is expected that persons with widely varying backgrounds will have use for varying amounts of the material contained in the book. The purpose of the first three chapters is to provide a basis for the remainder of the book.

The next two chapters go into the flow of the gate array design process itself from a top-level viewpoint. Different options available to the user are explained. The distinction between design with macros and design with transistors is made, and the pros and cons of each are given.

The next three chapters go into CAD and other tools used in gate array design. Included are manual options that can be used and discussions of when such a use is suitable. The subject of design for testability is included as a separate chapter because of its importance. Sample exercises of design with macros are provided.

The next two chapters deal with design of the macros themselves. Alternatively, these two chapters could be called Design with Transistors. Although many users will never design a macro, the author has found that knowledge of how macros are designed is extremely useful. The user who knows how to design with transistors is much more aware of subtleties in using such macros.

The last chapter deals with processing. Although most designers have little interest in this subject, knowledge in this area is helpful in vendor selection, and in being aware of some of the limitations and pitfalls that can occur in making the device, and is really necessary for a thorough understanding of the subject.

The book can be used in several different ways depending on the needs and backgrounds of the students and the amount of classroom time available. Chapters 1 and 4 and portions of Chapters 2, 3, 5, and 6, 7, and 8 are probably central to any discussion on gate arrays. The mix of the portions will depend on the needs and backgrounds of the audience and on the time available. A long design course would include Chapters 9, 10, and 11.

The general flow of the writing involves first answering the "what" and the "why is it important" questions before answering the "how to do it" questions.

ACKNOWLEDGMENTS

At one time or another, the author has had helpful conversations with many people. It would be very difficult to acknowledge all of them, and the author hopes that those not mentioned will not feel offended.

The first group to be mentioned are the author's students. The author has greatly benefitted over the years by having had some very bright students in classes that he has taught both at Sanders Associates and in the State-of-the-Art program at Northeastern University. The level of the students has ranged from design engineer to vice president and chief engineer, with a few company presidents thrown in for good measure. Their questions and comments have been very useful in shaping this book.

The author was extremely fortunate to have done the early work on gate arrays and custom logic in what eventually became the Sanders Associates, Inc. Microelectronics Center (MEC) under the direction of Ken C. Chin. Helpful discussions were held with Dr. David Meharry, manager of Monolithic Microwave IC Design; Dr. John Heaton, manager of Gallium Arsenide Processing; George Norris, and Andrew Moysenko; Ralph Stevens, MEC operations manager; Frederick Hinchliffe II, manager of CAD/CAE; and Robert Boerstler, all of Sanders Associates, Inc., MEC. Arthur Berlin and the library staff at Sanders procured many needed papers and articles.

A lot of encouragement to write the book came from Thomas E. Woodruff, vice-president of science and technology for Sanders, and William R. Hutchins, manager of Sanders phase 0 VHSIC (Very High Speed Integrated Circuits) program.

The author has also benefitted from discussions at one time or another with a wide variety of people, including Orhan Tozun, vice-president of engineering, International Microcircuits, Inc.; Darouche Samani and Thomas Bonica, GE/Calma; Tim Hammond and Dr. Victor Twaddell, General Dynamics-Pomona; David Goldstein, Stromberg-Carlson; Martin Jurich, Silicon Systems, Inc.; Anthony Anderson and Gary Griffin, AMCC; Douglas Greetham, Raytheon; Charles Stern, Silicon Compilers; Dr. Don McClennan and David Hightower, GE/Intersil; and Dr. Paul Greiling, Hughes Research Laboratory. Rob Walker, vice-president of engineering, and Wilfred Corrigan, chairman of the board of LSI Logic, kindly shared their views on where the gate array industry is heading.

Richard Offerdahl, president, and Alan Gorlick, Zycad, Inc., reviewed the material on the Zycad special-purpose processor. John Bastion of Rockwell reviewed an early version of the circuit simulator section. Ken Lai of Sanders provided the FET model, which the author modified to make the SPICE simulation runs (other than those made at Mentor). Maurice Dumont of Sanders developed the grid of the International Microcircuits, Inc. background layers.

The many organizations that supplied material are acknowledged in the figure captions. A number of groups and individuals made special efforts to supply material. In particular, David Coelho, CAD manager, Silvar-Lisco; Michael P. Gagliardi, design manager, GE/Intersil; Frank Deverse, president, International Microcircuits, Inc.; Robert Lipp, chairman of the board, California Devices, Inc.; Stephen Swerling, vice-president and general manager, Mentor Graphics, CAE Systems, Inc.; Alan Metalak, Universal Semiconductor, Inc.; John Stockdale, Digital Equipment Corp.; and Dr. Satish M. Thatte, Central Research Laboratories of Texas Instruments, Inc., made special photos to either support the book or supplied information, or both, in a timely fashion.

Dr. Jonathan Allen of MIT and Dr. Glen G. Langdon of IBM Research Division reviewed early chapters of the book and made many helpful suggestions as to content, format, and style. Dr. Alan V. Oppenheim of MIT, Harold R. Ward of Raytheon, and Vartan Vartanian of Sanders, authors of books in other fields, gave much useful advice on the writing process itself. Bernard M. Goodwin, executive editor, and Sophie Papanikolaou, production editor, both at Prentice-Hall, were very helpful in innumerable ways in getting the book published.

On a more personal level, the encouragement of my two sons, Richard and David, has kept the effort alive at times when the time demands of other needs might have crushed it. Richard reviewed the first two chapters for general flow and made a number of useful comments. The book is dedicated to them, to my wife, Susan M. Hollis, and to the memory of my beloved Dad, Mr. Ray C. Hollis, Sr., with whom I was privileged to work for many years.

All of these contributions are gratefully acknowledged.

Chapter 1

Introduction

1.0 INTRODUCTION

Gate arrays are an important branch of custom VLSI (Very Large Scale Integration). By 1990 it is estimated that more than half of all semiconductors sold will be semi-custom designs [1] of which gate arrays are a major part. Today, gate arrays outsell standard cell ICs by a 4-to-1 margin. [2] Over 70 vendors currently offer gate arrays, and this number is constantly increasing. Hence, the gate array methodology is and will continue to be a major force in the electronics industry.

Applications of gate arrays cover the gamut from high performance computers, such as the Data General MV 20000 [3] and IBM 3084, to industrial controls, instrumentation, communications, and aerospace/military. [4]

Although gate arrays have been used in high-volume applications, a lot of their usefulness occurs in applications where the quantities are so low as not to be practical for full custom. Although semiconductor manufacturers often compare the costs of a gate array to those of a full custom design, gate array costs are probably most usefully compared to those of the one or more printed circuit boards (PCBs) that the gate array replaces. It is true that the die cost of a gate array is much higher than that of a full custom die. However, in low-volume applications, the nonrecurring development costs are the dominating factors, not the differential piece-part costs. It is here that the gate array has a clear edge over the full custom approach.

The purpose of this chapter is to give an overview of gate arrays and to introduce terms that will be used throughout the book. The meaning of these will become clear in subsequent chapters. Those familiar with semiconductors in general and gate arrays in particular may wish to skip this chapter.

Figure 1.1(a) shows a gate array *die* mounted in a standard 24-pin dual-in-line package (DIP). The gate array die is the small square in the middle of the package *cavity*. The cavity is the larger square to which the horizontal and vertical (in the picture) *fingers* butt. The small wires connect the gate array input/outputs (IOs) to the metal fingers of the cavity. The fingers are connected inside the package to the package pins. This is the mechanism by which the gate array die (or any other semiconductor die) connects to the outside world.

Figure 1.1(b) shows the same package with a lid on. The important message is that it is the gate array die that is being customized, *not* the package. To the outside world, there is no visually discernible difference between a gate array IC and any other custom or standard IC (except for the name of the user organization on the package). Gate arrays are made with standard processes on standard semiconductor manufacturing lines.

One significant difference between a standard IC and a gate array is that the gate array wafer is not completed at the time of initial manufacture. Rather, all the steps which are common to that particular member or to that particular gate array family are done. The gate array wafer is then put to one side to await a customer's design, from which it will then be customized. The latter is done with the last few processing steps.

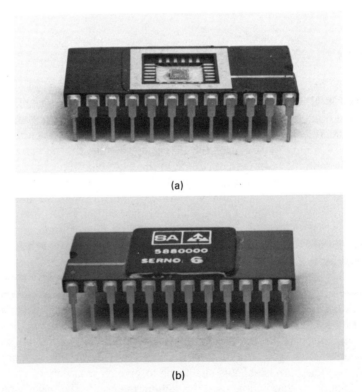

(a)

(b)

Figure 1.1 (a) Gate array die mounted in a 24-pin dip; (b) same package with lid on. Except for company name, the package is indistinguishable from any other standard 24-pin DIP.

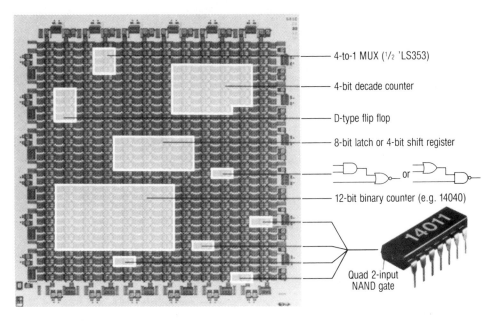

Figure 1.2 Map of very small gate array chip indicating that (a) different macros use different amounts of space, (b) different macros can be placed almost anywhere on the chip as opposed to being in fixed positions, and (c) the transistors in a given part of a gate array can be configured to a wide (almost unlimited) variety of functions. For clarity in presentation, a *very small* (360-cell) array was used. This is the smallest member of a family that includes members with over 5000 cells. (Courtesy of International Microcircuits, Inc.)

An example of the versatility of the gate array is shown in Fig. 1.2, which depicts one of the smallest members of one of IMI's gate array families, with the sizes taken by typical circuit functions outlined in block form. The reader will appreciate that (1) an almost endless variety of functions can be implemented on a gate array, (2) any portion of the array can be configured to make any of an innumerable variety of random logic functions, and (3) the circuit designer has a wide latitude in terms of where the functions can be placed on the chip.

Moreover, because the background structure of the gate array is generally the same for all members of a given gate array *family,* except for size and numbers of elements the same functional elements or even the entire circuit can be used on other arrays in the family. This permits some or all of the design effort for one gate array design to be used for another design, which in turn helps amortize the cost of design.

To describe gate arrays, it is necessary first to give a few definitions. These definitions are expanded on later in the book. The reader should not get the impression that this is a book on processing of semiconductors. It is *not*. It is a book on design. Most logic designers have little interest in processing. However, to fully appreciate the gate array methodology and to become familiar with its use, it is necessary to understand a few basic concepts. Chapter 11 is provided for those who *do* have an interest in how gate array wafers are converted to finished gate array ICs.

A *wafer* contains many *dies*. Figure 1.3 is a picture of a wafer; the small

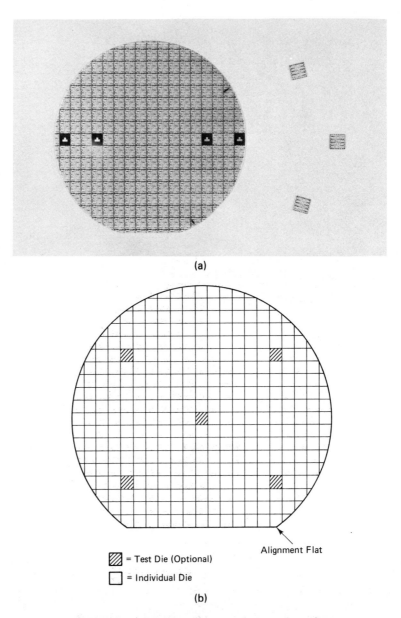

Figure 1.3 (a) Photograph and (b) diagram of a wafer.

rectangles are the dies. A die (sometimes called a chip), when packaged, constitutes an *integrated circuit* (IC). The term *chip* is also used (somewhat loosely but very commonly) to denote a packaged die. Which meaning is intended is generally clear from the text.

A wafer, and, hence, a die, consists of layers of conducting, partly conducting, and nonconducting material. The layers are used to form the active (transistors, diodes) and passive (resistors, capacitors, conductors) components of the die and, hence, of the IC. Definition of each layer is done using a *mask*. The mask contains the *pattern* for that layer replicated for each die on the wafer. This permits each layer to be made at the same time for all dies on the wafer. Whatever is done to the wafer is done simultaneously to all dies on the wafer. The above holds true for any integrated circuit, whether digital or analog, whether gate array or other type of IC.

Figure 1.4 shows ultraviolet light passing through one of the masks to form an image on the wafer. Material such as impurities or oxide, is added or removed after each such masking step. The process is replicated until the wafer is completed to the desired point in the manufacturing cycle. For parts other than *mask-programmable* ICs (e.g., gate arrays and ROMs), the wafer is made entirely. For mask-programmable devices, it is made to the point where metallization will define its function.

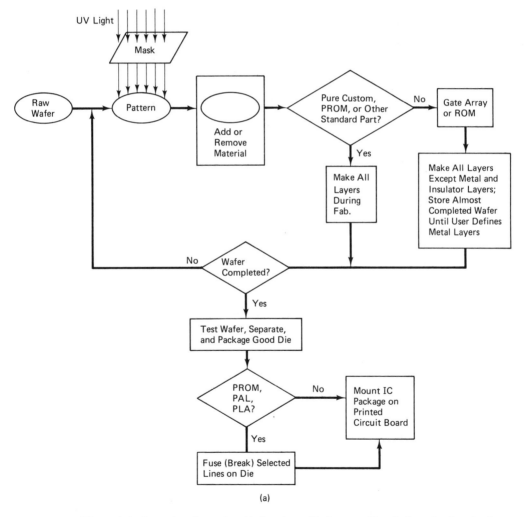

Figure 1.4 Patterning the wafer: (a) flowchart; (b) diagram. (For clarity only ultra-simple patterns are used.)

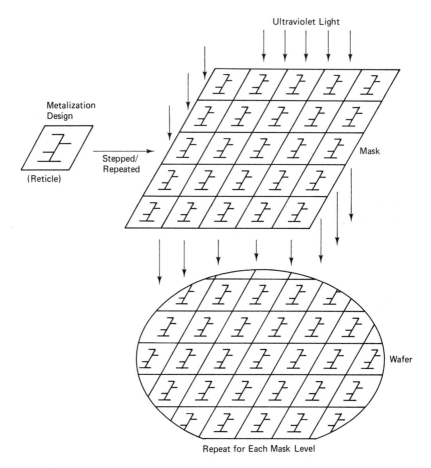

Figure 1.4 (*cont.*)

Two important points are that the patterns of the different mask layers are different and the patterns are aligned with one another.

Figure 1.5 is a sequence of cross-sectional views of the layers of a single transistor pair on a die. The important points of Fig. 1.5 are (1) the layered structure and (2) the metalization layer, the final major layer to be added to the wafer/die. (A protective coating is added after the last metalization layer, but the characteristics of this layer are far less critical than those of the preceding layers.) The reader should not be concerned at this point about not understanding the details of the process itself.

With the foregoing definitions of wafer, die, masks, and layers on a die in hand, gate arrays can now be defined.

1.1 WHAT ARE GATE ARRAYS?

Gate arrays are *semicustom digital integrated circuits,* which are mostly made ahead of time and which are customized to the user's needs by defining one or more layers of metal (via the appropriate masks) on the die itself. The word "customized" means

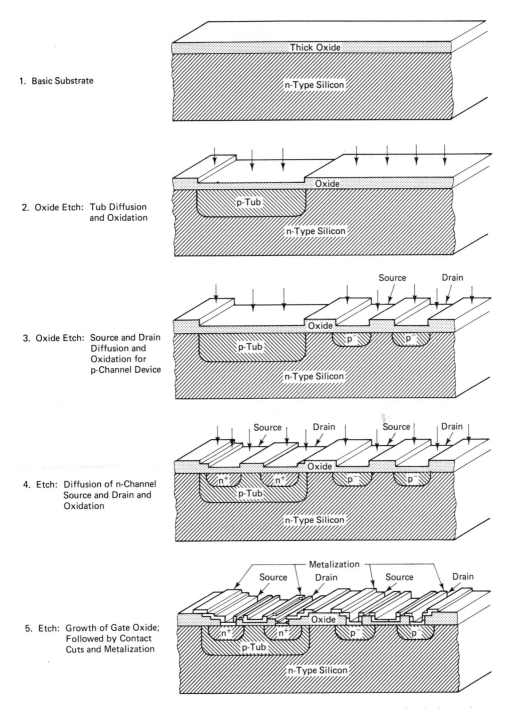

Figure 1.5 Sequence of cross-sectional views of the layers of a single transistor pair on a die. (Courtesy of Integrated Circuit Engineering, Corp.)

that the gate array is configured to perform the same function as that of part or all of a user's circuit. The term "mostly" is used to indicate that the wafer is processed up through the first (and possibly only) layer of metallization. The definitions of these layers are the last steps in the processing of the die itself (except for the addition of a protective coating to the wafer).

Gate array dies are collections of premade transistors placed in fixed positions on the dies together with suitable general-purpose peripheral I/O (input/output) structures and associated bonding pads. A characteristic feature of a gate array is the very large number (over 70,000 for a popular medium-sized array) of *potential* internal contact points. The word "potential" is used because not all elements will be used in any given design and because most of the internal circuit elements have at least two spatially separated contact points. The two or more contact points are usually electrically the same. They are redundant to aid in routing the interconnecting metal lines. Interconnection among the transistors using a subset of these contact points effects the customization.

The term "semicustom" means that the IC can be made into many custom designs (by interconnecting the transistors in different ways) as opposed to being designed specifically for a given user's circuit. It refers to the fact that the structure of the active (transistors, diodes) and passive (resistors, capacitors, diffused or polysilicon underpasses, contact points, bonding pads, power and ground buses, etc.) elements is fixed, and that circuit definition is achieved by routing metal interconnects among the potential circuit element contact points.

Some writers (Hively [5]) include standard cells in the definition of semicustom. Others (Eklund [6]) do not. Hence, the terminology is not uniform at this point.

Semicustom circuits include not only gate arrays but also linear circuits. They also include the *masterslice* category, which may have linear, digital, and a variety of other circuits, such as voltage regulators on chip. The latter definition is not universal, however.

Another term used extensively to denote the class of semicustom circuits is *Application Specific ICs* (ASICs). This is an umbrella term that includes gate arrays, standard cells, PLAs, PALs, and so forth.

Figure 1.6 shows the background metallization layer of a typical gate array die (the IMI 71960 die). The background metallization, which is present in all gate array dies, includes the bonding pads that the fine wires from the package are attached to; the distributed power and ground buses; and the most important item in this discussion, the potential contact points (the small squares) of one section of this typical gate array. The contacts are to the transistors and to the polysilicon underpasses which are described later in the chapter. The very large squares are I/O "bonding pads" (see Chapter 11). The relatively wide lines running in parallel are power and ground buses. These will be discussed in greater detail later.

The important feature from Fig. 1.6 that the reader should note is the *repetitive nature of the structure*. It is not expected at this point that the structure itself has meaning.

Figure 1.7 shows a generic map of a gate array die. Its purpose is to exhibit the key elements mentioned above. Not shown in Fig. 1.7 are the array cells and under-

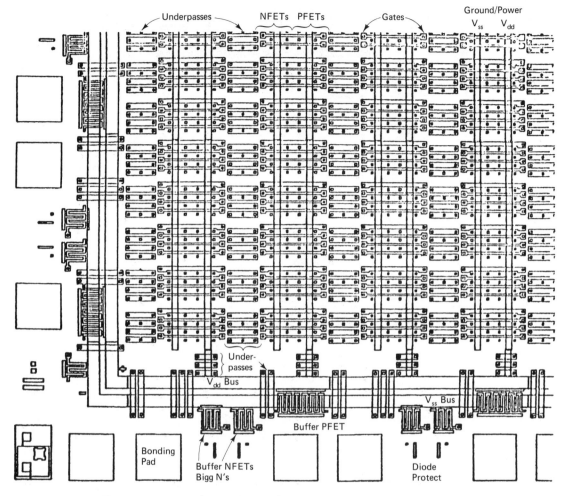

Figure 1.6 Corner of gate array die. (Courtesy of International Microcircuits, Inc.)

passes (if used). The latter are shown in Fig. 1.6. This subject is amplified in Chapter 3.

The structure itself will differ from manufacturer to manufacturer and from technology to technology. However, most manufacturers offer "families" of gate arrays. The members of a given family generally have the same repetitive structure of transistors and underpasses and differ only in the number of elements on a die. This means that a design done for one family member can be placed on any other family member that has a large enough number of elements. Techniques for doing this will be delineated in subsequent chapters.

In a given technology, for example, single-layer metal HS (high-speed) CMOS, there is marked similarity among the vendors in terms of the background structure. After becoming familiar with the structure of one such vendor, one can

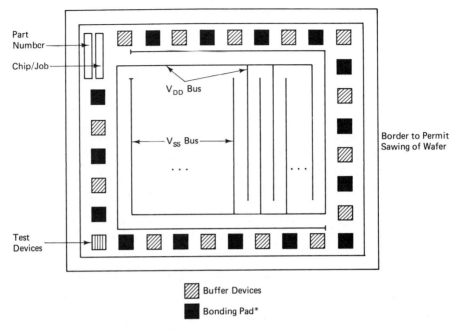

Figure 1.7 Generic map of individual gate array die. V_{dd} and V_{ss} are the power and ground buses, respectively.

generally recognize very rapidly structures and patterns in the offerings of other gate array vendors, which are in the same technology grouping. This topic is discussed more fully in Chapter 3.

One or more metal layers and sometimes one or more layers of partly conducting polysilicon are used for the interconnect. The *pattern* of the metallization layers interconnecting the transistors and underpasses constitutes the definition of these layers and determines the logical functions that the gate array will perform. By varying the interconnect pattern, a gate array can be configured to virtually an unlimited number of circuits.

Stated another way, the transistors on the gate array die can be hooked up in an almost limitless variety of patterns. Each pattern constitutes a combination of logical elements ranging from simple gates to complex ALUs, registers, and so on. The amount of such circuitry will depend, of course, on the number of transistors available on the gate array selected. Macros can be designed (using techniques shown in this book) or can be accessed or acquired from gate array vendors. More will be said about this later in Chapters 4, 5, 8, 9, and 10.

Unlike the metallization layers, the polysilicon layer if used is spatially fixed in terms of the number and placement of the polysilicon underpasses that constitute that layer. Each of the underpasses has two or more contact points to which metal from the metallization layer may be run. This structure is described in more detail below.

At least two conducting layers separated by an insulator such as glass (SiO_2), whether they are polysilicon and metal or both metal, are needed to permit two conducting lines to cross without contacting.

1.1.1 Macros

The reader should not be left with the impression that gate array design is currently at the level of interconnecting transistors. Most gate array design currently involves *macros*. These are predefined patterns of interconnects (among transistors) that perform a logical function and can be used repeatedly. An example of such a macro used to make a common digital function [in this case, an eight-input AND-OR-INVERT (AOI) circuit] is shown in Fig. 1.8. The small squares represent potential interconnect points to the transistors and underpasses. The dots represent allowable paths for the interconnect metallization lines.

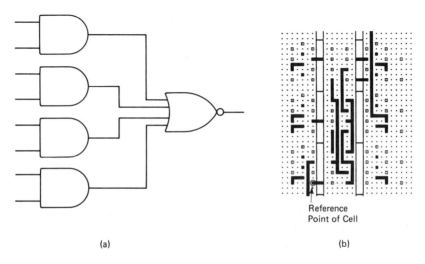

Figure 1.8 Macro (in this case it is an eight-input AOI, one column by three rows): (a) schematic; (b) layout on gate array die.

The important points about Fig. 1.8 are that (1) digital functions are formed by the patterns of interconnects, (2) not all of the potential interconnect points are used, and (3) by using a larger or smaller number of array "cells" (see below) and interconnecting them differently, a wide variety of logic functions can be made. This topic is developed more in the chapters on design. The reader of this introductory chapter should not be concerned about the details of the interconnect itself and the structure of the grid on which the interconnect is done.

The eight-input AOI shown above takes up three of the array cells. In this particular family, the largest chip has almost 2000 such cells. The smallest member of this family has 250 such cells. The same pattern, replicated if desired, can be used on any of the family members.

Other names for gate arrays are *uncommitted logic arrays* (ULAs), *configurable gate arrays* (CGAs), *master slices,* and *logic arrays.* Motorola uses the trade name Macrocell Array to denote that the user can design only with macros and cannot design with individually connected transistors (see the discussion in Chapter 4). The most popular term by far, however, is *gate array*; and this term will be used in this book.

1.2 PURPOSE OF CUSTOMIZATION

The purpose of customization is to enable the gate array to replace a number of standard (off-the-shelf) integrated circuits. The number of standard ICs that can be replaced by a single gate array varies widely. Current CMOS gate arrays can replace over 400 small scale integration (SSI) and medium scale integration (MSI) ICs. This is equivalent to replacing more than eight 8 in. × 6 in. printed circuit boards (PCBs). Gate arrays in development [7], [8], [9] will double and quadruple this number.

Indeed, every improvement in speed and density (number of functions per IC) of standard ICs is also reflected in gate arrays. The reason is that gate arrays are made with standard IC manufacturing processes. Gate arrays are available in almost every technology (see Chapter 2).

1.3 OTHER FUNCTIONS

1.3.1 Memories and Microprocessors on Gate Arrays

As of this writing, gate arrays are used almost exclusively to replace random logic functions. They are really not suitable vehicles for replacing substantial amounts of memory or implementing microprocessors, although both have been done. The reason is that both memories and microprocessors are better designed using full custom techniques wherein each transistor (or more likely each group of transistors) is designed and positioned specifically for the function served by that group of transistors. It *is* fairly common, however, to implement small amounts of "scratch-pad" memory on gate arrays. The advantage is that both pin-out and number of IC packages can sometimes be reduced if the system is partitioned correctly.

A second reason is to provide the content addressable memory (CAM) part of CAM/CACHE techniques. (In this technique, a small amount of memory in the CAM keeps track of what memory addresses are stored in the CACHE memory. Data in the latter comes from nonvolatile storage, such as disks.)

The situation is changing, however. NEC has reported on a developmental gate array which includes eight blocks of configurable static random access memory (RAM). Each block is organized as 32 by 9 bits, for a total of 2.25 K of storage. IBM has also reported on a gate array that can have a static RAM section. To obtain reasonable die size, the IBM die uses four layers of metal.

Motorola offers a version of its 2500-cell ECL array configured as 1000 bits of RAM and 1500 gates of logic. AMCC also has a version of its 3500-gate ECL (see Chapter 2) array that contains memory.

Nonvolatility. The static memory of the AMCC array (the QM1600S) can be made nonvolatile by tying the bases of the cross-coupled flip flops used to the appropriate voltages. In this sense it functions as a ROM (Read Only Memory). Typical access time for the high speed memory macro is 5 ns. Chip size is 360 × 360 mils, and the chip contains 1600 equivalent gates in addition to the 1280 bits of RAM.

AMD has a series of ECL arrays with 5000 gate equivalent densities. One of these, the Am3525 has 3718 gates PLUS 1152 bits of RAM. Hitachi has under development a gate array that is suitable for up to 12,000 logic gates, up to 10,000 bits of memory, or some combination of the two. NEC also has under development a CMOS gate array whose cells can be made into static RAM. Integrated Logic Systems, Inc.'s CMOS Programmable Performance Logic array has an architecture that enables both ROM and PLA to be combined with normal gate array functions.

Until recently, microprocessors were generally too complex to be put on a gate array chip, although IBM had done it. With current gate array densities of 7000 to 20,000 gates per chip, it is possible to design 4-bit slices of a microprocessor (together with perhaps a 2-bit slice of the remaining data path, including memory) on a single gate array chip.

An example is the class of "structured array" designs of LSI Logic. These are metal mask-programmed devices that contain transistors that can be interconnected as with any other gate array to provide gates and other logic functions. However, unlike a gate array, the background layer contains what LSI terms *megacells* (to distinguish them from the macros or macrocells of their gate arrays). Megacells include functions such as 4-bit slice microprocessors, RAMs, ROMs, ALUs, and multipliers. Different members of the structured array family have different combinations of megacells. Unlike a gate array, the transistors in the megacells are designed to be optimum for those individual megacells. Moreover, the megacells are fixed in position in a given array family member (because all the layers are designed uniquely for each megacell).

As the densities of gate arrays increase, it will be possible to implement even more such combinations. Alternatively, in the future one may find gate arrays on a portion of the die of other mask-programmable devices, such as ROMs. Another development by LSI logic in conjunction with Toshiba is a CMOS gate array with more than 500,000 transistors. This is the equivalent of 130,000 (2 input) logic gates. Because it utilizes a *sea of gates* (see Chapter 3), which has no routing channels, not all gates can be used. Typically about 50,000 of the 130,000 gates are used. Significantly, LSI has developed layout software which will accommodate what they term compacted arrays [7].

1.3.2 Mixed Analog and Digital Functions in Gate Arrays

A number of companies offer gate arrays with mixed analog and digital function capability. These include Interdesign and its parent Ferranti, GE/Intersil, Telmos,

Holt, Array Technology, Micro-Circuit Engineering, Silicon Systems, Universal Semiconductor Corp., and Exar. Some of these, such as the arrays offered by Telmos, have a high-voltage (up to 300 V) capability built in. Most of the manufacturers offer kits containing discrete parts which when breadboarded will replicate the performance of the actual chip.

1.4 SPECTRUM OF POSSIBILITIES

Gate arrays started out in life as a quick and inexpensive way for users to obtain low- to moderate-performance custom ICs. The first gate array with which the writer was involved many years ago cost $3000, took five weeks (by IMI) to develop, and replaced 11 SSI and MSI ICs. Since that time the emphasis has shifted to larger and more complex arrays in the range of 5000 to 50,000 cells.

The key question is not necessarily "What is the largest circuit that can be put on a gate array die?" but rather "What is the size of the gate array required to adequately meet the need at hand?" Finally, while major gate array companies with their sophisticated computer-aided design (CAD) tools have made it possible to design gate arrays with almost no knowledge whatsoever of the gate array methodology, users of CAD tools will find it highly useful to understand some of the basic principles underlying the design and layout of gate arrays. For all these reasons, the *spectrum of possibilities* of gate array design and implementation from the simple to the complex is given in this book. Because the trend is toward use of CAD in the design of gate arrays, including the use of CAD workstations, this book is oriented in the latter direction.

1.4.1 Computer-Aided Design

CAD refers to the use of computers and appropriate software to aid the design process. The brethren of this term are computer-aided test (CAT), computer-aided engineering (CAE), computer-aided manufacturing (CAM), and sometimes facetiously "computer-aided everything" (CAEVY).

There is a measure of overlap among some of the terms, particularly CAD and CAE. Often the latter term refers to a system design, whereas the former refers to a component design.

Some people have proposed that the term CAD should really be HAD for "human-aided design" (and sometimes for "I've HAD it with this software package."). The reason is that in some cases, the human being is really aiding the computer to do its job rather than vice versa.

Although some computer programs are now at the stage where only minimal intervention is needed by the designer, the author personally prefers the term CAD to the term HAD. The reason is that the term CAD properly implies the need for a designer to *think* and the responsiblity for the designer to use the computer-assisting tools intelligently. The term CAD is used almost universally and will be used here.

If artificial intelligence (AI) fulfills some of its many promises, the term HAD may become more appropriate in the future.

The discussion of CAD tools proliferates throughout this book. A major purpose of the discussions is to show where the current tools can most successfully be used, some of the subtle benefits of using them, and to indicate some of the newer tools that are coming along.

On occasion, someone questions what the definitions of VLSI (very large scale integration) and LSI (large-scale integration) are in terms of the number of gates and whether gate arrays are truly VLSI. Different speakers and writers use different numbers of gates as the break point between LSI and VLSI. The author believes that any IC that replaces five or more printed circuit boards of MSI/SSI ICs (as current-generation gate arrays do) is very definitely VLSI in any sense of the word.

1.5 GENERATIONS OF GATE ARRAYS

Writers and speakers sometimes refer to "generations of gate arrays." In this introductory chapter, it is appropriate to give what are of necessity inexact definitions of these terms.

CMOS gate arrays have gone through three fairly distinct generations. The first was that of *metal gate CMOS* (see Chapter 2). Gate arrays in this category were single-layer metal and generally had no CAD tools available to aid in design. As such, they tended to be labor intensive but have relatively low nonrecurring engineering (NRE) costs. *Drawn gate length* (the length of the gate that is actually on the mask) is typically 7.5 micrometers (μm; also called "microns").

The second generation is that of *silicon gate CMOS*, which is oxide isolated but which is also single-layer metal. Drawn gate lengths on these arrays are in the range of 4 to 5 μm. They are an order of magnitude faster and roughly four times the density of metal gate CMOS arrays. International Microcircuits, Inc. (IMI) was the first gate array company to offer such arrays commercially in the United States.

The third generation of CMOS gate arrays is characterized by silicon gate and oxide isolation, by *two* layers of metal, and by more extensive CAD tools. Drawn gate lengths are currently in the range of $1\frac{1}{2}$ to 3 μm.

Bipolar gate arrays do not break down as easily into generations. Most, if not all, have always been two-layer metal. The reasons are that the voltage swings of bipolar technologies, such as ISL/STL and ECL, are much lower than that of CMOS; at the same time, the currents that flow are much larger. The higher resistance of the polysilicon used as the second conducting layer in single-layer metal systems would cause unacceptably high voltage drops in bipolar arrays.

ECL gate arrays can be classified according to the method of isolation among the transistors. First generation arrays are junction isolated (see Chapter 2) whereas second generation arrays are oxide isolated.

Although not a firm line of demarcation, the later ECL gate arrays generally have TTL-compatible buffers around the periphery for use in interfacing to CMOS and TTL. (They also have ECL-compatible buffers.)

1.6 TWO FREQUENTLY USED TERMS THAT HAVE MULTIPLE MEANINGS

Two terms used frequently throughout this book need to be discussed. These are the terms "cell" and "gate."

1.6.1 Cell

The term *cell* has at least two different meanings. First, it denotes a group of active elements (transistors) on a gate array. It is sometimes called an *array cell* when used for this purpose. The author prefers the term "cell" to the word "gate," which is often used for this purpose because there is no current standardization on what a gate means in terms of numbers of inputs and outputs. (Some firms have tried to standardize on the definition of a gate used in this context as a two-input NAND or NOR gate with a fan-out of 2. However, neither this definition nor any other is universal. The result is that the IMI 2000-cell array, which contains three-input cells, is in reality a 3000-gate array by the above definition. However, most people (including the author on occasion) loosely call it a 2000-*gate* array.

The word *cell* is also used to denote a functional element that is predefined in terms of the layout of the interconnects among the transistors. In terms of size, it could be as small as a logic gate or as large as the entire chip but generally is somewhere in between. Cells used in this context are generally called *macros* or sometimes *macrocells* or *function cells* (the two terms are interchangeable). Of late, especially in the U.S. Department of Defense (DOD) VHSIC program, a distinction is made between small and large function cells. The former, which are referred to as *microcells,* contain a small number of gates and would be simple multiplexers (MUXs), D flip-flops, and so on. The larger function cells are referred to as macrocells. The gate count for the latter is 2500 or more, although the reader can understand that there is no hard-and-fast break point between microcells and macrocells.

1.6.2 Gate

The word *gate* has three different meanings. First, it denotes the physical input to a MOS transistor or, in the case of CMOS, a CMOS transistor pair (unless otherwise stated in the text). Second, it refers to a logic gate, such as a NAND or NOR gate. Third, it sometimes loosely denotes the number of repetitive active element groups of a gate array (e.g., "a 2000-gate array"). In the case of ISL and STL gate arrays (see Chapters 2 and 10), each transistor represents a gate, and there is little confusion. In the case of arrays in other technologies (see Section 1.6.1) considerable ambiguity is often the result of this lack of standardization.

1.7 OTHER PROGRAMMABLE METHODOLOGIES

Semicustom circuits are "mask programmable" as opposed to being "fuse-link" programmable, electrically programmable, or software programmable. Examples of *fuse-link programmable* devices are programmable read-only memories (PROMs), programmable array logics (PALs), and field-programmable logic arrays (FPLAs). Examples of *electrically programmable* devices are electrically erasable programmable read-only memories (EEPROMs) and EPROMs. The best example of a device that can be *software programmed* to perform logic functions is, of course, the ubiquitous microprocessor.

A PAL is a collection of *premade logic functions* (typically simple gates and flip-flops) on a chip, as contrasted to a gate array, which, as noted, is a collection of premade transistors. On a PAL, the interconnect lines among the logic functions are *fixed in place*. The user's only choice is to *break or not break* (via a fusible link) a given line. Unlike a gate array user, the PAL user can neither redefine the logic function nor change the routing of the interconnect lines among the logic functions.

Although the user has fewer options when using PALs than when using gate arrays, the simplicity of design coupled with the ability to get working parts rapidly have made PALs extremely popular for relatively low-density applications.

Gate arrays provide many more functions per chip than do PALs but are more difficult to program and take longer. (There are mask-programmable versions of the PAL called HALs which are used for high-volume applications. They have the same densities as those of the PALs. Newer PALs with much higher densities have recently been introduced.)

Gate arrays can provide a wider variety of circuit functions than PROMs or FPLAs can but are volatile. PROMs, FPLAs, and EEPROMs are really aimed at the memory applications for which gate arrays are not suited. Gate arrays are faster than microprocessors but generally take longer (because of the need for a mask) to reprogram *once the problem has been found*. However, finding the problem associated with either a microprocessor or a gate array may take a considerable amount of time.

It is extremely dangerous and difficult to make comparisons of costs between PALs and gate arrays. A great deal depends on the vendor, the technology, the individual circuit, and so on. (See Chapter 5). However, *on a per gate basis*, the cost of a gate array is often much less than that of a PAL. The cost per gate of National's SCX6200 family of CMOS gate arrays ranges from $0.01 to $0.015. In some cases the cost of the HAL version of the PAL is comparable to that of the gate array.

Xilinx has developed what they call a logic cell array. [10], [11], [12] It has the unique feature of being software programmable. This means that the custom design can be changed by means of a program stored elsewhere in some memory device, such as a PROM, EPROM, DRAM, or SRAM. The time to change the configuration is about 12 msec (milliseconds).

Each of the 64 logic cells contains a D type flip-flop and a combinatorial block. The configuration is set by the six, two-and three-input switches in each cell. Xilinx indicates that the complexity is about that of a 1000 to 1500 gate gate array.

Harris Microwave offers an array in GaAs (see Chapter 2) which is midway between a gate array and a PAL. Like the PAL, each of the family members has functions such as D flip-flops and combinatorial gates which are predefined and fixed in place on that particular array family member. Unlike a PAL, however, the routing of the interconnect lines is determined by the mask design similar to that on a true gate array. The array is specifically aimed at replacing standard data book functions with GaAs chips.

Bell Telephone Laboratories have developed what they term the "polycell" approach to customization. This is midway between a gate array and a standard cell. Like the gate array, the wafers are partly built ahead of time. It differs from a true gate array in that less of the wafer is prebuilt meaning that the user has more flexibility to determine some but not all of the transistor and connection parameters. However, as always, more flexibility means more responsibility on the designer's part. (CIC and to a lesser extent Signetics have also used this approach.)

Lattice Semiconductor, Beaverton, Oregon, has developed a CMOS array based on electrically erasable fuse technology. These emulate 20 pin PAL devices. Because the chip is electrically reconfigurable, only one chip type is needed to perform a variety of functions. The GAL (generic array logic) 16V8 is a 64×32 programmable AND array. Data retention is said to exceed 10 years.

1.8 GATE ARRAYS VERSUS PURE CUSTOM AND STANDARD CELLS; SILICON COMPILATION

1.8.1 Standard Cell Approach

The standard cell approach to customization is one in which designs of macros are stored on a computer and are interconnected to form a customized circuit, just as similar macros are placed and interconnected to customize a gate array.

One difference between the two methodologies is that in the standard cell approach, the wafers are not built ahead of time. Therefore, unlike a gate array, the designs of *all* the mask levels of each macro have to be stored on a computer. There may be 14 or more such mask levels to be stored, as contrasted to generally one such mask level for gate array macros. (Even in the two-layer metal gate arrays to be discussed, only one of the metal layers is used for an interconnect among the transistors to form the macro.) This means that the tooling cost for a standard cell library is considerably higher than that of a similar library of macros for gate arrays. (Silicon compilers, discussed elswhere in this chapter, may largely eliminate this difference if CPU charges for running the compiler are ignored.)

1.8.2 Relative Die Sizes between Standard Cell and Gate Arrays

Because the wafers of a standard cell are not made ahead of time, the dies of standard cells do not have to accommodate unused routing channels and transistors (except for transistors that exist in unused portions of macros).

There are some unused portions of routing channels even with the standard cell approach. These occur because the routing channels are made wide enough to accommodate all the lines that will transit a given area. Because most of the lines will not transit the entire length of the routing channel, some space will be lost. However, such lost space is generally minimal compared to that of a gate array.

In the gate array approach, not all of the array cells are used. Although current placement and routing programs for gate arrays permit percent utilizations of the cells that are greater than 90%, the percent utilization is often much less. (See Section 1.8.3.) Also, there is still an area penalty associated with the fixed sizes of the transistors of the gate array.

Because the routing channels and transistor sizes are minimal and the utilization of cell areas is maximum in the standard cell approach, the standard cell die is generally smaller than that of the gate array (see the discussion on large numbers of buffers for exceptions to this). The size difference depends on the exact circuitry being implemented as well as on the arrays being used, but is typically 20 to 30% smaller.

The difference in die size is generally touted by standard cell vendors as enabling lower costs to be obtained with their products. The author believes this argument is often fallacious, for several reasons. First, in the VLSI era, the major cost drivers are often the costs of packaging and testing, not the costs of the silicon itself. VLSI packages can often cost $20 or more in the relatively low quantities characteristic of the gate array market. The cost to develop and run test programs on expensive multimillion dollar testers also contributes.

NRE (nonrecurring engineering) costs vary so much among standard cell and gate array manufacturers that general statements about such costs are not possible. In any specific situation, however, the cost of NRE amortized over each chip produced by the two methods also has to be factored in.

In cases where very large numbers (greater than 125) of I/Os have to be accommodated on the chip, there may be relatively little (less than 15%) size difference between the two methodologies. This is especially true when (1) the circuit is mostly random logic or the gate array is of the type that uses minimum area for the type of logic used (see the GE study mentioned in Chapter 3); (2) the gate array size closely approximates that of the circuit being implemented; (3) placement and routing is able to achieve 95% utilization of the array cells; and (4) the circuit itself is relatively small (a few hundred gates). The reason is that a pad takes up about 16 square mils of area. The associated driver circuitry typically takes very roughly about the same area. The spacing between pads is also about the same area. The net result may be a chip with a lot of empty space regardless of which method is used.

1.8.3 Die Size versus Yield

Finally, the argument about the cost of silicon itself being less for the standard cell deserves a few words. This argument is based on a given die being uniformly packed such that all areas are used. (The so-called "Seeds" and other models mentioned in Chapter 11 are based on this assumption.) This means that a given defect, be it a dust particle that acts as a mask or a crystalline defect, will make the die unusable.

Although increasing numbers of gate arrays are being designed with over 90% of their internal array cells being used, many designs will use only 60 to 80% of a given die. The reason is that gate arrays typically come with four to six members in a given gate array family. Each family member is of a different size. A given circuit that is too large for one family member will be put on the next largest family member. On the larger member, the utilization may be in the range 60 to 80%. This makes for a lot of open space on the die. Even in gate array designs in which the percent utilization is high, there may still be a lot of open space just because of the space taken up by unused routing areas. (Remember, in the standard cell approach, these unused routing areas will be there but will be very minimal.)

Also, in many cases, a given design will be put on a larger member of a given gate array family in order to get the larger number of I/Os of the larger chip.

All of the above means that there is often a lot of empty space on a gate array die that has been processed. The probability of a given dust particle or defect being in a spot on the gate array die where it can damage the functionality of the die is much lower in general for a gate array than for a standard cell chip. Therefore, the usual models that relate the die cost to die size are really not valid for the gate array die.

Larger dies for a given degree of functionality do mean fewer dies per wafer. However, silicon is relatively cheap. A silicon wafer starts out in life costing about $10. By the time it is processed, the cost will be in the range of $100 to $700; and it will contain one hundred to more than several hundred potentially good dies.

All of the above reinforces the author's earlier statement that the size of the die is really not the cost driver in the era of VLSI.

1.8.4 Benefits of Standard Cells

There *are* a number of major arguments in favor of the standard cell approach. First, the ability to put an extra 20 to 30% or more of functionality on a die often enables the system to be partitioned in such a way as to minimize the connections to the world outside the chip. Because the bulk of the power dissipated is sometimes, but *not always,* in the I/O buffers, minimization of their numbers can drastically cut the power dissipated by the chip. Speed will also be enhanced by the elimination of the capacitance associated with the no-longer-needed I/Os.

Moreover, because the background structure of the chip does not have to be designed to accommodate an unlimited number of designs (as it does on a gate array), elements such as microprocessors and memory elements, which have been laid out with minimal areas, can be more readily accommodated on a standard cell design than they can on a gate array design. (As noted elsewhere, however, memory is being incorporated on some gate arrays.)

The ability to incorporate elements such as microprocessors and RAMs is one reason the standard cell approach is so popular. Today, over 70 vendors offer this method.

1.8.5 Converting from Gate Array to Standard Cell

GE/Intersil offers the capability to transform a gate array design into a standard cell design. It will accept a Tegas, Daisy, or P-CAD netlist. [13] Therefore, the design does not have to be done originally on a GE/Intersil gate array.

1.8.6 Gate Arrays versus Standard ICs (Not Standard Cell ICs) and Full Custom ICs

A full custom IC is one in which, theoretically, every transistor is designed and sized for exactly its place in the circuit. (In the VLSI era with as many as two million or more transistors on a chip, a relatively small number of transistor designs will be used over and over again; and groups of circuitry will be replicated where possible.)

Gate arrays versus full custom ICs. Gate arrays differ from standard ICs and from full custom ICs in that (1) the transistors of a gate array are unconnected before customization, (2) the transistors are of fixed sizes and are in fixed positions for a given gate array, and (3) the gate array is not completely made until customization is complete. A gate array wafer may be stored for many months until a design comes along that enables it to be personalized (customized) to a particular application.

By contrast, the transistors of standard and full custom ICs are sized and placed appropriately for the function (e.g., shift register, arithmetic-logic unit, logic gate) that particular IC performs and the IC IS completely made with no interruption in the manufacturing cycle. (In this book, the term *standard IC* will be used to denote an integrated circuit which is offered as a standard "jelly-bean" product by a semiconductor manufacturer. Examples of standard ICs are such well-known products as the SN74LS74 D flip-flop and the MC68000 microprocessor ICs.)

Because the gate array is mostly made ahead of time, it offers significant savings in both cost and time. Large volumes of wafers can be made up through the final stages of metallization. This provides the cost benefits of the learning curve with the attendant improvements in reliability.

1.8.7 Silicon Compilers

Recently, several firms have introduced commercial "silicon compilers" to the commercial marketplace. The firms are Silicon Systems Technology (SST) of Seattle, Washington, which offers the Concorde software; Silicon Compilers, Inc. of Los Gatos, California, which offers the Genesil compiler; Silicon Design Labs of Liberty Corner, New Jersey; and Metalogic of Cambridge, Massachusetts. In Europe, Lattice Logic, Inc. of Scotland offers the Chipsmith compiler. IBM has the Yorktown Silicon Compiler for its own internal use. [14]

A silicon compiler is an extensive set of software that transforms the behavioral or gate-level description of a logic structure into a set of pattern generator tapes from which the chip can actually be fabricated. Thus, they enable pure custom circuits to be designed by system designers. The foundation for this methodology was laid by Carver Mead and others at the California Institute of Technology and Lynn Conway at the Palo Alto Research Center (PARC) of the Xerox Corp. The reader is referred to [15], [16] plus the excellent books by Mead and Conway, [17] and by Ronald Ayres. [18]

At this stage, it is too early to tell what impact the availability of pure custom circuits designed using silicon compilers will have on the gate array market. The author's guess is that silicon compilation will become another (but major) CAD tool available to the gate array designer. There is no fundamental reason why some variation of silicon compilation cannot be used to synthesize gate arrays from libraries of macros or even to create special macros for such arrays.

A key problem in designing gate arrays with more than 2000 or 3000 gates is managing the "data base" from which the design was created to assure "auditability" and traceability of the final design. Properly designed silicon compilers can contribute to the auditability and traceability of chip designs.

When such silicon compilation software is developed for gate arrays (and there is little likelihood that it will not because of the impact it can have), systems designers will have the options of having either a pure custom or a semicustom chip designed from the set of system and behavioral specifications rather than from a set of chip component specifications and a schematic. Developments are already in motion for the forerunner of these techniques. The SST software mentioned above is now part of the software package offered on the Valid workstation. (See Chapter 8 for a discussion of workstations.) Silicon Compiler's Genesil system is offered on the Daisy workstation.

Aside from their uses in designing custom and semicustom chips, silicon compilers have enormous potential for simplifying the design of logic systems regardless of whether they will ultimately be put on silicon (or GaAs). The reason is the ease with which the designer can specify and encode the blocks to be used for simulation. In the very impressive Genesil system, for example, the designer can specify very large functional blocks simply by answering a few questions asked interactively and changing default values of other parameters. Typical parameters on a register might be, for example, the number of inputs and outputs, whether all the resets (if any) are to be synchronous, and so on. On a counter, some of the parameters might be the number of bits, the count value, whether the counter is to be able to count both up and down, and so on.

The Genesil software will also automatically configure the internal logic to perform the specified function. It will also change the symbols used to reflect the numbers of I/Os, resets, clocks, and so on. Therefore, the schematic so created is automatically *documented* upon command. Those who have had to create schematic libraries using conventional means (although not a really big problem on most workstations) will nevertheless appreciate this time saving feature.

Silicon Compilers also offers the Genesis Silicon Development System [19] which enables sophisticated users to generate their own macro generators. Sophisticated users might want to design their own ALU generators, for example.

1.9 A GUESS AT THE FUTURE

The author believes that the future will see combinations of customization techniques on one chip. The customization techniques may include fuse-programmable devices (PROMs, PLAs, and PALs) as well as mask-programmable devices (gate arrays, ROMs), and a few microprocessors, and software-programmable devices. (As noted above, both analog and memory devices are also to be expected.)

These combinations will be made possible by four factors. The first is the larger degree of funtionality that can be achieved on chips because of the shrinking feature sizes and to a lesser extent, the somewhat larger sizes of the dies.

The second reason is that mask programmability will have become accepted as a tool by both designers and managers much more so than it currently is.

A third reason (which will actually help positively influence the second reason) is that CAD tools for going from a logic design to a mask will become much more widespread and efficient. A fourth reason is that "back-end" (i.e., patterning and etching metal and insulator layers as opposed to doing all of the processing steps) fabrication facilities will become more widespread. Many companies already have their own such facilities. These are far less expensive both to build and to maintain than are full processing facilities.

Today, standard cells and gate arrays sometimes compete "head to head" for a given customization project. Yet there is nothing (except perhaps denying that a competing method is sometimes useful) to prevent standard cells from incorporating gate arrays. One or more of the standard cells in the library could simply be a gate array cell of a given size.

The advantage of such an arrangement is that it would permit a chip to be designed that could be made into a number of permutations. The permutations could either be to accommodate a rapidly changing market or (in the case of a company that is "testing the waters" with a product) to accommodate a low-entry-cost product.

1.10 WHY USE GATE ARRAYS?

The assumption in this book is that a decision has already been made to use gate arrays, and now the question is how to do it. Nevertheless, the question of why to use gate arrays deserves a modicum of attention.

To those digital designers who feel comfortable with the gate array methodology, the freedoms of being able to add lower level functions to a design without having to add several discrete packages of standard parts are real plusses.

Special ancillary circuits, which could not be implemented with standard ICs, can often be put on a gate array at little additional cost in time, space, and dollars. If these were implemented with standard ICs, additional packages would generally be required which would take up additional space. Using standard IC's, a quadruple two-input NAND gate takes the same space as does a 64-kilobit dynamic RAM IC. While there are often unused gates and other functions residing in standard ICs on PCBs, these are often geographically in the wrong places on PCBs or, equally important, are of the wrong type. A two-input NAND gate is just that and no more. An unused cell on a gate array can be made into a variety of functions.

An irony of the VLSI era is that the standard parts being produced have so many features to make them widely applicable that designers wind up using only relatively small (30%, 50%) fractions of the chip in many cases. Yet the entire chip has to be powered. This is in sharp contrast to a gate array, in which *only the portions of the chip being used are powered.* System logic designers using gate arrays design in *only* and *exactly* what is needed to do the task at hand instead of being forced to work a design around what is available.

1.10.1 Benefits of Gate Arrays

The most visible benefit of using a gate array is overall size reduction. Figure 1.9 shows an example of such size reduction. In this figure, a given circuit has been built using 133 SSI ICs (top left view). The same circuit has also been built using 45 MSI ICs (top right view). In both cases, the entire circuit has been replaced by just one gate array IC (lower left). Therefore, size reduction is certainly a valid consideration.

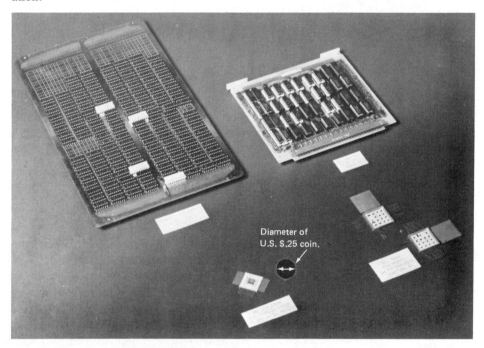

Figure 1.9 The same circuit implemented four different ways: in SSI (upper left), in MSI (upper right), two hybrids (lower right), gate array (lower left).

However, there are many other benefits of using gate arrays. These will be discussed below. The benefits cover a wide range of technical and economic parameters. Not all of them apply to any given circuit. In some cases none of them apply. No two organizations have the same objectives; competitive positions; and human, material, and capital resources. Moreover, both gate arrays and the CAD tools to produce them are rapidly changing. Each organization must select for itself that path that is most viable in terms of its own resources and its own expectations of markets and future resources. For that reason, this book delineates a *spectrum of possibilities* in chips and design methods for gate arrays.

1.10.2 Technical Advantages of Gate Arrays

From a technical standpoint, a key reason for using custom LSI and gate arrays in particular is often to minimize the number of interconnects to the outside world. Doing this impacts many, many other areas, as shown in Fig. 1.10. The rationale is that every interconnect outside the IC, regardless of whether it is an off-the-shelf or some form of custom circuit, involves significant amounts of capacitance both in the I/Os of the ICs themselves and the lines of the printed circuit board (PCB) to which they connect.

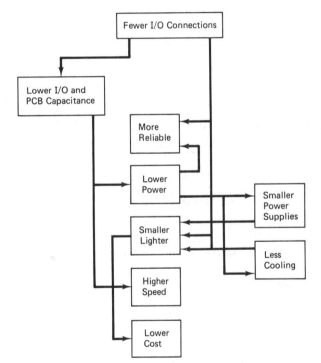

Figure 1.10 Diagram showing how reducing the number of interconnects to the outside world (i.e., outside the chip) improves size, weight, power, reliability, and so on.

This capacitance must be charged and discharged with displacement current, which in turn must ultimately be generated by a power supply. The flow of displacement current through resistors creates heat, which must be dissipated. The power supplies which generate the current must be cooled. The cooling requires physical structure for the coolant paths. Physical structure is also required for the power supplies, which are larger than they would be if less displacement current were generated, which would happen if fewer interconnects outside the chip were used. If there is less heat, the system is more reliable because of the lower temperatures involved.

It is also lower in cost because of smaller physical structure; smaller, less costly power supplies; and fewer parts to inventory.

The interelectrode capacitance of the ICs also slows down the signal according to the relation $I = C\,(dV/dT)$, where I is the displacement current, C the total capacitance due to all sources, dV the voltage change between logic levels, and dT the increment of time. For a given amount of displacement current available (and, of course, logic-level swings, which are fixed by the technology involved), the greater the total capacitance, the longer the time required to switch, all else being equal.

To obtain any respectable speed using standard parts or gate arrays, vendors reduce the driving resistance so that the RC time constant of the interconnect remains low. However, decreasing the resistive portion of the RC time constant forces some generally undesirable trade-offs. In MOS devices, decreasing the driving resistance can be done either by doping the "channel" more heavily or by increasing the size of the driving device itself. Aside from the space penalty, increasing the size of the driving device forces the transistors inside the chip, which are driving it, to be physically larger. The result is that the space on the chip required for drivers to outside the chip multiplies rapidly.

Doping the channel more heavily to decrease the driving resistance has the drawback that the "punch-through" voltage decreases as the transistor dimensions get smaller. Therefore, there are limits to how much channel doping can be increased while keeping the voltage compatibile with existing logic families.

No matter how it is done, the lower resistance value of the driving source will cause higher power dissipation for a given power supply voltage. Therefore, either higher speed or lower power is one result of decreasing the number of interconnects outside the chip by replacing many standard ICs with a gate array.

Hence, there is a very real advantage, which is not easily quantifiable in dollars, to reducing the number of interconnects. For example, if a gate array replaces 100 16-pin SSI/MSI parts with one, say, 40-pin package, then the number of interconnects has been reduced by a factor of 40. The capacitance of a CMOS gate inside a gate array chip is typically 0.12 to 0.15 pF (twice that for the pair of CMOS transistors). This contrasts sharply with the 3 to 20 pF commonly measured on the inputs to a packaged IC. Therefore, the capacitance per on-chip I/O is 20 to 133 times the off-chip capacitance. The overall interelectrode capacitance has been reduced by a factor of 40 times this range or a resultant factor of 800, to over 5300, by replacing the PCB with a gate array. This factor is exclusive of the additional capacitance of the PCB runs relative to the runs on a chip. The former are typically 50 to 100 pF/ft, as shown in Fig. 1.11. The on-chip capacitances are much higher on a per foot basis, but the run lengths are much shorter. The latter are discussed more fully in subsequent chapters. However, typical numbers are 1.9^{-17} pF/μm^2 for aluminum separated from a ground plane by 1 μm of insulator. Of course, the resistance must also be figured into the speed calculation. This is done in subsequent chapters.

In the example above, the power supply requirement dropped from roughly 1.4 W to 35 mW worst case for the circuit shown in Fig. 1.9. Although most of this decrease in power is attributable to replacing LSTTL parts with CMOS, some of it is due to the decrease in the number of interconnects.

Fairchild has made the point in a slightly different fashion. Their point is that it

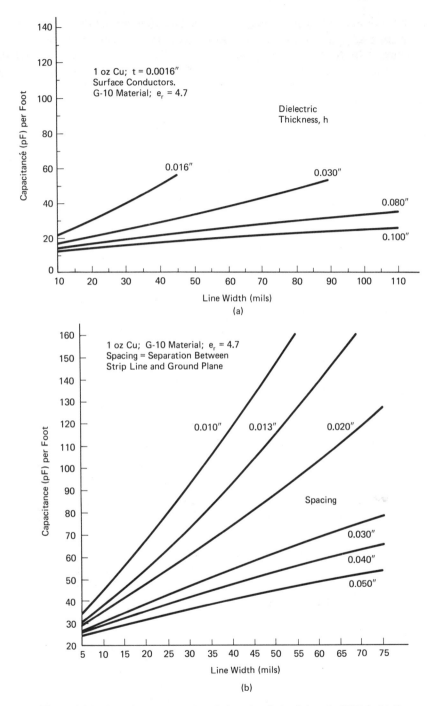

Figure 1.11 Capacitance versus length for printed circuit boards (PCBs): (a) Capacitance versus line width and dielectric thickness for microstrip lines; (b) capacitance versus line width and spacing for strip lines. (Courtesy of Motorola.)

Sec. 1.10 Why Use Gate Arrays?

does little good to keep reducing the propagation delay of the individual gates if the resultant speed improvement is lost on the printed circuit board and interconnects among the ICs. Figure 1.12 shows that as the logic stage delay decreases, the percent of the total stage delay attributable to interconnects increases very rapidly.

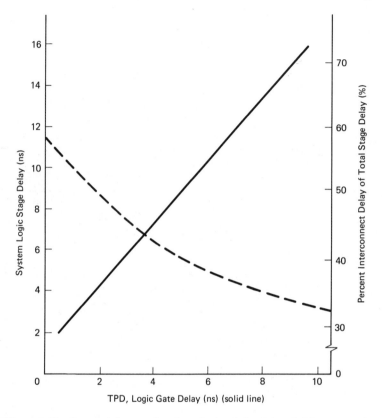

Figure 1.12 System delay as a function of printed circuit board delay as gate delay decreases. (Courtesy of Fairchild.)

1.10.3 Economic Advantages of Gate Arrays

Economic considerations of gate arrays involve much more than simply cost. First and foremost, such considerations must include an answer to the question of what is required to become or remain competitive in one's chosen marketplace and what resources are available to effect such a competitive posture. The costs involved in comparing a gate array to a PCB can be classified as nonrecurring and recurring, sometimes abbreviated as NRE and RE, respectively. A continuing problem in trying to compare the NRE of a gate array with that of a PCB is that few program managers know the true cost of the latter. The reason is that a variety of costs involved in the design, layout, test, debugging, and rework of a PCB are spread across a variety of task codes, jobs, and personnel in the typical engineering organization.

By contrast, the design of a gate array includes costs that are much more quantifiable and, more important, visible. Generally, one or two engineers are involved with a gate array design, but are dedicated to the task until it is finished. The simplest case is where the gate array is purchased from a gate array vendor which starts with the schematic and produces finished, tested parts. To the vendor's cost must be added the labor charges of the user organization to interface to the vendor, but these generally involve only one engineer part time and hence can be tracked quite easily.

A user organization that uses gate arrays a lot often builds up a core group of designers with appropriate design equipment which "bids" on gate array design jobs. Here, too, the costs are quite definable.

The economic benefits of gate arrays are much harder to define quantitatively than are the technical benefits. As noted, few companies know precisely what their costs of PCB design, development, fabrication, test, debug, and rework are. Some of these tasks, such as design of the basic digital circuit and test of the finished product, are common to both gate arrays and PCBs. When making a judgment about relative costs, however, the user must be careful to compare the cost of the gate array with the cost of the PCB(s) or portions thereof which it replaces, as well as these costs can be determined. For example, if a gate array replaces three PCBs of standard logic, the cost of three completed "stuffed" PCBs should be totaled. However, the cost of the edge connectors, mother board, and harness wiring (if any), which are no longer necessary, should also be included in the recurring cost of the PCBs, which will be compared against the cost of the gate array.

One of the few in-depth studies of the cost advantages of gate arrays was done by Texas Instruments. This is shown in Fig. 1.13, which indicates that even though

Component	1 — 40-Pin Array 377 Utilized Gates		59 — LS SSI/MSI Devices 477 Utilized Gates	
Recurring Cost				
Component Insertion		25.00		14.10
Procurement		0.10	0.03/unit	1.77
Goods-In Test		0.25	0.08/unit	4.72
Inventory		0.10	0.10/unit	5.90
Assembly		0.75	0.40/unit	23.60
Board Test		0.25	0.10/unit	5.90
Board Rework		0.55	0.09/unit	5.31
Insertion Total		2.00		47.20
PC Board	3 in.^{2} @ 0.50	1.50	88.5 in.^{2} @ 0.238	21.12
Power	0.5 W @ 3$	1.50	1.0 W @ 1$	1.00
Board/Power Total		3.00		22.12
Reliability 35K hr (= 5 yr × 80%)	0.007%K hr F.R./ Unit $400 Service Cost	1.00	0.001%K hr F.R./ Unit $400 Service Cost	8.20
		31.00		91.62

Figure 1.13 Comparison of costs of using standard ICs on a printed circuit board versus using gate arrays. (Courtesy of Texas Instruments, Inc.)

the gate array initially costs more, the overall cost compared to using standard parts is lower by a factor of almost 3:1. Figure 1.14 shows that even the smallest gate array (500 gates) in TI's STL (see Chapter 2) family can make a very significant impact. Power is reduced by a factor of 2, and the PCB area is reduced by a factor of almost 30. Equally significantly, the number of I/O connections from/to the outside world has decreased by a factor of 21. Different users with different circuits would have different costs. However, for a true comparison, the elements in Fig. 1.13 should be used.

Category	STL A500 Array	Low-Power Schottky TTL
Implementation Hardware	1-TAT004 500-Gate Array 40-Pin Package	12-74LS00 7-74LS04 11-74LS09 1-74LS10 1-74LS20 10-74LS74 2-74LS126 3-74LS136 2-74LS221 59-Total
Total Utilized Gate Count	377	477
Total Power Dissipation (typ.)	521 mW	1032 mW
Board Area (in.^{2})	3	88.5
Total Board/Pin Connections	40	842

Figure 1.14 Comparison of space and power on a printed circuit board versus that of a gate array. (Courtesy of Texas Instruments, Inc.)

1.10.4 Other Benefits of Using Custom LSI in General and Gate Arrays in Particular [20]

There are several other reasons for using custom LSI in its various forms. First is that for a given maturity of technology, the reliability increases. A study made by the quality assurance group of the author's company showed an increase of 11.5 in the MTBF (mean time between failure) when a group of SSI, MSI, and LSI parts were replaced by a single gate array. A study at RADC showed that as the number of gates on a die increases, the reliability per gate increases dramatically.

A second reason is the competitive advantage of a proprietary part which is tailored to the needs of the system and which is unavailable to the competition. Closely coupled with this is the ability to get on-chip specific functions not otherwise available. These might include (to mention just a few widely divergent examples) an on-chip built-in test (BIT), an arithmetic unit of some kind (such as the equivalent of an LS182 ALU), a small amount of scratch-pad memory (memory is not efficiently put on a gate array, but small amounts of it can be and can often same pin counts thereby), perhaps a FIFO (first in, first out), and so on. All of these functions can be done with discrete ICs. One of the experiences of logic designers using off-the-shelf parts is that they often use only some fractional percentage of off-the-shelf VLSI ICs (but have to use them in order to get certain desired functions if means of customizing chips are not available).

Another reason for using a custom part is that standard parts are often withdrawn from the market. Nearly every company has its own "horror story" of this happening. In the author's own experience, a well-known semiconductor manufacturer stipulated a minimum buy of nearly $200,000 for a part which was being withdrawn but which was crucial to a system. To make the situation even more difficult, the part is often withdrawn at a time late in a product's manufacturing cycle, when it is questionable whether funds committed to procuring a custom part could be recouped by additional sales. If the part had been customized earlier in the life cycle of the product, more functions could have been put on the custom chip to defray all or some of the development costs, as noted above. Owning the mask set to one's own custom part does not cure this problem entirely. One still has to be assured that a vendor can be found to make wafers using this mask set. It is here that the gate array is somewhat better than a pure custom circuit. In the case of the gate array, wafers *usable for many designs* can be stockpiled either by the user or by the gate array vendor. Because the wafers can be used to create *many* custom designs by use of different metal masks, the costs of the wafers can be spread over several programs instead of just one.

All the reasons for using pure custom circuits delineated above apply to using gate arrays. A question of interest, then, is: Why use gate arrays in preference to some of the other forms of custom VLSI?

The answer to this question is not as clear as it was even four to five years ago, indicating how fast the industry and technology are changing. Just four to five years ago, the answer was relatively simple. If one had high-production volumes of 100,000 pieces or more per year or low-production volumes of 5000 to 10,000 pieces per year but could pass on the added costs of a pure custom circuit because of dominance in the marketplace, then pure custom was the way to go. The user would have to be prepared to put a great deal of money ($150,000 to $250,000 or more) "up-front," wait a year or more, and take the attendant risks of a pure custom development. For the investment, the user got a high-performance IC that could replace 100 or more SSI ICs and perhaps have special nondigital functions on it.

By contrast, if one had a handful of low-performance (less than a 1-MHz clock rate) digital ICs to replace and wanted rapid turnaround without a lot of up-front investment, one chose gate arrays. A few companies, notably General Dynamics in Pomona and Raytheon in Bedford, Massachusetts, did have high-performance 300-gate LSTTL gate arrays back then. Also, Fairchild had a 300-gate ECL gate array with a 750-ps internal gate delay. MITs Lincoln Laboratory also developed a superfast ECL gate array together with a set of macros to enable logic to be rapidly converted to gate array form. However, these were the exceptions. Back then, the only other possibility was the standard cell approach. In the United States only two companies, Harris and Signetics, offered real products for sale. A number of other companies, such as TRW, had their own standard cell families, but these were for internal use only and were not for sale outside.

Today, the situation is considerably more complex. Gate arrays span a very broad spectrum in both capabilities and costs. Newer methodologies, such as the

Mead–Conway technique and symbolic layout plus a host of improved CAD tools, have made the design and layout of pure custom circuits both easier and less risky. More vendors have entered the standard cell arena with excellent products and tools. Partly as a result of the DOD VHSIC program, others may follow suit. (All six major VHSIC phase Ia and Ib contractors are developing standard cell libraries.) PALs, offered by MMI, National Semiconductor, and others, have taken over a lot of the lower-performance, lower-density (number of chips replaced) market.

Muddying the waters still further (if "muddying" is the correct term for a technical accomplishment) is the gate array of CIC. This uses five programmable mask layers and begins to get closer still to a standard cell approach, even though the wafers are still made ahead of time. The polycell approach of Bell Laboratories is somewhat similar to this.

The author hopes that the discussions provided within the pages of this book will enable the reader to ask intelligent questions and make intelligent decisions about the complex issues involved.

1.11 WHY NOT USE GATE ARRAYS?

This question may sound like heresy in a book on gate arrays, but it needs to be honestly addressed. A number of reasons have been given above. If there is one salient reason that crops up repeatedly, it is that it is an unfamiliar methodology to first-time users which has some additional real and *perceived* drawbacks:

1. A gate array chip cannot be probed internally with a scope probe in the conventional fashion. [However, it will be seen in Chapter 11 that (a) test points can be brought out to the edge of the die, and (b) limited probing can be done inside the chip using special techniques and equipment.]
2. The circuit designer has to hand over at least the chip fabrication part of the design to others and hence loses control to a certain extent. By contrast, if a microprocessor is used, the designer can "do it himself or herself." Most system logic designers are adept at using the various microprocessor development systems (MDSs), such as the Intel MDS 220/230/240 series, for software development.

Other, more general reasons for reluctance to use gate arrays are:

3. The designer must be more sure of his or her design before committing to a gate array than to a PCB. With the latter, traces can be cut and jumpers put in and the part shipped. The rebuttal, of course, is that in the VLSI era, the sheer complexity of circuits is forcing usage of large-scale circuit simulation programs to verify the logic before it is built, *regardless* of whether it is implemented in standard ICs, gate arrays, or some other form of CLSI.
4. From a cost standpoint, funds are often not allocated ahead of time for the gate array. Because the *perceived* cost of a gate array is considerably greater than that of a PCB, program managers who have not previously used gate arrays are

reluctant to allocate funds (again the unfamiliarity aspect) to what appears to be a risky unknown. The author's experience is that this attitude undergoes a dramatic change after a program uses its first gate array.

5. Second sourcing is still something of a problem, although significant strides have been made. Most gate array vendors today have at least one other second source, at least for wafers. Moreover, some of the second sourcing often claimed sometimes leaves something to be desired. Some industry observers will state that unless wafers are being built on a line in which several thousand wafer starts per week are being made, there is not sufficient volume to really keep the line "in gear." [The wafers do not all have to be gate array wafers; rather, they just have to be built on the same line by the same process to satisfy this criterion. One major gate array vendor for example, currently uses wafers built on the Toshiba DRAM (dynamic random access memory) line.]

6. Lack of commonality of tools and to a lesser extent terminology is probably a bigger problem at this time than is second sourcing. A user tends to select a small number of gate array vendors (if doing the designs in-house) because of this factor. (Even though CAD programs and background layer information are supplied by the gate array vendor, the user who does his or her own design still makes an investment.) TI is trying to change the above by making its TIDAL (Transportable Integrated Design Automation Language) set of CAD programs available to others.

7. The gate array structure is inflexible, especially with regard to the number of routing channels. The author (whose organization has routed some rather complex arrays) has not found this to be the problem that it is touted to be with the second-generation and beyond gate arrays. With rare exceptions, the routing channels are adequate for the purpose. (Techniques for placement and routing are given in subsequent chapters.)

1.12 EXAMPLES OF GATE ARRAY USE

Four examples may help clarify the spectrum of possibilities available to the gate array user to which the author alluded earlier. The examples are purposely chosen to illustrate the *wide range of possibilities available with gate arrays* (and perhaps to defeat the notion that they are useful only when they replace three of four PCBs with one gate array chip). The examples range from the ultrasimple to the relatively complex.

Example 1: Standard Part Being Withdrawn by Semiconductor Manufacturer

(This is not the same case alluded to earlier in the chapter.) This is an example which illustrates that even though the gate array represents significant overkill, it is nevertheless the best solution to the problem. This application involved a system that had been produced many years ago following a lengthy period of development. A SUHL I (an old form of TTL) part which had just four EX-NORs feeding a four-input NAND gate was being withdrawn. The problem was complicated by the fact that the system had gone through very extensive and very expensive military qualification many years earlier. Any change that required requalification of the PCB containing the part was unacceptable from the standpoint of cost. The possible options were:

1. Make a very substantial up-front investment in a minimum buy of parts before the parts were withdrawn. However, the number of parts in the minimum buy exceeded the number that would probably be used by such a factor that the amortized cost per part of the parts that would actually be used was unacceptably large.
2. Use a PAL. For such a simple part, this normally would be the solution indicated. A problem in usage of PALs was that at that time every PAL containing EX-NORs had a register on the output. This would have required changing the printed circuit board to obtain the additional clocking and clear signals and was unacceptable for that reason alone.
3. Use a 38510 (a military-quality assurance specification) qualified die mounted in a hybrid package. The die in this case was a 5485 comparator, which has the same truth table for the functions of interest that the part being withdrawn has. The problems in this case were that (a) the pinouts were not the same as the SUHL part nor could they be made the same by mounting the die on a hybrid substrate to reroute the connections. (b) Even if the pinouts could have been changed using the hybrid substrate, the resultant package would not fit the "footprint" of the original package because of the extra space taken by the ceramic substrate of the hybrid. (c) It was not clear that requalification would not have been necessary for at least the hybrid. (The latter problem was, of course, academic because of the first two problems.)
4. Use a gate array (there are a few) which has already been qualified to 38510. This was the route ultimately chosen even though the circuitry involved used only a minute part of the gate array die.

Example 2: Reducing the Manufacturing Cost of a Hybrid that Contains Both Analog and Digital Functions

Hybrids are very useful for relatively short runs and where both analog and digital functions have to be put in the same package. A problem with hybrids is that they are very labor intensive unless the volumes involved are large enough to warrant "pick and place" or other automated assembly methods. In the case in point, one hybrid contained 11 dies which were of a digital nature, including a digital phase-locked loop containing a voltage-controlled oscillator (VCO). By replacing the 11 chips with just one, significant savings were achieved in the manufacture of the hybrid. (An added plus is that higher-frequency operation was achieved.) All parts of the 11 chips (including two small noncritical resistors) were made on the gate array chip.

Example 3: Lowering the Cost of a Printed Circuit Board

It is not necessary to replace an entire PCB in order to justify the cost of a gate array. In the case in point, replacing just 15 ICs on a PCB changed the PCB from an expensive and sometimes unreliable several-layer PCB to a simple low-cost double-sided PCB. The elimination of the NRE cost of the multilayer PCB paid for the gate array. Added bonuses occurred in subsequent manufacturing of the PCBs and in reliability.

Example 4: Replacing Large Amounts of Circuitry

Nowadays all that one hears about in the literature is how many printed circuit boards of logic can be replaced with a single gate array chip. In this case, 133 SSI ICs were replaced with one gate array. Speed of operation was slightly higher with the CMOS gate array than with the LSTTL parts replaced with one gate array. However, the worst-case power used by the CMOS gate array was 35 mW as contrasted to over 1 W for the LSTTL devices *when operating at the rated clock frequency of* 12 MHz. The latter point is emphasized to avoid any charge of "specmanship" which could occur if the power of the CMOS were not taken at the frequency of operation of the circuit. (The power dissipated by CMOS increases linearly with the frequency of operation.)

1.13 VHSIC GATE ARRAYS

The U.S. DOD (Department of Defense) sponsors the VHSIC (Very High Speed Integrated Circuits) program. Most of the information on this program is restricted. The following information came from an unclassified trade magazine article. The inclusion of the material below does not confirm its content. Those with proper credentials can contact DOD for more information.

There are six phase I VHSIC contractors. Each is building a number of chips. Typical geometries are in the 1.25-μm range. Most are teamed with other contractors.

One parameter used is the *functional throughput rate* (FTR). This takes into consideration the trade-offs among speed, power, and density. For the phase I VHSIC program, the FTR is 5×10^{11} gate-Hz/cm^2. This specification could be met, for example, by a chip operating at 100 MHz with 3750 gates on a 300-mil-on-a-side chip. Alternatively, it could be met by a chip operating one-fourth as fast but with four times as many gates on the same-size chip.

The FTR for phase II of the program is 5×10^{13} gate-Hz/cm^2. Feature sizes in the vicinity of $\frac{1}{2}$ μm are used in phase II.

Gate arrays are some of the chips being designed and built. These are listed in Table 1.1.

TABLE 1.1 VHSIC GATE ARRAYS

Vendor	Technology	Number of gates	Chip size (mils^2)
		Phase 1 chips	
TI	TTL	10,900	100
Westinghouse	CMOS	10,000	
Hughes	CMOS/SOS	8,000	370
Motorola/TRW	CMOS	6,500	270
Honeywell	Bipolar	7,000 (not part of original set of VHSIC chips but is used to support the VHSIC chips)	
		Phase II chip	
Honeywell	CML[a]	66,000	300

[a]Similar to ECL (see Chapter 2).

1.14 STANDARDS FOR GATE ARRAYS AND FOR CAD DESIGN TOOLS

Currently, no standards exist for gate arrays. There is, however, one Joint Electronics Design Engineering Council (JEDEC) subcommittee (JC-44) which is working on such standards. The group has split into two subgroups. One group is working on standards for gate arrays; the other is working on standards for standard cells. Activities of the groups include defining terms and definitions, proposing voltage and other interface standards (including those needed to transfer test vectors among various CAD systems), and recommending performance benchmarks. They are also generating a macro set.

No standards exist for the design software. At least six languages have been proposed as standards. These are Texas Instrument's TIDAL (Transportable Integrated Design Automation Language), University of California at Berkley's CDIF (common interchange description format), VHDL (VHSIC high-level design language), workstation maker Daisy's GAIL (gate array interface language), IGES (Initial Graphics Exchange Specification), and EDIF (electronic design interface format). [21, 22] Each of the six VHSIC phase I teams is developing a set of chips using different CAD tools.

An IEEE Standards Subcommittee (IEEE SC20 ATPG) is developing a Hardware Description Language (HDL) and is considering the foregoing possibilities.

1.15 SUMMARY

In this chapter basic definitions pertinent to gate arrays were introduced. Gate arrays were compared to other methodologies. Future developments in gate arrays were mentioned. The what and why and why not questions were answered. The chapter concluded with four examples of gate array use.

REFERENCES

1. Gold, M., "Established chip vendors shift their strategy to the semicustom market," *Electronic Design,* Oct. 31, 1985, p.29.
2. VLSI Update, Dec. 1985, p.1.
3. Ohr, S., "Minicomputer turns to ECL gate arrays, reaches 8 MIPs," *Electronic Design,* Nov. 28, 1985, pp. 37–40.
4. Beresford, Roderick, "A profile of current applications of gate arrays and standard-cell ICs," *VLSI Systems Design,* Sept., 1985, pp. 62–66.
5. Hively, J., "The case for standard cell semicustom ICs," *Electronic Products,* Feb. 7, 1983, pp. 67–69.
6. Eklund, M., "Semicustom high density business," *ICECAP Report,* Aug. 28, 1981, Integrated Circuit Engineering Corp., Scottsdale, AZ.
7. VLSI Systems Design Staff, "Survey of custom and semicustom ICs," *VLSI Systems Design,* Dec. 1985, pp. 54–90.

8. Cole, B., "How gate arrays are keeping ahead," Electronics, Sept. 23, 1985, pp. 48–52.
9. "Channel-less CMOS gate array yields 50K gates," *Electronic Engineering Times,* Oct. 21, 1985, pp. 1, 12.
10. "Startup Xilinx puts its faith in reconfigurable logic array chips," *EE Times,* Dec. 9, 1985, pp. 39, 42.
11. "Logic array reconfigures itself on the fly," *Electronic Products,* Nov. 15, 1985, pp. 27–30.
12. "A simple device to cut your gate array costs by 50%," *EE Times,* Nov. 4, 1985, pp. 18–19.
13. Brayton, R., et.al., "A microprocessor design using the Yorktown Silicon Compiler," IEEE International Conference on Computer Design: VLSI in computers," *IEEE publication 85CH2223-6,* Oct. 7–10, 1985.
14. Gabay, J., "Gate arrays boast highest bipolar densities, subnanosecond speed," *Electronic Products,* Jan. 2, 1986, pp. 25–26.
15. Collett, Ronald, "Managing the VLSI explosion with silicon compilation," *Digital Design,* Sept. 1985, pp. 30–40.
16. Burich, Misha, "Fledgling silicon compilers will get best use in hands of chip designers," *Electronic Design,* Oct. 31, 1985, p. 67.
17. Mead, Carver and Conway, Lynn, *Introduction to VLSI Systems,* Reading M.A.: Addison-Wesley, 1980.
18. Ayres, Ronald P., *VLSI: Silicon Compilation and the Art of Automatic Microchip Design,* Englewood Cliffs, N.J.: Prentice-Hall, Inc., 1983.
19. "SCI's Genesis: A silicon compiler development system," *VLSI Systems Design,* Nov. 1985, p. 108.
20. "ASICS," *ICECAP Report,* March, 1986, issue 6-1, Integrated Circuit Engineering Corp., Scottsdale, AZ.
21. Nedbal, R., "EDIF standard lacks the discipline to keep its offspring compatible," *Electronic Design,* Nov. 18, 1985, p. 60.
22. Hampton, L., "EDIF breaks CAE/CAD compatibility barrier," *Digital Design,* Oct. 1985, pp. 69–72.

Chapter 2

Gate Array Technologies

2.0 INTRODUCTION

By the term *technologies of gate arrays,* the author means the underlying semiconductor technologies, which are used to fabricate the devices themselves.

Although most designers have little interest in semiconductor structures, a modicum of information is necessary to enable designers to understand what is happening inside the device. It is important for the reader to understand something of these processes in order to better appreciate and use the gate array methodology. In particular, the gate array user needs to know the characteristics of the pertinent gate array technologies, what happens when they are scaled, how they compare to one another, and something about how they are made.

It is true that noncomplex gate arrays can be designed without such information. It is also true that CAD tools available from some gate array vendors have sometimes made it possible to design more complex gate arrays without such information. However, the author's experience is that feasibility assessments, technology conversion details (such as from LSTTL to CMOS), and other design decisions can be made more intelligently with the factors enunciated in this chapter in mind. Such information is also needed to understand models in circuit simulation programs, such as SPICE and TRACAP. Systems designers will find some of the information on scaling useful for predicting what can happen when device parameters change under different manufacturing conditions and with different scalings. They will also find it useful for dealing with gate array vendors.

The treatment below is not intended to be an inclusive tutorial on semiconductor processing and fabrication. Rather, it is intended to give the reader a reasonable background and be complete enough for the purpose. Additional information will be found in the discussion of circuit simulation programs and in a variety of very readable and excellent books.

The treatment below is organized into 10 sections. Section 2.1 is a brief overview of silicon technologies used for either VLSI or standard parts or both. The purpose of that subdivision is to show how the silicon semiconductor technologies relate to one another on a family tree. In Section 2.2, gate arrays and standard parts are treated as subsets of this family tree. The dominant technologies are stated.

From the high-level overview, the discussion then jumps to a very basic level in order to give meaning to some of the terms just used. Sections 2.3, 2.4, and 2.5 present detailed discussions of bipolar, MOS, and CMOS technologies, respectively. The technologies useful for gate arrays are discussed as subsets of the bipolar and MOS subdivisions. Each type is individually described from the standpoint of the silicon structure. The aspects important to the gate array methodology are then pointed out and explained in the context of the earlier material.

Section 2.6 is the technology of interconnects. Primarily, this relates to the resistance and capacitance of the metal and silicon interconnect paths. The conditions under which this resistance and capacitance must be considered are pointed out. In subsequent chapters the rationales of routing such interconnects will be developed. Problems that contacts pose to scaling are pointed out.

Section 2.7 presents a comparison of technologies. Although this might logically be placed earlier in the chapter, a lot of its meaning would be lost without the detailed background embedded in the earlier subdivisions.

Major isolation methods are then treated in Section 2.8. Because of the importance of this topic, it is worthwhile to have it in one place. A structure called a ring oscillator is described in Section 2.9. It is important for the reader to be aware of this structure because many propagation delays of circuit elements are based on it.

Section 2.10 gives a brief description of a few new developments that hold considerable potential for VLSI. These include silicon-on-insulator (SOI) and vertical CMOS techniques. Although these have not yet been applied to gate arrays, it is worthwhile to discuss them briefly because gate arrays can be built in any technology. This subdivision is purposely kept minimal in size. There are many other fine and promising techniques that might equally well have been discussed.

For those who need to brush up on basic terms and definitions, the rudiments of semiconductors are given in Appendix 2A at the end of this chapter and for the benefit of those who have not been exposed to this material before.

The presentation assumes a reader background equivalent to an introductory semiconductor course; that is, it is assumed that the reader is familiar with the following concepts: holes and electrons, p- and n-type materials, depletion regions of semiconductor junctions, and so on.

2.1 OVERVIEW OF SILICON TECHNOLOGIES USED IN LOGIC

Although gate arrays are beginning to appear in gallium arsenide from companies such as Honeywell and Tektronix, the present discussion will be restricted to the silicon technologies. [Gallium arsenide (GaAs) is a material, as is silicon, in which many types of integrated-circuit structures can be built.] GaAs is discussed in Section 2.10 and where appropriate elsewhere.

Figure 2.1 shows the relations among the silicon technologies. The terms and definitions are expanded on later in the text.

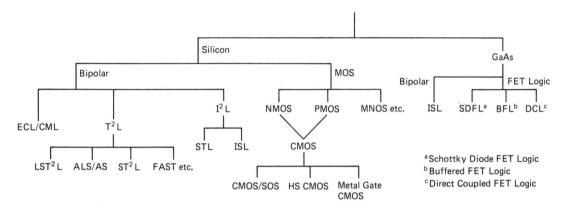

Figure 2.1 Relations among silicon technologies: SDFL, Schottky diode FET logic; BFL, buffered FET logic; DCL, direct-coupled FET logic.

It will be seen from Fig. 2.1 that two major classes of silicon structures exist. These are called *bipolar* (meaning that both majority and minority carriers participate in the transistor action) and *unipolar*. The most important class of unipolar devices for VLSI is that of MOS (metal–oxide semiconductor) as opposed to JFET (junction FET).

The term MOS itself is something of an outdated term, although it is still almost universally used. The metal in this case refers to the gate metal, and the oxide refers to the insulation between the gate metal and the silicon (semiconductor) channel being controlled by the gate. The dominant unipolar processes today, however, make use of polysilicon rather than metal for the gate material. Details are given later in the chapter.

Figure 2.1 applies to silicon technologies used for standard logic ICs. Other VLSI technologies, such as MNOS (metal–nitride oxide semiconductor) used in some nonvolatile memories, also exist.

The dominant standard IC technology family for random logic (as opposed to memories and microprocessors) used for standard parts (as opposed to gate arrays) is currently TTL (transistor-transistor logic). TTL has several permutations. These are discussed in the bipolar subdivision.

2.1.1 TTL De Facto Standard

The important point about TTL is that its voltages constitute a de facto standard to which other semiconductor technologies must interface. (Voltage values are given in Chapter 9.) Data sheets of most current standard parts advertise that the I/Os are

"TTL compatible" or sometimes "LSTTL compatible." Generally, this means that the given device is capable of driving at least one TTL load of the type specified to the required voltages or operating with the voltages from a given TTL device.

This interface standard, however, is being somewhat permuted to apply only when TTL is not fully loaded with the maximum fan-out possible (normally 10). (When driving fewer than the maximum allowed fan-outs, the voltage on TTL devices rises slightly but significantly.) The reason is that as feature sizes of MOS, used so extensively in memories and in CMOS, are scaled downward, it becomes increasingly difficult to maintain the voltage standard without seriously degrading the performances of the devices.

Moreover, studies have repeatedly shown that rarely is the maximum fan-out used. Although no two circuits are alike, most circuits exhibit pronounced peaking in the fan-out versus number of gates having that fan-out curve at a fan-out of 2. The number of gates having higher-order fan-outs generally drop off precipitously as the number of fan-outs increases. The gate array technologies form a subset of the technologies used for random logic and other standard parts.

2.2 OVERVIEW OF GATE ARRAY TECHNOLOGIES

Although gate arrays exist in nearly every technology, there are three to five (depending on which vendor one talks to) dominant technologies. Some of these, such as CMOS, have two or more permutations.

The major technologies are clearly CMOS in its three different forms, ISL/STL (integrated Schottky logic/Schottky transistor logic) and ECL (emitter-coupled logic), all discussed below; TTL in its various permutations; and IIL (integrated injection logic).

Gallium arsenide (GaAs), shows promise of being a major substrate material for gate arrays. At least five companies (TI, Honeywell, Toshiba, Harris, and Tektronix) have GaAs gate arrays or standard cell arrays operating in the laboratory. Those from Honeywell, Harris, and Tektronix are being offered commercially.

2.2 Combinations of Technologies— BIMOS and GaAs/Si

There are also combinations of technologies. Both Motorola and Hitachi are developing gate arrays in a combination of CMOS and bipolar. [1,2] In a typical configuration, most of the logic in a macro is CMOS except for the output drivers of the macro or of a peripheral device. The latter are made from a bipolar process.

The advantage of the bipolar is that it is capable of driving much higher loads without significant sacrifice of speed. (Its source resistance is much lower than that of CMOS.) CMOS offers the advantages of both low power and high density. Motorola calls its process BICMOS and is developing a 6000 equivalent gate gate array in a 1.5-μm (see Section 2.4.7) feature size.

Both Motorola and AMCC offer arrays that have TTL interfaces and require only the standard TTL voltage of 5 V and ground but which consist entirely of ECL inside (except for the ECL to TTL interfaces) and run at ECL speeds. (There is a slowdown at the TTL to ECL interfaces.) AMCC's arrays accommodate both ECL and TTL interfaces on the same chip. (In the latter case, the chip requires both ECL and TTL power supplies.)

There are also advantages to combining substrate materials. See Section 2.10.4 for a discussion of GaAs on silicon.

Of the above, CMOS is clearly the dominant technology. Nearly every gate array vendor offers it in at least one of its three forms and with good reason. CMOS offers the highest densities and the lowest powers of any of the major gate array technologies; at the same time it is able to operate at very respectable speeds.

Strictly speaking, ISL/STL grew out of and are forms of IIL. However, the performance of the former is so much different than that of the latter that a separate category is justified. Today, the main selling point of IIL is its hardness to radiation and its promise of very high bipolar circuit densities. (One manufacturer, CIC, offers an IIL array with 9000 gates.)

NMOS [of which HMOS (registered trademark of Intel Corp.) is a permutation], which is used so extensively in the large dynamic random access memories (DRAMs), is used very little in gate arrays. Interdesign offers gate arrays in this technology commercially, and Tektronix has gate arrays for internal use in NMOS; however, it is certainly not a major technology for gate arrays.

The reasons are that gate arrays are aimed at the relatively low volume uses for which NRE charges amortized over the number of chips built often dominate the chip cost. The somewhat smaller dies of NMOS with their attendant lower (exclusive of NRE) die costs are not sufficiently less costly to make a mass market. Moreover, CMOS offers significant advantages in power and speed. Finally, the process improvements to NMOS required to keep it competitive have caused its complexity to equal or even exceed that of CMOS.

The technologies described above can be built using different techniques. Two major classes of such techniques are the methods used to isolate one transistor from another and the method used to achieve alignment of the transistor geometries. Different combinations of the techniques are used by the different vendors. However, the mainstream combination currently in use is that of oxide isolation and self-alignment especially for MOS devices.

The first category includes junction, oxide, and gate isolation as well as the use of insulating substrates (specifically SOS—silicon-on-sapphire) and three-dimensional isolation techniques.

Junction isolation was the technique almost universally used until the advent of oxide isolation. Junction isolation techniques include collector diffused isolation (CDI). Reports by Ferranti indicated that bipolar gate densities of 100,000 gates per gate array chip are possible.

Three-dimensional isolation techniques include silicon-on-insulator and the dielectric isolation technique of Harris. The second category includes the important case of self-alignment using polysilicon gates.

These topics are discussed in separate sections below.

2.2.2 The Term "Substrate"

One term that will be used frequently in the discussions of both MOS and bipolar circuits is the word *substrate*. This refers to the material on which or in which other structures are to be built. It may refer to the entire wafer itself or only to that portion of the wafer being used for a particular purpose. For example, in CMOS (see Section 2.5), the NFETs are built into *p wells*. The latter are local substrates implanted/diffused into what is generally an *n*-type wafer.

2.3 BIPOLAR TECHNOLOGIES

As shown in Fig. 2.1, major bipolar technologies are ECL (emitter-coupled logic) and its variant CML (current mode logic), and IIL (integrated injection logic) and its important permutations ISL (integrated Schottky logic) and STL (Schottky transistor logic). In Europe, where it originated, IIL is called MTL (merged transistor logic).

The most widely used technology family is, of course, TTL (transistor-transistor logic). This has many permutations. Some have disappeared from the marketplace to be replaced by other permutations. Two long-standing major permutations are LSTTL (low-power Schottky TTL) and STTL (Schottky TTL). The latter is faster but more power hungry than the former. These two technologies are now being replaced by versions called, respectively, ALS (advanced low-power Schottky TTL) and AS (advanced Schottky TTL). The term TTL will be used to denote all of the permutations of TTL discussed above unless otherwise noted.

ALS and AS plus Fairchild's FAST (Fairchild advanced Schottky TTL) are all "oxide isolated." Oxide isolation is a technique employing glass (silicon dioxide) to isolate transistors from each other. It is also used with MOS devices. In both cases it increases circuit densities (transistors per unit area).

Bipolar transistors are characterized by lower "on" resistances (the resistance through which current must flow to charge or discharge an output capacitance) and higher current capabilities for a given size of device than are MOS devices. They also often exhibit lower capacitance to the driving source than do MOS transistors. Because of this, the allowable fan-outs (number of outputs that a given logic gate can drive or be output to without seriously degrading its performance) is much higher for bipolar devices than for MOS devices. The term *device* in this context refers to a logic gate. (It will also sometimes be used in this book to denote a transistor.) For example, an LSTTL logic gate has an allowable fan-out of 10 LSTTL loads. By contrast, CMOS logic gates are generally restricted to fan-outs of 4 to gates of the same size.

2.3.1 Bipolar Structures

A bipolar transistor is a three-terminal device consisting of the collector, base, and emitter. In digital designs, the emitter is often, if not generally, grounded; and the input control signal is applied to the base. An important exception is the "emitter-follower" configuration, which is widely used to provide isolation of the driving source as well as additional drive capability. It can also be used to provide level-shifting capability on the chip.

In this configuration, the load resistance is attached to the emitter lead. The emitter voltage thus "follows" the input signal applied to the base. It is assumed that the reader is familiar with basic bipolar transistor operation.

The symbolism for an *npn* bipolar transistor is shown in Fig. 2.2. In Chapter 1 it was mentioned that a wafer contains many dies, that the wafer is made from layers of conducting, semiconducting, and nonconducting material, that the layers acting in concert form the active and passive regions of the die, and that each layer of each die is formed simultaneously as the same layer of all other dies on the wafer. These concepts will be the starting point for the discussion. Definitions of common semiconductor terms are given in Appendix 2A.

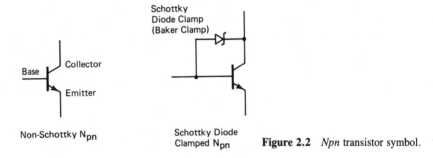

Figure 2.2 *Npn* transistor symbol.

In what follows, a number of cross-sectional views of a wafer are given. These are somewhat simplified for clarity.

The cross section of a bipolar transistor is shown in Fig. 2.3. The cross section shown is that of an *npn* transistor. This type is more widely used than the *pnp* type because of the better "beta" (current gain). The cross section shown is somewhat

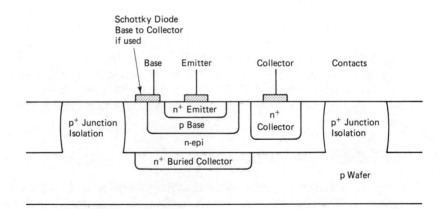

Figure 2.3 Cross section of bipolar transistor (junction isolated).

generic to the important bipolar technologies, ECL, ISL/STL, and TTL in its various forms. The process parameters (dopings, spacings of elements, depths of diffusions, numbers of elements, and placement of metal to produce Schottky junctions) will vary among the various technologies and vendors. However, the reader can be introduced to the subject; and the important features can be explained with the aid of Fig. 2.3.

The important features of Fig. 2.3 are as follows. First, there is a "buried layer" embedded in the silicon starting material. The purpose of this layer is to provide a low-impedance path from the collector to the emitter and hence to provide low collector resistance. This collector resistance is the "on" resistance referred to earlier, which is used to charge and discharge the capacitance of the succeeding stages. The current path involved from collector to emitter is shown in Fig. 2.4.

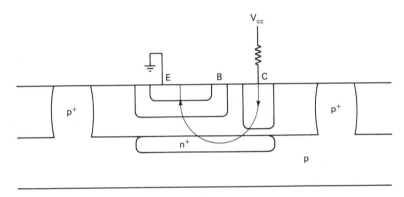

Figure 2.4 Current path from collector to emitter.

Epitaxial region. The layer of material on top of the buried contact and the basic silicon starting material (i.e., the wafer itself) is the *epitaxial layer*. This is crystalline silicon which is grown to provide a highly pure structure for implanting or diffusing the base, emitter, and collector.

A major difference between MOS and bipolar technologies is the use of an epitaxial layer in the latter. Such a layer is rarely used in current bulk MOS (although TI does use it in their 64K DRAM process). It *is* used in a number of newer MOS processes. It is also used in the form of CMOS called CMOS/SOS (which is not a bulk CMOS process). The layer is somewhat difficult to put down, and its characteristics are crucial.

The epitaxial region is needed in bipolar because two successive changes of semiconductor type must be made. This double change of type is not needed in MOS devices of only one type (i.e., all PMOS or all NMOS) and accounts for some of the simplicity of the MOS manufacturing process. (There *is* a double change of type in CMOS, in the "*p*-well" structure. However, the requirements delineated below for both low impedance and high breakdown are not as critical for CMOS as they are for bipolar. See Section 2.5.)

The two changes are as follows. First, the epitaxial region is doped with *p*-typed impurities (which have "holes") to create the base region. Then part of the base region is doped to the opposite polarity with *n*-type impurities to create the emitter inside the base region. The collector is formed in one or more steps. Because its depth into the epitaxial region is greater than that of the emitter, a separate masking and processing step is sometimes used to provide a "deep collector." The second step fills in the top of the collector region and takes place at the same time as the emitter diffusion. Reference to Fig. 2.3 shows this. As geometries shrink, some of this information will change. The reader will also appreciate that this is a short, simple explanation of a complex process.

Current flows from the collector through the base region to the buried collector through the epitaxial region to the collector diffusion. The purpose of the epitaxial region is to provide a low-resistance path which has a reasonably high breakdown voltage. Control of the epitaxial thickness is therefore essential. The thin epitaxial region between the buried collector and the base–collector junction has a low enough donor concentration to provide a high breakdown voltage but offers very little collector series resistance.

The epitaxial region is formed in several ways. An older method is by passing a gas, such as silane (silicon tetrachloride), over silicon at a very high temperature (1200°C). Newer methods are LPE (liquid-phase epitaxy), MBE (molecular-beam epitaxy), VLE (vapor levitation epitaxy)[3], MOVPE (metal-organic vapor phase epitaxy), and the use of LPCVD (low-pressure chemical vapor deposition). The reader is referred to the literature for more information. [3, 4, 5]

Epitaxy is called "single crystal" even though there are normally a few defects in the otherwise crystalline matrix. Epitaxy has long been an important part of bipolar fabrication and is being used increasingly in MOS fabrication.

Typical thicknesses of epitaxial layers are in the range $\frac{1}{2}$ to $1\frac{1}{2}$ μm for bipolar processes and $\frac{1}{2}$ μm for CMOS/SOS. Typical growth rates are in the range of 0.2 to $1\frac{1}{2}$ μm/min in the CVD (chemical vapor deposition) processes currently used in production. Another process, MBE [5], allows very precise control of the epitaxial layers.

Ion implantation and diffusion of impurities. Ion implantion is used to provide what is known as a "predep" (predeposition) of impurities at the surface of the material. It is used because the dosage can be controlled with far greater precision than the now-almost-ancient ways of passing gases containing such impurities over wafers. The impurities are then "driven in" (interstitially diffused) to the material at high temperatures.

The reason that drive-in diffusion is needed is that the drive-in depths are greater than those which can be accomplished using ion implantation alone without very large accelerating voltages of the ion implanter and consequent excessive damage to the silicon crystal lattice. For a given accelerating voltage, the penetration depth is inversely proportional to the atomic weight.

As an example, at an accelerating voltage of 300 keV (kilo electron volts), boron with an atomic weight of 10.8 will be implanted to a depth of about 1 μm in silicon. By contrast, phosphorus, arsenic, and antimony, with atomic weights of about 31, 75, and 121.8, respectively, will be implanted to mean depths of only about 0.39, 0.16, and 0.115 μm, respectively. Almost all current bipolar structures have requirements that exceed all except the first figure. Laboratory devices are being built which have shallow enough dimensions to permit the use of all ion implantation.

An annealing step at a relatively high temperature is needed to repair the damage to the crystal lattice caused by the ion implant.

The next important feature that the reader should note in Fig. 2.3 are the contacts to the emitter, base, and collector. They are of two types, ohmic and Schottky. The Schottky contact is the metal overlapping the base and the collector. This provides the very low voltage (100 mV) "Baker clamp" used to prevent the transistor from going into saturation. This feature is present in Schottky permutations of TTL and in ISL/STL but is not present in ECL.

Nature of contacts. As noted above, two main types of semiconductor contacts are Schottky barrier and ohmic or resistive. As the name implies, the latter breaks down at zero voltage and hence provides a voltage that varies linearly with current.

The Schottky contact is in reality a Schottky diode. It clamps the voltage at a level that increases very little with increase in current once the "knee" of the curve is passed. Its normal purpose is to prevent bipolar transistors from going into saturation (and increasing the times required to turn them off and, hence, their propagation times). The Schottky contact is also used to provide isolation. Examples are the Schottky contacts at the outputs of the collectors in IS1/STL gate arrays.

The clamp voltage of a Schottky diode depends on the "barrier height" of the materials used. Different metals have different Schottky clamp levels. In the STL technology, this factor is of critical importance. There, the output voltage swing is dependent on the *difference* between the barrier heights and hence clamp voltages of two different kinds of Schottky diodes fabricated on the same chip.

Isolation methods will be discussed in a separate section. The most common isolation method currently is oxide isolation. It replaces the junction isolation techniques formerly used.

Characteristics of two major bipolar technologies, ISL/STL and ECL, will now be discussed. Because of its extremely wide use in standard logic families, a great deal of information is readily available on TTL and its permutations. Therefore, the discussion will be restricted to ISL/STL and to ECL.

2.3.2 ISL/STL Technology

In the ISL/STL technology, each transistor is used as an inverter in the grounded emitter configuration. ISL and STL are very similar to one another, differing primarily in the way in which the collector is clamped to the base of the *npn* transistor. In ISL this clamp is via the forward voltage drop of a *pnp* transistor, as shown in Fig. 2.5.

In STL, a Schottky diode is used between base and collector in a manner similar to that of the Baker clamp of standard LSTTL. Figure 2.6 shows this. The object in all cases is to keep the *npn* transistor out of saturation. The voltage swings of

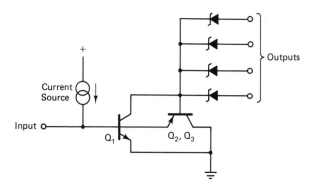

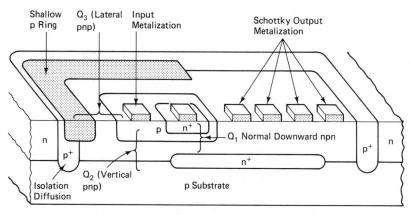

Figure 2.5 ISL structure. Note transistor clamp to base. (Courtesy of Signetics.)

these devices is very small, being typically about 150 to 200 mV. For this reason some rather wide metalization lines are used for the power and ground buses. In the Signetics first-generation version of ISL, for example, the top metalization layer is 12 μm wide and the first layer is 6 μm wide. (Signals are also routed on these layers.)

The voltage swing of STL is the difference between the barrier heights of the clamp diodes and the output diodes. Proponents of ISL claim that building two different types of Schottky diodes with two different barrier heights on the same chip is very difficult. However, TI and other manufacturers seem to have mastered it.

Both ISL and STL are "collector fan-out" logics as opposed to the "emitter fan-in" logic of TTL and its permutations. Each *npn* transistor has five collectors (in the versions of ISL and STL offered by Signetics and TI, respectively) of which any four can be used. The five are there for flexibility in layout. Each of the five is isolated from the others by a Schottky diode and has a fan-out of 1. Figures 2.5 and 2.6 show typical structures. In these diagrams, one of the output diodes is connected to the base input.

ISL/STL have been also been called AND/NAND logic. This is discussed more fully in Chapter 10, but a brief explanation will be given here.

Unlike CMOS, wire ANDing is fundamental to ISL/STL design. The outputs

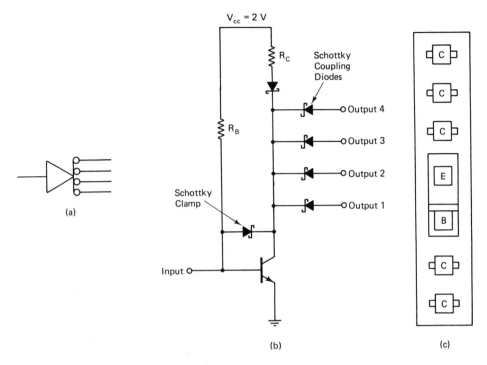

Figure 2.6 STL structure: (a) logic; (b) schematic; (c) topography. Note diode clamp to base. (Courtesy of Texas Instruments, Inc.)

being isolated by Schottky diodes permit this. Any inversion in the signal requires use of a gate, which in this case is one transistor. Therefore, if a four-input gate is considered, an AND gate would require no transistors, a NAND gate would require one transistor (wire AND of four inputs with the transistor being used for the inversion), a NOR gate would require four transistors (four inversions followed by a wire ANDing), and an OR gate would require five transistors. This is discussed more fully in Chapter 10.

Variation of speed with injection current. The speed of any IIL (of which ISL/STL are permutations) device can be changed by altering its injection current. This is the current injected into the base. Figure 2.7 shows this for STL. The same curve shows that by proper selection of the operating point, temperature variations can be minimized or eliminated. The injection current required to effect such minimization may not yield a high enough speed. Hence there is a trade-off among the parameters of speed, power dissipation, and temperature dependence.

2.3.3 ECL Technology

The third major gate array technology is ECL (emitter-coupled logic). This is the highest-speed silicon technology, offering gate speeds in the subnanosecond range. It is also the highest-powered silicon technology.

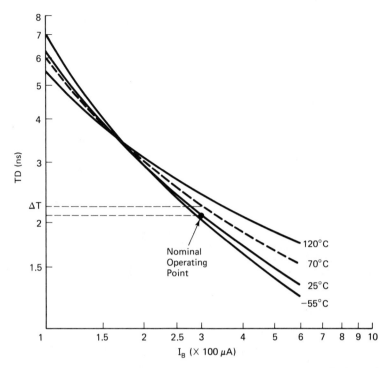

Figure 2.7 Variation of speed of STL with injection current (F.O. = 4). (Courtesy of Texas Instruments, Inc.)

The basic structure of ECL is the differential pair driven by constant-current sources, as shown in Fig. 10.9. When the input signal crosses (becomes greater than) the reference signal threshold, the current is suddenly switched from the reference transistor to the signal transistor, which in turn raises the voltage of the collector resistor of the reference transistor. Monitoring the latter produces the desired voltage switching from one logic state to the next. Monitoring the collector resistor of the other member of the differential pair produces the inverse of the first signal. Hence both phases of the signal are available. This is one of the advantages of ECL.

Just as ISL/STL can be thought of as AND/NAND-type logic, ECL can be thought of to some extent as OR/NOR-type logic. However, ECL is not as strongly OR/NOR as ISL/STL is AND/NAND (see Chapter 10).

Power versus speed tradeoffs. Some ECL gate arrays offer user selectability of speed and gate power. The Fairchild GE1000 series is an example of this. The speed and power is selectable at the macro level. Trade-offs are (per equivalent gate)

0.4 ns at 4 mW

0.6 ns at 2 mW

1.1 ns at 1 mW

Internal resistors are selected by the wiring configuration of the macro to effect the selection. Typical propagation delays and power dissipations for common functional elements are given in Table 2.1.

TABLE 2.1 PARAMETERS FOR COMMON MACRO ELEMENTS

Macro	Delay(ns)[a]	Power dissipation (mW)[b]
High-speed, high-power mode		
D flip-flop	0.9	25
4 in XOR	0.8	18
4-to-1 MUX	0.7	20
Full adder	1.1	27
5 in OR/NOR	0.5	14
Low-speed, low-power mode		
D flip-flop	2.5	10
4 in XOR	2.2	7
4-to-1 MUX	2.0	7
Full adder	3.2	11
5 in OR/NOR	1.4	6

[a] Propagation delay for worst signal edge.
[b] Typical for fan-out 3.

2.4 FUNDAMENTALS OF MOSFETS

MOS transistors are known as FETs (field-effect transistors) and are referred to as MOSFETs or sometimes simply as FETs. Major MOS technologies are NMOS and PMOS. The corresponding transistors are called NFETS and PFETS, respectively, or less commonly as *n*-channel and *p*-channel transistors. CMOS is comprised of both NFETs and PFETs.

MOSFETs have the opposite properties of the bipolar devices discussed in Section 2.3; that is, their "on" resistances and circuit densities are higher and their fan-outs and current-handling capabilities are lower for a given size of device.

The above are generalizations. By scaling and doping the semiconductor to enable higher-speed, lower-voltage operation, or higher circuit density, the generalizations above can be reversed in a given specific instance.

Because CMOS is the dominant technology of gate arrays and because there are virtually no NMOS or PMOS gate arrays, this discussion is heavily slanted toward CMOS.

To fully appreciate CMOS, it is necessary to understand the properties of MOS (metal–oxide semiconductor; the term is still used even though the gates are currently generally polysilicon) in general. The information presented below will be used in several places in the remainder of the book, and the reader is advised to be sure that he or she clearly understands it before moving on. Of particular importance are the topics of *sharing of* MOS *transistor source/drains* and *sizes of* MOS *transis-*

tors and how they affect performance. The intent is to present only as much information as is needed for the task at hand. For other information, the reader is referred to any of the several excellent books on MOS [6], [7], [8], [9].

2.4.1 Terminal Connections

A MOS transistor is a four-terminal device. The four terminals are the source, drain, gate, and substrate. The term "gate" in this case means the physical input to the transistor. As noted in Chapter 1, the term "gate" is also used in two other contexts; (1) logic gate, NOR, etc.) and (2) to denote the number of cells in a given gate array.

The *source* and *drain* are physically completely interchangable. A distinction that is sometimes made is that current flows out of the drain. For this reason, the terminal, which becomes the output, is sometimes labeled the drain. The terminal connected to ground or to the supply voltage is usually labeled the source. When FETs are connected in series, the more positive source/drain of PFETs and the source/drain closer to ground of NFETs are sometimes denoted the sources of those elements. This convention will generally be followed in this book. The term *source/drain* will be used to denote terminals that are connected to neither an output nor a power supply terminal and/or for which nothing is gained by making a distinction.

The *gate* of the MOSFET is the physical input to it of the control signal. The gate represents almost a purely capacitive load to the driving source. This point will be important in the discussion of logic structures and feasibility analysis.

V_{dd} and V_{ss}. Repeatedly throughout this book, the terms V_{dd} and V_{ss} will be used. These refer respectively to the most positive and most negative connections of the MOSFET. For purposes of this book, V_{ss} will always be at ground potential, and V_{dd} generally will be in the range 3 to 8 V unless otherwise noted. This is the way in which most CMOS devices operate. (The voltage range of V_{dd} is sometimes greater or less depending on the feature sizes involved.)

Substrate bias. An NFET usually has its substrate connected to V_{ss}, and a PFET usually has its substrate connected to V_{dd}. However, substrates can be connected to a voltage which is neither V_{dd} nor V_{ss}. In a technique called *back gate bias*, the substrate is purposely reverse biased to speed up the FET and to raise the absolute value of the threshold voltage. This technique is currently not used in CMOS gate arrays, however, and will not be discussed further. The interested reader is referred to reference [9], pp. 56–57.

2.4.2 Enhancement versus Depletion Mode

Both PFETs and NFETs can be made to be either *enhancement-mode* or *depletion-mode* devices, depending on, among other parameters, the work function of the gate. This is set during the manufacturing process and is not alterable by the user. The simplest definition of depletion- and enhancement-mode devices is that the former are "on" until turned "off," whereas the opposite is true for enhancement-mode devices. All FETs used in CMOS gate arrays are enhancement-mode devices, and it is these that will be discussed exclusively in this book.

2.4.3 Symbols Used

The IEEE standard symbol for an enhancement-mode FET is shown in Fig. 2.8(a). However, because the only interest in MOSFETs in this book is in those used in CMOS gate arrays, the simpler symbol shown on the right in Fig. 2.8(c, d) will be used. Most gate array manufacturers use this also, for the same reasons. A mental "crutch" helpful in remembering which way the arrow (which denotes the substrate connection) points is the word "pout," which means that for a PFET, the arrow points out.

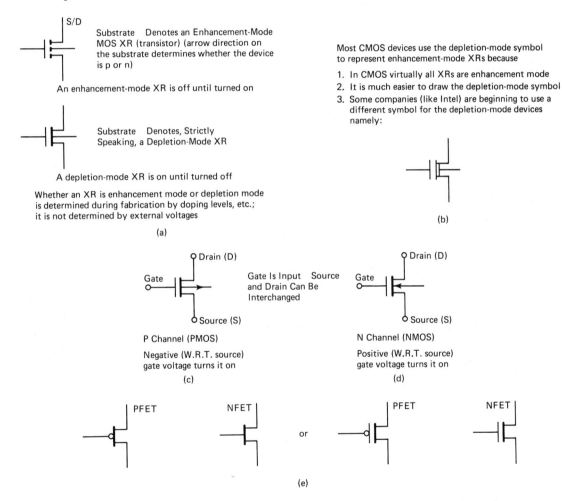

Figure 2.8 Symbols for FETs: (a) IEEE standard symbol; (b) simplified symbol sometimes used to distinguish enhancement from depletion; (c), (d) a more common representation (to be used in this book); and (e) additional alternatives.

Sec. 2.4 Fundamentals of MOSFETS

Another symbology used by some vendors, such as LSI Logic, is that shown in Fig. 2.8(c). In this symbology the bubble denotes the PFET, as it does in the symbology in Fig. 2.8(d). Figure 2.9 gives the properties of MOS transistors and defines a "string."

- Four-terminal device: source (the reference point), drain, and gate (the input)
- Source and drain are completely interchangeable physically
- Source and drain exist at opposite ends of the "channel"
- Whether the channel conducts or not depends on the gate (input) voltage relative to the source voltage
- Source and drain can be shared with adjoining XRs if done properly
- V_{dd}, drain voltage; V_{ss}, source voltage; V_{ds}, drain-to-source voltage; S, source; D, drain; XR, transistor

(a)

- "n-Channel string": the series connection of two or more NMOS transistors (XRs)

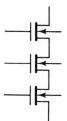

- Analogous statement holds for "p-channel string"

(b)

Figure 2.9 (a) Properties of MOS transistors; (b) the term "string."

2.4.4 Cross-Sectional View

Figure 2.10(a) is a cross-sectional view of a MOSFET. An NFET has source/drains which are doped with *n*-type impurities that have excess electrons. Electrons are therefore the carrier type used for conduction between the source and the drain when the transistor is on. Unlike bipolar transistors, the other type of carrier (in this case the hole) plays no part in the conduction between source and drain. For this reason,

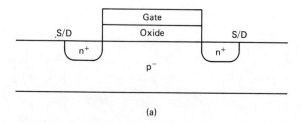

(a)

Figure 2.10 (a) Cross-sectional view of NFET; (b) V–I curve of NFET. Note: Scale is expanded near the origin. Actual values will vary with size and processing.

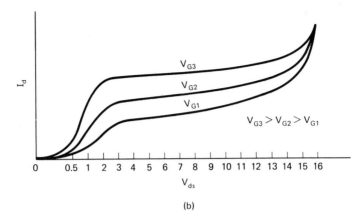

(b)

Figure 2.10 (cont.)

MOS transistors are called "unipolar." For a PFET, the source/drains are of *p* material; and the conducting carriers are holes.

The gate is physically located between the source and the drain and is separated from the substrate by means of *gate oxide*. This is silicon dioxide (glass), whose characteristics are especially important to the operation of the device. These characteristics include freedom from pinholes and uniformity of thickness and of dielectric constant.

2.4.5 Transistor Action

The V–I curve of an NFET is given in Fig. 2.10(b). In this case, I is the drain current and V is the drain-to-source voltage. The parameter is the gate voltage.

To make a useful device, a load must be connected to the drain of the NFET to make an inverter. Later, this resistive load will be replaced with a switchable load in the form of a PFET. For purposes of discussion, the drain-to-source voltage, V_{ds}, has been made greater than the range of gate voltages.

Threshold voltage V_t. As the gate voltage is increased in the positive direction, electrons from the *p*-type substrate are attracted to the interface between the gate oxide and the substrate, although they remain in the substrate because the gate oxide is an insulator. At a given value of voltage, conduction measured in picoamperes begins to occur between source and drain. The voltage at which this occurs is called the *threshold voltage* and is a very important parameter of the transistor. Threshold voltage depends in a complex fashion on such factors as substrate channel doping, gate oxide thickness, gate material, and the charge in the gate oxide. (For those steeped in semiconductor theory, the threshold voltage is the gate voltage that produces "strong inversion." The latter definition is beyond the scope of this book, however.) Typical values of threshold voltage are in the range 0.6 to 1.2 V for the silicon gate (see below) transistors used in gate arrays.

The scale in Fig. 2.10(b) has been expanded around the origin to show the threshold effect. As the gate voltage is increased still further at a constant value of drain-to-source voltage, the drain current increases linearly.

Pinch-off. The magnitude of the current at any given gate voltage also depends on the drain-to-source voltage. At any given value of gate voltage, there is a value of drain-to-source voltage at which all the electrons attracted by the positive gate voltage are collected by the drain. This condition is known as *saturation* or *pinch-off*, and the region to the right of the dashed line in Fig. 2.10(b) is known as the *saturation region*. Increasing the drain-to-source voltage still further produces essentially no further increase in drain current, until a point is reached at which the drain current will once again begin to increase sharply.

The resistance between source and drain varies greatly with gate voltage. Below threshold, the resistance is on the order of 10^{14} Ω. Above threshold, the resistance (the "on" resistance) of the channel between source and drain will decrease to values in the hundreds of ohms to thousands of ohms, depending on the size of the transistor, the feature size of the technology (e.g., 2 μm), the resultant dopant and other levels, and the gate and drain-to-source voltage.

2.4.6 Logic Voltages Pertinent to CMOS Devices

For purposes of the discussion on CMOS that follows, a positive (with respect to the source) voltage on the gate of an NFET equal to 70% or greater of V_{dd} will turn the device on and one less than 30% of V_{dd} will turn it off. The same signal that turns the NFET on will turn the PFET off, and vice versa. This assumes that the input signal goes to both a PFET and to an NFET. This is normally the case.

Strictly speaking, the types of PFETs used in bulk CMOS gate arrays require a negative voltage to turn them on. The source of the PFET in CMOS structures is at or close to V_{dd}. A logic low of 30% or less of V_{dd} is actually negative with respect to the source voltage and hence turns the device on. Conversely, a high voltage on the gate that is roughly equal to V_{dd} is essentially zero voltage with respect to the PFET source and hence turns the PFET off.

2.4.7 Sizes of MOS Transistors

An important parameter of FETs is the *W/L* or *width-over-length ratio*. The larger this ratio, the lower is the "on" resistance of the FET. Figure 2.11 is a three-dimensional view of a MOSFET showing the pertinent parameters. These are set in the design of the background layers by the designer of such layers and are not under the control of the logic designer who designs a logic circuit on such a gate array.

Length. The source and drain are interchangeable. The *length* of the device is the distance between the source and the drain. The *width* of the device is the dimension of the source and drain that is perpendicular to the length. To the uninitiated this seems completely backwards.

There are three lengths that are of interest: the electrical gate length, the effective gate length, and the drawn gate length, which is set by the lithography. In a

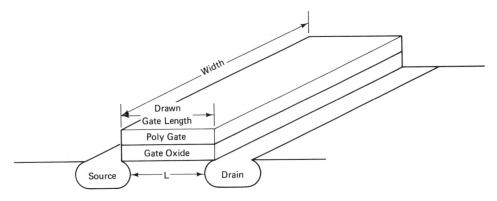

Figure 2.11 Three-dimensional picture of NFET showing width and lengths. L, Electrical length. Effective gate length (EGL, not shown) is a mathematical artifice used to account for parasitic capacitance. For each FET described in TRACAP, W, L, and EGL are specified. EGL depends on W and gate length. Current flows from source to drain if gate voltage is proper for that type of MOS transmitter. In current usage, effective gate length normally denotes electrical length.

self-aligned (sometimes called "silicon gate," although the gate could be made of a refractory metal, such as molybdenum, instead of polysilicon) gate MOSFET, this is the feature size generally 2 μm for the current generation of CMOS gate arrays.

The term "feature size of X_1 μm" (where X_1 is a number like 2 or 3) generally, although not always, implies a drawn gate length of X_1 μm. It is far better to specify exactly what is being referred to by the feature size, but this is not always done.

The drawn gate length differs from that on the mask. Because some photoresists (see below) shrink and others expand when exposed and developed, the mask-drawn lengths and widths must be adjusted accordingly. (See Chapter 11.)

Photoresist is a coating put over the wafer and patterned with an image (interconnects among transistors or its reverse in the gate array case generally) that protects the wafer surface from etchants and dopants. More will be said about this in Chapter 11.

The second length is the *electrical gate length*, frequently called *effective gate length* (but see discussion that follows). It will be noted that the source and drain bulge toward each other in Fig. 2.11. The reason is *lateral diffusion*. The normal manufacturing process is to use ion implantation to do a "predep" (predeposition of dopants) and then to drive the impurity into the source and drain at a very high temperature, such as 1050 or 1100°C, to the desired depth. The elevated temperature allows what is primarily interstitial [10] diffusion to take place.

Unfortunately, it also allows lateral (sideways) diffusion to take place, which, in effect, narrows the distance between source and drain. The deeper the diffusion, the greater is the narrowing. If an ion implant could be used for drive-in, a lot of the narrowing would be eliminated. Production single-ion implanters are not capable of doing this for the current 2-μm technologies. A 300-keV (kilo-electron volt) ion implant of phosphorus in silicon goes in only 0.3812 μm, for example. [11]

The narrowing in the gate length can be significant. In one of the more popular 2-μm processes in use today, the electrical gate length is 1.6 μm.

The problem with the narrower gate is that it limits the maximum voltage that can be applied between source and drain. A popular 3-μm process is limited to 6 V between source and drain unless wafers are selected which have larger gate lengths or breakdowns. (The narrowing stated takes into account the manufacturing tolerances. If the voltage is increased above a certain value, an effect called *punch-through* (see below) will occur between source and drain. When this happens, the depletion region of the reverse-biased drain extends to that of the source, allowing charge carriers to be injected directly from the source into the drain depletion region, where they are swept directly by the electric field over to the drain. Because there is a plentiful supply of carriers in the source, the current will increase until it is limited by the external circuit.

If the source and drain are doped more heavily, the punch-through voltage is increased; but the transconductance, G_m, and hence the ratio of G_m to the input capacitance of the device, are decreased because the mobility is decreased. This in turn decreases the speed–power product of the MOSFET. [12]

The last length is a fictitious parameter called the *effective gate length* (EGL), which is sometimes used to take into account the change in parasitic and edge capacitances with increase in gate width. The parasitic and edge capacitances *increase* with gate width but *decrease* as a percentage of the total gate capacitance as the gate *width* (not length) is increased.

An expression for the EGL is

$$EGL = GL + k_1 + \frac{k_0}{W} \qquad (2.1)$$

where GL is the gate length, W the width, and k_0 and k_1 are parameters associated with capacitive overlap effects that depend on GL. The load capacitance that the gate presents to a driver is also given as a function of the same parameters. The EGL decreases as W increases because the overall capacitance of the gate increases so much. As noted above, EGL is an artifice to account for the parasitic capacitances of the gate.

Width. The width is a variable that the designer of gate array background layers (i.e., not the user) uses to adjust the "on" resistance of the device. Typical W/L values for CMOS gate arrays in the 3- to 5-μm feature sizes are 8.5:1 to 13:1. Corresponding "on" resistances are in the vicinity of 6000 Ω for 5-μm PFETs and in the vicinity of 400 Ω for 3-μm NFETs. These are for the array FETs as opposed to the buffer FETs.

The reader will appreciate that three W/L values (corresponding to the three lengths) are possible. Normally, the gate length used is either the drawn gate length or the electrical gate length. The user should always ascertain which is being referred to. The W/L is an approximate value used to "ballpark" the size of a device.

2.4.8 Punch-Through and Avalanche

The next few paragraphs describe two important phenomena that can occur in MOSFETs. The discussion assumes an understanding of basic semiconductor theory.

Those who do not have this background are referred to the several excellent books on the subject listed elsewhere. The topics to be discussed are drain-to-source punch-through and avalanche breakdown.

Punch-through. Around both the source and the drain are *depletion regions* in which there are no free (i.e., mobile) charges. Because the drain is reverse biased, the depletion region of the drain may extend into that of the source if the drain-to-source voltage is made sufficiently high. When this happens, charge carriers (electrons in the case of an NFET) can be injected directly into the drain depletion region and thence to the drain without being under control of the gate. This is called punch-through.

The importance of punch-through is that it places an upper limit on the drain-to-source voltage that can be used between source and drain for a given feature size, doping, and so on (see the scaling rules below). Because the speed of a MOS device is directly related to the drain-to-source voltage, the speed is thus limited by the need to operate at low-enough voltages to avoid punch-through.

Avalanche Breakdown. The second type of breakdown that can occur is avalanche breakdown. This occurs when the junction depletion region is large enough to accelerate charge carriers with enough energy to ionize formerly neutral silicon atoms and, thus, generate new carriers. The new carriers themselves generate carriers, and the process continues until limited by the external circuit.

2.4.9 Scaling of MOS Transistors

Unlike bipolar circuits, MOS transistors were for many years scaled according to a fairly standard set of scaling rules first published by IBM researchers. [13] Because of the impact that scaling can have on device performance, it is useful for the reader to be aware of its effects. The IBM rules are given in Fig. 2.12. They are based on keeping the power density (actually the electric field, but this term means little to most digital designers) constant as the devices are scaled.

One potential problem with the scaling rules which has led many, if not most, manufacturers to use modified versions of them is that the voltage must be reduced by the scaling factor. Because it is still to a certain extent a "TTL world," this is unacceptable. Also note that the channel doping concentration must increase by the scaling factor. One deleterious effect of this is that the G_m (transconductance) of the FET decreases because the mobility decreases as the doping increases (see Appendix 2A). This factor has also led to a modification of the scaling rules. A typical modification is also given in Fig. 2.12.

The shortened channel length due to the scaling can also produce punch-through, as noted above, unless the channel doping is increased and/or the voltage decreased. A second problem is "hot electron" emission into the gate oxide. Hot electron emission occurs when electrons composing the channel current are accelerated by the the large electric field in the drain depletion region of a FET operating in the saturation region. If the kinetic energy gained by such electrons is large enough

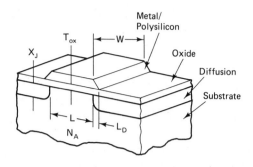

	Classical	Modified ($\alpha > 1$)
Substrate Doping, NA	S	S
Device Dimensions, T_{ox}, L, L_D, W, X_J	1/S	1/S
Supply Voltage, V	1/S	1
Supply Current, I	1/S	−1
Parasitic Capacitance, WL/T_{ox}	1/S	$1/\alpha S$
Gate Delay, VC/I	1/S	$1/\alpha S$
Power Dissipation, VI	$1/S^2$	1
Critical Charge, Q = CV	$1/S^2$	$1/\alpha S$
Power-Delay Product	$-1/S^3$	$1/\alpha S$
Power Density, VI/A	1	S^2

Figure 2.12 MOS scaling rules.

to overcome the barrier activation energy at the Si–SiO₂ interface of 3.2 eV (electron volts; 1 eV = 1.6×10^{-12} erg), they will be injected into the gate oxide.

2.4.10 Self-Alignment and Silicon Gate

Self-alignment. In the literature on ICs, one often reads the term *self-alignment*. This refers to using a gate of a MOS transistor as a mask for the source and drain diffusions. (The technique can be generalized to more than just the gates of MOSFETS, although this is where the greatest usage lies.) To be usable, however, the material used for the gate must be capable of withstanding high temperatures used to "drive in" the impuritites of the source and drain. These temperatures are on the order of 1000 to 1100°C. Because aluminum has a melting point of 660° C, it cannot be used. Currently, polysilicon is used almost universally for the gate material. As noted elsewhere, refractory metals, such as molybdenum and titanium/tungsten, are being used for this purpose in the laboratory and to a small extent in non-gate-array CMOS devices.

A benefit of doped polysilicon is that its "work function" (see below) enables enhancement-mode NFETs to be formed on 100 (Miller indices of crystal orientation) silicon. This, in turn, allows the higher mobilities possible in this orientation to be used. Other materials considered for the gate material have to meet the same conditions. Figure 2.13(a) shows the sequence of steps involved in silicon gate process-

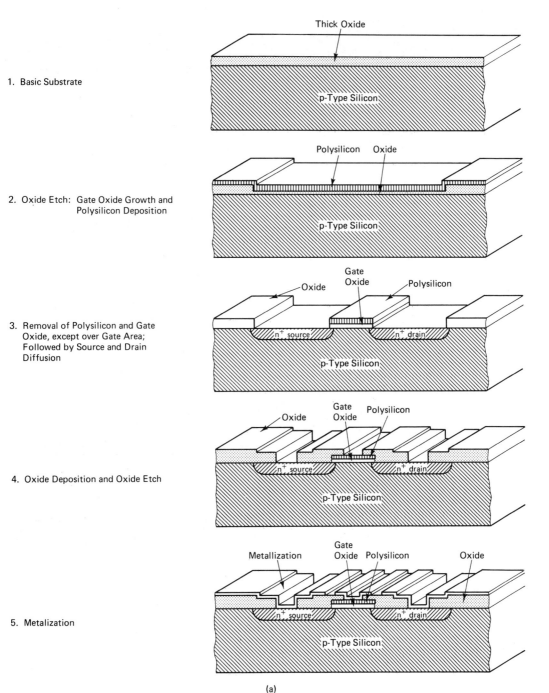

Figure 2.13 (a) Sequence of steps in silicon gate processing; (b) ISO-CMOS process of MITEL; (c) design and analysis of self-aligned gate MOSFETs, and terminology. (a) Courtesy of Integrated Circuits Engineering Corp. (b) Courtesy of MITEL.]

The diagrams show the process stages for Mitel's ISO-CMOS process. This process uses a total of 9 masks. P-type transistors are formed directly in the n-type substrate which will be biassed to the most positive voltage used (V_{dd}). N-type transistors are formed in p-type wells formed by inverting the n-type substrate with a high concentration of p-type dopant. The p well will be biassed to the most negative supply voltage (V_{ss}). The process has self-aligning gates, as the polysilicon gate can be used to mask the source/drain diffusion areas. Interconnect between the transistors is primarily with metal. However, the polysilicon is used as a second interconnection medium, greatly facilitating interconnect design. Isolation to prevent the formation of spurious transistors in the field-oxide region is achieved by using a very thick, recessed oxide.

The 5-μm design rules coupled with oxide-isolation and self-aligning gates gives the process major speed and packing-density advantages over metal gate CMOS. While maintaining similar power dissipation characteristics to metal gate CMOS, packing densities approach those of NMOS in similar geometries and speeds are comparable with low-power-Schottky technology.

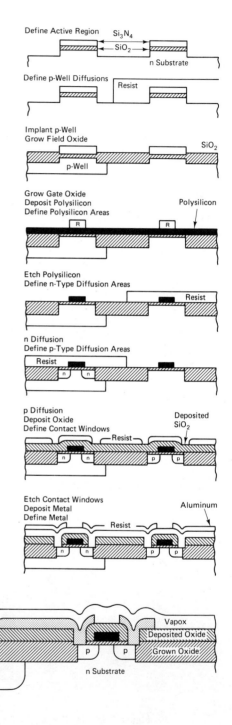

(b)

Figure 2.13 (*cont.*)

- W/L (width over length) ratio sets "on" resistance and drive requirements for a given process; Process includes gate length specification
- W/L is made >1 by increasing width of channel
- W/L is made <1 (rare cases for high-Z devices) by increasing L; otherwise, L is normally at its process-defined minimum
- For self-aligned gates, L is set by gate length
- Increasing W/L gives faster response (lower "on" resistance) of that FET but requires higher drive

Note: For gate arrays, W/L of the individual transistors (XRs) is fixed. However, paralleling of XRs in gate arrays can be used to increase the effective W/L at the expense of increasing the load on the driver of the paraleled XRs.

Si gate CMOS = silicon gate CMOS
 = self-aligned (gate) CMOS = poly gate CMOS
 = polysilicon gate CMOS

HS (high-speed) CMOS (semistandard term) is both Si gate and oxide isolated. It is bulk CMOS as opposed to CMOS/SOS.

But . . . a Si gate device does not have to be oxide isolated:
- Si gate structures are also used extensively in NMOS and CMOS memories
- A Si gate has a metal contact to it (to permit interconnect), but this does not make it a metal gate device
- A self-aligned gate does not necessarily have to be made of polysilicon

Crossover block = poly block = crossunder block
 = nonmetallic conductors on the chip which enable lines to cross without touching

(c)

Figure 2.13 (*cont.*)

ing. Figure 2.13(b) shows the silicon gate process of MITEL, called the ISO-CMOS process, which has been widely licensed.

The work function of a material is the energy required to remove an electron with energy equal to the Fermi energy from the material to vacuum.

As noted elsewhere, virtually all CMOS ICs use enhancement-mode FETs. The electron mobility on the 100 crystal orientation n-type inversion layers is almost three times the hole mobility of p-type inversion layers formed on crystalline silicon with 111 Miller indices. The latter is the orientation that historically has been used for bipolar devices for years.

The polysilicon itself enables self-aligned "silicon gate" structures to be formed. Hence, the benefits of both higher electron mobility and silicon gate can be obtained by using 100 crystalline indices.

It might be suspected that a totally ion-implanted fabrication sequence would allow lower-melting-point-temperature materials to be used. Certainly as feature sizes decrease, ion implantation of the source/drain becomes much more feasible because the depths to which the impurities are driven are less. (Even in 3-μm CMOS, an ion implant is used only to provide *predeposition*, which is the controlled amount of impurities that are to be driven in.)

The temperatures *are* lower with an all-ion-implanted process. However, the gate material must still be able to withstand temperatures of several hundred degrees Celsius even in cases where no diffusion and all-ion implantation are used. The reason is that ion implantation causes damage to the silicon lattice which must be annealed out at temperatures in this range.

Decrease of Miller effect capacitances. Self-alignment significantly decreases the Miller feedback capacitance due to the gate overlapping the drain. The gate-to-source overlap capacitance is essentially eliminated as well. The elimination of these capacitances, in turn, speeds up the signal. If self-alignment were not used, then a greater overlap, and hence greater capacitances, would have to be allowed to accommodate mask alignment tolerances.

Other names for self-aligned gates are *poly gate* and *silicon* or *SI gate*, although as noted above, the gate does not have to be polysilicon.

Transconductance of MOS devices. Transconductance (G_m) is the incremental change in drain current that results from an incremental change in gate voltage. [12] An expression for transconductance is

$$G_m = \frac{U_s W C_{ox}(V_g - V_t)}{L} \qquad (2.2)$$

where U_s is the mobility in the substrate. For PFETs, $U_s = U_p$, and for NFETs, $U_s = U_n$. V_g is the gate voltage, C_{ox} the oxide capacitance per unit area, V_t is the threshold voltage, and W and L are the width and length of the transistor. C_{ox} can be calculated from [14]. (Transconductance G_m is normally couched in terms of milli-Siemens per millimeter of gate *width*. This is abbreviated mS/mm. For those readers not familiar with the unit, a milli-Siemen is a change of one milli-amp of drain current per one volt change of gate bias. The unit tends to be used more in GaAs than in silicon.)

$$C_{ox} = \frac{K_{ox} E_o}{T_{ox}} \qquad (2.3)$$

where K_{ox} is the dielectric constant for the gate oxide (3.9), E_o the permittivity of free space (8.85×10^{-14} F/cm), and T_{ox} the gate oxide thickness.

Speeds of MOS devices. The ratio of transconductance to the input gate capacitance is a measure of the speed of the device. The latter depends not only on the capacitance calculated from the expression above but also on the capacitances due to the overlap of the gate and source/drains. Decreasing the gate capacitance will increase the speed of the device.

One of the benefits of the self-alignment feature discussed in this chapter is that the overlap capacitances are essentially eliminated. This is one reason that silicon gate structures are so much faster than the older metal gate structures. In the latter, the gate had to be put down *after* the source/drains had been formed. This, in turn, required a gate overlap tolerance in the manufacturing process.

The ratio of transconductance to input gate capacitance can also be increased in the following ways:

1. Decreasing the drain-to-source spacing. This is one effect of scaling.
2. Increasing the mobility of the carriers in the inversion layer. This requires decreasing the channel doping. (See the discussion of "on" resistance below.) Decreasing the doping, however, requires that the maximum operating (drain-to-source) voltage be decreased to avoid punch-through.
3. Increasing the gate-to-source voltage. However, in CMOS devices this carries the risk of inducing latch-up unless the resultant gate-to-source voltage is less than the drain-to-source voltage.

As noted above, however, the maximum drain-to-source voltage must be decreased if the mobility of the carriers is to be increased. Therefore, there is a trade-off between items 2 and 3.

Equations of operation. For the linear region of a MOSFET V–I curve in which the gate voltage V_g is greater than the threshold voltage V_g and for which the drain-to-source voltage V_{ds} is less than $V_g - V_t$, an expression for the drain current I_d is

$$I_d = KV_d[2(V_g - V_t) - V_d] \qquad (2.4)$$

where K is the conduction factor:

$$K = C_{ox}U_oW_{eff}\,2L_{eff} \qquad (2.5)$$

U_o is the zero bias surface mobility; W_{eff} and L_{eff} are the effective width and length, respectively, of the MOSFET; and C_{ox} is the oxide capacitance per unit area. C_{ox} can be calculated from Eq. (2.3). An assumption made in the foregoing equations is that the channel has a potential of $V_{ds}/2$.

In the saturation region, an expression for the drain current is

$$I_d = K(V_g - V_t)_2 \qquad (2.6)$$

where all symbols have been defined previously. Expression (2.6) shows that the drain current is independent of V_{ds} to the first-order approximation used in the equation.

"On" resistance, lateral diffusion, and mobility. "On" resistance R_{on} is the resistance between source and drain of the FET when it is on (i.e., when the gate voltage is above threshold). It can be calculated as the ratio of V_{ds} (drain-to-source voltage) to I_d.

It is noted that R_{on} depends on surface mobility among other parameters. Mobility in turn, depends on several parameters. Splinter, for example, has shown that mobility depends on film thickness, and that for film thicknesses of 2300Å (angstroms) or less, the mobility of the NFETs is less than that of the PFETs. For most work, however, the PFET mobility is less than that of the NFET mobility by a factor of typically 3. A subsequent figure in this chapter shows how the mobility varies with dopant concentration. As doping increases, the mean free path of the carriers is effectively decreased. This gives rise to the *reduction* in mobility with increase in dopant concentration as shown in the figure.

Therefore, for a large PFET, such as a buffer on the chip periphery, to have the same R_{on} as an NFET, it must have roughly three times the *W/L* ratio. This, indeed, is what typically happens. Figures 3.1 and 3.2 of the inside of the chip show the much longer sepentined PFET (labeled "BIG P" in Fig. 3.2) buffer on the periphery. The source is on one side of the PFET and the drain is on the other. The thin narrow ribbon in the middle is the gate. The width of the FET is the thin dimension. The width is the "long" dimension. (It sounds backward, but it is true.) The high *W/L* gives the low "on" resistance.

High *W/L*, and hence low "on" resistance, can also be obtained from the effects of lateral diffusion. This makes the effective gate length less than it otherwise would be. The deeper the source/drains are driven in, the greater will be this effect all else being equal. (Because different impurities diffuse at different rates at a given temperature, the results will vary with the impurity which is used.)

One gate array manufacturer reportedly has used this difference to aid in making resistances of the PFETs more nearly equal to those of the NFETs without using substantially more area on the chip. Of course, making the effective gate length less increases the possiblities of punch-through at a given voltage.

Mobility variations with temperature. Mobility also varies with temperature. Increasing the temperature has somewhat the same effect as increasing the dopant concentration in the sense that the mean free path of the carriers is reduced. This gives rise to the slowdown as temperature increases, noted in Chapter 5. Interestingly, Gaensslen [15] has found that below threshold, just the opposite effect occurs (i.e., mobility increases in that regime as temperature increases). However, because the interest in this book is in digital logic circuits that operate above threshold, the latter effect will be ignored. It is mentioned for the sake of completeness and because a small number of readers who do extensive circuit modeling may wish to use the results. The reader is referred to Gaensslen's paper for more information.

Lateral diffusion, its effect, and its minimization. The effective gate length is narrower than the drawn gate length because of lateral (sideways) diffusion of the source/drains during "drive-in" as noted earlier (see the Appendix 2A). The source/drains diffuse laterally as well as vertically. This shortening of the gate length limits the allowable drain-to-source voltage that can be used. The effect of punch-through has already been mentioned. Figure 2.11 shows this.

A way of minimizing lateral diffusion is the stepped source/drain shown in Fig. 2.14. In this technique, the portion of the source/drain immediately adjacent to the gate is made considerably shallower than is the remainder of the source/drain. Although this portion of the source/drain has a much higher diffusion capacitance per unit area than does the rest of the source/drain, its overall area is kept relatively

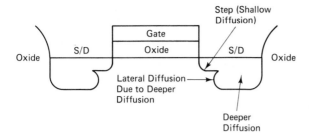

Figure 2.14 Stepped source/drain used to minimize lateral diffusion.

small. The shallowness of the step is created by driving in that portion of the source/drain diffusion for a much shorter period of time. The lateral diffusion, which is time dependent, is therefore also minimized. The remainder of the source/drain is driven in to the normal depth (for that particular scaling). The process is more complex because of the extra steps involved.

The process takes advantage of the fact that the MOSFET transistor action takes place in a very small depth of the channel between source and drain.

2.4.11 Simple Logic Structures Using MOS Transistors

The purpose of this section is to show how simple logic devices can be made from MOSFETs, in this case NFETs. In the following section, these same logic structures are built from both PFETs and NFETs (i.e., CMOS). Introducing the subject here enables other topics, to be described in this section.

The simplest type of logic structure is the inverter. The FET structure of this is shown in Fig. 2.15(a). This has one NFET with a load resistor. (The load resistor is often a depletion-mode NFET. Bear in mind that the discussion here is on NMOS only as part of the general treatment of MOS. It is not on CMOS, for which depletion-mode loads are not used.)

A high on the input to the NFET yields a low on the output. By stacking three such NFETs in series, a three-input NAND gate can be made. In this case the gates to all three NFETs must be high in order to pull the output low. However, note that if the *"on"* resistance of the NFETs in Fig. 2.15(b) is the same as that in Fig. 2.15(a), the discharge path to ground (i.e., a high-to-low transition) will be three times that of the inverter, and the appropriate propagation time would be expected to be about three times greater, all else being equal. This is approximately what actually happens.

The ratio is not exactly 3:1, for a variety of complex reasons, including the fact that the gate potential of each transistor *with respect to that transistor's source* depends partly on the relative position of each transistor in the series "string."

A three-input NMOS NOR gate has three NFETs in parallel instead of three in series. This, of course, will lower the RC time constant for a high-to-low transition of the load capacitance by a factor of 9 over that of the three-input NAND gate if the "on" resistances are the same in both cases. For this reason, NMOS is sometimes called a NOR-gate technology. One can appreciate that if NOR gates are to be used

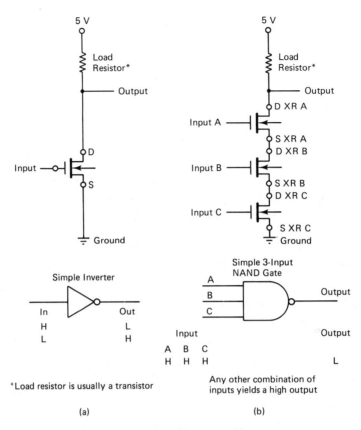

Figure 2.15 NFET logic structure: (a) inverter; (b) three-input NAND gate.

extensively in a given non-gate-array part, the FETs used in such gates can be designed with much smaller W/L ratios than for the case where NAND gates were used. This means that the gate capacitance which they present to a driving source is much less, and so on. The overall die becomes much smaller primarily using NOR gates if this is possible for the given circuit design. SDFL of GaAs has a similar property.

Sharing of MOS transistor source/drains. The simple three-input NAND gate shown above uses three NFETs in series. If three separate NFETs were used, there would be 3 × 2 or 6 source/drain diffusions. If the reader could see the inside of a gate array, custom, or standard chip, only four such source/drains would be seen. (A two-input cell gate array would have more.) The reason for the difference lies in the ability of MOS devices to share source/drains. This means that the actual source diffusion of one FET can be the drain diffusion for the immediately adjoining FET. Remember that the sources and drains of a FET are completely interchangable and a source/drain diffusion can act as either a source or a drain.

This is unlike most bipolar processes. In the latter, the collector of one device cannot be the emitter of the next adjacent device. On occasion two LSTTL (low-power Schottky TTL) transistors may share the same collector. The latter cases are rare, however, and the collector so shared is generally double width in the instances the writer has seen. This is also different from the case that occurs in IIL-like structures, in which there are multiple emitters (or collectors in the case of ISL/STL) embedded in the base region (epitaxial region for ISL/STL). Although the latter *regions* are being shared, the actual emitters or collectors are not. Therefore, the larger the number of emitters, the larger must be the area of the base.

Sharing of MOS source/drains has several benefits and helps account for the considerable circuit density advantage that MOS has over bipolar. On gate arrays, it allows the design of regular structures that can be used to make many types of logic gate configurations, which are quite efficient in their usage of chip area.

An example of sharing MOS transistor source/drains is shown in the Fig. 2.16.

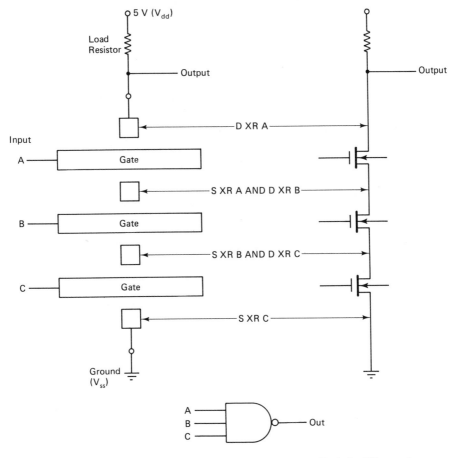

Figure 2.16 (a) Sharing of MOS source/drains: S, source; D, drain; XR, transistor; signal designations A, B, and C are arbitrary; (b) three-transistor pair structure.

Sec. 2.4   Fundamentals of MOSFETS

It can also be seen from Fig. 2.15(b), which is a pictorial representation of a three-transistor pair structure. Figure 2.15(b) also shows that if one of the three transistors in the series NFET string is low, the output will still be high, as it should be for a three-input NAND gate.

Pure custom versus gate array design—getting high W/L ratios. If the NFETs of the three-input NAND gate described above could be custom designed for that portion of the circuit, they could in theory be made three times larger, so that their resultant "on" resistances are three times lower. The combined resistance of the three in series would then equal that of the inverter. In theory, therefore, logic gates with an unlimited number of inputs can be made and the "on" resistance of the combination kept small simply by making the transistors with larger W/L values.

Gate array FETs are of fixed size. Because they must be capable of being used in either series or parallel structures, it is not desirable to make them of different sizes. Therefore, in theory they suffer a substantial disadvantage compared to their pure custom counterparts.

The preceding theory is not 100% true for a variety of reasons. First, even pure custom FETs cannot be made as large as desired. The reason is that if the size of a FET is increased by a large factor, of say 27, to decrease its "on" resistance, the load that it presents to its driver is also increased by a factor of roughly 27. Therefore, to avoid severely degrading the propagation time of the driver, the driver FET structure would also have to increase in size by a large factor, say 9, 3, and so on, as would its driver. The result is that the size increases "ripple back."

The problem is compounded even further in the case of CMOS by the desire to try to equalize rise and fall times. (In CMOS NAND gates, the NFETs are in series and the PFETs are in parallel. Just the opposite is true for NOR gates.) It can be shown that the optimum W/L under this criterion increases as the square of the number of inputs. For these reasons CMOS logic gates in even pure custom circuits (in which every transistor can be theoretically designed for its exact place in the circuit) are generally limited to about the same four or five inputs that logic gates built using gate arrays are restricted to normally (but not always).

The second reason that gate arrays are not in as disadvantageous a position with regard to size as might be thought is that array FETs *can be paralleled* to achieve a similar effect to increasing the W/L ratio of a pure custom circuit. Paralleling is discussed in Chapter 9.

A third reason why even pure custom circuits cannot be designed with arbitrarily large numbers of inputs and hence W/L ratios is that to achieve a reasonable form factor of such devices, the overall structure must be "wrapped around" or "interdigitated" in some fashion. The buffer devices on the periphery of the gate array chip are examples of high-W/L-value devices.

Consider the three-input FET structure shown in Fig. 2.17. The source/drains and gates are denoted by dashed and solid lines, respectively. The structure could be a NOR gate, in which case every other source/drain would be connected together to form the paralleled structure, or it could be part of a NAND gate in which the FETs are in series.

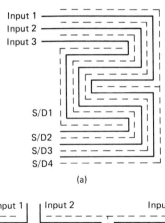
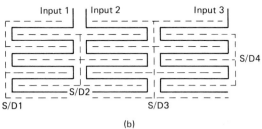
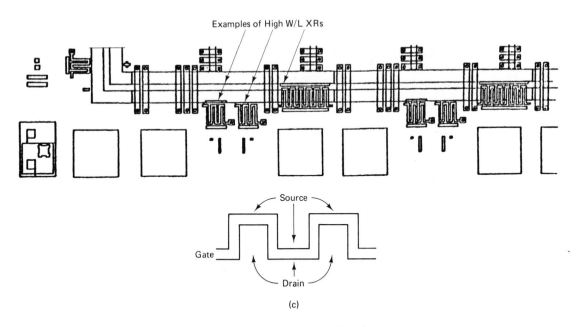

Figure 2.17 (a) Serpentine structure with three inputs (gates). Lines are shown for clarity to denote gate (inputs) and source/drains. In reality they are of nonzero dimension, of course. Source/drains, S/D, are shared. Note the change in width/length ratio between the inner and outer transistors. (b) Interdigitated structure with three inputs. (c) Examples of single transistor serpentine structures used on gate arrays. In this case they are the buffer PFET and NFET.

The structure in Fig. 2.17(a) is a "serpentine" structure, so called because of the way in which it wraps around. Although it is not drawn to scale, the reader can appreciate that to achieve a good form factor, a fair amount of space is wasted due to the fact that the corners use up a lot of the common area between successive source/drains in the structure. The designer and user, however, are "paying" for the entire capacitance of each of the gates, including in those areas where no useful transistor action is taking place because of the lack of adjacent *common* source drains.

An alternative to the serpentine structure is the interdigitated structure shown in Fig 2.17(b). Again the source/drains and gates are denoted by dashed and solid lines, respectively. The interdigitated structure overcomes some of the problems of the serpentine (especially low overlap for relatively small W/L's, where there is just one turn) but has problems of its own. The chief problem of the interdigitated structure is that there are no inner source/drains to be shared. This makes it less area efficient, in general, than the serpentine under *some* conditions. The conditions include form factors in which there are minimal turns to the serpentine.

It is possible to conceive of hybrid structures employing combinations of serpentine structures interleaved in an interdigitated structure.

For all of the foregoing reasons, gate arrays with their fixed W/L ratios are not in as disadvantageous a position as might be imagined.

Having introduced the reader to MOS devices in general, the discussion now turns to CMOS, the dominant technology of gate arrays.

2.5 CMOS AND ITS PERMUTATIONS

CMOS has been called the "technology of the 80s" with good reason. All variants of CMOS offer the following properties.

- Operation over a wide range of voltages. Typical values are 3 to 15 V for metal gate CMOS, 3 to 8 for 5-μm (feature size) HS CMOS, and 3 to 5.5 V for 3-μm HS CMOS.
- Low-power operation. CMOS uses power only when switching. At all other times, there is virtually an open circuit to either V_{dd} or to V_{ss}. (V_{ss} is generally ground, and V_{dd} is the operating voltage.) Because only a fraction of the logic gates in a circuit are switching at any given time, this factor can be significant. The actual power is $CVVf$, where C is the total load capacitance, VV the voltage squared, and f the switching frequency.
- High density.
- A somewhat simpler manufacturing process than most bipolar devices. This is reflected in lower manufacturing costs due to higher yields (see Section 2.7). As goemetries shrink, this advantage over bipolar becomes very minimal.

In addition, CMOS offers freedom from power supply regulation, good noise margins, and can be TTL compatible. The high-speed versions are comparable in speed to even the newer versions of TTL. State-of-the-art experimental CMOS transistors fabricated by Honeywell have exhibited gate delays of only 90 ps (0.09 ns) at 3 V in the laboratory. (This speed should not be interpreted as that which is realizable with practical devices in even the near future.)

As shown in Fig. 2.1, CMOS is made up of NFETs and PFETs. Except in the case of buffer FETs on the periphery of the die, the NFETs and PFETs are normally paired in such a way that a given signal is input to both a PFET and an NFET. This pairing is responsible for the ultra-low-power usage of CMOS. One of the two transistors is always on and the other is always off. When the input signal reverses polarity, the FET that was "on" becomes "off," and vice versa. Only during the switching interval are both FETs of the pair on.

An important exception to the foregoing generalization is the element called a transmission gate (TG). These are used extensively in CMOS flip-flops and other circuit elements. (See Chapter 9.)

2.5.1 Variants of CMOS

CMOS exists in three primary forms. The three forms are categorized by the gate material (metal gate CMOS), the gate and isolation material [high-speed (HS) CMOS], and by the wafer substrate material (CMOS/SOS).

Metal gate CMOS is the oldest of the CMOS technologies. It presently offers the gate array user the advantages of relatively high voltage operation (up to 15 V), relatively low prices, and high reliability resulting from being high on the manufacturing learning curve, generally compatibility with the standard "4000B" CMOS logic family offered by many major semiconductor manufacturers, and availability in respectable numbers of gates per chip (up to 550). It is multisourced by many vendors, although few of the versions are mask compatible with one another.

The primary disadvantages of metal gate CMOS are that it is an order of magnitude less than HS CMOS in both density and speed and that it has largely been phased out by gate array vendors. As competition intensifies, metal gate CMOS will increasingly get squeezed out except for replacement parts of the same gender. Manufacturing yields of the high-speed version are now comparable to those of the metal gate process at many gate array manufacturers; and this, coupled with a larger volume of HS CMOS version business, is putting pressure on metal gate CMOS.

Figure 2.18 is a cross-sectional view of a metal gate CMOS transistor pair. The fabrication sequence in terms of mask levels is shown. The important feature for the reader to note is that the metal gate is defined after the source/drains (see below) are fabricated. By contrast, the polysilicon gate of HS CMOS is fabricated before the source/drains are fabricated.

The reader will note that there are really two substrates used in CMOS. The n wafer is the first. The "p tub" or "p well" into which the NFETs are diffused is the second. The two substrates are also used in other types of "bulk" (see below) CMOS.

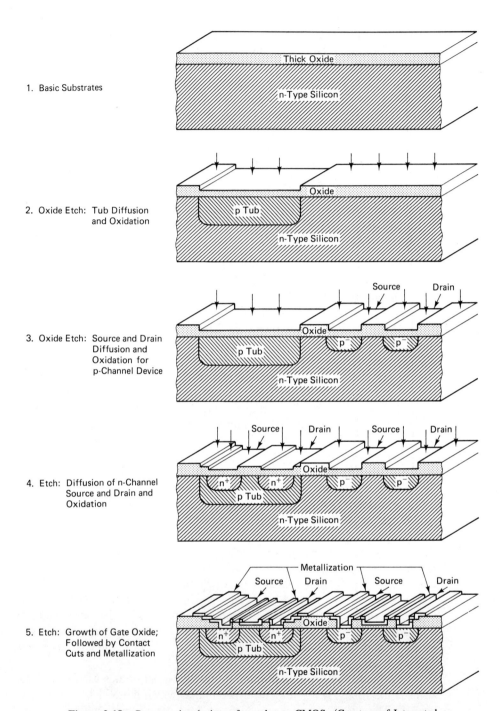

Figure 2.18 Cross-sectional view of metal gate CMOS. (Courtesy of Integrated Circuit Engineering Corp.)

HS CMOS. HS CMOS is currently the variant in which most of the new gate array designs are being done. HS CMOS is both oxide isolated and uses a silicon (polysilicon) gate. There currently is no standard term to denote this form of bulk CMOS. Some people have used the terms SO CMOS (selective oxidation CMOS) and LO CMOS (local oxidation CMOS) for this combination. The term HS CMOS is used by several major semiconductor manufacturers and is the term that will be used in the remainder of this book.

HS CMOS is fast, dense, and of low power. Current gate arrays from National Semiconductor, for example, offer up to 6090 equivalent two-input gates. Internal propagation delays are 0.85 ns under no-load conditions at 4.5 V = V_{dd}, and 100°C junction temperature. (Specifying propagation delays under conditions of no load is reasonably common among gate array manufacturers such as National and CDC. The propagation delay of the load (fan-out plus metal interconnect) is added to this baseline propagation delay.)

Although both oxide isolation and silicon gate are used in virtually all current high-speed bulk CMOS processes, the two features do not *have* to be used simultaneously. Moreover, refractory metals, such as molybdenum, are being experimented with in the laboratory. These may eventually replace the polysilicon gate. The advantage of a refractory metal gate is the lower (than polysilicon) sheet resistance. The former has a sheet resistance on the order of 5Ω/square (see Section 2.7), which while over two orders of magnitude greater than aluminum is nevertheless considerably lower than the 40 to 60 Ω/square of polysilicon.

Gate arrays with two layers of metallization are sometimes called *dual metal arrays*. Neophytes sometimes confuse this (somehow) with metal gate CMOS. The reality is that almost all dual-*layer* (which is what the term implies) metal CMOS arrays are silicon gate, not metal gate.

CMOS/SOS. Both metal gate CMOS and HS CMOS are *bulk* device, meaning that the transistor action takes place in the substrate material. Actually, it takes place in an infinitesimal depth of the bulk (substrate) material at the interface between the gate oxide and the substrate.

By contrast, the third variant of CMOS, CMOS/SOS, is not a bulk device. The transistor action takes place in a crystalline layer (the epitaxial layer) of silicon placed on top of an insulating substrate (sapphire). This gives rise to the term SOS, which stands for "silicon-on-sapphire."

Figure 2.19 is a three-dimensional view of a CMOS/SOS transistor. The epitaxial layer is the region between the gate insulator and the sapphire.

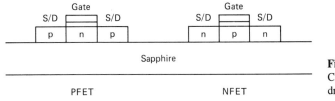

Figure 2.19 Two-dimensional view of CMOS/SOS transistor: S/D, source/drain.

CMOS/SOS is still restricted primarily to military applications because of high price. Only a handful of manufacturers (such as Rockwell, Raytheon, RCA, and Hughes) have it, and it is mainly for in-house use at these facilities. [RCA has an 800-gate CMOS/SOS gate array which was developed under U.S. Army (ERADCOM, Ft. Monmouth, New Jersey) sponsorship.]

Two other variants of CMOS are worthy of mention. The first is called C-HMOS. It differs from the forms of bulk CMOS discussed previously in that it makes use of an *n* well instead of a *p* well. The starting material for the substrate is made from the same *n* material as that used for the newer NMOS DRAMs. This and other similarities to the NMOS processes make possible the sharing of some of the process steps used in DRAM manufacture. Figure 2.20 shows a cross section of the process. The reader will note the absence of the *p* well under the NFETs, the presence of the *n* well under the PFETs, and the *p* substrate.

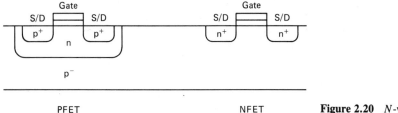

Figure 2.20 *N*-well CMOS.

The second variant is called "twin-tub" CMOS and makes use of both an *n* tub and a *p* tub. Its purpose is to try to eliminate latch-up entirely.

With this introduction, CMOS will now be discussed in detail.

2.5.2 Latch-Up

Latch-up is a property of CMOS in which the output signal stays "latched" (stuck) in its high state. It is caused by the two substrates of CMOS creating what is in effect an SCR (silicon-controlled rectifier) in the CMOS device. An SCR can be regarded as a combination of an *npn* and a *pnp* transistor.

Because an SCR is a bipolar device, some readers may have trouble understanding how a bipolar device can exist simultaneously with a MOS device. The answer lies in the fact that bipolar devices are formed from semiconductor junctions which are inherent in the CMOS structure. Figure 2.21 shows this. The NFET source/drain diffusions (*n* type, as noted earlier), the *p* tub, and the *n*-type substrate of the PFETs (the wafer) form an *npn* transistor. The resistance of the wafer is the collector resistor R_2 in the figure. The *pnp* transistor is formed from the source/drain diffusion of the PFETs, the *n*-type wafer itself, and again the *p* tub (substrate of the NFETs). The resistor R_3 is the resistance of the *p* tub.

These two structures are hooked together as shown in Fig. 2.21. From Fig. 2.21 it can be seen that there are three ways in which latch-up can occur. Two of the ways involve currents flowing in R_2 and R_3, which in turn bias either of the transistors on. If either is biased on, it will bias the other on. Hence the structure latches up.

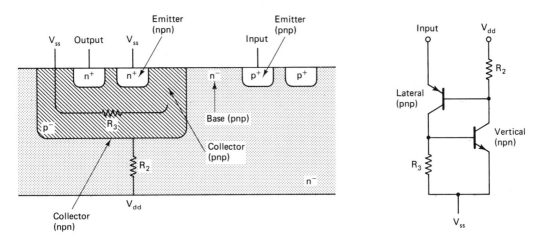

Figure 2.21 Latch-up in CMOS. (Courtesy of Harris Corp.)

The third way in which latch-up occurs is when the input signal voltage is greater than V_{dd}. This is often the most common way.

To minimize the chances of latch-up, most CMOS manufacturers periodically tie the two substrates to ground (the *p* tub) and to V_{dd} (the wafer). A common way to do this is to have one connection from the distributed power and ground buses at each cell location. The figure showing the close-up view of the chip shows this (though just barely). The small ovals at the top of each cell are the steps of the contacts.

Because CMOS/SOS structures utilize doped epitaxial regions in place of the substrates for the NFETs and PFETs, they are essentially immune from latch-up. The reason is that there is effectively an open circuit (the sapphire substrate) between the *n* epitaxial region of the PFETs and the *p* epitaxial region and *n* source/drains of the NFETs. Hence no *npn* transistor can be effectively formed as it can in bulk CMOS. For a similar reason, no *pnp* transistor is formed.

2.5.3 Propagation Delays

Propagation delays of any MOS structure depend on the type of logic structure (e.g., NAND or NOR gate) and on the "on" resistance of the transistors used in that structure. In gate arrays, the "on" resistances are not user selectable (except by means of a technique called "paralleling" described in Chapter 9). However, different gate array manufacturers use different values of "on" resistance. The "on" resistance values tend to be somewhat similar among manufacturers for a given feature size (i.e., the $5/\mu$m technologies and the 3-μm technologies).

Propagation delays can be calculated in several different ways. The methods differ in the degree of sophistication of the model, the amount of CPU time (if any) required, and the approximations used. Two major methods are discussed below. More advanced work done at MIT and Cal Tech will be discussed briefly in Chapter 5. The first way is amenable to machine calculation; the second is useful for hand calculation of critical paths. Both ways will be presented.

Machine calculation. Propagation delays in the first way are calculated as the product of the "on" resistance of the driving stage, the total capacitance being driven, and a suitable constant. The latter represents the number of RC time constants required for the circuit voltage to reach switching levels.

The total capacitance being driven is the sum of three capacitances. These are:

- Output capacitance of the driving stage, C_{out}
- Capacitance of the wiring channel, C_w
- Input capacitance of the stages being driven, C_{intot}

where C_{intot} is the sum of all the gate capacitances to which the output signal is fanning.

In CMOS, the output capacitance is that of the drains of the PFETs and NFETs that are connected together. Although it varies with gate array manufacturer, this capacitance is typically on the order of 0.15 to 0.6 pF per drain, with most values falling in the range 0.2 to 0.4 pF. (These numbers are representative but are not guaranteed to be worst case for all manufacturers.) The numbers quoted are for the internal array cell transistors. The values also vary with the position of the drain in a given manufacturer's cell structure and will normally be different for the NFETs and PFETs of a given manufacturer.

A large portion of the drain capacitance is that of the depletion layer of the drain. As such, its value is voltage dependent. This is a subtlety that is almost universally ignored for ordinary calculations except those involved in using a circuit simulator program (see Chapter 6).

The capacitances of the wiring channels are discussed in a separate section of this chapter. Some systems, such as that developed by LSI Logic, Inc., will automatically extract the wire lengths of each path in the layout and calculate propagation delays from it using stored values of the constants.

The gate capacitance likewise varies from manufacturer to manufacturer but is on the order of 0.115 to 0.2 pF NFET or PFET gate (i.e., given that the signal normally goes to both a PFET and to an NFET, the total gate capacitance for fan-out of 1 would be double these figures).

This brings up another point.

Fan-out in CMOS. Because each signal in CMOS normally goes to both one PFET and one NFET, each such combination (of one PFET and one NFET) gate capacitance is almost universally (among the gate array manufacturers) defined to be equal to one fan-out.

Hand calculation—the second method. In the second method, either tables or graphs of propagation delay versus fan-out are used. Examples of the latter are shown in Fig. 2.22.

The reader will note that the high-to-low propagation time of the NAND gate is greater than the low-to-high propagation time of such a gate and that the opposite is true for a NOR gate. These relations can be varied by changing the W/L ratio and the lateral diffusion, as noted earlier.

The values obtained from the graphs and tables have to be combined with an estimate of the wire (metal) lengths for those runs that are quite long. Delays due to polysilicon underpasses should always be accounted for. An example of such accounting is given in Chapter 5.

Variation of driving-point resistance. A NAND gate has one NFET in series and one PFET in parallel for each of its inputs. A NOR gate has just the opposite configuration. This is dealt with in detail in Chapter 9.

Constancy of the Driving-Point Resistance. The calculations above assume that the load capacitances are being charged or discharged through a constant-resistance value equal to the net on resistance of the device under consideration. This is a reasonable assumption for most calculations. Strictly speaking, some switching action begins to occur before the driving transistors are fully turned on (i.e., before the "on" resistance is at its minimum value). Fortuitously, very little switching action occurs until the gate voltages of the driving transistors cross the switching thresholds for those devices. At that point, the channel resistance drops dramatically, as noted in the expanded V–I curves of transistors shown earlier. Further amplification of this point is given in Section 9.16.

Number of RC Time Constants. The constant Kr above is the number of RC time constants required to effect the switching between the switching voltage levels. This number can be calculated using the assumption that the driving-point resistance is a constant as follows.

For low-to-high transitions, the load capacitance charges exponentially toward V_{dd} through the driving-point resistance starting at the V_{ol} (output low voltage of the driving device) according to the relation

$$V_o = V_{ol} + (V_{dd} - V_{ol})\left[1 - \exp\left(\frac{t}{RC}\right)\right] \quad (2.7)$$

where V_o is the output voltage in question. Note that at $t = 0$, $V_o = V_{ol}$.

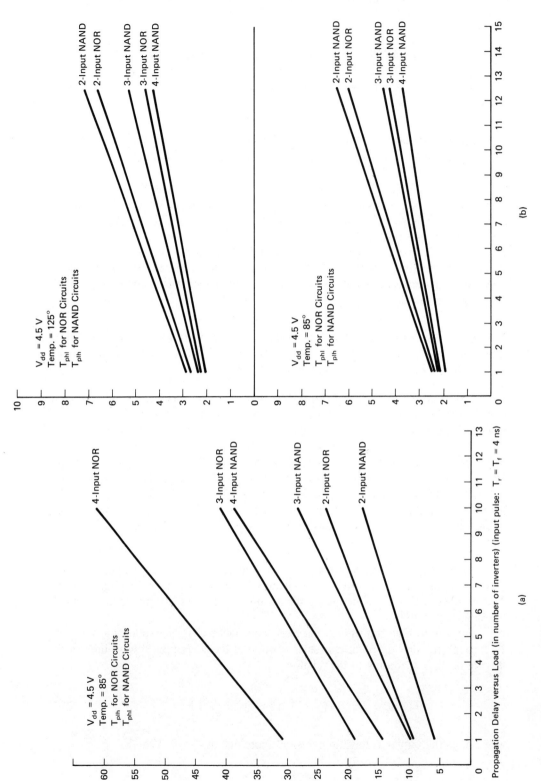

Figure 2.22 Examples of propagation delay versus fan-out. 5M CMOS (Courtesy of International Microcircuits, Inc.)

If the succeeding stage is guaranteed to switch at (or before) 70% of V_{dd}, then the number of RC time constants can be found as

$$\frac{t}{RC} = -\ln\left(\frac{1 - 0.7V_{dd} - V_{ol}}{V_{dd} - V_{ol}}\right) \quad (2.8)$$

As an example, suppose that V_{ol} is 0.4 V and V_{dd} is 5 V; then t/RC will be 1.12. If R is 6000 Ω and the total capacitance is 300 fF, the low-to-high transition will be 2.01 ns.

The high-to-low transition time can be found similarly from the relation.

$$V_o = V_o\left[\exp\left(\frac{-t}{RC}\right)\right] \quad (2.9)$$

Penfield and others at MIT [16] have shown that when the resistances of the lines being driven are comparable to that of the driving-point resistance, there is generally no closed-form solution to the problem. They give a way to bound the problem. This situation occurs most frequently in single-layer metal CMOS, in which logic gates drive significant amounts of polysilicon interconnect.

Temperature variations of CMOS. In the preceding section of this chapter, it was mentioned that there are four ways to speed up MOS devices by increasing the ratio of transconductance to input capacitance. One of the ways is to increase the mobility of the carriers in the channel.

Propagation-delay variations with temperature of MOS devices are directly related to mobility variations. These in turn are directly related to temperature. At lower temperatures, the mean free path of the charge carriers (electrons for NFETs, holes for PFETs) is greater; therefore, the mobility is greater. The mean free path is the average distance between collisions of the hole or electron. A simple explanation is that there are two competing processes in the channel. One is the random motion of the charge carriers due to collisions. The other is the forces exerted by the electric field gradient due to the drain-to-source voltage. At lower temperatures, there is less random motion; and the electric field gradient can act more effectively. Therefore, MOS devices operate much faster at low temperatures than at high temperatures.

It is possible to give some "ballpark" numbers for the variations in propagation delay over temperature. Although the values will vary somewhat, surprisingly good correlation has been found among a variety of manufacturers. The values that are given in Chapter 5 are factors by which the propagation delays (in either direction, i.e., high to low or low to high) at the reference temperature should be multiplied to find the propagation delays at the higher temperature.

Propagation delay values can be obtained from each individual manufacturer. A word of caution is in order. A number of gate array manufacturers are very conservative in the values that they promulgate for propagation delay. There is virtually no chance that the logic devices will run more slowly than the numbers presented by such manufacturers. Actual chips may run at clock rates twice what the data would indicate. These manufacturers take into consideration all of the layout and processing tolerance to make sure that any gate array IC from any production lot will run at the desired clock frequency.

Other gate array manufacturers are not so conservative or sometimes simply do not have their processes that well characterized. The propagation delays of gate arrays of a given feature size from most of the manufacturers are in the same "ballpark." The user should strongly question claims of operating frequencies that are out of line with those of other manufacturers at a given feature size.

The propagation delays above are only those for the active elements. The propagation delays of the polysilicon and metal interconnects are essentially temperature invariant. However, with very rare exceptions, the sum of the propagation delays of the active elements vastly outweighs those of the interconnect. Therefore, one is normally justified in scaling the overall propagation delays (including those of the embedded interconnect) by the ratios above. Except in very rare instances, if the error made in such calculations (by scaling the interconnects as well as the active elements) is important, the user is probably operating too close to the frequency limits of the given manufacturer's technology and is probably within the manufacturing tolerances of such technology.

Voltage variations of propagation delays of CMOS. Increasing the drain-to-source voltage will increase the speed of MOS devices. The curve of such variation is nonlinear and shows a clear saturation effect. A few typical values are given in Table 2.2. Some of these numbers are based on operation of silicon-gate CMOS FETs in regions where they would not normally be operated (because of punch-through). This region includes voltages greater than about 7 or 8 V for the 5-μm feature sizes and is something of a composite. The user should consult the gate array manufacturer for specific information on a given gate array. Values are referenced to 4.5 V, which is labeled V_{ref}. The other voltage values are labled V_{new}. The corresponding propagation values are labeled PV_{ref} and PV_{new}, respectively.

TABLE 2.2 VOLTAGE VARIATIONS

V_{ref}	V_{new}	PV_{new}/PV_{ref}
4.5	4.75	0.93
4.5	5.0	0.87
4.5	5.25	0.82
4.5	5.5	0.77

Source: LSI Logic 5000 series data sheet. Courtesy of LSI Logic, Inc.

Wafer-processing effects on propagation delay. Wafer-processing tolerances can yield large variations in propagation delays of the resultant gate array wafers. Tolerances of dopant levels and thicknesses of the various layers can yield variations in such factors as channel mobility and capacitances of the various terminals. Often the largest changes are seen from one run of wafers to the next. (A run is a *multiple* of typically 24 or 48, sometimes as few as 12, wafers. Forty-eight wafers will fit in the "flat zone" of the normal diffusion furnace. Fewer will fit in the carousel of an ion implanter.) Smaller changes may be seen from wafer to wafer.

Estimating the aggregate change in propagation delay from all of the tolerances is quite difficult. Different manufacturers use different values. However, a value of 1.4 for the ratio of longest to shortest propagation time of a given function is a typical number for this parameter. Other gate array vendors use as much as 2:1.

The static high and low voltages of CMOS. The statement is often made that CMOS outputs will "pull (approximately) to within one diode drop of the power supply rails." The reason for this statement can be seen from a diagram of the cross-sectional view of a MOS transistor shown earlier and repeated with modifications as Fig. 2.23. The source diffusion and the substrate form a diode. The substrate is tied (normally) to V_{dd} for PFETs. Under static conditions, this diode will clamp the output voltage to one diode drop greater than the substrate voltage. For NFETs, a similar diode exists; and the voltage is clamped to one diode drop less than V_{ss}.

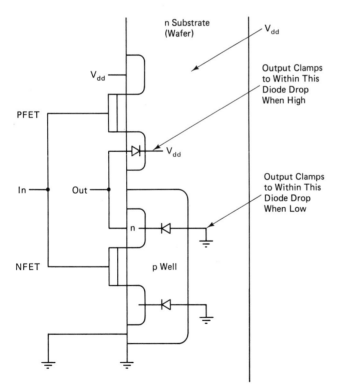

Figure 2.23 Why CMOS pulls to within one diode drop of the rails. Formation of diodes is shown.

Sec. 2.5 CMOS and Its Permutations

2.5.4 Protection of Inputs from Electrostatic Discharge and Other Voltages

A person walking across a wool rug can develop as much as 10,000 V of static electricity if the air is dry enough. Similar voltages can be developed in a variety of other circumstances.

The effects on MOS devices of such voltages are disasterous. The gate oxides of 5-μm (drawn) gate length devices are only typically 750 Å thick. Those of 2-μm gate-length devices range from 150 to 250 Å in thickness. The tetrahederal radius of the silicon atom is 1.18 Å. Thus the gate oxides of current CMOS and other MOS devices are only 100 or more atoms thick and can easily be severely damaged by electrostatic discharge (ESD).

For this reason, virtually all MOS devices have some form of ESD protection built into them, Typically, this takes the form of diode voltage clamps in conjunction with a resistance of some kind. A cross-sectional view and a schematic of a typical input protection are shown in Fig. 2.24(a) and (b), respectively.

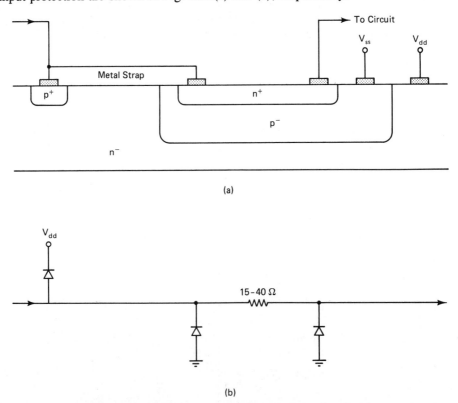

Figure 2.24 Input protection circuit: (a) cross section; (b) schematic.

The diodes typically limit the input voltages to one diode drop (0.7 to 0.8 V) above V_{dd} and one diode drop below below V_{ss}. Voltages greater than these cause the diodes to conduct and clamp the signal to V_{dd} and V_{ss}, respectively. The resistance helps act as a current limiter and in conjuction with the capacitance of the junctions helps act as something akin to a low-pass filter to integrate the energy in high-voltage spikes and spread the effect out over a long period of time. Many ESD occurrences involve high voltages but relatively small amounts of actual energy (as anyone who has petted a cat on a dry day knows; what the cat thinks, however, has never been determined!).

The clamp to V_{dd} is formed from a p^+ diffusion or implant over the N^- substrate (wafer). The anode of the diode is, of course, the p^+ diffusion [see Fig. 9.24(a) and (b)].

The clamp to V_{ss} is formed by an n^+ diffusion/implant over the p^- well discussed earlier. The resistance is that of the n^+ diffusion over the p^- well.

One way of implementing this structure is to use what is in reality a transmission gate (see Chapter 9). This is shown in Fig. 9.55(a) and (b), and its voltage response is as shown in Fig. 9.55(c). By connecting the gates of the PFETs and NFETs to V_{ss} and V_{dd}, respectively, both FETs are turned on. This yields the resistance shown. From the preceding discussions about diodes, the reader should be able to readily understand how the diode clamps to V_{dd} and V_{ss} are formed.

2.5.5 A Useful Feature of Battery-Operated CMOS

Most batteries produce lower voltages at lower temperatures. Because the lower voltage (which would make CMOS run more slowly as noted above) occurs in the temperature regions where CMOS runs fast, the two parameters tend to compensate for one another. The same effect occurs at higher temperatures where the battery voltages are normally greater but where CMOS would slow down in the absence of increased voltage. The result is that battery-operated CMOS often has a fairly constant (over temperature) maximum clock rate. The reader should be aware, however, that a battery discharging at a high rate (because of perhaps powering other circuits) may develop internal temperatures which are much different from those of an adjacent CMOS chip.

2.5.6 Interfacing CMOS to TTL

The propagation delays of TTL do not vary with temperature the same way that they do for CMOS. A circuit that runs perfectly well at room temperature may not work or may work only marginally at other temperatures because the propagation delays of CMOS have changed so much. The shorter delays of CMOS at $-55°C$, for example, may aggravate race conditions that did not show up in a PCB tested at room temperature. The designer must take these differences into account in the early phases of design.

2.5.7 Transition to CMOS Logic

The FET structures in Fig. 2.15 represent NFET logic. It will be noted that the load resistor (which may be a depletion-mode transistor with its gate tied to an appropriate voltage) allows current to pass through the NFET series string whenever the latter is "on" (i.e., whenever the output is low). If the load resistor could be replaced with a switchable load such that the overall structure drew current only (except for leakage currents) when the device was switching, a considerable savings in power would result. This is exactly what happens in CMOS technology. A CMOS inverter is shown in Fig. 2.25(a). Figure 2.25 shows a three-input NAND gate in CMOS technology. It will be noted that the load structure of Fig. 2.15 has been replaced by three PFETs in parallel. The reason for 3 (instead of perhaps just 1) is to ensure that the structure will pull the output high if *any* input is high. This brings up an important point that neophyte CMOS logic structure developers sometimes overlook:

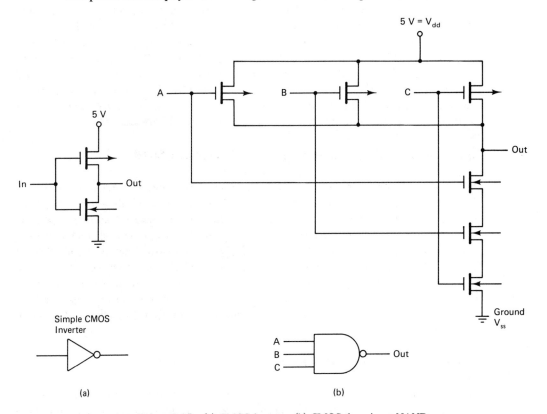

Figure 2.25 (a) CMOS inverter; (b) CMOS three-input NAND gate.

IN ORDER *To have a valid low on the output, there must be not only a low-impedance path to ground, but also a high-impedance path to V_{dd}. The opposite is true for a valid high on the output.* A result of this point is that "wired AND" configurations which are readily doable in both ISL/STL and ECL are not generally doable in CMOS except when using either three-state devices or transmission gates. This is explained in Chapter 9.

NOR gates. NOR gates in CMOS are the opposite of NAND gates in the sense that they have one NFET in parallel (with other NFETs) and one PFET in series for each input. This is shown diagrammatically in Fig. 2.26.

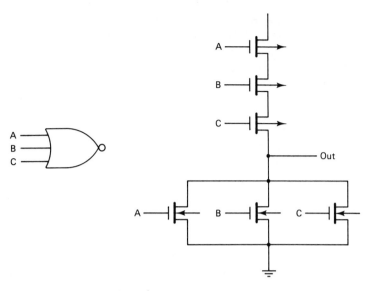

Figure 2.26 CMOS NOR gates.

2.6 TECHNOLOGY OF INTERCONNECTS

This section covers the topics of:

- Characteristics of the materials used for interconnects
- When the resistance and capacitance of the interconnects is of concern to the designer
- Comparison of the abilities of the various gate array technologies to drive interconnect paths

The reader is reminded that what is being interconnected are the transistors or more likely the macros selected by the user to provide the functional performance required.

On a gate array there are two levels of interconnect. At least two levels are needed to permit signal and power lines from crossing without contacting. The two levels can both be metallic or one level can be metallic and the other semiconductor.

If semiconductor interconnect is used as one level, it will be fixed in place and not definable by the user. On each end and sometimes in the middle of such interconnects are contact points to which can be connected signal lines where required. By contrast, both layers of metal are user definable. This includes the cases where automatic routing programs do the actual configuration of the routes. Some vendors having two layers of metal require that all horizontal routes be done in one level and all vertical routes be done in the other level.

Current semiconductor (as opposed to metal) interconnects are mainly made from doped polysilicon, the same material used for gates of MOS transistors. On occasion, the diffusions of CMOS transistors are also used to cross without contacting one or both bus bars of single-layer metal CMOS gate arrays. However, such cases are rare and are not amenable to automatic routing techniques.

From a systems viewpoint, the important parameters of a interconnect are its resistance and capacitance. From a manufacturing standpoint, the important parameters are adhesion to the surface of the silicon or dielectric, stability with time, uniformity of the thickness of the film, etchability, ability to withstand processing temperatures, and compatibility with metals used for Schottky diodes if such are used on the chip. Hence, what may seem like an ideal metal from a systems standpoint may not be useful because of processing problems.

The predominant metal used in gate arrays is aluminum. It may be composed of a few percent of either silicon or copper. The latter is useful in reducing the *grain size*. The grain size of aluminum is typically about 2 μm and has the property of increasing slightly the resistance of the metal at narrow line widths.

Some vendors use a "sandwich" of metals of *each* layer of metal interconnect. This is especially true in bipolar technology. A typical sandwich consists of platinum (to form Schottky diodes) followed by "Ti-Tung" (titanium tungsten, Ti-W), on top of which is overlaid either gold or aluminum to provide good conductivity. The titanium is used because it adheres well; the tungsten is used to provide a barrier. Note that unlike the case of two-layer metal, there is *no insulator* between the metal layers of the sandwich.

A common term used to describe interconnect paths is *pitch*. Pitch is the sum of the width of the line and the spacing between lines. For example, if the pitch is μm and the metal width is 2 μm, the spacing between metal lines is 3 μm.

Another term used extensively in describing interconnects is the number of *squares* in a given length of interconnect. The length of each square is equal to the width of the line. Ten squares of a 2-μm-wide line have a length of 20 μm. If the resistance is 0.05 Ω/square, then the total resistance is 0.5 Ω.

The resistance value, ohms/square, is the *sheet resistance* of that material. It is derived from the resistivity (couched in units of Ω-cm) divided by the thickness of the film. Aluminum, which is 10,000 Å (i.e., 1μm) thick has a sheet resistance of roughly 0.03 Ω/square. Tungsten has a value of roughly 0.05 for the same thickness. The sheet resistance of polysilicon depends on the doping but is typically in the range 15 to 100 Ω/square. Table 2.3 gives the sheet resistance and capacitance of the polysilicon underpasses of a popular single-layer metal process.

TABLE 2.3 INTERCONNECT CHARACTERISTICS OF POLY UNDERPASSES

	Length (μm)	Width (μm)	Area (μm^2)	Capacitance (pF)	Length (no. of squares)	Resistance (Ω min./ max.)	RC min./max.) (10^{-12} s)
Poly A	113.5	13	1475.5	3.5×10^{-2}	8.73	131/349	4.6/12.2
Poly B	53	13	689	1.7×10^{-2}	4.08	61.2/163	1.04/2.8
Poly C	73	13	949	2.3×10^{-2}	5.62	84.3/225	1.9/5.2
Poly D			1750	4.2×10^{-2}	17.5	263/700	11.0/29.4

Contrast: Aluminum 0.5 Ω/☐; 1.9×10^{-17} F/μm^2.

As line widths become smaller, a number of problems arise. One of these is that the line sizes become comparable to the grain sizes mentioned earlier. A second is that the *aspect ratio* of the metal becomes undesirable unless the thickness is changed. The aspect ratio is the ratio of the width to the thickness. If these two dimensions are similar and if the spacing between lines is comparable to those of the other two, it becomes difficult to etch away the line material properly. Figure 2.27 shows this.

Figure 2.27 Why narrowing line spacing makes it difficult to etch to necessary depth. Cross section of two parallel lines on chip.

Decreasing the thickness improves the aspect ratio. However, the sheet resistance increases as the thickness is decreased. The problem is very significant for all types of VLSI, and major efforts are under way to develop newer and better interconnect materials. A promising group is the silicides of many of the refractory metals (e.g., molybdenum silicide, etc.).

Another method of providing good (low-ohmic) conductors with low thickness is that of polycides. These are sandwiches made from refractory metal (see below) silicides overlaid on polysilicon (again with no insulator in between). The combination has a much lower sheet resistance in a given thickness than does polysilicon by itself. (The latter has a sheet resistance of 15 to 40 Ω/square for typical realizable thicknesses.) The resulting sandwich is sometimes called a refractory metal *shunt* when used for gates. The problem, of course, is the increased processing difficulties and resultant costs.

A group of metals that has been investigated for a number of years is the refractory metals group. The major property of the metals in this group is that they have melting points significantly greater than the processing temperatures used in wafer processing. Tungsten and molybdenum, for example, have melting points of 3410 and 2610°C, respectively. This is in sharp contrast to that of aluminum, which has a melting point of 660°C. Normal processing temperatures for silicon are in the vicinity of 1000 to 1110°C for diffusions and 600 to 900°C for "annealing" (after ion implantation, for example).

2.6.1 Distribution of Interconnects

Figure 2.28 shows the distribution of interconnect path lengths in a 5000-gate logic gate array normalized to the length L_c of one edge of the die. This figure shows that the average interconnect path is roughly 20% of the die size. The figure will obviously vary with the type of circuit.

Figure 2.29 shows the wiring statistics of a microprocessor made from a gate array. Very little information is available in the open literature on wiring statistics and fan-out characteristics. Hence the information in Figs. 2.28 and 2.29 is relatively rare.

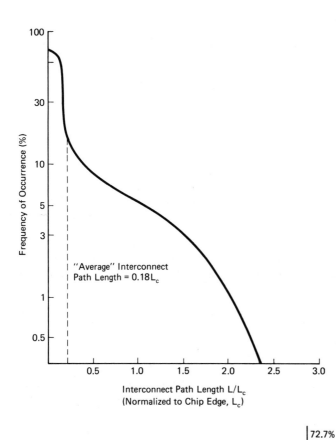

Figure 2.28 Distribution of interconnects in a 5000 logic gate array using two levels of metal. From D. J. McGreivy and K. A. Pickar, "VLSI technologies through the '80's and beyond"; registered copyright 1982 by IEEE.

Chip Size	7 mm
Number of Internal Circuits Wired	4705
Number of Off-Chip Drivers	96
Number of Receivers	122
Number of Signal Pads Used	200
Total First-Metal Wiring Length	2.40 m
Total Second-Metal Wiring Length	3.40 m
Total Number of Wiring Vias	33,516
Number of Nets	4437
Number of Connections	10,605

(a)

Figure 2.29 Data for microprocessor made from gate array: (a) wiring statistics; (b) histogram of total net wiring lengths; (c) histogram of wiring capacitances; (d) fan-out distribution (total number of nets = 4437). (From A. H. Dansky, "Bipolar circuit design for a 5000 circuit VLSI gate array," *IBM J. Res. Develop.*, May 1981, pp. 116–125. Copyright 1981 by International Business Machines Corp.; reprinted with permission.)

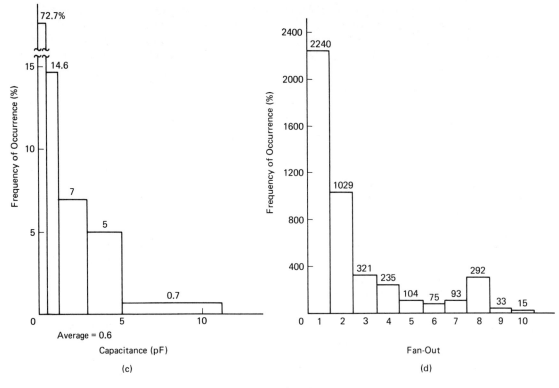

Figure 2.29 (cont.)

A summary of the chip is given in Fig. 2.29(a). Figure 2.29(b) is a histogram of the wiring lengths. It shows that the average length is 1.3 mm. Figure 2.29(c) is a histogram of the wiring capacitances of the microprocessor/gate array. It shows that the average wiring capacitance is 0.6 pF.

The distribution of fan-outs is also of interest. Figure 2.29(d) gives this information and shows that for this particular circuit, the overwhelming majority of gates have fan-outs of 2 or less. The distribution is monotonic through fan-out 6, at which time it begins to rise. However, the maximum rise is less than 10% of the total fan-outs. Later, it will be stated that normally one likes to keep the fan-out of CMOS to less than 4. Figure 2.29(d) shows that *for this application,* such cases occur 86% of the time.

2.7 GENERAL COMPARISON AMONG TECHNOLOGIES

The dominant gate array technology is clearly CMOS. Nearly every manufacturer has or is coming out with one or more families of CMOS gate arrays. The dominant permutation of CMOS is the bulk [as opposed to the silicon-on-sapphire (SOS)] version, which is both oxide isolated and silicon gate (see below). This is called by a variety of names, including HS CMOS (which will be used in this book), SO CMOS (selective oxidation CMOS), and LO CMOS (local oxidation CMOS).

HS CMOS began life at MITEL Corp. in Bromont, Quebec, Canada, as a 5-μm (drawn gate length) process. This process was widely licensed and adapted. A few companies did photolithographic scaling to 4-μm gate lengths. However, the next real jump was to 3 μm and below. This variant of HS CMOS is called HCMOS to denote the HMOS process to which it is similar. On occasion, the term "HCMOS" is used to refer to CMOS, which uses n tubs or n wells (as opposed to p tubs or p wells) to be compatible with the HMOS process. However, this usage is not universal.

Next most popular in terms of both number of vendors offering it and devices built is probably metal gate CMOS. This was one of the first gate array technologies offered and is still being used, although the costs of HS CMOS on a per gate basis are rapidly approaching it as competition increases and manufacturing techniques improve. Its second-place position (in terms of numbers of designs implemented as opposed to number of vendors offering it) is rapidly going to a high-performance technology, ECL.

Metal gate CMOS will probably continue to be manufactured for a period of time simply because of the very large number of designs that have been done in it. It is probably safe to say that otherwise the process has been largely phased out insofar as new designs are concerned.

Figure 2.30 is a comparison of IIL, STL, ISL, and CMOS with regard to power and switching speeds. These numbers are subject to change as technologies change but are representative.

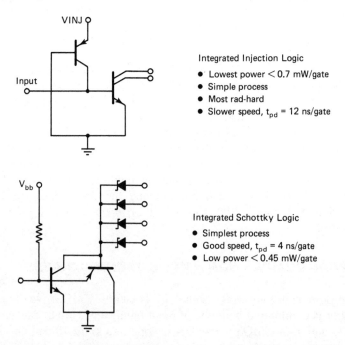

Integrated Injection Logic
- Lowest power < 0.7 mW/gate
- Simple process
- Most rad-hard
- Slower speed, t_{pd} = 12 ns/gate

Integrated Schottky Logic
- Simplest process
- Good speed, t_{pd} = 4 ns/gate
- Low power < 0.45 mW/gate

Figure 2.30 Comparison of technologies circa 1981. The *relative* numbers are still valid. (Courtesy of Harris Corp.)

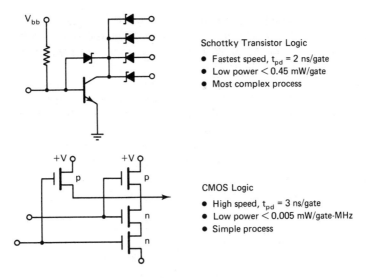

Figure 2.30 (*cont.*)

Oxide isolated ISL gate arrays, such as the Raytheon 5040 gate CGA 50L15 fit in between CMOS and ECL in terms of frequency and power. It is reportedly twice as fast as CMOS without the speed degradation of high fanout loads of CMOS and yet is lower power than ECL. Its speed power product is less than 0.2 picojoule when operating at a propagation delay of 1.2 ns. It is also radiation hard.

Because ECL and CMOS are on opposite ends of the silicon power, speed, and density curves, it is useful to make a few comparisons. Exact details will be given in the respective chapters on macro design with these technologies.

2.7.1 Speed-Power Product: ECL versus CMOS

ECL (emitter-coupled logic) and its variant CML (current-mode logic) are the highest-speed silicon technologies, offering gate speeds in the subnanosecond range. However, on a per gate basis, HS CMOS for low fan-outs has a speed–power product surprisingly close to that of ECL; both its power and its speed are very similar. Indeed, some people have conjectured that if the voltage swing of HS CMOS were made equal to the few hundred millivolts of ECL, HS CMOS would actually be faster. The threshold voltages of MOS devices, which are in the 0.6 to 1.1 V range currently prohibit this. Different materials with different Fermi levels and work functions would have to be used.

Two classes of factors conspire to make the per gate speed–power product results invalid for comparison purposes. First is that CMOS uses power only when switching. In any given circuit, typically only 10 to 30% of the gates are switching at any given time. All else being equal, this means that CMOS will use only $\frac{1}{10}$ to $\frac{1}{3}$ of the power of ECL.

The second class of factors is that in ECL there are five different ways of getting multiple levels of gating with essentially just one level of propagation delay.

These are series gating, emitter dotting (wire ORing), and collector dotting (wire ANDing), paralleling transistors to obtain the NOR function, and the inherent existence of the inverse of the given function due to the differential structure of ECL. These are discussed in Chapter 10.

By contrast, in CMOS there is only one class of structures that enable multiple levels of gating to be combined into fewer levels of delay. This class of structures is the AOI (AND-OR-INVERT) and its cousin, the OAI (OR-AND-INVERT). These are extensively discussed in Chapter 9.

In between HS CMOS and ECL in both speed and power is STL (Schottky transistor logic). This is a variant of IIL (integrated injection logic, sometimes called MTL, merged transistor logic, in Europe). STL is very similar to ISL (integrated Schottky logic), differing primarily in the way in which the base of each transistor is clamped to the collector.

In ISL, the clamping is achieved by the base-to-emitter drop of a *pnp* transistor. In STL, the clamping is done by a Schottky diode. This Schottky diode must be different (must have a different barrier height) from the Schottky diodes used on the collector outputs. (The output voltage swing is the difference between the two barrier heights.) This has caused some (mainly the competitors of STL, i.e., the vendors offering ISL) to believe that manufacturing STL devices is too difficult. Results would seem to prove them wrong. DEC uses STL gate arrays in their VAX 11/750 and other computers.

CMOS, STL, and ECL occupy roughly three different power dissipation and signal frequency regimes. As device feature sizes shrink, there are increasing amounts of overlap among these regimes. Currently, 5-μm-feature-size HS CMOS is good to about 25 MHz at 5 V V_{dd}. (The 2-μm double-layer metal version of LSI Logic/Toshiba is good for roughly twice that frequency.) STL of TI and Harris is good up to about 80 MHz in the newer versions. ECL is good to about 200 MHz to over 500 MHz depending on vendor and temperature range. The three technologies also rank the same way with regard to power dissipation, with CMOS being the lowest. These are simply very rough guidelines which should be used with a great deal of caution by the reader. Individual user circuits vary greatly in throughput rates, and vendors are constantly making improvements to their technologies.

2.7.2 Power Dissipation: CMOS versus Bipolar

The power dissipation of bipolar devices is [17]

$$\frac{CV_1 V_{cc}}{t} N \qquad (2.10)$$

where C is the load capacitance, V_1 the logic swing, V_{cc} the power supply voltage, t the propagation time per gate, and N the number of gates.

By comparison, the power dissipation per gate of CMOS devices is very similar,

$$(CV_1)^2 f N P \qquad (2.11)$$

where all terms are the same as for the bipolar case, and P is the percentage of gates that are switching at any one time.

Of course, if not all gates of either type have the same load and other characteristics, then to obtain the total power dissipation for that technology, the effects of the separate elements will have to be calculated and then summed.

Equations (2.10) and (2.11) back up what has been stated previously; that is, the power of CMOS on a per gate basis can be similar to that of bipolar (except for the switching duty cycle), and the lower logic swing of bipolar (which CMOS currently cannot match) contributes to lower-power operation.

2.7.3 Drive Capabilities of Gate Array Technologies

An important consideration in technology selection is the ability to drive the lines on a chip. Different technologies have differing abilities to drive the interconnect. Figure 2.31 shows the wide variations in such capabilities. Most of the differences among the technologies are due to the differences of the "on" resistances (i.e., the driving-point resistances) among the technologies. Hence ECL, in addition to being the fastest silicon gate array technology, is able to drive lines on the chip better. The wide variations in the drive capabilities of CMOS are due to (1) the logic configuration (NAND or NOR gate), (2) the relative dimensions of the given transistors (5-μm geometries in this case), and (3) the relative sizes of the transistors (W/L ratio). (This parameter is not under control of the gate array user.) By changing this parameter, certain CMOS logic gate configurations can be made to have better drive characteristics in one direction of propagation than in others.

	T_{plh} (ps/mil)	T_{phl} (ps/mil)
ECL (Motorola)	0.6	3.8
HS CMOS (5 μm) 4-Input NAND	7.5 (3.8) Calc.	120 (60) Calc.
2-Input NOR	60 (30)	15 (7.5)
LST^{2}L (Fujitsu)	5	1.75
ISL (Signetics)		Avg. 52.3
STL (TI)		Avg. 20

Figure 2.31 Ability of different technologies to drive aluminum interconnect lines on a chip, based on 0.005 pF/mil (Motorola number). Numbers in parens are 0.0025 pF/mil, which is actual for 5-μm process.

2.7.4 Differences between Designing with CMOS and ECL Technologies and Gate Arrays in Particular

There are a few important differences between designing with ECL gate arrays and designing with CMOS gate arrays.

1. Fan-outs of ECL macros are greater than those of CMOS.
2. The propagation delays of ECL change very little with increase in fan-out. A typical number is 0.1 ns per added fan-out.

3. The wire OR IS allowed in ECL. As with any wire OR, care must be taken not to wire OR signals that have fan-outs of greater than 1. Otherwise, the other leg of the wire OR can feed a false signal to the second fan-out of the first signal in addition to wire ORing it with the first signal.
4. Both the signal and its inverse are commonly available at most ECL macro outputs.
5. The higher fan-out of ECL applies to both the signal and its inverse (i.e., if the fan-out of a given macro is 9 for the noninverted signal, the fan-out of the inverted output will also be 9). Properties 4 and 5, in conjunction with the fact that the propagation delay of the signal and its inverse are the same or nearly the same, give rise to a very important technique for reducing propagation delay and to a lesser extent power when using EXORs and EXNORs. Two input EXORs and EXNORs have the property that if the polarities of *both* input signals are reversed, the truth table remains unchanged. This is equivalent to putting an inverter on *each* of the two inputs. (The reader is invited to try this conversion.) Stated another way, a two-input EXOR with an inverter placed on each of its inputs is still an EXOR.

   Suppose that one has high fan-out signals (say of fan-out 9) going into many EXORs. An example might be a Hamming error detection scheme. In ECL although typical fanouts for the high speed versions of macros are 9, they must be derated by 40% in effect leaving an allowable fan-out of a little more than 5. Normally, this would dictate use of a high fan-out driver, one which has a fan-out of at least 15. Such a driver may introduce 2 to 3 ns of delay (including that due to wiring) in the critical path, however, and would be unacceptable.

   A nice way to work around the problem is to let roughly half of the input data signals to the EXORs come from the input signals and the other half from the inverse of the input signals. That way both the signal and its inverse sees no more than the properly derated fan-outs, *and* the fan-out driver with its delay and (power) requirements is eliminated.
6. When both the signal and its inverse are used to differentially drive the same device (such as an output macro), care must be taken in the fault analysis not to try to put faults on the signals inside the common driver and that which is being driven. The effect of this is shown in Chapter 7.
7. While extra power and ground pins are sometimes desirable in CMOS, they are often mandatory in ECL, depending on how many outputs are switching at a given time. A typical rule of thumb is that an extra power and ground pin must be added for every eight outputs that switch on that side of the chip. If this is not done, a few signals will potentially suffer differential propagation delays.
8. The designer often must try to equalize the power in the quadrants of an ECL chip within limits. This factor is almost never a consideration in CMOS.
9. In ECL, the designer must consciously invert a string of signals to enable signal restoration and edge sharpness and to avoid pulse width shrinkage or ex-

pansion. In CMOS, this problem does not occur (except potentially in the case of AOI) because any noninverter is made from the series connection of an inverting element (such as a NAND or NOR gate) along with an inverter. Therefore, the signal is generally restored automatically in CMOS.

10. In an AND or NAND gate in CMOS, propagation time from high to low (but not from low to high) is largely (neglecting third-order effects) independent of which of the inputs is activated. (The other inputs are assumed to be enabled so that the propagation time of a given input can be discussed.)

    In an ECL AND/NAND gate made using the very common series gating technique discussed elsewhere in the book there will be a significant difference in propagation delays of the inputs. Those in the lower part of the series "tree" will have longer propagation times than those higher up in the tree. At first glance, the difference of typically 0.6 or 0.7 ns (for a two-level series gated structure) between the highest and next-to-highest inputs in the tree may seem small. It is not small, however, compared to the tpyical propagation time of 1.4 or 1.6 ns for the inputs that are highest in the tree.

11. ECL gate arrays have a significant overhead power, which CMOS gate arrays do not. This means that if power and ground connections are made to an ECL gate array, which has *no functionality whatsoever* (not even one input or output defined), the chip will still draw substantial power (often as much as over 900 mW worst case). The power is used internally in reference voltage and current generators.

12. ECL gate arrays have many more distributed buses than do CMOS arrays. The latter generally have just power and ground. In addition to power and ground, ECL arrays have buses for (a) reference voltages of the current sources, (b) DC returns from emitter followers, etc., and (c) reference voltages for up to three levels of series gating (one bus per level). In addition, in arrays which have both TTL and ECL I/Os, there will be peripheral buses for both 5 V as well as V_{ee} and generally two separate grounds for the TTL and ECL I/Os.

    The reader should not get the impression especially from points 11 and 12 that ECL is a difficult technology to design with. It is not. Speaking as a person who has designed with all three kinds of CMOS, ISL, two kinds of ECL, and GaAs, the author believes that ECL is a beautiful technology with which to design. The flexibility inherent in the five methods of achieving essentially one level of propagation delay when using multiple levels of logic (see Chapters 2 and 10) makes ECL one of the most powerful logic forms available.

2.8 ISOLATION METHODS

Isolation methods are techniques used to prevent transistors on the chip from interacting with one another except under control of valid signals. Isolation methods have been referred to in earlier sections of this chapter and will be referred to in subsequent chapters. Moreover, when dealing with gate array vendors or when reading the literature on gate arrays, the designer will come across terms relating to such methods. Hence, it is highly useful for the reader to be given a *qualitative* and rudimentary overview of such methods at the wafer/die level.

What is presented here is intended to be enough that the user can rapidly obtain a qualitative understanding of such techniques without being swamped by a lot of detail which is not necessary for gate array design. The reader should regard this material as being slightly more than a dictionary or glossary. The object is to make the reader aware without making the reader an expert.

Several types of structures are used on wafers to obtain isolation as defined above. These are broadly classified as junction, oxide, and gate isolation, the use of insulating substrates, such as SOS, and three-dimensional isolation techniques, such as silicon-on-isolator and dielectric isolation.

The most widely used isolation technique currently is oxide isolation. This has largely supplanted the older junction isolation methods which were used for so many years. Hence, it is instructive to compare the two methods for both CMOS and bipolar transistor structures.

2.8.1 Oxide Isolation versus Junction Isolation for Bipolar Structures

Silicon dioxide, or glass, is the oxide referred to in the term *oxide isolation*. Junction isolation refers to isolation using back-biased diodes. The diodes in this case are formed from an isolation diffusion (of the *p* type except in the case of the CDI process described below) in contact with an *n*-type epitaxial region.

An example of a bipolar process that uses oxide isolation is Motorola's MOSAIC II process. A cross-sectional view of this process is shown in Fig. 2.32. Compare this to the older junction-isolated transistor in Fig 2.3. Both transistors are of the *npn* type.

From the discussion in Section 2.3 plus the labels on Fig. 2.32, the reader should be able to identify the particular circuit elements of the transistor.

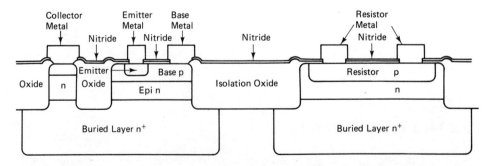

Figure 2.32 Cross-sectional view of oxide-isolated bipolar. (Courtesy of Motorola.)

Several benefits accrue from using oxide isolation. First, because the oxide is inert, alignment tolerances are considerably lessened. The oxide can be in direct contact with the buried collector referred to in Section 2.3.

This is considerably different from the junction-isolated case, in which physical separation must be maintained between the p^+ region used as part of the part of the isolation region and the buried layer. Additional tolerances had to be allowed for the lateral diffusion of the buried collector during subsequent high-temperature processing steps, such as diffusion of the vertical collector and the base regions. The tighter tolerances enable greater packing densities to be achieved.

Because of the lateral diffusion that occurred during subsequent processing steps, an impurity that diffuses slowly, such as antimony, was generally used for the buried collector. Oxide isolation largely removes this requirement and permits greater flexibility in the selection of the material.

Another problem with the junction isolation method is that the reverse leakage current increases substantially with increase in junction temperature. This not only added to the heat and power supply requirements on large chips, it also degraded the value of the isolation.

Yet a third improvement of the oxide isolation method over that of junction isolation is the ability to run conductors among the transistors in the oxide itself. This was never possible with junction isolation. Because 50% or more of the area of a given die may be taken up with such interconnects, this property is of considerable value. A cross-sectional view showing such interconnects within the oxide is given below in the comparison of the two techniques for CMOS.

Oxide isolation also eliminates a lot of the parasitic capacitances of the diodes used in the junction isolation. For all of these reasons, power is reduced and density and speed are increased via the use of the local oxidation.

2.8.2 CDI Technique

The collector diffused isolation technique and the prediction of Ferranti that it would produce gate arrays with 100,000 gates were mentioned earlier. Because CDI is a subset of the junction isolation technique, it is instructive to compare it in this section with both oxide and the normal junction isolation.

In the CDI technique, the starting material is a p-type wafer. On top of this is grown a p-type (instead of an n-type as in the usual junction-isolated process) epitaxial region. The isolation regions are themselves the n^+-type collectors of the pertinent transistors from which the process gets its name. Because the isolation regions are n type rather than p type, they can be in contact with the buried collector, thus enabling reductions in alignment tolerance similar to those discussed for the oxide isolation process.

The junction isolation is between the n^+ collector and the p-type epitaxial region. The process uses 5 masks instead of the 10 to 14 masks normally required and incorporates self-alignment for the isolation regions. Ferranti makes some other very telling arguments for this method. The reader is referred to the excellent paper referenced earlier. [17]

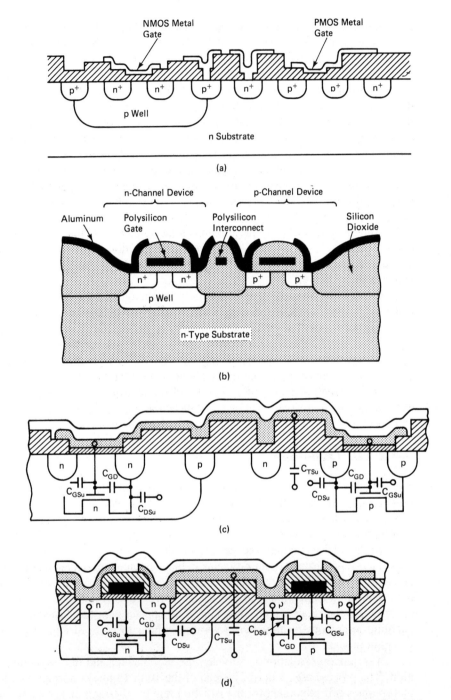

Figure 2.33 Comparison of (a) junction and (b) oxide-isolated CMOS; (c), (d) performance-limiting capacitances of the same structures: C_{GSu}, gate-substrate; C_{GD}, gate-drain (Miller); C_{DSu}, drain-substrate; C_{TSu}, track-substrate (not to scale). [(a) Courtesy of Interdesign; (b), (c), and (d) courtesy of MiTEL.]

2.8.3 Oxide Isolation versus Junction Isolation for CMOS Structures

It is also instructive to compare the two major types of isolation for CMOS. This comparison is shown in Fig. 2.33. From the preceding sections plus the figure itself, the reader should be able to identify the pertinent elements. Note the conductor running through the oxide isolation. The value of this was discussed above. The guard rings present in the junction-isolated case are no longer present, thus enabling higher circuit densities.

The same benefits of oxide isolation over junction isolation mentioned under bipolar structures also apply to the MOS case.

Bird's beaking. A problem with oxide isolation is that it causes a lifting of the silicon nitride used as a mask. This phenomenon, called *bird's beaking* or *encroachment*, makes it difficult to scale devices because it increases the required tolerances. Figure 2.34 is an example of bird's beaking applied to MOS devices.

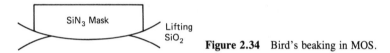

Figure 2.34 Bird's beaking in MOS.

The HP SWAMI technique. Hewlett-Packard has come up with an interesting variant of oxide isolation known as the SWAMI (side-wall masked isolation) which eliminates bird's beaking. This process has no ecroachment at all, and hence has considerable promise for the future. The interested reader is referred to the literature. [18]

2.8.4 Use of Insulating Substrates

The predominant insulating substrate material is sapphire (Al_2O_3). On top of this material is grown an epitaxial layer to a thickness of roughly $\frac{1}{2}$ μm. Mesa structures are created by etching the epitaxial layer. Isolation is obtained from the mesa structures on the nonconducting substrate. A *mesa* is a protrusion with sharply defined edges. The earliest transistors were built with mesa structures. Mesa structures today are utilized in CMOS/SOS and in some GaAs (another type of material as opposed to silicon) transistor types. The source and drain regions are then implanted and/or diffused using the gate as a mask in the usual case where self-alignment is employed. For an example of a mesa structure, see the discussion of CMOS/SOS in Section 2.5.1.

2.8.5 Dielectric Isolation

Two distinct methods of dielectric isolation exist. The first is a refinement of oxide isolation in which the transistors are isolated on the bottom as well as on the sides either by oxide or by silicon nitride. Doing this eliminates the bottom capacitance as well as the sidewall capacitance with resultant additonal improvements in speed, density, and power. The silicon-on-isolator technique discussed in Secton 2.10.1 is one such method.

The second form (which actually preceded all or most of the refinements mentioned above) is that used by Harris Semiconductor to provide not only the other benefits of density and low leakage currents, but also hardness to radiation. It differs from the others in the sense of being a much more extensive process. The reader is referred to the literature for a description [19].

2.9 RING OSCILLATORS

Propagation delay measurements are often quoted as being measured on a ring oscillator. Hence, it is useful for the reader to know what this structure is.

A *ring oscillator* is a logic structure built on a die consisting of several to many logic elements (generally inverters, NAND, or NOR gates) connected in series such that the output of one logic gate feeds the input to the next gate. There is no standard ring oscillator.

A ring oscillator is a very idealized structure for several reasons. First, generally no line length is used between the output of one gate and the input of the next. Second, the fan-out is generally 1. Third, because of the previous two items, the layout is considerably simplified. Fourth, the output buffers are generally designed to be custom matched to the output being driven. Finally, the output load of the chip itself is generally kept to a minimum.

With all these restrictions, why are ring oscillator measurements so popular? The answer is that they provide a *lower limit* to the propagation delays of the devices being measured. Because such measurements do eliminate the factors that will vary from circuit to circuit, they provide baseline measurement data.

Another reason for the use of ring oscillators is that they enable the measurement of signal propagation times which would otherwise be difficult to measure because of the short duration of such signals. The propagation delays of internal gates are also difficult to measure because the input capacitances of scope probes and package pins are so much larger than the capacitances of internal gates and on-chip lines that the measurement is distorted.

Still another reason is that the time delays of active elements are generally temperature dependent, while those of the interconnect are temperature *independent* for all intensive purposes. Elimination of the latter aids in obtaining baseline data on the devices themselves at various temperatures.

These problems are overcome by dividing the net propagation time of the composite signal by the number of stages through which the signal must pass to produce the output. The result is the propagation delay attributed to each of the gates. The gates are all made the same generally to enable the divided-up delay to be attributed to each. There are generally enough propagation stages in the circuit to minimize the effects of loadings of the output buffer which does drive to the outside world (i.e., outside the chip).

Figure 2.35(a) is a diagram of a ring oscillator. The pulse input is initially low, which puts a high on the other side of the NAND gate. (The reader should verify this.)

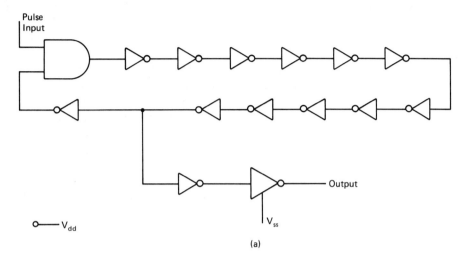

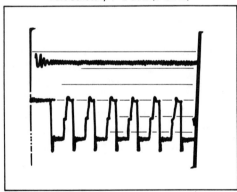

- < 3 ns/CMOS Gate
- CMOS is 1/100 power of pure Schottky T²L
- 2000 gates (40 to 100 ICs) on one chip
- Twenty-seven levels of CMOS logic

(b)

Figure 2.35 (a) Ring oscillator; (b) resultant output.

In order for the signal to appear (go high) at the output, it must make two complete passes through the network. There are 11 inverters plus one NAND gate through which the signal must pass twice before the output signal goes high for a high on the input. This is a total of 24 delays. There are also three inverters, labeled

IV_1, IV_2, and IV_3, which the signal passes through once before the output goes high. This is a total of 27 stage delays. All but three of the 27 are due to propagation through the same-size inverter. The remaining two are two passes through the NAND gate and one pass through buffer inverter IV_3. Hence, the total propagation delay divided by 27 gives a slightly conservative (because the delays through the NAND gate and IV_3 are longer) estimate of the propagation delay of the 13 inverters that are the same size.

The top of Fig. 2.35(a) is a photograph of both the input trace and the resultant output. In this case, the output load is a pure Schottky TTL device, a 54S140.

The reader may have noted that the propagation delays produced by this circuit arrangement monitor only the average propagation delay of the inverters. The average propagation delay is defined as the sum of the high-to-low and low-to-high propagation times divided by 2. If the circuit element being measured has propagation times in the two directions that are approximately equal to one another, the measurements are representative of the delays in either direction. If such is not the case, the user must be alert that actual circuit performance may differ markedly depending on the signal edge being used.

It is not possible in the foregoing circuit to measure the propagation time in one direction (i.e., in the low-to-high or high-to-low directions) only. It is left as an exercise for the reader to figure out such a circuit.

2.10 NEW DEVELOPMENTS

In the rapidly changing world of microelectronics, new developments or refinements of existing processes happen almost daily. In this section two of the more promising developments in CMOS are discussed. A few words are also said about GaAs gate arrays.

The choice is the author's own and does not imply that any of the many other developments that are happening are less significant. For example, GE/Intersil has developed a "retrograde" CMOS process that has exhibited propagation delays in the few-hundred-picosecond range. Workers at Honeywell and University of Dortmund, Federal Republic of Germany, have also developed CMOS structures which have exhibited similar delays. However, there is not a great deal of information on these developments as of this writing.

The author believes that the developments briefly mentioned below hold considerable promise in terms of making significant impacts on the technologies used to fabricate VLSI in general and gate arrays in particular.

2.10.1 Silicon-on-Insulator Techniques

Silicon-on-insulator (SOI) techniques imply the use of a nonconducting substrate in contrast to the semiconductor substrates used in all the technologies discussed in this chapter except SOS. Unlike SOS, in which the substrate is the entire wafer, SOI techniques build a nonconducting substrate (in effect, a floor) into the wafer. The nonconductor can be either silicon dioxide or silicon nitride. The major benefits derived from a nonconducting substrate are lower leakage currents, less parasitic capacitance (and hence faster speeds), and much higher packing densities.

2.10.2 Vertically Stacked CMOS

As noted earlier, in CMOS, a given signal goes to both a PFET and to an NFET in the majority of applications. This property can be taken advantage of to build vertically stacked transistor pairs. Typically, the PFET is on the bottom, the NFET is on the top, and the common gate is in the middle. Figure 2.36 shows this arrangement (called *joint gate* CMOS) in work done at MIT's Lincoln Laboratory. Note that in order to make contact with the bottom device, the channel length must be greater for it. One might expect that because of the higher mobility of the electrons, the NFET would be on the bottom. This has been done in later work done at Texas Instruments.

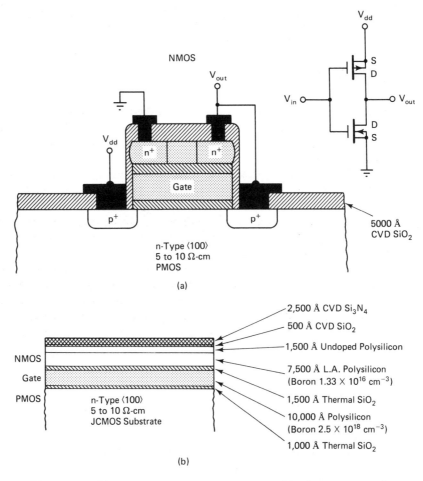

Figure 2.36 (a) Vertically stacked CMOS structure; (b) substrate cross section. (Courtesy of MIT Lincoln Lab.)

The advantages of such an arrangement are several. Most obvious is the increased density. This occurs not only because two transistors have been built in the space of one, but also because one gate serves both FETs. A second advantage which is not so obvious is the elimination of latch-up. There is no p tub or n tub.

Matsushita and Futjitsu have also developed vertically stacked CMOS circuits. Matsushita has achieved a 1-MHz operation in a 10-bit dynamic shift register. Fujitsu has obtained propagation delays of 450 ps with 2-μm gate lengths. In both cases, lasers are used to recrystallize silicon on insulators without thermally disturbing substrate circuits.

A problem with vertically stacked structures is that their use increases problems in making the wafer planar and hence makes it more difficult to achieve good step coverage.

A related development is that of "trench isolation" used by TI and Nippon Telephone and Telegraph (NTT). [20] The TI work enables the cell size of a 4-Mbit DRAM (dynamic RAM) to be half the size of a nontrench isolated 1-Mbit DRAM cell. The transistors in the NTT work (which NTT calls TMOS or TFETs) are applicable to many types of digital VLSI.

2.10.3 GaAs [21, 22, 23]

Another material in which gate arrays are being developed is GaAs (gallium arsenide). The principal advantage of GaAs is that it has much greater mobility than silicon has. This enables faster-than-silicon devices to be made or a trade-off of speed for power.

Significant progress has been made in overcoming the materials and fabrication problems that in the past have plagued GaAs. For example, unlike silicon, GaAs has no native oxide to act as an insulator. Another problem is that GaAs is a compound semiconductor as opposed to the single-base material silicon. Unless special precautions are taken, arsenic will outdiffuse from the gallium because of its low vapor pressure. This causes changes in the material.

The impetus to develop gate arrays in GaAs comes from both the private and the military sectors. In the private sector, several companies have developed GaAs gate arrays in the lab. A quick summary is given in Table 2.4 in terms of the developer, the size of the array, the speed and power parameters mentioned. The reader should note that some of these are one-of-a-kind laboratory items which are the forerunners of future products.

TABLE 2.4 GaAs GATE ARRAYS

Honeywell: 2000 equivalent gates; 250 ps/gate; 4-5N power (typical)
Tektronix: 1224 gates, 200 ps/gate; 250 μW per gate
TI: 4000 gates
Toshiba: 1050 gates, 350 mW and 10.6 ns when implemented as a 6 × 6 parallel multiplier.
OKI Electronics: 1000 gates, 390 ps
Rockwell: 306- and 732-cell arrays
Lockheed: 320-cell array; 184-ps delays

In addition to the above, Harris offers a PAL-like device which has preconfigured gates. Unlike a normal PAL, however, the interconnections among the gates are not already in place with fusible links. Rather, the user can define the interconnects but not the gate structures.

A small GaAs gate array from Tektronix. The GaAs gate array from Tektronix is one of the very first commercial products of its kind. As noted above, other GaAs gate arrays in the laboratory (including those in the laboratory at Tektronix) are much larger. The delay of a NOR gate with a fan-in and fan-out of 3 of the current offering is about 190 ps with 20 mW of power dissipation. Up to 14 D-type flip-flops of 28 latches can be made on the chip, and clock frequencies of 1.4 GHz can be achieved with it. Tektronix calls this device a *cell array*. Like most gate arrays it is in reality a collection of groups of transistors that can be connected to form logic structures. (The fact that the designer is allowed to use only the macros in the Tektronix library to do the interconnect among the transistors does not obviate the fact that it is a collection of transistors.)

One of the many uses of devices possessing such speeds is to enable counter circuits to be made which can divide the frequency into ranges where it can be handled by other technologies, such as ECL, which are currently more mature, less expensive, and have higher densities.

Differences between silicon and GaAs. One of the differences between GaAs and silicon fabrication is that while most silicon devices in production are in the range 2 to 3 μm or above, GaAs FETs have for many years been fabricated almost universally in the range 1 μm and below. The 1 μm refers to the length of the gate itself, not to the spacing between source and drain. The latter is often 2 to 3 μm. There is, hence, a resistance between the channel and the drain and a second resistance between the channel and the source. This is shown diagrammatically in Fig. 2.37(a).

The reader will also note that the gate is recessed. The reason for this is the fact that the surface of the channel is depleted even with no material covering it and thus has a high resistance. By going even 500 Å below the surface, the resistances between channel and source and drain are greatly decreased.

Another difference is that because of the lack of a native oxide, isolation among GaAs transistors is achieved by damaging the lattice with an ion implant and then not activating the implant with an anneal step.

GaAs structures. FETs in GaAs are almost universally NFETs. The very large mobility advantage of GaAs over silicon comes from using electrons as the majority carrier. (The type of FET known as JFET is used by one or more manufacturers and does have a poor but usable PFET. This is the exception to the rule, however.) Because of the lack of a PFET, complementary structures are not used in GaAs.

NFETs are formed by ion-implanting silicon ions into GaAs. The depth of the implant determines whether the device will be a depletion-mode or an enhancement-mode device. The implant depth of enhancement-mode devices is typically 500 to

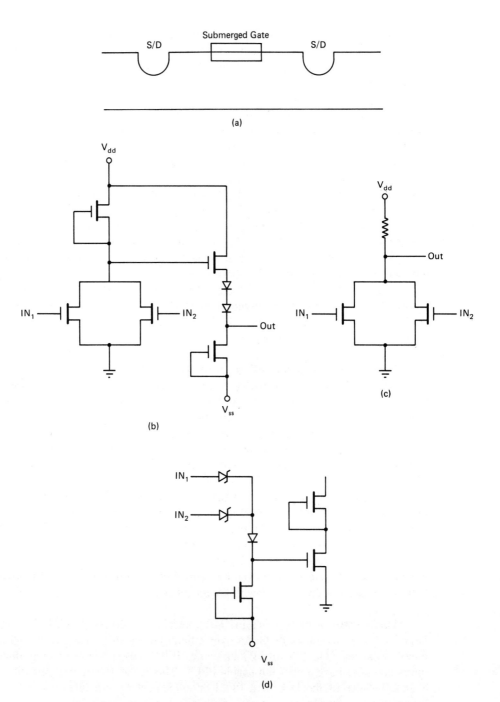

Figure 2.37 (a) GaAs gate structure; (b) BFL, buffered FET logic; (c) DCL, direct-coupled FET logic; (d) SDFL, Schottky diode FET logic.

1000 Å; that of depletion-mode devices is typically 1000 to 2000 Å. The depth determines whether the device is pinched off (see Section 2.4) at zero voltage or not. The larger depth of depletion-mode devices means that at zero volts, the built-in depletion region due to the gate is not sufficient to pinch off the device.

Gates. The gates of GaAs devices are metal, usually aluminum. Because, unlike silicon, GaAs has no native oxide, the metal gate is deposited directly over the implanted GaAs structure. It thus forms a SBD (Schottky barrier diode). It is this SBD that gives rise to the depletion region mentioned above. This gives rise to the term MESFET (MEtal Semiconductor FET) for GaAs devices.

The gate capacitance is typically 1 fF/μm of width for a gate length of 1μm. For a *very* crude and rough comparison to MOS, consider an NMOS gate that is 3 μm by 20 μm and which exists over 250 Å of gate oxide. A typical value is 140 fF. This is an extremely crude comparision but an interesting comparison, nevertheless. First, the value for the GaAs device is at zero potential. Obviously as the voltage varies, so will the depletion capacitance. Second, by the time the gate area of the MOS device is scaled to the 1 μm^2 of the GaAs device, effects such as parasitics and the effects of fringing fields, which are second and third order at 3 μm, will play a much larger role at the lower scaling. Third, as noted in Section 2.4, scaling to 1 μm would also entail a scaling (in this case a decrease) of the gate oxide of the MOS device with an attendant increase in capacitance. Despite the somewhat "apples to oranges" comparison, the comparison is interesting.

Several types of logic structures are used by different GaAs manufacturers. One of the most widely used is BFL (buffered FET logic). This structure is shown in Fig. 2.37(b). The benefit of this structure is that it has a very high drive capability, which thus enables the inherent speed advantage of GaAs over silicon to be taken full advantage of. Its disadvantages are: (1) it requires a level-shifting output stage; (2) the level shifter is constantly dissipating power (unlike a CMOS FET structure), typically two-thirds of the power dissipated by the logic structure to which the level is attached; and (3) the additional space taken up by the level shifter.

Three other types of logic structures used with GaAs are the SDFL (Schottky diode FET logic) of Rockwell and Honeywell, the CDFL (capacitor diode FET logic) of Gigabit Logic, and DCFL (direct-coupled FET logic). The first two use depletion-mode transistors (as does BFL), while the third uses both depletion mode and enhancement-mode transistors. Examples of BFL, SDFL, and DCFL are given in Fig. 2.37. CDFL is quite similar to SDFL. The basic structure is shown and a configuration of either OAI (OR-AND-INVERT) or AOI (AND-OR-INVERT (see Chapter 9). The reader will note that a "dual-gate configuration" is shown. Although not unique to GaAs, it is far more widely used than in MOS.

BFL is the highest-powered, highest-speed, and most mature of the three structures. DCFL is the newest, slowest, but also most dense and lowest powered of the three. Newer structures containing both enhancement-mode and depletion-mode devices are being made in the laboratories.

Indium phosphide is another compound semiconductor material on which laboratory devices have been built for a number of years. Its advantage is that unlike GaAs, it does have a good native oxide. Its mobility is also superior to that of silicon. Its disadvantages are that it is an extremely difficult material with which to work and (partly because of the lack of economic incentives), material uniformities are even worse than with GaAs.

A common method used to achieve isolation among the transistors in GaAs is the air bridge. Because the propagation delay varies inversely as the square root of the dielectric constant and because air has about the best dielectric constant of any material, its use degrades the speed of the interconnect lines the least. The problem, of course, is the lack of support mechanisms, which in turn limit the lengths of the paths so formed. Harris has built chips with air bridges which are as long as 50 to 80 μm using metals which are electroplated to 3- to 5-μm thicknesses.

HEMTs and HBTs. GaAs has an electron mobility three to five times that of silicon. The mobility in either case is limited by the collisions the electrons experience. It is possible to make submicron GaAs structures in which the electrons traverse a quasi-ballistic path giving rise to velocities of 3×10^7 cm/s (as opposed to 1×10^7 cm/s or so depending upon temperature, etc., for silicon). Two types of GaAs structures that exhibit this phenomenon are the HEMT and the HBT. HEMT stands for "high electron mobility transistor" and also for "heterostructure enhanced mobility transistor." (Both terms are used.) These are FET devices. Another name for the HEMT is MODFET.

An example of a recent HEMT device is the Bell Laboratories 4×4 multiplier, which has a multiplier, which has a multiply time of just 1.6 ns. [24]

The HBT (heterojunction bipolar transistor) is a bipolar device. Because the current flow in a bipolar device is vertical, essentially the entire area can be used. Therefore, relatively high currents can be obtained in relatively small devices. This is in contrast to FET devices, in which the current flow is confined to a small depth adjacent to the surface. HBT logic structures include both ECL (being done by Rockwell and others) and IIL (TI).

2.10.4 GaAs on Silicon

Researchers at the University of Illinois led by Hadis Morkoc [25] have successfully produced MESFETS, MODFETS, and HBTs by growing GaAs on silicon substrates using MBE (Section 2.3). The advantages of this method are that

- Thinner wafers can be used because silicon is more mechanically rugged than GaAs.
- Larger GaAs chips can be produced again due to the mechanical ruggedness of the silicon. (This is just one of the many factors involved in making large GaAs chips of course.)
- Silicon has better thermal conductivity than does GaAs. This factor helps enable better cooling.
- Silicon wafers cost much less than GaAs wafers. Therefore, there is promise that the overall costs can be lower than with pure GaAs wafers.

This technique is not to be confused with the technique of bonding a GaAs chip to a silicon substrate, which is then bonded to the cavity of a package. The latter technique is used by companies such as Gigabit Logic. It enables the bonding wires to the *decaps* (decoupling capacitors) to be much shorter than would be the case if the GaAs chip were simply bonded to the package cavity. (The decaps are fabricated directly on the silicon.) This in turn means that far fewer package pins have to be used. (Typically, every high speed signal pin has an adjacent ground pin.)

Interconnecting GaAs ICs. Interconnecting GaAs ICs so as to not lose the inherent speed of the GaAs is a major problem. One method that has been proposed is to use very tiny coaxial lines. [26]

2.11 SUMMARY

One of the purposes of this book is to enable the reader to read the literature. In this chapter, the different technologies, the principles of their fabrication, and how they differ were introduced. Comparisons among them were made. Permutations of the technologies, such as the three forms of CMOS, were described. A structure on which many delay measurements are quoted, the ring oscillator, was mentioned. The important subjects of isolation methods and interconnections were discussed. Gate arrays made from GaAs were mentioned. Finally, developments that show promise for future gate array use were noted. These included the JCMOS (joint-gate CMOS) developments at MITs/Lincoln Lab and elsewhere.

APPENDIX 2A: RUDIMENTS OF SEMICONDUCTORS [10, 11, 27]

The purpose of this brief overview is to provide the reader with a handy reference for some of the terms used earlier in the chapter. It is *not* an attempt to act as a first course in semiconductor theory for those who have not been exposed to the latter. Nevertheless, the nonspecialist will learn the meanings if not the full implications of basic semiconductor terms and concepts. The reader is referred to any of the excellent books already in existence on this subject for more information.

Semiconductors utilize two basic types of carriers, holes and electrons, known, respectively, as *p*- and *n*-type impurities or acceptor and donor impurities. Material that has an excess of donor over acceptor impurities is said to be "*n* type" and that with an excess of acceptor over donor impurities is said to be "*p* type." At any given temperature, the product of the impurity concentrations of the two impurities must equal the square of the *intrinsic concentration*.

2A.1 Doping the Semiconductor

Impurities are embedded into the silicon wafer by means of one or a combination of two methods. The two methods are ion implant and diffusion. The process of embedding the impurities is called *doping* the wafer. Typically, ion implant is used to provide a precise amount of impurity at the surface of the silicon. This is called a *predep* or *predeposition*. Diffusion is then used to "drive" the impurity into the silicon. The latter operation takes place at very high temperatures (1000 to 1100°C). When much larger quantities of impurities are needed, they are embedded using diffusion only. When smaller quantities are acceptable, ion implant alone is used.

Doping "profiles" refer to impurity distributions versus distance into the semiconductor surface. For ion implant acting alone, the impurity distribution is Gaussian. The depth of the mean of the Gaussian depends on a combination of the mass of the atom being implanted, the accelerating voltage of the ion implanter, and the material being doped by the implantation. For example, boron (which with an atomic mass of 10.8 is the lightest of the impurities used to dope silicon) will penetrate (mean of the distribution) to depths of roughly 0.52 μm in silicon, 0.56 μm in silicon dioxide, 0.44 μm in silicon nitride, and 0.55 μm in aluminum. The accelerating voltage of the ion implanter in all cases is 200 kV. By contrast, phosphorus (a much heavier atom with a mass of 30.1) at the same accelerating voltage penetrates only roughly half as far, or 0.25, 0.21, 0.16, and 0.23 μm respectively, for silicon, silicon dioxide, silicon nitride, and aluminum, respectively.

For diffusion, the doping profile as a function of depth x into the silicon is C_1 erfc $[x/(4DT)^{1/2}]$, where erfc is the *complementary error function*, C_1 is the solid solubility (see below) of the impurity to be diffused under the given conditions, D is the diffusion coefficient at the given temperature, and T is time.

From the discussion above, the reader will note that most of the action takes place in a depth into the wafer of a few micrometers. By contrast, the wafer is typically 14 mils (thousandths of an inch) thick or roughly 350 μm. The rest of the wafer is used for mechanical rigidity during processing. (Also, it is difficult to saw the raw wafer off the "boule" if it is too thin.)

For silicon wafers, the range of doping concentrations is roughly 10^{14} to 10^{19} impurities per cubic centimeter. Most doping falls below 10^{17} or 10^{18} per cubic centimeter. The true upper limit of doping concentration depends on the actual impurity itself and temperature but is in the range of 10^{19} for aluminum at 700°C to about 2×10^{21} for arsenic at 1100°C. This is called the *solid solubility limit* for that impurity at the given temperature.

Regions of wafers with impurity concentrations at the low end of the scale are said to be doped n^- or p^- depending on the impurity type. Those doped at the higher end of the scale are said to be n^+ or p^+, respectively. Sometimes the terms n^{++} and p^{++} are used to denote very heavy doping. 10^{14} impurities per cubic centimeter is roughly the residual doping concentration of an undoped wafer.

A material that has been doped with acceptor impurities can be changed to one that is *n*-type material by doping with sufficient donor impurities. The concentration of the donor impurities must be great enough first to neutralize the acceptor impurities and then to dope the silicon to the required *n*-type impurity concentration. This is the process that takes place when a portion of the base region of a bipolar transistor is converted to the emitter region. The dominant impurities are called the *majority carriers* and the other type (the holes in this case) are called the *minority carriers*.

2A.2 Mobility

An important parameter of semiconductors is mobility. Transconductance, and hence the speed–power product of MOS devices, depends on mobility, for example. Mobility refers to the velocity of the carriers as a function of electric field gradiant (V/cm). The units of mobility are cm/s per V/cm, generally seen as $cm^2/V\text{-}s$.

For non-polar semiconductors, such as Si, two scattering mechanisms affect the mobility. These are scattering due to acoustic phonons and due to ionized impurities. [28] The variation of mobility with impurity concentration for donor and acceptor impurities is given in Fig. 2.38 which shows that as doping concentration increases, mobility actually decreases. The reason is that the "mean free path" of each type of carrier is decreased. There are simply more carriers into which to bump in the course of the carriers trying to traverse a given distance.

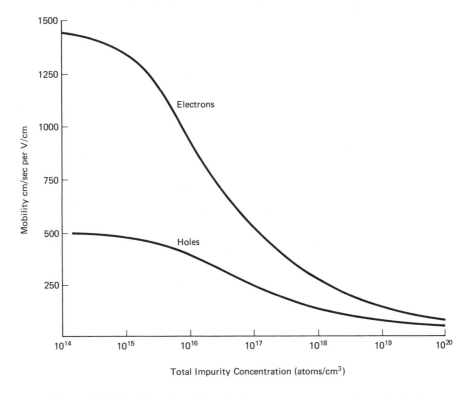

Figure 2.38 Variation of mobility with impurity concentration. (From the Semiconductor Technology Handbook; courtesy of Technology Associates, Portola Valley, Calif.; all rights reserved.)

An important point is that the mobility also decreases when temperature is increased for a similar (but not the same reason). Again, the mean free path is decreased. This time, however, the decrease is partly due to the thermal agitation of the crystalline structure.

Mobility also depends upon the *effective electron masses* (sometimes called "electron-effective masses.") [28] (These are not to be confused with the gain in mass of the electron due to relativistic effects.) For GaAs, the effective electron mass is only 7% that of silicon. [29] Because mobility varies inversely with the electron-effective masses, the mobility of GaAs can be much higher than that of silicon. The situation is actually more complex than is represented here, and the reader is referred to Sze's excellent book [28] for more explanation.

Splinter and others [30] have shown that mobility is a function of film thickness in SOS (silicon-on-sapphire). The mobility of electrons in *n*-type material is much greater than that of holes in *p*-type material if the film thickness is greater than about 2300 Å (0.23 μm) (*in* CVD/SOS). Because of the variation of mobility with film thickness, two important types of mobility are often distinguished. These are bulk mobility and surface mobility. The latter is often the dominant type in MOS transistor action for gate lengths greater than about 2 μm. For gate lengths less than about 2 μm, device current gain becomes dominated by saturation rather than by mobility. [31]

2A.3 Effects of Doping on Resistivity and Sheet Resistance

Increasing the doping decreases the resistivity of the silicon in accordance with the classical curve shown in Fig. 2.39. The equations of the curves in Fig. 2.39 are

$$r = \begin{cases} \dfrac{1}{q(uN \times n)} & \text{for } n\text{-type silicon} \quad (2A.1) \\ \dfrac{1}{q(uP \times p)} & \text{for } p\text{-type silicon} \quad (2A.2) \end{cases}$$

In Eqs. (2A.1) and (2A.2), q is the charge on the electron, uN is the electron mobility, n is the number of conduction electrons/cm^3, uP is the hole mobility, and p is the number of holes/cm^3.

The curve in Fig. 2.39 is based on uniform doping of the semiconductor. As noted above, neither of the two prevalent doping methods provides uniform doping. Therefore, to use Fig. 2.39, an average doping is calculated. This gives rise to an average conductivity. Alternatively, differentials can be used to incorporate the doping profiles in the expressions above. This is left as an exercise for the reader.

It was noted earlier that sheet resistance is resistivity (in Ω-cm) divided by thickness of the doped layer.

2A.4 Semiconductor Junctions

When two types of oppositely doped silicon are brought together (or more likely as in the illustration of the emitter being made from a portion of the base, *made* from one another), the result is a semiconductor junction.

Two forces result from such a juxtaposition of oppositely doped material. The first is due to a mechanical concentration of the excess carriers on each side of the junction. The excess electrons, for example, in the *n*-type material physically diffuse

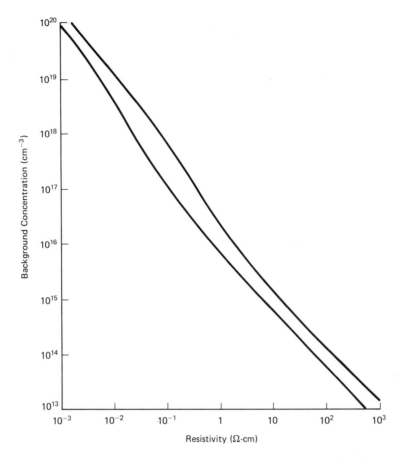

Figure 2.39 Resistivity versus dopant concentration for silicon at 300° K. (From *Semiconductor Technology Handbook*; courtesy of Technology Associates, Portola Valley, Calif.; all rights reserved.)

across the junction into the *p* material, where the mean free path between collisions is initially lower. This gives rise to a net positive charge in the *n* material in the vicinity of the junction. The result is the second force, which is an electrical force that tends to limit the process. In the vicinity of the junction, there is a *depletion region*, which is a region devoid of all mobile charge. The depths of the depletion region in the *n* and *p* materials will vary inversely as the relative dopings in the two regions. Specifically, if X_n and X_p are the depletion-layer penetration distances into the *n* and *p* sides and N_n and N_p are the concentrations of the *n*- and *p*-type impurities, respectively, then

$$N_n X_p = N_p X_n \qquad (2A.3)$$

has the effect of widening the depletion region. Forward biasing it has the opposite effect. This concept is important in the consideration of punch-through in MOS transistors.

Appendix 2.A: Rudiments of Semiconductors [10, 11, 27]

2A.5 Contacts

Two major types of contacts exist between the silicon and a metal. The first is an "ohmic" (resistive) type in which the current from the contact varies linearly with voltage. The second type is a Schottky type, which exhibits a diode-like behavior. The current remains relatively low and independent of the voltage until a threshold voltage is reached. Thereafter, it increases rapidly. The threshold voltage of the Schottky diode is called the *barrier voltage*. In STL the output voltage swing is the difference between the barrier heights of two different types of Schottky diodes. Schottky diodes formed from Schottky contacts are widely used in several silicon and GaAs devices.

(2A1) *Semiconductor Technology Handbook,* Technology Associates, Palo Alto, Calif., 1978, p. DIF 9

REFERENCES

1. Bambrick, R., "Motorola, Hitachi describe new processes," *Electronic News,* Oct. 14, 1985, p. 2.
2. Alvarez, et. al., "Technology considerations in BICMOS ICs," *IEEE Conf. on Computer Design,* Oct. 9–10, 1985.
3. Checkovich, P., "New VLE technique," *EE Times,* May 20, 1985, pp. 27–28.
4. Checkovich, P., "Other methods of growing epitaxial layers: pluses and minuses," *EE Times,* May 20, 1985, pp. 27–28.
5. Wang, K., "Novel devices by Si based molecular beam epitaxy," *Solid State Technology,* Oct. 1985, pp. 137–46.
6. Glasser, Lance, and Dobberpuhl, Daniel, *Design and Analysis of VLSI Circuits,* Reading MA: Addison-Wesley, 1985.
7. Mavor, J., M.A. Jack, and P.B. Denyer, *Introduction to MOS LSI Design,* London: Addison-Wesley, 1983.
8. Mead, Carver and Lynn Conway, *Introduction to VLSI Systems,* Reading, MA: Addison-Wesley, 1980.
9. Carr, Jack and William Mize, *MOS/LSI Design and Application,* New York: McGraw-Hill, 1972.
10. Ghandhi, Sorab, *VLSI Fabrication Principles,* New York: John Wiley, 1983.
11. Trapp, O.D., R.A. Blachard, and W.H. Shepherd, *Semiconductor Technology Handbook,* San Mateo, Calif.: Bofors, 1982.
12. Richman, Paul, *MOS Field Effect Transistors and Integrated Circuits,* New York: John Wiley, 1973.
13. Dennard, R., et. al., "Scaling of MOS Transistors," *IEEE J. Solid State Circuits,* Vol. SC-9, Oct. 1974, pp. 256–68.
14. *Design Manual for HC CMOS Gate Arrays,* San Jose, Calif.: California Devices, Inc., 1982.

15. Ganensslen, F., "MOS devices and ICs at liquid nitrogen temperature," *IEEE International Conf. on Circuits and Computers,* Oct. 1–3, 1980, *IEEE publication 80CH1511-5.*
16. Rubinstein, J., P. Penfield, and M. Horowitz, "Signal delay in RC tree networks," *IEEE Transactions on Computer-Aided Design,* July 1983, pp. 202–11.
17. Grundy, L., and C. Bruchez, "100,000 gate gate array," *Electronics,* July 14, 1983, pp. 27–29.
18. Chiu, K., "SWAMI, a zero encroachment local oxidation process," *HP Journal,* Aug. 1982, pp. 31–33.
19. Hamilton, Douglas, and William Howard, *Basic Integrated Circuit Engineering,* New York: McGraw-Hill, 1975.
20. Myrvaagnes, R., "Planting transistors in holes fits more on a chip," *Electronic Products,* Jan. 2, 1986, pp. 17–18.
21. DiLorenzo, James, and Deen Khandelwal, *GaAs FET Principles and Technology,* Dedham, MA, Artech House, 1982.
22. "Here comes GaAs," *Electronics,* Dec. 2, 1985, pp. 39–44.
23. Wilson, D. "High performance ICs fuel up with GaAs," *Digital Design,* Oct. 1985, pp. 41–46.
24. Brown, C., "HEMT GaAs IC may begin new era," *Electronic Engineering Times,* Dec. 2, 1985, pp. 33, 47.
25. "GaAs on silicon technology," *Semiconductor International,* Oct. 1985, pp. 16–18.
26. "Linking fast chips solved by tiny coaxial wiring," *Electronics,* Oct. 28, 1985, p. 14.
27. Sze, S.,ed., *VLSI Technology,* New York: McGraw-Hill, 1983.
28. Sze, S.M., *Physics of Semiconductor Devices,* New York: John Wiley, 1981.
29. Private communication, Gigabit Logic.
30. Splinter, M.R., "A 2-μm silicon gate CMOS/SOS technology," *IEEE Transactions on Electronic Devices,* Aug. 1978, pp. 996–1004.
31. Chwang, T., and L. Hu, "CMOS," *VLSI Design,* fourth qtr., 1981, p. 42.

Chapter 3

Gate Array Grid Systems

3.0 INTRODUCTION

Chapter 2 described the dominant gate array technologies in detail. With this information as background, the array grid structures of gate arrays can be described. It is necessary for the reader to understand array grid systems in order to fully appreciate the material in the next chapters. In the process of teaching gate arrays to a wide variety of students, the writer has found that only a superficial understanding of these structures by the student is not sufficient.

It is appropriate to devote a separate chapter to this material instead of scattering it around in a variety of chapters because of the similarities that exist among the different gate array grid structures. Putting it in one chapter avoids undue repetition of comments. Moreover, it enables ready comparisons to be made.

3.1 DEFINITION OF GATE ARRAY GRID SYSTEMS

The reader will recall from Chapter 1 that the gate array die is made up of several layers and that the customization involves using the one or two metallization layers to interconnect among the contact points of the underlying layers which form the underpasses, transistors, and so on.

The grid system of a given gate array is a representation of the metallization layers which defines the allowable interconnect points that can *potentially* be used. It may or may not be the actual background of the metal layer itself which shows the interconnect points to the underlying transistors. In some cases, a simplified representation is easier to use, especially when layout is to be done manually and then "digitized" (converted to the data base of an interactive graphics system by line tracing). A layout sheet is a prime example of a gate array grid system. The interconnect points go to transistor "terminals" and to underpasses, if used, on the array. For arrays from manufacturers which allow only macros to be used, the interconnect points are to I/Os of the macro. All the current ECL arrays, for example, have this restriction (i.e., allow only macros to be used).

To allow for a wide variety of circuits to be built on the same gate array background, most active elements have redundant terminals. For example, the source/drain of each array transistor in most second- and third-generation CMOS arrays comes out in two places, one on each side of the appropriate power bus. The IMI 2000 (actually 1960) cell array has approximately 71,000 such contact points among its approximately 12,000 transistors and approximately 10,500 underpasses.

The reader will also recall from Chapter 1 that all gate arrays, regardless of technology and manufacturer, consist of several basic elements:

- Pads to enable connection from the "outside world" (package or hybrid)
- Buffer devices to drive to the higher capacitance of the outside world
- Distributed power and ground buses
- Transistors and diodes
- Underpasses to cross under the power and ground buses without contacting them or more commonly a second metal layer

Many array dies have in addition to the above:

- Test devices
- Voltage-level shifters to enable operation from only one external voltage source (as in the case of ECL and ISL/STL) or to make the devices compatible with TTL logic levels

Most gate array grid systems define the spatial positions of the interconnects to all of the elements above, and most also define the allowable routing channels. Most gate array manufacturers offer families of gate arrays. Each family may have several members. The structure of the gate array grid systems is generally the same for all members of a given family, although the numbers of elements, pads, buffers, and so on, will change.

Figure 3.1 is a composite picture of three layers, including the metal layer and polysilicon underpass layer of a single-layer metal gate array. Note! The gate array shown in Fig. 3.1 is NOT an actual chip. It differs from an actual chip only in that the numbers of array cells is 40 instead of 1440 on the actual chip, and the numbers of I/Os and pads are correspondingly reduced. The purpose is to exhibit with reasonable clarity all the elements mentioned above.

The next sections develop a grid structure for one of the popular single-layer metal gate arrays. Then grid structures from a variety of manufacturers are presented. This is followed by discussions of two relatively new developments: gate

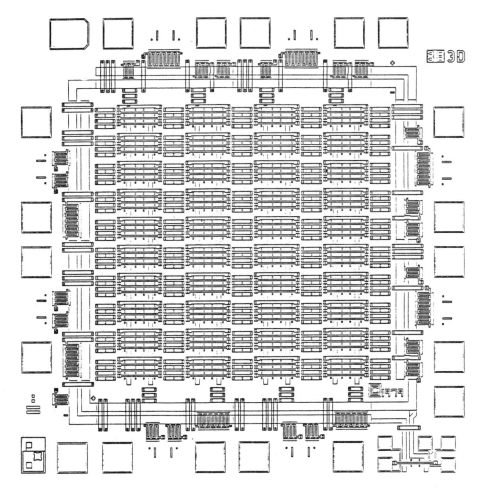

Figure 3.1 *Fabricated* picture of gate array die showing circuit elements. The figure was *created* by *removing* 1400 of the 1440 cells which are present in the layout of the *actual* die and similarly *reducing the number of I/Os so that a* complete die could be shown. *This is not a picture of an actual die.*

isolation and the "sea of gates" approach. Finally, an important study done at the General Electric Microelectronic Center is discussed.

All of the CMOS arrays discussed below are both oxide isolated and silicon gate (i.e., they are HS CMOS arrays).

3.2 SINGLE-LAYER METAL SILICON GATE CMOS ARRAYS

3.2.1 Detailed Example

There is a measure of similarity among gate array grid systems, even among those of different technologies. The reason is that most contain the elements listed above. By going through an example of one such gate array grid system in great detail, the reader will be prepared to make observations about others. Full appreciation of these systems will come after the reader finishes the chapters on design using the different technologies.

The design chosen is that of the IMI HS CMOS array. There are currently eight members in this gate array family, ranging in size from 200 *three*-input cells to almost 2000 (1960 to be exact) *three*-input cells. When it came out, the latter was the largest gate array available at the time. Note the emphasis on three-input cells. Sizes of newer arrays are often couched in terms of numbers of gates, with a two-input gate generally being implied. The IMI 2000-cell array is equivalent to 3000 such *two*-input gates. If the reader is not familiar with CMOS, it will be helpful to review Chapter 2.

The lower left corner of this array was shown in Chapter 1 and is repeated in Fig. 3.2. Each of the small squares represents a contact point and a potential interconnect. From this figure the reader can also observe which of the contacts are electrically connected together via polysilicon (in this case) gates and underpass. Not shown are the diffusions representing the source/drains of the transistors.

The way to get oriented on Fig. 3.2 and on the grid structure itself is as follows.

1. Find the "railroad tracks." These are the pairs of vertical lines which are the power (V_{dd}) and ground (V_{ss}) buses. This establishes the rotational orientation of the structure to within 180°.
2. Find the cells containing the transistors. These are clustered around the V_{dd} and V_{ss} buses, as contrasted to the "poly" (for polysilicon) blocks, which are in alternate columns from the transistor cells.

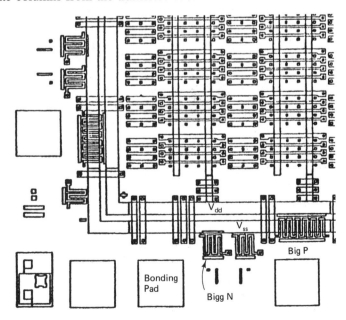

Figure 3.2 Corner of gate array. (Courtesy of International Microcircuits, Inc.)

Sec. 3.2   Single-Layer Metal Silicon Gate CMOS Arrays

3. Find the "horns" of the cells containing the transistors. In Fig. 3.2 these are the end contact points of the single thick horizontal bar adjacent to three thin horizontal bars. By orienting the grid so that the thick bar is above (instead of below) the three thin bars, the up-down orientation is established.

   The thick bar in the transistor cell represents an underpass. In Chapter 9 it will be seen that this underpass (arbitrarily called "poly D") is the preferred method of getting under the railroad tracks. The three thin horizontal bars are the gates of the transistor pairs.
4. Locate the "poly block" containing the poly C (arbitrary designation which will be used in the book) underpasses to the left and right of the array cells.

The reader will recall from Chapter 2 that the word "gate" has three different meanings, that one of them is the physical input to a MOS transistor, and that 99.9% of the time CMOS uses transistors in pairs consisting of one PFET and one NFET. (But, as will be shown below, it is sometimes advantageous to be able to make contact to the gate inside the V_{dd}/V_{ss} bus. For this reason, the transistor gates in the array connect both transistors.

Some manufacturers, such as CDI, have some gates that are broken in the middle and have separate contact points. These have advantages when building the logic device known as a transmission gate. To a limited extent they also enable interconnect lines which have been routed between the railroad tracks to be attached to transistor gates without using a transistor source/drain to get outside the railroad tracks.

The contact points of the source/drains of the NFETs and PFETs are clustered around the railroad tracks. The NFETs contact points are on either side of the V_{ss} bus and the PFET contact points are on either side of the V_{dd} bus. The source/drain contact points for a given transistor which are electrically the same (except for negligible resistance) are horizontally opposite each other.

Figure 3.3 is a gridded representation of a portion of the internal array structure in Fig. 3.2. Readers should use the clues above to orient themselves on this grid (i.e., first find the railroad tracks and then find the horns of the poly D. The latter has no source/drain contact above it.)

The small squares in Fig. 3.3 represent potential contact points just as they do in Fig. 3.2. The columns of small squares closest to the railroad tracks are contacts to the source/drains. The squares in the next columns (going outward from the railroad tracks) are contact points to the poly underpass and to the three gates in each cell. The squares in the latter have a slightly heavier dot in the middle.

The dots in Fig. 3.3 represent the allowable routing channels. This is an advantage of this representation over that in Fig. 3.2. For example, two vertical lines can be routed between each of the contact points of the poly Cs. A total of 21 such vertical lines are possible in each repetition of the cell and poly block. Of these, about 11 have some restrictions applied to them.

Reference to Fig. 3.4 may help the reader to better understand the above. Figure 3.4(a) is the diagram of a grid structure which spatially shows the locations of the contact points. The contact labeled underpass is the poly D referred to above. The three contacts labeled input A, B, C, respectively, are the three gates to the three CMOS transistor pairs. Figure 3.4(b) shows horizontal lines crossing the rail-

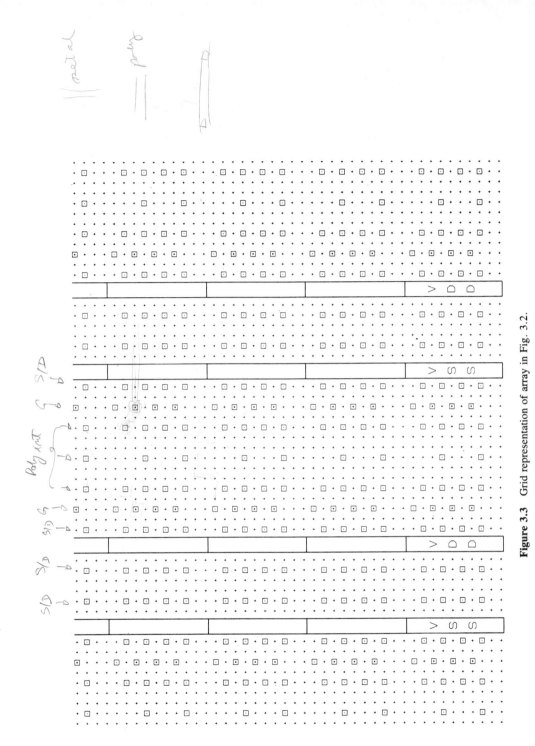

Figure 3.3 Grid representation of array in Fig. 3.2.

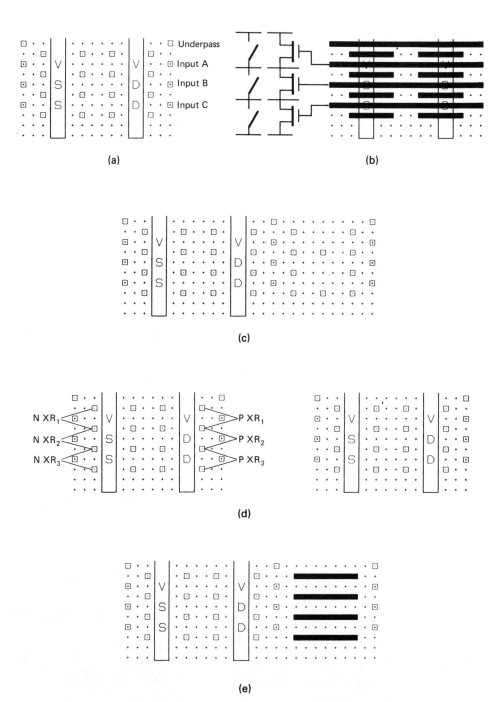

Figure 3.4 Explanations of grid: (a) basic cell; (b) internal interconnects; (c) basic cell and crossover block = poly block; (d) XR = transistor, N XR1 = n-channel transistor 1, P XR1 = p-channel transistor 1, etc.; (e) internal interconnects (horizontal lines) of crossover block.

road tracks. These represent gates and diffusions and clearly show what contact points are electrically the same. The polysilicon gates and underpasses and the diffusions are, of course, electrically isolated from one another and from the railroad tracks. Also shown in Fig. 3.4(b) are representations of the FETs as switches. The FET diagram is shown in Fig. 3.4(c).

One point that always causes a great deal of confusion is the sharing of the MOS transistor source/drains. This was discussed Chapter 2, and the reader should review that material if there is confusion. Each of the transistors can be regarded as a switch, as shown in Fig. 3.4(b). The control to the switch is the gate of the transistor.

The points labeled P XR_1, P XR_2, ... , N XR_1, ... in Fig. 3.4(d) are the source/drains of the array. Outputs are taken from source/drains as noted in Chapter 2; but, of course, not every source/drain is used as an output. (Strictly speaking, the outputs come off the drains, but the two are physically the same.)

The internal interconnects of the poly block are shown in Fig. 3.4(e). Every other such poly can be accessed from the middle as well as from either end.

Figure 3.5 is a diagram that may further help clarify the internal connects of the array cell. The underpass and the sources, drains, and gates of the FETs are all labeled in terms of their relative spatial positions in the array.

Figure 3.6 is grid structure which represents the bottom I/Os of the chip. Instead of having a long thin figure, the structure is repeated across the bottom. The reader should be able to readily identify the bonding pads. Other elements are as labeled. The elements labeled "Big PCH" and "Big NCH" are the buffer FETs on the periphery of the chip. These can be used for both input and output (see Chapter 9).

Also in Fig. 3.6 are poly underpasses labeled "poly A" and "poly B." The former is an underpass under the V_{dd} bus and is about half the length of the poly D. (At

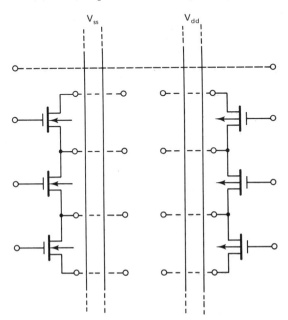

Figure 3.5 Another representation of the cell structure of the previous figures. (Courtesy of Frank Wolfenden, Sanders Associates.)

Sec. 3.2   Single-Layer Metal Silicon Gate CMOS Arrays

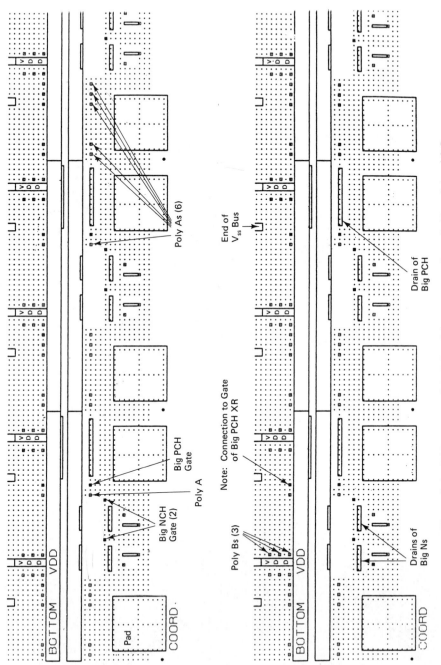

Figure 3.6 Grid structure of the I/Os of the gate array shown in Figs. 3.4 and 3.5.

the top where the V_{ss} makes contact to the peripheral V_{ss} bus, the poly B goes under it. The V_{dd} and V_{ss} buses are interleaved in such a way that the former is closest to the internal portion of the array on the bottom and the latter is closest on the top.

The poly A is an underpass under the peripheral V_{dd} and V_{ss} buses. In some cases the poly A contacts directly as shown to the gate of a PFET. Not apparent in Fig. 3.6 is that the sources of the buffer PFETs and NFETs are connected directly to the V_{dd} and V_{ss} buses, respectively.

The reader should also note that the gates of the buffer PFETS are not connected to those of the buffer NFETs as are the gates of the internal FETs.

Note that for a given gate array family, the same grid structure is used. The smaller members simply accommodate fewer replications of the grid sheets. The grid background can be transferred to Chronoflex sheets to permit remote design. The design on the sheets is then "digitized." Now that the reader is familiar with a representative gate array grid structure, other such structures can be introduced.

3.2.2 CDI's Grid Structure

CDI's single-layer metal grid structure (Fig. 3.7) differs from that of IMI's in that a given array cell is made up of two subcells. The first subcell has three inputs and is somewhat similar to that of IMI. The second subcell contains two inputs and is aimed primarily at building an important class of elements in CMOS called transmission gates (TGs). In the latter subcell, the PFET inputs are not collinear (i.e., not on the same straight line) with those of the NFETs. This permits the source/drains of the PFET/NFET pairs that make up the TG to be directly opposite one another. This technique and other comparisons are discussed in Chapter 9.

In the CDI grid representation, the squares without a larger square around them are the routing channels. Those with such a square are the contact points.

The reader will note the poly block and the poly underpasses in each cell. The cell underpasses have a connection in the middle, however.

An important note: In the discussions in this book it will be assumed that the chip background layers are arranged in such a fashion that the distributed power and ground buses run vertically unless otherwise noted. An important exception is that of the ISL arrays, where many of the buses do run horizontally. These will be discussed in the appropriate sections.

CDI does have space for one more vertical line crossing the polysilicon underpasses *adjacent to the array cells* than does IMI. However, one more line can run vertically inside the IMI V_{dd}/V_{ss} buses than can be run inside the CDI buses because the contacts to the gates of the latter force routing transiting signals around the contacts.

3.2.3 SPI's Grid Structure

SPI's grid structure (Fig. 3.8) has some of the elements of both IMI's and CDI's. Like CDI's, it is a "3/2 structure," meaning that the cell contains a three-input subcell and a two-input subcell. Like IMI's, none of the gates are broken in the middle;

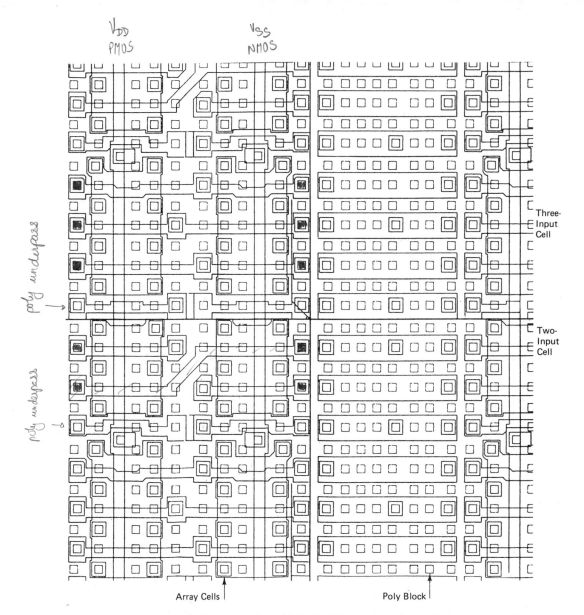

Figure 3.7 CDI's single-layer metal HS CMOS gate array structure (not to be confused with their newer "sea of gates" structure). (Courtesy of CDI.)

that is, each gate *always* connects to both an NFET and to a PFET. The polysilicon underpass of the two-input subcell does have a contact point in the middle, however. As with both IMI and CDI, every other polysilicon underpass *in the poly block adjacent to every array cell not in the array cell itself* has a contact in the middle. However, unlike either CDI or IMI, the contact point is offset on every poly that has such a contact point. A macro drawn on SPI's grid structure is shown in Fig. 3.8(d) to illustrate its use. See Chapter 9 for more detailed explanations.

3.2.4 MITEL's Grid Structure

Figure 3.9 shows the internal structure of the MITEL gate array. All elements are labeled. This is an example of a four-input cell array. It also has the interesting feature that poly interconnects run between but are not connected to source/drains of PFETs and NFETs which are adjacent to one another. This permits local routing within the cell wihtout interfering with global routing passing through the cell.

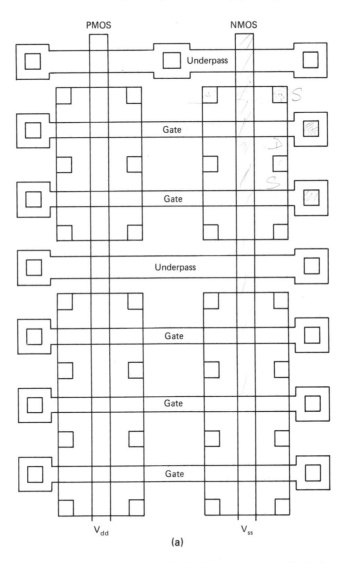

Figure 3.8 SPI's single-layer metal HS CMOS grid structure: (a) device cell; (b) crossover cell; (c) section of array at 200×; (d) double inverter. (Courtesy of Semi Processes, Inc.)

Figure 3.8 (*cont.*)

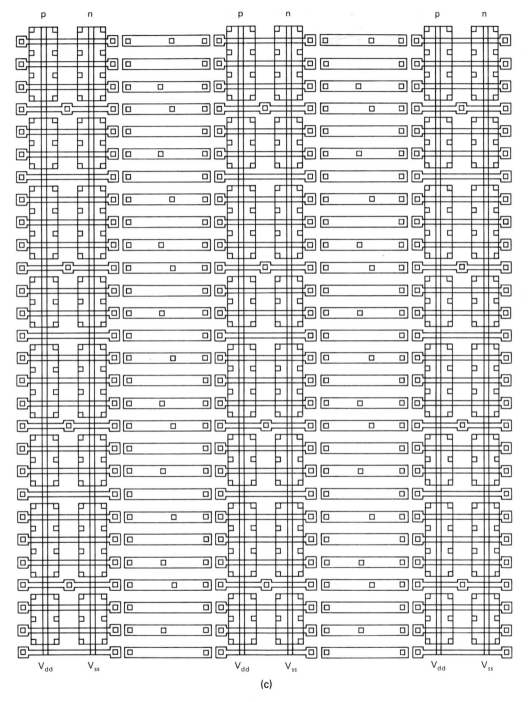

Figure 3.8 (*cont.*)

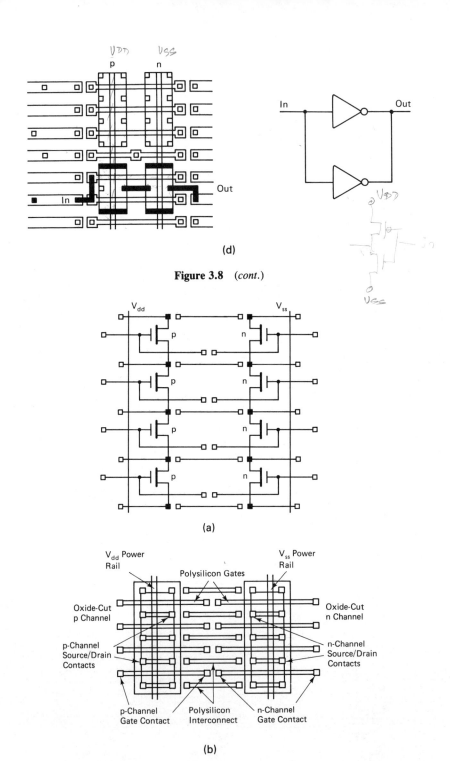

Figure 3.8 (cont.)

Figure 3.9 MITEL CMOS gate array structures: (a) array cell mask layout (unprogrammed); (b) a layout of basic array cell; (c) schematic of basic array cell. (Courtesy of MITEL.)

3.3 TWO-LAYER METAL CMOS ARRAYS

The first generation (see Chapter 1 for a definition of generations) CMOS gate arrays were not designed (in general) for use with IGSs. The second-generation CMOS gate arrays were designed for IGSs but not for "auto P&R" (automatic placement and routing of macro cells). The third-generation CMOS gate arrays were designed for both but especially for auto P&R.

A major difference between the one- and two-layer metal gate arrays is, of course, the lack of the underpass blocks in the latter. The space of the poly blocks is used to route the first-layer metal. A major advantage of the two-layer metal is that good speed can be obtained in both horizontal and vertical directions over long runs of interconnect. (Metal can run in both directions instead of just in the direction parallel to the V_{dd} and V_{ss} buses.)

In a few two-layer metal arrays, the cells in the center are spread farther apart than those nearer the periphery. This is based on recognition that the greatest density of signal lines is likely to occur in the center of the array.

3.3.1 STC's Grid Structure

STC's grid structure (Fig. 3.10) is representative of two-layer metal CMOS arrays. It, too, is a four-input structure. Note the source/drains under the power and ground buses (which in this case are labeled V_{cc} and Gnd). This is possible because the power and ground buses are on second-layer metal. (As with the other arrays, the source/drains are adjacent to the buses.) In this structure, the Gs are gates.

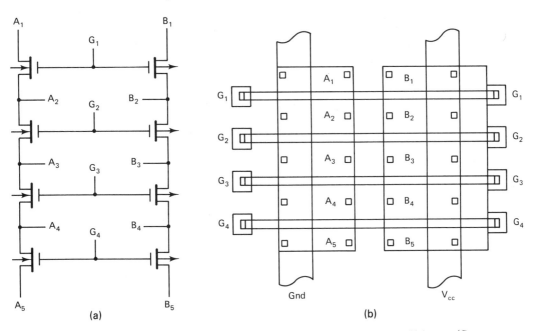

Figure 3.10 STC two-layer metal HS CMOS structure: (a) schematic; (b) layout. (Courtesy of STC Microtechnology.)

3.4 VALUE OF THE BREAK IN THE MIDDLE OF THE GATE

CMOS gate arrays also differ in whether the gate inputs to the PFET/NFET pairs have a break in the middle. In the author's opinion, the "jury is still out" on the value of doing this. (By a break in the middle, we mean that each PFET gate is not connected internally to an adjacent NFET gate.)

Some of the pros and cons of this method are given in Chapter 9. Briefly stated, the advantages accrue from being able to route lines between the distributed V_{dd} and V_{ss} buses to connect to the the inputs of the next logic structure. Because the outputs are available inside the V_{dd} and V_{ss} buses, this would seem to be a good method (and very definitely is a good method for some designs.)

A second advantage is that three-state gates can be made much more easily and with less wastage of transistors. (Note that the buffer FETs of gate arrays which do not have the break in the middle of the array FETs almost invariably are not connected. This permits three-state devices to be made using the peripheral FETs with no wastage of transistors.)

A problem, however, is that the extra contacts (two per gate that is so configured) take up a fair amount of space and in general force round-about routing of what would otherwise be straight lines. Examples of both methods are given in Chapter 9, and readers can judge for themselves.

3.5 ISL GRID STRUCTURE

Signetics's ISL gate arrays are two-layer metal devices. As much of the routing as possible is done on the first layer. The first-layer metal also contains three parallel buses running in horizontal rows. The outer two of these buses are both ground (illustrating the importance of this parameter in a technology where the voltage swing is a couple of hundred millivolts). The middle bus is V_{bb}. The three horizontal buses separate the array into groups of transistors.

The transistors are in groups of four. The base inputs of the four are closest to the buses with the bases of two members of the quad adjacent to the upper set of buses and two adjacent to the lower set of buses. The second layer is used to route over the first layer power and ground buses and to route what cannot be routed on the first layer. This especially would be the interconnects from row to row of the array.

The arrays consist of various numbers (depending on the size of the array) of *npn* transistors surrounded by buffer cells on the periphery. A somewhat unusual but highly useful feature is the column of Schottky transistors in the middle of the array. These are used to drive high fan-out lines on the array. The remainder of this section describes the Signetics 8A1200 array. Except for different numbers of elements, the information is generally representative of the other members of the family.

Figure 3.11 shows an overview of the chip. There are 40 bonding pads for the 36 I/Os. The remainder are used for the three power and ground buses. ISL uses two power supplies. One of these supplies V_{cc} (5 V). The other supplies V_{bb} (1.5 V) for the internal structure itself. It determines the injection current and in conjunction

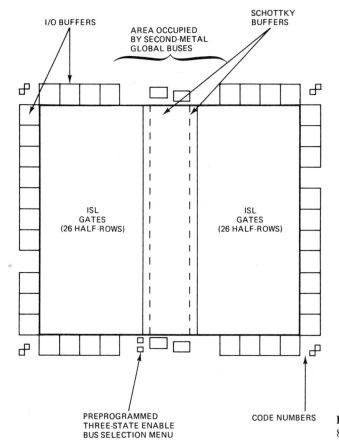

Figure 3.11 Overview of Signetics 8A1200 chip. (Courtesy of Signetics.)

with the pull-up resistor on each transistor and hence the speed of each device (subject to the loading, of course).

Figure 3.12 gives the circuit diagram of an ISL gate. It is discussed more fully in Chapters 2 and 10. The reader will note the single input and the four outputs.

Figure 3.13 shows the grid structure of the internal cells. A detailed explanation of the structure is given in Chapter 10.

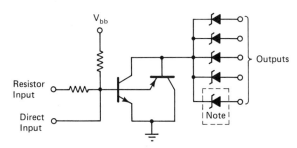

Note: Any four of the five outputs can be used but not all five at the same time.

Figure 3.12 ISL schematic. Note: Any four of the five outputs can be used but not all five at the same time. (Courtesy of Signetics.)

Sec. 3.5   ISL Grid Structure   135

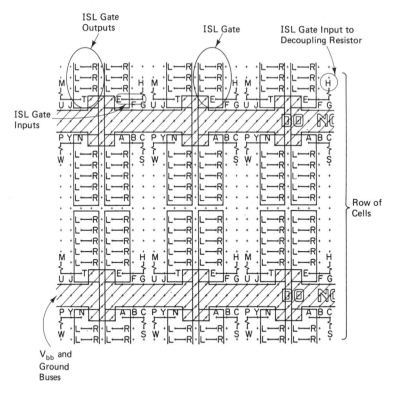

Figure 3.13 Signetics ISL structure. (Courtesy of Signetics.)

Each cell contains one transistor with its five collector outputs (four of which can be used). There are four possible inputs to each cell. Three of these are electrically the same. The fourth is a resistive input used to prevent current hogging.

3.6 ECL GATE ARRAY STRUCTURE

ECL gate arrays generally employ spatially fixed cell sites where macros can be located. The fixed cell sites are generally also fixed in size. This means that the macros have to be kept to a given number of transistors. Details of the transistor makeup of the cells are given in Chapter 10.

Figure 3.14 shows Motorola's array. A typical macro is also shown. Figure 3.15 shows two of AMCC's arrays, the Q700 and the Q720. Figure 3.16 show the structures of Fairchild's GE2000 arrays. The latter figures show separately the various elements that make up a typical ECL gate array. These include not only the usual internal cells and output drivers but also cells to translate from TTL to ECL levels as well as bias generators to provide the reference voltages for "series gating" (see Chapter 10) within the array.

The grid structures of ECL gate arrays contain many more bus lines than do those of CMOS gate arrays. The reason is the larger number of such buses. In

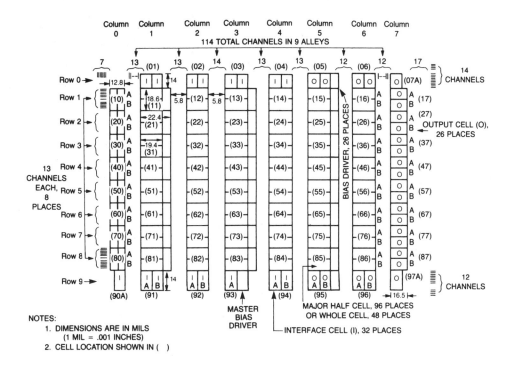

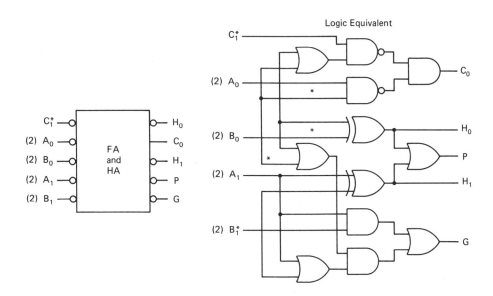

Figure 3.14 Structure of Motorola ECL gate array chip. (Courtesy of Motorola.)

Sec. 3.6　ECL Gate Array Structure

Truth Tables

A_1	B_1	A_0	B_0	G	P	H_0	H_1
L	L	L	L	L	H	H	H
L	L	L	H	L	H	L	H
L	L	H	L	L	H	L	H
L	L	H	H	L	H	H	H
L	H	L	L	L	H	H	L
L	H	L	H	H	L	L	L
L	H	H	L	H	L	L	L
L	H	H	H	H	H	H	L
H	L	L	L	L	H	H	L
H	L	L	H	H	L	L	L
H	L	H	L	H	L	L	L
H	L	H	H	H	H	H	L
H	H	L	L	H	H	H	H
H	H	L	H	H	H	L	H
H	H	H	L	H	H	L	H
H	H	H	H	H	H	H	H

A_0	B_0	C_1	C_0
L	L	L	H
L	L	H	H
L	H	L	H
L	H	H	L
H	L	L	H
H	L	H	L
H	H	L	L
H	H	H	L

IDC = 1.5 ns
IDC* = 1.8 ns
PD = 312 mW

Figure 3.14 (*cont.*)

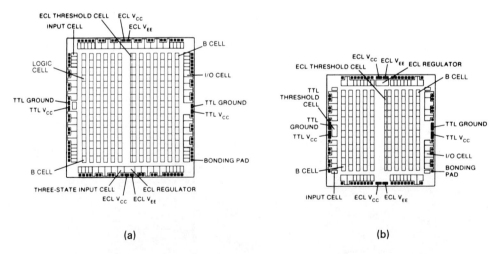

Figure 3.15 Structure of AMCC array chips: (a) 236 × 243 mils; (b) 195 × 197 mils. (Courtesy of AMCC.)

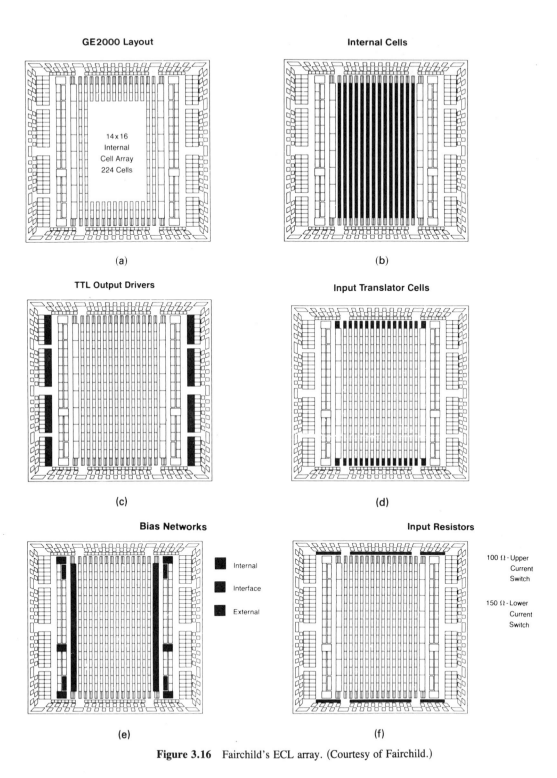

Figure 3.16 Fairchild's ECL array. (Courtesy of Fairchild.)

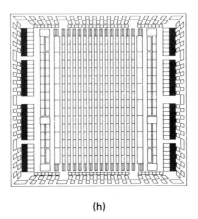

(g)	(h)

Figure 3.16 (*cont.*)

CMOS, there are only two such bus structures, V_{dd} and V_{ss} (power and ground). In ECL, in addition to power and ground (labeled -V_{ee} and V_{cc}, respectively), there is also a voltage reference for the current source plus one reference voltage bus for *each* level of "series gating" (see Chapter 10). Thus if there are three levels of series gating, an ECL chip will have four more buses than a CMOS chip. All of these additional buses must be distributed across the chip.

An example of such a structure is that of the AMCC Q3500 array. This is a two-layer metal oxide isolated array whose library contains both fixed- and variable-sized macros. (See Chapter 5 for a discussion of the differences between the two types of macros.)

3.7 GUIDED TOUR OF THE INSIDE OF A GATE ARRAY CHIP

Having given the reader several representative gate array structures, it is instructive to now take a "tour" of the inside of an actual gate array chip.

The reader is taken through a sequence of steps in which the magnification is increased with each step. The tour begins with Fig. 3.17, which shows a close-up of the gate array die mounted in a 64-pin package. Its purpose is to give a better overview of this particular die and to enable the reader to separate the parts of the die from the parts of the package.

Figure 3.18 shows a corner of the gate array. The reader will note the wires running to the bonding pads. In addition, the reader will be able to notice that some parts of the die are lighter colored than other parts. The light color represents the selective metallization of the array. Note that there are large peripheral areas which seem to have negligible metal. The reason is that not all array cells are used.

There are at least three reasons for not using all the array cells. First is that when there are a lot of interconnect lines, it is sometimes difficult to judge a priori how much space is needed. Second, as a general rule of thumb, only 80 to 90 of the

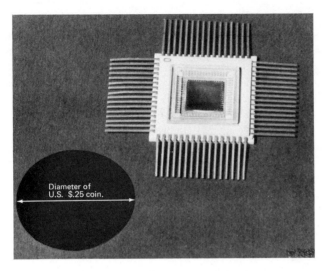

Figure 3.17 Gate array die in 64-pin package.

cells should be used, to promote ease of interconnect. This is true for both manual and automatic placement and routing.

Third, in the case of this particular circuit, the chip size was dictated by the need to run a number of high-speed 16-bit buses in and out of the chip with equal delays. This implied usage of the I/O pads at the top and bottom of the chip as opposed to the sides of the chip. The reason is that this particular array is a single-layer metal gate array. Consequently, signal buses which are almost (except for getting

Figure 3.18 Close-up of die in Fig. 3.17.

Sec. 3.7 Guided Tour of the Inside of a Gate Array Chip

under the *peripheral* power and ground buses) entirely metal can be run vertically. They can also be run horizontally until they have to connect to an underpass to dive under the vertical power and ground buses. However, running metal lines horizontally *across* the poly block underpasses (as opposed to connecting to each end of a given poly) eliminates the ability to run other signals vertically. All of this will be brought out more fully in Chapter 9.

The next figure in this sequence, Fig. 3.19, is an inside look at the die itself taken through a microscope. This chip is an IMI (International Microcircuits, Inc.) 2000-cell gate array. (The exact model is 71960.) Again, the light-colored areas and lines are metal (aluminum in this case). The very wide and moderately wide metal lines are part of the background layer and are present in every circuit. The narrow metal lines are those put in by the user to customize the chip to a particular circuit configuration.

Readers can orient themselves by noting first the railroad tracks running vertically, and second, the wires running to bonding pads. The latter are the squares with the ball ends of the wires attached to them. The very thick bars running horizontally are the power and ground buses. Running vertically off what will turn out to be the V_{ss} bus is the local V_{ss} bus. If the top part of the chip were shown, the reader would see the vertical V_{dd} buses connected to a similar thick horizontal bar. The V_{dd} and V_{ss} buses are interleaved in a clever fashion to permit this.

A map of Fig. 3.19 is given in Fig. 3.20. Although the grid structures of a number of high-speed CMOS gate arrays including this one are given in a succeeding chapter, it is appropriate to make a few observations. Readers unfamiliar with CMOS should consult Chapter 2.

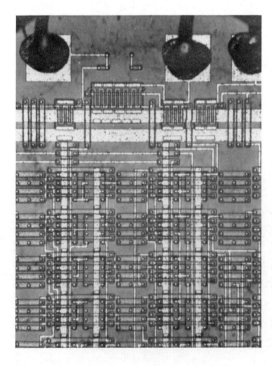

Figure 3.19 Inside look at gate array die of Figs. 3.17 and 3.18.

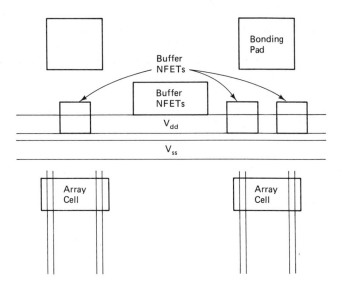

Figure 3.20 Map of die of Fig. 3.19.

The large serpentine structures in the vicinity of the square bonding pads are the buffer NFET and PFET. The serpentine structure (in traveling-wave tubes this is known as a *Karp structure*) enable transistors with large W/L ratios to be obtained in a convenient form factor.

These devices are much larger than the array PFET and NFET. This size difference illustrates a major point that will be made in Chapter 4—that reducing the *total* number of interconnects to the outside world of a given circuit via use of VLSI substantially reduces power, size, and weight, among other parameters. The rationale is that to interface to the outside (the chip) world requires larger transistors, which in turn consume more power and hence require larger, more expensive, and heavier power supplies, cooling mechanisms and so on. Readers should be able to relate the remaining elements with that of the diagrams shown in Sec 3.2.

3.7.1 Latch-Up Protection

Latch-up was discussed in Chapter 2. It was mentioned that two of the ways in which latch-up can occur involve currents flowing in the p well and in the substrate and that to combat this problem, the p^- well and the substrate are connected to the V_{ss} and V_{dd} buses, respectively, by a connection in each array cell. This helps eliminate the possibility of stray currents flowing in the well and the substrate and causing latch-up. The well and substrate connections can be seen as the small ovals at the top of each cell in Figs. 3.19 and 3.20. The ovals represent the coverage of the via (hole) connection to the well and substrate.

3.8 STRUCTURES OF GATE ARRAYS EMPLOYING THE GATE ISOLATION TECHNIQUE

Up until now the transistors of all the CMOS gate arrays have been clustered in two, three, and four (and sometimes combinations of two and three) transistor pair cells. In the case of HS CMOS, the cells are surrounded by oxide, which isolates each group of cells from its neighbors. The transistors within each group of cells share source/drains.

Whether a given transistor interacts with its immediately adjacent neighbor in the same cell is determined by the voltages on the gates of the transistors and also by the connections of the source/drains. An example is given in Chapter 9 of the layout of a three-input NAND gate using one of the cells discussed earlier in this chapter. The reader will note that the three PFETs are paralleled and that the three NFETs are in series. A high on any of the gates of the NFETs will turn on that NFET and produce a low-resistance path to its neighbor(s).

The effect produced will depend partly on where the NFET is in the series "string" of NFETs. If the NFET is the NFET closest to V_{ss}, the source of the next highest (spatially in the series string) NFETs will be put essentially at ground (V_{ss}). If it is the topmost NFET and the one or more of the other NFETs is off (input low) such that there is a high on the output, that high will be transmitted to (place on) the drain of the next NFET farther down in the string.

The same signal will produce a very high resistance path between source and drain of the PFET to which that gate is attached, and hence effectively remove that PFET from the paralleled structure by making it an open circuit.

A benefit of this arrangement of (in this case) three transistor pairs in a cell is that the transistors in the adjacent cell can be ignored because none of the source/drains of one cell are shared with those of an adjacent cell. Moreover, the cells are isolated from each other by means of the oxide isolation process described in Chapter 2. Therefore, the transistors in one cell do not interact with the transistors in an adjacent cell unless specific wiring connections are made. This is the way in which all HS CMOS gate arrays are currently made. (Note the "HS CMOS".)

A disadvantage of this arrangement is the space taken up by the oxide isolation and the two diffusions (one *p*-channel, the other *n*-channel), which are not shared with the PFET and NFET in the next cell.

Researchers at Honeywell and Mitsubishi have developed a number of gate arrays that eliminate the oxide isolation among transistors. The basic "cell" is one transistor pair. Every source/drain diffusion of every transistor (except those on the periphery of the chip) is shared with a transistor of the same type. This means that many more transistors can be put on a given-size chip than can be put on a chip with oxide or junction isolation between the cells. This is shown in Fig. 3.21.

Isolation is achieved by biasing the gates of individual FETs. No longer can there be a gate that is connected to both a PFET and an NFET. The reason is that the voltage that turns a PFET off (a high) turns an NFET on, and vice versa. To achieve isolation between two transistor pairs that are separated in the middle by a third transistor pair, the third transistor pair must be used and is not available for use as a signal element.

The key question is: What increase in circuit functionality for a given chip size can be achieved using this arrangement? The reader will correctly surmise that a lot depends on the individual circuit (as happens with most designs). The technique is obviously more useful for logic structures that employ relatively large numbers of transistors in series or in parallel. However, the maximum number of inputs to a

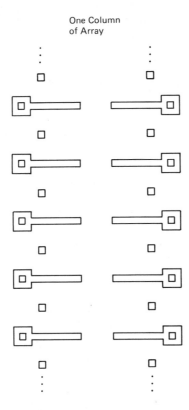

Figure 3.21 Gate isolation technique.

NAND or NOR gate in CMOS is typically 4, because of the addition of the "on" resistances involved in the circuit structure.

A technique that can help gate isolation to achieve its full potential is the use source/drains for isolation. This technique is described in Chapter 9. By using source/drains to isolate, the number of transistors that have to be used for isolation is greatly reduced. The source/drains involved in isolation would be those used for the power supply connections. This means that two two-input NAND gates, for example, would be laid out back to back with the power supply connections made common. For example, one source of one NFET would be used for both (one in each NAND gate) NFET strings.

One can also expect that logic design aimed at using logic structures which have relatively large strings or parallelings of PFETs and NFETs can be developed to optimize the use of this technique. The AOI (AND-OR-INVERT, which Mitsubishi calls an AND NOR) described in subsequent chapters is an example of one such method.

Mitsubishi has done analysis which indicates that the improvement in density to be expected using gate isolation is a factor of 1.68 over a gate array made from two-input transistor pair cells. It is to be expected that if the same circuit had been compared against an array of three- or four-input oxide isolated cells, the factor would have been somewhat lower. (Less space overall is taken up by the oxide.)

Nevertheless, the technique appears to be quite powerful and certainly is a valuable option to have in the gate array "arsenal."

3.9 STUDY OF YIELDS OF DIFFERENT GATE ARRAY CELL STRUCTURES

Michael Gagliardi and others at the GE Microelectronics Center studied the effects of 15 unique array cell types on yields of two-layer metal CMOS gate arrays with drawn gate lengths of 2 μm. [1] The 15 were grouped into five classes in order to exhibit differences among grid structures which have maximum W/L ratio, maximum routability, different placements of source/drain and gates and their connection points, and so on. The important *sea of gates* developed by Robert Lipp and others was one of the classes studied. Sea of gates as implemented by GM/Hughes and some others requires an extra mask level for the contacts.

FET widths in the study varied from 20 to 51.5 μm with resultant W/L values of 10 to 25.8. Accurate data for chip area was obtained by doing several layout iterations of each cell and grid structure. It was found that a cell which had a lower cell count when used in a flip-flop-intensive application will have a higher cell count when used with a combinatorial logic intensive circuit. This is based on the use of transmission gates (see Chapter 9) to implement the flip-flops. The area taken by the different cell types under different conditions of routability was also studied.

The five classes are shown in Fig. 3.22. Readers should remember that the process is two-layer metal. In cells where the V_{ss} and V_{dd} bus bars run across source/drain or (FET as opposed to logic, of course) gate contact points, the contact point is on first-layer metal and the bus bars are on second-layer metal. The s/ds contact points are isolated black squares in the figures. By contrast, the gates have crosshatched connections running to them. This is how the two structures can be distinguished from each other.

The class of structures in Fig. 3.22(a) is characterized by maximum device widths and common gates. Those in Fig. 3.22(b) have one diagonal gate. Figure 3.22(c) is that of GE's ICG20000 series, the design of which arose from the study. Figure 3.22(d) is a sea of uncommon (nonshared) gates. Figure 3.22(e) is a sea of gates with alternate gates shared.

Figures 3.23 and 3.24 show the results of the study. The curves in Figs. 3.23 and 3.24 labeled a to e refer to the cells labeled (a) to (e) in Fig. 3.22. Because the ICG20000 (curve c) was developed as a result of the study, it is a good for both combinatorial and sequential logic.

3.10 SUMMARY

The repetitive structures of the gate arrays of several different manufacturers were shown in this chapter. Examples were shown of both single- and double-layer metal and both CMOS and bipolar arrays. A grid system that will be used in subsequent

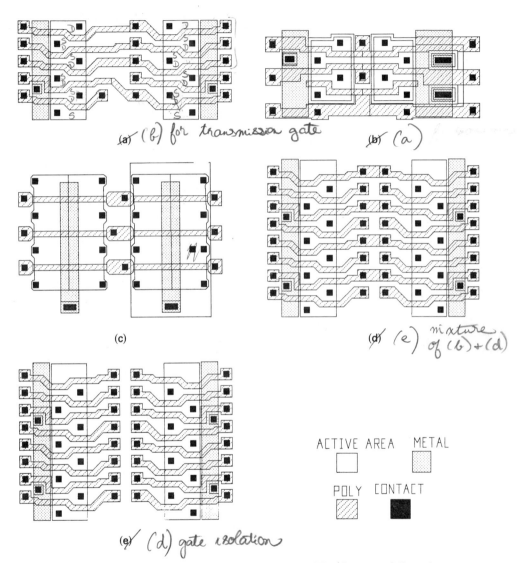

Figure 3.22 Cell classes studied by researchers at GE. (Courtesy of General Electric Microelectronics Center.)

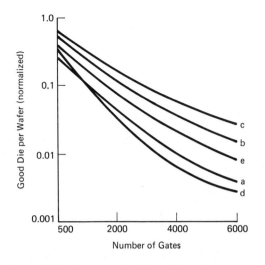

Figure 3.23 Layout results for the five cell classes showing predicted relative yield versus gate count. (Courtesy of General Electric Microelectronics Center.)

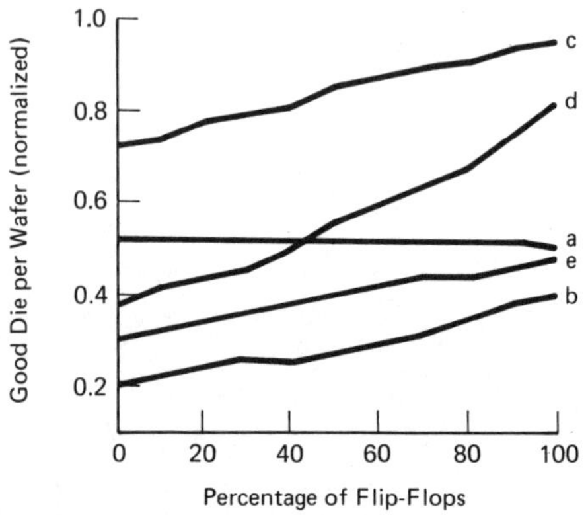

Figure 3.24 Layout results showing predicted relative yield versus fraction of sequential logic. (Courtesy of General Electric Microelectronics Center.)

chapters was introduced. Finally, a guided tour of the inside of an actual gate array chip was taken. The GE study of cell structures was discussed.

REFERENCE

1. Gagliardi, M., "Gate array cell types," *VLSI Design*, Feb. 1984, pp. 78–79.

Chapter 4

Overview of Gate Array Design

4.0 INTRODUCTION

Chapters 1 to 3 developed basic concepts and definitions. Technologies of gate arrays and their grid systems were discussed. With this background, the flow of gate array design can now be discussed. The purpose of this chapter is to give the reader an overview of the gate array design process and to mention the alternatives available to the designer at each step.

Also included in this chapter is a section on vendor interaction. The purpose of this section is to make the user aware of what can be expected from a gate array vendor and what a gate array vendor expects from the user.

Gate array design can be divided into five major categories. These are:

1. Feasibility analysis, circuit analysis, and partitioning
2. Technology conversion (if required) and design for testability
3. Logic coding (if used) and verification and testability analysis
4. Mask design
5. Processing, packaging, and test

A flow diagram showing the interrelationships of these various steps is shown in Fig. 4.1. Figure 4.1(d) shows the flow of gate array design in a somewhat more pictorial fashion.

All the topics in Fig. 4.1 except technology conversion will be discussed in detail in separate chapters. Technology conversion will be discussed in this chapter. Mask design will be discussed as a part of several chapters, including this one.

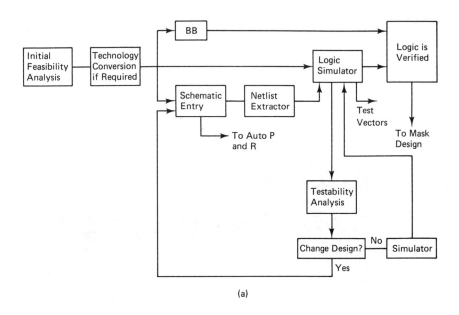

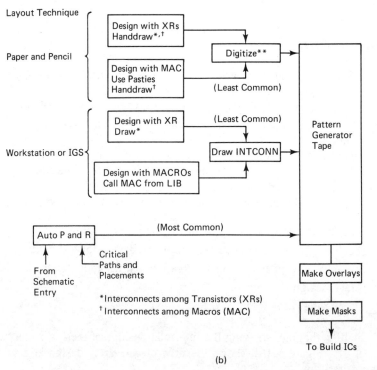

Figure 4.1 (a) Flow of gate array design; (b) mask design options; (c) building IC's; (d) pictorial view.

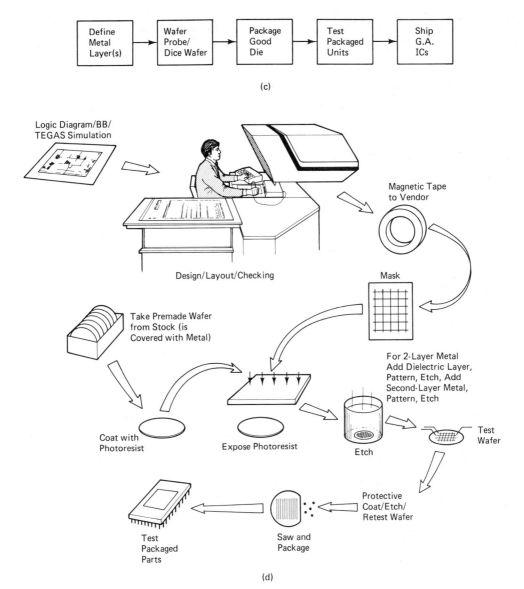

Figure 4.1 (cont.)

Although the divisions imply a sequential work flow, there are often substantial look-a-head and feedback operations going on simultaneously. The need to look at testing early is mentioned below. However, there are others. Certainly, one needs to know if a given macro is available on the layout system (mask design) before he or she proceeds with technology conversion, for example.

The initial partitioning is sometimes changed after the "floor plan" is done. Another case of nonsequential work flow is the iterations that occur during technology conversion. The feasibilty of one approach will often be traded off against that of another approach.

Another case involves the basic design as modified by the designer for testability approaches. When changes are made to permit better testability, the design must then be rechecked and reverified to ensure that the logic performs correctly. (This is why the author recommends against extensive simulations at speed before testability is established. By the term "at speed" is meant at the actual clock and signal change rates used, as opposed to slow clock rates used to simply check out the functionality without having to worry about clock overlap and other parameters.) It should be noted that the simulation being referred to is logic simulation as opposed to circuit simulation.

Yet another example is that during logic coding and verification, brief hand calculations will often be done to establish the *feasibility* of a small permutation to the original approach. (This is often faster than changing the coding of an innermost and all subsequent outer modules, resimulating, and examining the results.) A basic assumption is that a digital logic circuit has been partly or wholly or partly designed and that a good specification exists.

4.1 SUMMARY OF THE FLOW

4.1.1 The Starting Point—A Good Specification

The process begins with a good specification. A good specification is one which is complete and precise and which addresses the issue of testability of the device. (The term "complete," however, does not include specification of the pin-outs of the packaged unit. Such pin-outs are better left to the discretion of the designer whenever possible.)

Getting a good specification for a gate array is almost no different than getting a good specification for any other chip or, for that matter, the printed circuit board replaced by the gate array. An important feature of the specification is a set of tests that determine sell-off of initial prototypes and later on production quantities (see Section 4.5).

Other important features of a good specification are a description of how the unit works, a delineation of critical parameters and paths, the environmental conditions and interface parameters, and a block diagram of the major parts, showing the signals among the blocks. The last is an aid not only to help the gate array designer (if he or she is not the designer of the original logic to be implemented on a gate array) to rapidly get acquainted with the design, but also to partitioning of the circuit. It is also an aid to determining what functions should be made into what "macros" (see below) and how the macros should be "nested."

4.1.2 Definition of Other Tasks

Feasibility analysis, circuit analysis, and partitioning. Feasibility analysis is intended to show whether it is worthwhile to make the circuit in the form of one or more gate array chips. An integral part of such a decision for large and complex circuits is partitioning. Proper partitioning can sometimes enable a difficult circuit to be put on a single gate array which otherwise would require two or more gate arrays. Circuit analysis is part of feasibility analysis to the extent required to show that it is feasible to build the circuit on a gate array chip. Circuit analysis is also used to answer the questions of in what technology to build the gate array and what some of the compromises and risk areas are in building the chip in a given feature-sized technology from one vendor versus another feature-sized technology from another vendor. [1], [2]

Technology conversion and design for testability. Technology conversion is necessary when the original circuit design was for example, in some form of TTL and the gate array is to be built in CMOS. A new schematic is drawn to reflect the changes.

Design for testability involves adding features to the design to make it more testable. Examples range from simple test points to implementation of techniques, such as the Level-Sensitive Scan Design technique of IBM.

Logic coding and verification and testability analysis. After a basic design has been established, it is necessary to verify that it works as desired and also that it is testable. The two major methods of verifying logic are via breadboarding and via logic simulation.

If a major simulation program, such as Tegas5, is to be used, the circuit will have to be coded. Logic coding refers to defining what functional elements are to be used and how they are to be interconnected (i.e., the net list). Logic coding can come from a schematic entry program or it can be done manually. In either case, a program (e.g., the preprocessor in Tegas5) is run to create subfiles which are used for subsequent operations.

Although testability analysis can also take many forms, one of the best is that of the "trained eyeballs" of an experienced person who will be responsible for making sure that the gate array meets its performance specifications. This talent applied early, perhaps even during the feasibility and circuit analysis stages, can pay very large dividends.

As circuits become more complex or when organizations lack a person whose sole responsibility it is to make sure that gate array circuits meet specifications, it is highly desirable, if not absolutely necessary, to use a circuit analysis program. A typical circuit analysis program, such as COPTR (part of the Tegas family) or SCOAP from Sandia Laboratories, will yield a printout showing the degree of testability of each node in the circuit. For such purposes, testability of each node is defined as a combination of the controllability (to a 1 and to a 0) and the observability. More will be said about this later.

Methods of coding logic—Macros. There are several ways in which logic coding can be accomplished. In one method the designer uses schematic entry both to draw the schematic and to define the net list for the logic simulator. In a second method, the designer manually codes (enters) the net list. Schematic entry and logic simulators are both discussed in subsequent chapters.

Where possible, *macros* (sometimes called "modules" or less often "function cells") will normally be used for coding using either method. Macros are functional elements (D flip-flops, ALUs, multiplexers, counters, logic gates, etc.) which have predefined characteristics such as propagation delay versus fan-out. Where certain macros are not available, they can be created from primative elements (logic gates, flip-flops, etc.) and stored in memory for use in different circuits.

The macros used will be those of the gate array vendor whose chip is being used, those created and verified by the user organization for that particular gate array family, or a combination of the two. "Soft macros" are combinations of macros that form new and generally larger macros. In some cases, the unused portions of the macros so combined can be eliminated by another program. This saves layout space. An example of such a program is the Gate-Gobbler program of LSI Logic, Inc.

Usually the data base created from logic coding is used as an aid to mask design. In mask design, the term *macro* refers to the layout of a functional element (See below.)

4.1.3 Logic Verification

Verification of the static and dynamic properties of the logic design is the next step. This is more critical in the case of a gate array than in the case of a printed circuit board because it is generally harder to both test and modify the latter. Verification may be done with a breadboard or with logic simulation programs. *Very* simple circuits not yet built may be verified by an engineer making an independent (of the original designer) paper analysis. The latter technique is rarely satisfactory except for circuits which are noncritical with regard to speed and size.

Breadboarding versus logic simulation. For years the primary method of logic verification has been breadboarding even though logic simulators, such as Tegas in its various versions, have been around since 1972. As chip feature sizes shrink, however, breadboarding using discrete, standard ICs on a PCB becomes increasingly *less* representative of what is actually happening on the custom IC (whether gate array or other custom part).

The reasons are twofold. The primary reason is that the interconnect capacitance of both the PCB and the package pins is much greater than that of similar interconnects on the chip. Thus more displacement current is required to obtain a given speed of operation. More generally, signals will run faster on the die because of the much smaller on-chip capacitances.

It could be argued that if the capacitance scaled in the same ratio as the displacement current generated that there would be no net gain in speed obtained by replacing the PCB with a single chip. However, there *is* a substantial pickup in reduction of capacitance relative to driving capability. An example will illustrate the point. The internal gate capacitance of a typical 5-μm feature-size array CMOS FET pair is 0.15 pF. This is at least an order of magnitude less than the package pin capacitance of most packages and two to three orders of magnitude less than the pin capacitance of the same package when mounted on a typical PCB. The latter has run capacitances of up to 100 pF per foot of line length.

In simulating a logic circuit, the designer is substituting computer coding and debugging time for hardware building and debugging. Instead of drawing parts for a breadboard from an electronic stockroom, the designer calls macros or primitives from a library. However, unlike the parts in an electronic stockroom, which may or may not be there and which may or may not work, the macros and primitives of a simulation program are always "in stock" and always work the same way, "cockpit (operator) errors" not withstanding.

A wide variety of logic simulator programs exist. One of the most widely used programs, Tegas5, is discussed in Chapter 6.

Circuit simulator programs, such as SPICE or TRACAP, may be used to simulate specific parts of the logic which are critical with respect to speed. Such programs are not used as much as they are in the design of pure custom circuits for several reasons. First, the data to be input to such programs must come from the gate array vendor, who may or may not be reluctant to give it out. Second, the gate array vendor may have already done such simulations; and the vendor's results will almost invariably be better than (if not the same as) those obtained by a user, just by virtue of being closer to the process.

Third, process variations of 40% or more tend to prohibit "pushing" a given technology too hard unless one has the funds and the inclination to gamble on selecting dies that meet the speed requirements from a large sample of dies.

Nevertheless, there are times when it *is* appropriate for the user to use a *circuit* (as opposed to logic) simulator program. Not all vendors circuit simulate all cases of interest. Second, the user may desire to use a process which because of its newness has not yet been fully characterized by the gate array vendor.

Behavioral simulators, such as Helix of Silvar-Lisco, are a relatively new class of simulators. They enable the user to simulate a given circuit at a very high level of functionality as opposed to the gate level of most current logic simulators. (The Tegas5 library does contain a microprocessor, but it is modeled at the gate level. Boolean equations can be used with Tegas5, but Tegas5 is, nevertheless, a gate-level simulator.)

Although register-transfer-level simulators, such as Rockwell's Simstran program, have been around for many years, only in the past few years have true behavioral simulators become available. Moreover, algorithms involved in such simulators have improved to the point where such simulations are faster and occupy much less memory.

Mixed-mode simulators involve a combination of two or more of the foregoing classes. This means that part of a given circuit can be simulated at a very high block level with its gate structure undefined, while another part of the circuit is being simulated at the gate level, for example.

4.2 MASK DESIGN

Mask design is the process of creating a data base containing the interconnects among the transistors of a gate array. The data base is then used to create a *PG* (pattern generator) *tape* from which the actual masks are made.

The interconnects may involve two or more layers of metal. A separate mask is required for each such layer, together with a mask for each of the interlayer dielectric layers (which provide electrical insulation between the metal layers) and a mask for the protective dielectric or polyimide layer which coats the completed chip. The last may be common to all members of a given gate array family because the only openings in such a layer are to the bonding pads of the chip. (Some gate array vendors prefer to cover unused bonding pads with dielectric, but this requires a separate mask for the protective layer for almost every design. The reason is that rarely is the configuration of bonding pads the same for any two different gate array designs.) The interlayer dielectric mask defines the interconnects ("vias") between adjacent metal layers.

Mask design may be done completely manually, semimanually on an IGS or workstation, or automatically using automatic placement and routing (auto P&R) programs. Workstations and auto P&R programs use the data base created by either schematic entry or manual coding of the logic simulator program.

If semimanual methods are used, the data base may highlight the points in a given net to be interconnected (by means of a linking path, for example) and may prevent the designer from making an improper interconnect. This is called *correct by construction* and saves a great deal of very tedious line checking. If auto P&R is used, a small percentage (3 to 5%) of the routing will often have to be done manually if the circuit uses greater than about 80% of the array cells.

In many cases, the designer will want to manually place critical macros and route critical paths (those with very short propagation delays or those which must track other signals). This is possible with most auto P&R programs.

4.2.1 The "Floor Plan"

When manual and semiautomatic methods are used (and even on occasion when auto P&R is used and the user inputs critical macro placements and routings), the designer develops a "floor plan." This is a rough layout of the circuit intended to highlight routing problem areas and to serve as a basis for macro placement. As might be suspected, the better the placement, the less work the router has to do. Therefore, some P&R programs tend to stress one over the other.

It is a truism that a short line on a schematic can be a very long line on a layout. The reason is due to the form factors of the actual layout macros as contrasted to their respective shapes on the schematic. Another reason is the fixed numbers of rows and columns of cells of the array.

The floor plan is also an important part of the documentation of the chip. It can be a very important aid to locating functions when testing reveals potential flaws in the design.

A floor plan can be as simple or as complex as is needed for the task. Typically, a clear sheet of acetate is laid over a 100× pen plot of the background layer of the gate array being used. The acetate is much cheaper than the pen plot. Macros or other functional elements (designed with transistors perhaps) are represented simply as rectangles with the appropriate form factors on the acetate. Alcohol or water (depending on the type of pen used to do the drawing) can be used for small erasures. A new piece of acetate is used for each trial. Auto P&R systems generate floor plans automatically as part of the placement process.

Figure 4.2 shows a representation of a simple trial layout (floor plan). In this case the macros are of different sizes. A few lines are sketched in to indicate trial interconnects.

The first step in developing the floor plan is to decide how the chip is to be partitioned. The partitioning of the chip may or may not be similar to the way in which the circuit is partitioned for macro coding purposes. For simplicity, both partitionings will be similar whenever possible. However, some high-speed interconnects may dictate that some portions of the partitioning be different on the chip than for the coding used for simulation.

It may turn out that within some or all of the major blocks of the floor plan, it is necessary to develop sub-floor plans. Alternatively, it may be desirable to work with smaller (but more) blocks to begin with. No two people and no two circuits are alike in their ability to visualize and to be visualized, respectively. Regardless of

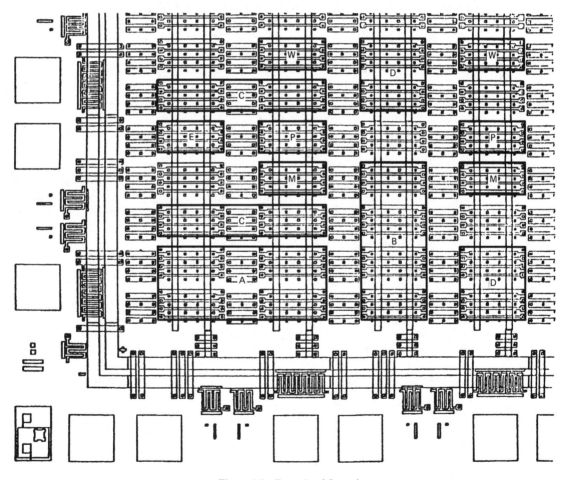

Figure 4.2 Example of floor plan.

Sec. 4.2 Mask Design

how it is done, the two key factors to be kept in mind are those mentioned previously, interconnects and I/Os.

A good starting point for the floor plan is to define the high-speed paths and the paths that must track each other in propagation delay. If the array used is single-layer metal, the high-speed paths will run parallel to the distributed power and ground buses as much as possible because they too will be metal. The same may or may not also be said of the paths that have to track each other in propagation delay.

If the array is double-layer metal, there is, of course, much more freedom. However, as line spacings shrink, the capacitance between parallel signal lines may become more significant than that of a given signal line to ground. For example, most metal lines are on the order of $\frac{1}{2}$ to 1 μm thick and are typically separated vertically from whatever they cross by roughly 1 μm or so of silicon dioxide or other insulator. Moreover, a 5-μm-wide line typically is spaced 5 μm from the next line. Therefore, the dominant capacitance is generally that to whatever the line is crossing over. As line widths and hence spacings decrease, the lateral capacitance between the "walls" of metal lines running side by side becomes much more important. (To keep the sheet resistance low, lines are kept as thick as possible, in the range $\frac{1}{2}$ to 1 μm, even at widths of 2 μm or so. Therefore, the wall capacitance at the smaller spacings can become important.)

If the high-speed interconnect paths involve I/Os, these will have to be planned for the top and bottom of the gate array if it is single-layer metal. In either double-layer or single-layer metal arrays, the circuitry feeding or being fed from the high-speed I/Os will have to be put as close to the periphery of the chip as possible. A common error among neophyte gate array designers is to forget to save the last inversion for the output buffers.

In either ECL or STL gate arrays, the voltage swing of the signal is much lower than for CMOS. It is in the neighborhood of 200 mV for ISL/STL. Although it is somewhat higher for ECL, the currents in ECL are considerably higher than for STL. The relatively high currents of these bipolar devices coupled with the low-voltage swing means that the signal lines must be kept as short as possible, to avoid decreasing the noise margins.

4.2.2 Methods of Mask Design

Mask design methods can be categorized as being transistor design methods or macro design methods in addition to being classified by the degree of automation involved. Normally, transistor design methods are used to design macros, and macros are used to design the chip. On occasion (if the gate array vendor allows it), a small portion (10 to 20%) of the chip may be designed with transistor design methods to obtain performance and/or density advantages of this method. See the discussion below on transistor design methods.

Macro design methods. A macro is a pattern of interconnects among transistors which form a functional element. The functional element can be very rudimentary, such as a logic gate, or it may be an entire circuit design. Libraries of macros generally encompass most of the common SSI and MSI functions available in data books of standard parts. Examples of macros are given in Fig. 4.3. The

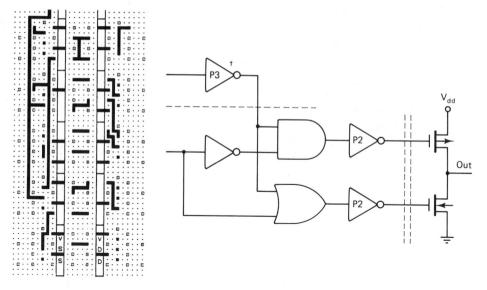

*Includes 1 row for an attached triple parallel inverter
†On function cell (Macro) but not necessary for TSB (Three-State Buffer)

Note: P2 and P3 mean parallel 2 and parallel 3 (inverters in this case), respectively

(a)

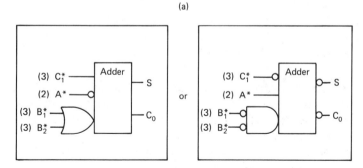

Note: Circle at Input or Output Indicates an Active "Low" Signal

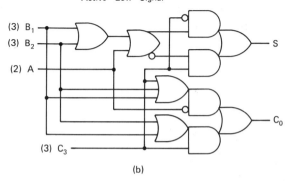

(b)

Figure 4.3 Examples of macro. [(a) Courtesy of Sanders Associates, Inc.; (b) courtesy of Motorola, Inc.]

Sec. 4.2 Mask Design

reader will note that the macro in Fig. 4.3(a) possesses a given number of rows (in this case 3) and a given number of columns (in this case 1).

Macros are predesigned, *checked* (layout-wise), hopefully characterized (by means of a circuit simulator program or by means of a test chip), and stored in computer memory for repeated use on many circuits.

Use of macros permits gate array designs to be done with lower engineering costs and times and increases the chances of "first-pass" success (getting the circuit correct in the first mask cycle).

The chief disadvantages of macros are that they are more wasteful of space than are transistor design methods and sometimes the performance of the resultant circuit is less. (See "Design with Transistors" below.) Gate array parts which are going into high-volume production are often designed wholly or partly with transistor design methods (if not with one of the methods that compete and complement the gate array methodology—standard and macro cell techniques, structured design, and less likely, pure custom).

However, many circuits do not go into high-volume production immediately. This is particularly true of chips incorporated into military products. Often, there is a fairly long "gestation" period between the time the first custom parts are utilized and the time the system goes into production. The gestation period may extend from one to three years or more. During this period, the circuit design may change substantially; or the entire project may be canceled. Therefore, a quick-turnaround, lower-cost gate array designed with macros may be preferable to a somewhat higher-performance, higher-cost chip designed with transistors that may take longer to get and which may not work on the first try. These are generalizations and, like any generalizations, are prone to exception being taken.

Size of library. An important question concerning the size of a given vendor's macro library is: How large does such a library have to be to be usable? By the term "usable" the author means that in the majority of cases, new macros will not have to be added to it to do a given circuit design in a gate array. The reader can appreciate that any number is very subjective and depends critically on the nature of the circuits being implemented. However, there is reasonable agreement that if a macro library contains roughly 100 (minimum) different functions *exclusive of permutations,* new macros will have to built in about 10% of the cases or logic will have to be redesigned to accommodate existing macros.

The term "exclusive of permutations" means that two macros which have the same functionality (e.g., a D flip-flop) but have, for example, different fan-outs or speeds in a given macro library would still be counted as just one element because the functionality is the same.

Fixed- versus variable-size macros. Macros can be categorized as being either of fixed size and shape or of variable size and shape. Fixed-size macros utilize the same number of rows and columns of the gate array cells for each macro type regardless of the function performed by the macro. Some vendors with fixed-size macros do allow half the fixed-size cell to be used for small macros. Figure 4.3(a) is an example of a variable-sized macro; Fig. 4.3(b) is an example of a fixed-size macro.

Each category has its advantages and disadvantages. Fixed-size macros are much easier to place and may be easier to route (depending on the circuit and the gate array grid structure).

However, fixed-size macros are wasteful of space internally. A two-input NAND gate or even a quad of such NAND gates does not approach the complexity of, for example, an ALU (arithmetic-logic unit). Moreover, if the fixed cell size is made large enough to accommodate the larger functions, many more replications of smaller functional elements will have to be placed in the fixed size of the macros.

The problem with this approach is that *all* of the many replications of the smaller functional elements in a given cell are really not that usable. The placement of a cell containing all of the replications forces some of them to be in the wrong place on the chip. For example, suppose that the fixed cell size is made large enough to accommodate an ALU or larger structure. The same cell size might accommodate 20 two-input NAND gates. To get one two-input NAND gate, the cell with the 20 (or perhaps a half-cell with 10—these are not hard numbers) will have to be placed. It is unlikely that the remaining 19 NAND gates are even semioptimally positioned in the array. A few might be; but the remainder will either be completely unusable or will cause severe routing problems if used. Therefore, there is a definite trade-off between the complexities of the macros that can be accommodated (and hence the fixed size of the macros) and the routability of the resultant design.

Smaller cell sizes used for the fixed cell size might seem to be a better solution than larger cell sizes. However, much of the compactness of a larger cell is lost when it is made from smaller elements. I/Os of the smaller macros used to fabricate the larger macros are often in the wrong places to build the larger macros. In the case of CMOS designs, sharing of transistor source/drains is often difficult or impossible when a larger macro is fabricated from many smaller elements.

Variable-size macros overcome some of the problems of fixed-size macros. The variable size permits fewer of the less complex functions to be incorporated and thus promotes greater layout efficiency from that standpoint.

The problem with variable-sized macros is that they make cell abutment more difficult. Suppose that three macros are of sizes (columns by rows) 1×3, 4×7, and 6×14. The reader can appreciate that figuring out how to interleave multiple repetitions of these three macros (without even considering the effects on line lengths and partitioning desires mentioned in Section 4.2.1) could be quite difficult.

A potential compromise solution is to have variable-sized macros with the dimensions of the small number of allowable shapes multiples of one another. Thus allowable (column $\times$ row) shapes might be 1×2, 2×4, 4×8, 8×16, 16×32. All macros would have to fit into one member of this or a similar set. There would be a higher probability that the shapes would fit together.

Pasties. Pasties are sometimes called "the poor man's macros." Like any other macros they represent predetermined layouts of functional elements. Unlike the macros discussed up to this point, they do not require the designer to have in his or her possession a terminal or computer system of any kind. The most sophisticated tool the pastie user needs is a pencil (and a big eraser).

A *pastie,* as the name implies, is an interconnect pattern (of transistors) on clear acetate with stickum on the back. Sets of pasties are obtained from appropriate gate array vendors, such as Interdesign. (Pasties are also used by the standard cell vendor Dumon Alphatron.) Figure 4.4 is an example of a pastie.

Normally, the pasties are put in several trial positions on the vendor-supplied layout sheet. The process is similar to doing a floor plan. The pasties can be thought of in terms of the rectangles mentioned under that topic. When using pasties, it is still useful to make out a floor plan for the different elements. It is also sometimes useful to (with the gate array vendor's permission) make Vu-Graph copies of the pasties to facilitate placement. These can be taped to the clear acetate sheet covering the layout sheet.

When the user is satisfied with the placement of the pasties, he or she peels off the backing to permit firm attachment to the layout sheet. Interconnect lines are then drawn to interconnect among the pasties. Sets of pasties include I/Os. The resultant sheet is then sent to the gate array vendor.

Upon receipt of the layout sheet, the gate array vendor (after doing some checking) calls up on an IGS or digitizer each of the macros represented by each of the pasties. The macro is then placed on the background grid structure, and the interconnect lines are either drawn in on an IGS or digitized. *Digitization* is the process of tracing a line in such a way that a data base is created. The pattern generator tape is then created by running a program on the IGS.

Interdesign was the first (to the author's knowledge) gate array vendor to offer the pastie approach some years ago. A few other companies, such as Mitel and MCE and standard cell vendor Dumont Alphatron, have since adopted it. Even in the era of low-cost terminals, it is still a feasible approach for circuits of less than about 1000 to perhaps 1500 gates. Increasingly, however, circuits in the several-hundred-gate complexity range are being done on the PAL methodolgy originated by MMI. Moreover, as workstations (Chapter 8) drop in price to under $10,000, this approach will become less attractive.

Design with transistors. With all the advantages just mentioned for design with macros, the reader may understandably wonder why information should be included on design with transistors. The reasons are several. First, the macro libraries offered by gate array vendors sometimes lack certain highly desirable (for a given circuit) macros. This is true not only of buffer cells for a particular configuration but also of internal cells as well.

Second, if one understands the internal structure of macros (and if the gate array vendor allows it), lines can be routed through the macro cells. Being able to route one or more critical lines, especially after a circuit change has been made, is sometimes crucial.

Third, designing a small portion of the chip using transistor design (and the rest with macros) may enable a smaller die to be used. The smaller die will cost less to produce and, perhaps more important, will use a smaller, *less expensive* package. Fourth, in some applications, package *size* is far more important than cost and time.

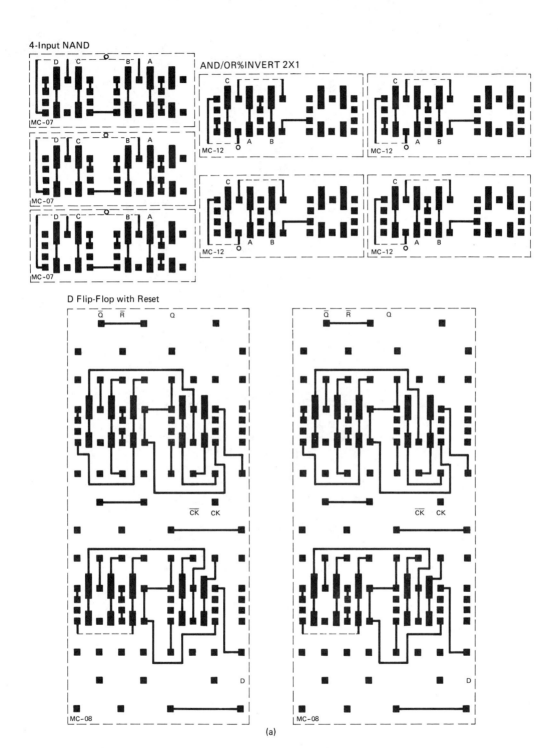

Figure 4.4 (a) Examples of pasties;. (b) placement of a pastie on a layout sheet. (Courtesy of Interdesign.)

Sec. 4.2 Mask Design

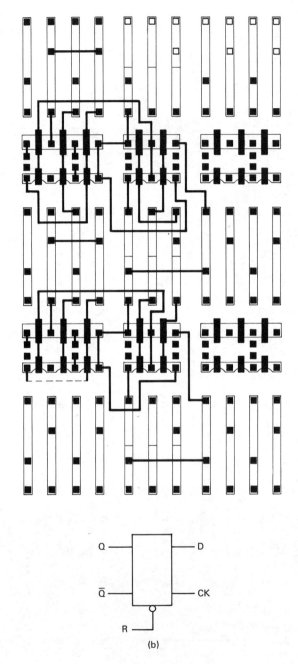

(b)

Figure 4.4 (cont.)

Fifth, being able to "cram" another 10 to 20% more circuitry on a given size die by using transistor or a combination of transistor and macro design techniques may enable the pin-outs of the chip to be drastically reduced. This will greatly aid the layout of the printed circuit board on which the chip is mounted.

Sixth, the performance (especially speed of operation) of some circuit designs can be improved using transistor design methods. Drive capabilities of certain functional elements can be *selectively* made closer to the optimum via paralleling techniques. "Ripple back," which results from such paralleling, can be better controlled and channeled using transistor design methods. These topics are discussed in the chapters on design with transistors.

Seventh, the *form factors* of either fixed- or variable-cell-size macros are often highly nonoptimum for a given circuit design. For example, the user's circuit might have many replications of D flip-flops. Because of I/O routing, the optimal layout might be horizontal. The optimum cell form factor might be one column by four rows. However, the form factor of the cells available from the gate array vendor might be just the opposite. In this case, being able to relayout the cell would pay great dividends. (Some companies offer macros in two or more configurations for the same function.)

Eighth, macro cells often contain elements which are redundant for the given application. These redundant elements are just so much space-consuming "excess baggage." A few companies have a program that automatically deletes unconnected portions of cells from overall mask design before layout begins. (The program of LSI Logic is appropriately called the Gate-Gobbler.)

Finally, to fully appreciate the gate array design process, the reader needs to understand how the macros themselves are designed. (They are designed using transistor design methods.)

Despite the advantages of transistor design, the preferable approach is to use macros whenever possible. The reasons are that speed, relatively low engineering costs, and probabilities of first-pass success generally outweigh other considerations. Densities of gate array chips are increasing at the same rates as those of standard products (the processes are the same). Therefore, chip size is often not as big an issue as it once was. Moreover, the user can count on the same design going on a smaller die (if the gate array vendor maintains the same grid arrangement) or being able to put more on a given die.

4.2.3 Checking the Mask Design

If some form of correct by construction or auto P&R is not used for mask design, layouts must generally be checked manually. (Even those done using correct construction and auto P&R methods are spot-checked, especially if line-search routing algorithms are used in the latter. The reason is that such algorithms can find bad paths.) A few programs exist which are capable of extracting the network list from the layout and comparing it against the database created by the schematic capture process. Examples of such programs are NCA Corporation's VGAC and Rockwell's MTRACE.

Line-by-line checking of layouts is extremely tedious and is error prone. The use of macros greatly reduces the need for such checking because only the interconnects among the macros need to be checked. Various CAD programs are used by some of the vendors.

One of the best ways to manually check layouts is to reconstruct the functionality of the device from the layout instead of checking the layout against the schematic (i.e., it is more effective to go from layout to schematic than vice versa, even though it takes considerably longer). The result is then compared to the original schematic and any differences noted.

The checking process itself is error prone if done by fallible human beings. A couple of "war stories" will illustrate the point. On one gate array in the author's experience, the design encompassed over 17,000 individual lines. Of these, five did not get digitized. After the part had been built, testing of the prototypes revealed that it was not functioning correctly.

Finding the first missing line took only 30 seconds starting with a description of the test results. However, finding the *last* one took five days with two people working nearly full time on the problem. [One of the lines was in a macro cell that had been paralleled (see Chapter 9). Thus it degraded the circuit speedwise but not catastrophically and hence was harder to find.]

This brings up another important point about checking. If an error is made that does not get picked up until after the mask and chip are made and built, respectively, the tendency is to "overkill" the checking process to ensure that no other errors in addition to those causing the problems determined by testing to date are in existence (i.e., generally the entire mask is rechecked once errors of this type have been found). One reason for this is that seldom are tests of complex chips anywhere near complete. There is generally always an area of uncertainty arising from this impracticality of testing every single possible mode of operation of the device. Moreover, the logic simulation program, if used, will be used to generate test vectors for less than 100% of the circuit nodes. (Rarely can all nodes be made testable and rarely is there both time and funding available to develop the test vectors even if all were testable.)

On another gate array, which was one of the first with which the author was involved, all lines were checked *extremely* thoroughly and a PG tape had been generated to make the mask and sent to the vendor. Fortunately, neither the mask nor the gate array had been built. The author had the embarrassment of having to call the vendor to state that "You'll never in this world guess what we did." The response came back: "You left out the connections to the chip of V_{dd} and V_{ss}." (power and ground). Author: "How did you know?" Vendor: "Everybody does it the first time." Moral of the story: Don't overlook the obvious; chips don't run very well without power and ground. (However, it certainly makes testing them rather easy; they're all dead!)

Even in hand layout, checking can be minimized to some extent. Pasties and macros are a major way because once the design of the macros is proven, only the interconnects among the macros or pasties need to be checked. (Sometimes the I/Os will also have to be checked because these are often specifically configured to the particular circuit.)

4.3 PROCESSING, PACKAGING, AND TESTING

Processing, packaging, and testing of a gate array differ little from the same functions for a standard part. The gate array *does* require far fewer *custom* processing steps than a standard or full custom part, as noted earlier. Moreover, it may be possible to enhance testability of a gate array by bringing out particular test points to enable testing of one or more particular modes of operation of the gate array.

On the negative side, however, runs of specific gate array designs are rarely long enough to build up the test histories that some popular standard parts have. Also, because a gate array involves active devices of fixed sizes, it may be difficult or impossible to bring out desirable test points without unduly loading the circuit. This is especially true when operating close to the frequency limits of CMOS.

A common practice is to use "drop-ins." These are functions placed on either every gate array die either by users who do many designs of a given gate array family or by the vendor. They may also be test dies placed at strategic locations on each wafer. History accumulated on the speed performance of such parts enables given wafers to be characterized (in terms of leakage currents, speed, etc.) and also serves to a limited extent as a process monitor. Different areas of a given wafer sometimes give better yields than do other parts of the wafer. One reason for this is tolerance buildup (called "runout"). This can be a problem when the metallization masks are not made on the same machines that made the masks for the other layers of the gate array. A second reason is that different parts of wafers stacked vertically in a "boat" in a diffusion tube receive slightly varying amounts of dopants.

4.4 TIMES REQUIRED FOR THE DIFFERENT STEPS

It is extremely difficult to make generalizations about either the elapsed or actual working times required to do a gate array design through to finished parts. Moreover, any times given can be very misleading when compared against, for example, a printed circuit board which performs the same function as the gate array that replaces it.

The reasons for this difficulty are several. No two circuits are alike in their degree of difficulty. No two gate array designers are alike in their degree of expertise. No two gate array vendors are alike with regard to the tools available and in their current work loads. Nevertheless, the author, having been through the gate array design process countless times with a wide variety of tools and techniques and a variety of gate array vendors, can offer a few generalizations.

The feasibility and analysis portion may take a few days to a few to several weeks. The longer time is generally time well spent if the circuit is complex. Technology conversion may take a day or two in addition to that which is done as part of feasibility analysis and partitioning and as a result of logic coding and verification.

The time to do logic coding and verification can be a few days to one or two months or longer. This includes iterations and design reviews. If the technology is being pushed, the program may make compromises or changes based on the simulation results, which will, of course, lengthen the time. (Drawing the schematic via schematic entry will diminish the coding time and make it more error free.) Similarly, any problems revealed by the testability analysis may force another iteration.

It is true that not all circuits undergo these iterations. A lot depends on the factors enumerated above as well as how hard the technology is being pushed. Similar numbers apply to mask design. In both cases, the bulk of the time is spent in *checking* rather than in the actual design or coding. Computers do marvelous things, but human beings still have to check simulation printouts, determine whether circuits are functioning correctly, and if so, how much margin there is just as one example. These are not trivial tasks.

Two major time elements in the category of processing, packaging, and testing are the time required to obtain the mask and the time required to test and, if necessary, debug the circuit. The former can be as short as a few days, but is generally at least a couple of weeks (see Chapter 11 for details). Almost no gate array vendor has its own "mask shop," except for those which are part of large semiconductor corporations.

The latter is even more highly variable and depends partly on how testable the gate array was made to begin with and on the equipment and people used to do the testing. The time also depends on the definition of testing. Sell-off testing (see Section 4.5.3) is different from tests done in the system. The latter may exercise more modes of operation than are tested during sell-off testing or may exercise them in a different fashion. (The user is often restricted to a fixed number, such as 10,000, clock cycles of a vendor's VLSI tester for sell-off testing.)

4.5 VENDOR INTERACTION

Vendor interaction has two phases. The first is before the gate array design is done by a given gate array vendor (if the user organization is not doing the design itself). The second is during the design, fabrication, and test of the chip.

Probably the most important advice that can be given during the first phase (when vendor selection is being done) is to be as truthful (about quantities and dates) and complete as possible. If you believe strongly that the circuit design will change before a gate array is completed, you have an obligation to inform your potential gate array vendor. Some vendors will allow this. Most will not; but it should be brought out "up front." The cardinal rule is "no surprises unless totally unavoidable."

You should expect the same candor in return from any reputable gate array vendor. You should be informed of peaks in the vendor's work load which may affect your design and the timing of such peaks. You should also expect a frank answer as to the probability of success on the first try for circuits of roughly the same complexity as yours. If your design pushes the frequency of the technology selected, you should be told how much margin exists and whether parts might have to be selected from different wafer lots to get a few parts meeting the frequency of operation.

The greater the portion of the design being done by the user, the more amenable a given vendor generally will be toward changes in timing, specifications, and perhaps quantities.

Queue time generally is much greater than actual working time in a gate array manufacturing operation of any size. Queue time is the time that a given part is in transit between work stations or is waiting to be worked on. The gate array designer must take this into account and not expect his or her gate array to be taken out of sequence unless arrangements have been made with the gate array vendor ahead of time.

4.5.1 User Involvement in the Design

This book is aimed at giving the reader a detailed understanding of how gate arrays are designed with the multitude of tools and techniques available. The question naturally arises as to where the user can become involved in the gate array design process. There is no uniform answer to this question. The answer depends almost entirely on the vendor's policies and the user's capabilities with regard to equipment, software, and experience. Some gate array vendors will allow the user to get into the gate array design process almost anywhere in the design cycle. Others will not.

Personnel availabiity also plays a key role, as does circuit complexity and the technology involved. Some companies use primarily one technology and are staffed to design gate arrays in that technology. For one or two unique designs in another technology, it may not be worth the expense to go through the learning curve for the second technology. The alternative is to let a gate array vendor offering the second technology do it all or to use the services of a gate array design house which has experience in that technology.

4.5.2 The Vendor's Point of View

It is to the user's advantage to be aware of the vendor's point of view even when it indicates that the user is not particularly desirable as a customer. The first tenet is that person power is precious in the semiconductor industry. This is one reason for the emergence of semicustom techniques, such as gate arrays that do not demand such a high usage of skilled labor in the design.

Although gate arrays are useful in low-volume applications, every gate array vendor will choose a high-volume application over a low-volume one. It will choose a less demanding design over a design which is critical in one or more parameters. It will choose a user which has several designs to be done over one which has only one unless that one has one of the other attributes of high volume or simplicity of design. In short, it will use its precious person power where analysis indicates that the greatest financial return occurs. It will not take on a very risky design so as to be able to "crow about it" unless other business is not available. Gate array vendors almost without exception are led by innovators who have a keen eye for maximizing return on investment (ROI). High ROI does not come by taking jobs which absorb undue amounts of person power. The same persons doing and possibly redoing *one* very difficult design can produce more income by doing several less complicated designs in the same period of time. It is an axiom that there is no money in "specials."

Those designers used to dealing with major semiconductor houses with lots of application notes and many field application engineers eager to answer questions are in for a bit of a surprise. While a few gate array vendors offer some form of field application engineering, may do not. Application notes are not nearly as widespread as they are, for example, in the field of microprocessors.

What then is the user with a difficult design to do? Having painted the foregoing bleak picture for those users with very special needs, the author should now "talk out of the other side of his mouth." First, with over 70 (currently) gate array vendors, there is a great deal of competition. Deals *can* be made. (But beware that yours is not the last deal that your gate array vendor makes before it goes out of business.)

Second, the user organization can develop its own staff of trained designers and, if necessary, its own gate array processing facilities. The latter method is not useful unless the user organization expects to have many gate array designs, including a large number that are critical in one or more parameters, including timeliness of the circuit.

Being able to control the priority of gate array designs by having one's own facilities is expensive. Moreover, not all gate array vendors will sell unpersonalized wafers (i.e., wafers which are complete except for customization). Even fewer gate array vendors want their processing people "hounded" by users trying to process wafers bought from them with questions about subtleties in the processing. Users contemplating developing such a facility should take into account the initial costs and the costs of maintenance and accommodating process changes by the appropriate gate array vendors.

To do one's own designs with a given manufacturer's gate array family requires information on parameters (as well as the background layers of the gate arrays being used). This information is often hard to come by, depending on the vendor. Most gate array vendors do not like to have their skilled people tied up on the telephone answering questions about the innards of a given gate array or worse, writing letters containing circuit simulation or other parameters, for example.

Some companies do offer one-week design courses for a fee in the neighborhood of (currently) $2000 per student. The student may or may not do a design as part of such a course. Normally, three to five days are required for the student to learn to use the vendor's CAD (computer-aided design) tools. The student turned designer determines for himself or herself whether the circuit will function as planned based on the simulation results.

Another possibility for the small user that cannot afford a workstation is to use the facilities of a distributor, such as Schweber, which is staffed and geared to providing such assistance.

For those users who do not wish to design their own gate arrays or have a vendor design them, independent design houses exist. A benefit of such a design house is that it can sometimes offer a somewhat more impartial view of what technology and which vendor's product to use.

Typical of such design houses is Custom Silicon of Lowell, MA. It offers a complete range of design and testing services in both standard cells and gate arrays backed up by a suite of integrated design tools.

4.5.3 Sell-Off

An important feature of a contract with a gate array vendor is the definition of what constitutes "sell-off." Sell-off is the dividing line between the gate array belonging to the gate array vendor and belonging to the user. Most gate array vendors (or at least the ones still in business) have quite definite criteria as to what constitutes sell-off. Generally, the sell-off criterion is based on tests that can be performed on the vendor's test equipment. Smaller gate array houses may accept a test fixture built by the user on which to perform the tests, but this is rare.

The reader should be aware that there are two types of tests which are sometimes used for very complex gate arrays (and other custom chips). The first type is design approval tests. The second type is production tests. The second may or may not be a subset of the first but is generally simpler in any case. The first is used to prove that the design works and functions in all modes of interest. For relatively short production runs, this may be the only type of test performed. For longer runs, relatively simple go–no go tests are performed, with sample devices from each lot being subjected to more extensive tests. These concepts and procedures are little different from any other device specification and will not be discussed further.

4.5.4 Where Can Changes Be Made?

Ideally, no changes are made to a gate array design once the design cycle has been started. This assumes that every other part of the system employing the gate array has been completely designed and debugged and that there is time to wait for the gate array to be designed and fabricated. Rarely is this the case. Therefore, changes must sometimes be accommodated. The earlier in the design cycle that a change is made, the lower is the cost of the change. The cost of a given change will depend somewhat on the equipment being used and on the type of circuit. For example, a given circuit may require a very long simulation run to exhibit the necessary properties needed to show its validity. Any changes to such a circuit will be more expensive than comparable changes to a circuit that can be proven valid with lower computer cost, *all else being equal*.

Making a given change to a design is often the least costly and time-consuming part of the change. *Checking* and *revalidating* the circuit design often outweighs the cost of making the change. Computer simulation programs help, but the user must remember that any simulation produces outputs that must be studied sometimes laboriously to avoid overlooking an error. One of the biggest dangers in making changes after a design has been validated in one fashion or another is that something was changed unintentionally. Checking only those portions of the circuit believed to have been changed will not pick up such errors.

Step function increases in cost due to changes occur after the masks have been made and after the parts have been fabricated. This is probably obvious to the reader. What may not be so obvious is that step function increases also occur after the design has been validated but before the PG tape (which makes the masks) has been made. Unless the changes are extremely simple, the requisite rechecking can be quite costly.

Although changes can be made to the gate array design, it is often, if not generally, to the user's advantage to treat the specification of the gate array as being "frozen in ice" unless there is a clear-cut error in the gate array specification. It is often easier to fix other parts of a PCB than to go through the redesign of the gate array. Nevertheless, gate arrays can be redesigned; and sometimes the redesign is less costly in both time and money than the redesign of other parts of the circuit.

4.6 CONFIGURATION CONTROL OF THE DESIGN

Because of the difficulty in making changes to a gate array once it is built, configuration control is probably more important when designing gate arrays than when designing printed circuit boards. Standard methods available at most companies involving ECOs (engineering change orders), CCBs (change control boards which for a small company might be just the designer and superior), and so on, can be used. In addition, there are two other opportunities for documenting changes. The first is on the chip itself. Some gate array vendors will allow the user to place a number and revision of the user's own choosing alongside the vendor's part number and revision letter. An example of this is shown in the corner of a gate array chip in Fig. 4.5.

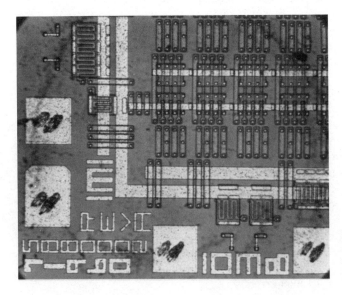

Figure 4.5 Example of user part number and revision letter on chip. Technique is used by users who create their own PG tapes.

The second method, which is useful when manually coding a simulator data base from which the masks will be designed, is to place dated comments in the coding. These can be referenced to more detailed explanations in notebook pages. The changes in this case are generally those the designer uses in doing the technology conversion and answer the question "Why did I do that?" The author has found such comments extremely useful when coming back to a project much later.

4.7 EXAMPLE TO BE USED IN THIS BOOK

To illustrate the principles involved in gate array design, the circuit described in this section will be used wherever possible. (In many instances other circuits will be used for greater clarity.) The circuit has several features that make it useful. First, it is a common structure useful in many applications. Second, it is a "nested" structure, which thus enables the principles of replication and modification of macros to be illustrated. In some cases the innermost modules will be used in place of the overall structure.

The basic structure is that of the 2-bit "slice" shown in Fig. 4.6. This consists of two D flip-flops arranged in a counter chain, two registers which latch a specific pattern of the counter stages upon command of the latch signal, and two EX-NORs which compare the value stored in each register against the current value of the counter stages. The 2-bit slice is used to make a 4-bit slice with essentially double the elements. This is shown in Fig. 4.7.

The overall module, designated a 10-bit slice, contains two replications of the 4-bit slice and one of the 2-bit slice. Figure 4.8(a) is a diagram showing how the modules are nested. (The nesting of the 2-bit slice inside the 4-bit slice is not shown explicitly.) All of the EX-NOR outputs of the nested modules feed into a 10-input AND gate, such as the 12-input TTL parts SN54S134. Outputs from stages 9 and 10 are compared in an EX-NOR and fed back to the input of the first stage. The circuit is thus a shift register sequence generator.

The reader will appreciate that more replications of the different modules could be used to produce even larger generators. For example, five replications of m_0 with the feedback still from the last two stages would yield a *maximal*-length feedback shift register sequence generator of length 22. The pseudorandom code pattern of this generator would not repeat for approximately 5 million clock cycles.

Figure 4.8(b) is the same 10-bit slice "flattened" to the primitive level.

4.8 TECHNOLOGY CONVERSION

Frequently, a circuit is designed in one technology and implemented in another. Because TTL and its permutations are the most common logic family, the two most common conversions are probably from circuits originally designed in TTL to CMOS or ISL/STL gate arrays. These two are discussed in this section. Unless otherwise noted, the term "TTL" will denote all the permutations listed in Chapter 2.

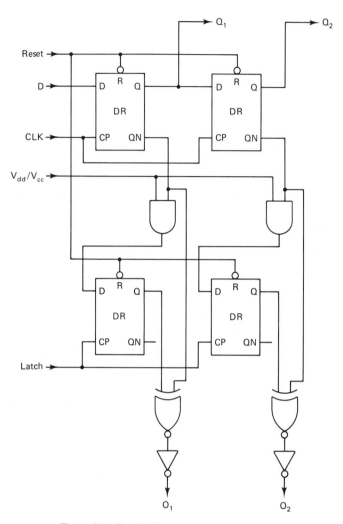

Figure 4.6 Two-bit slice used as example in book.

Technology conversion does not refer in this section to interfacing *between* technologies. The latter subject is dealt with in the chapters on design with the individual technologies. It affects technology conversion only in that additional stages must sometimes be used to drive the level shifters at the outputs, and the delays of such devices need to be accommodated in the feasibility analysis.

4.8.1 Conversion of TTL LOGIC to CMOS LOGIC

The salient difference between CMOS and TTL *insofar as logic conversion is concerned* is the much lower fan-in and fan-out of the latter. TTL devices will normally fan-out to 10 (sometimes 20) ULs (unit loads).

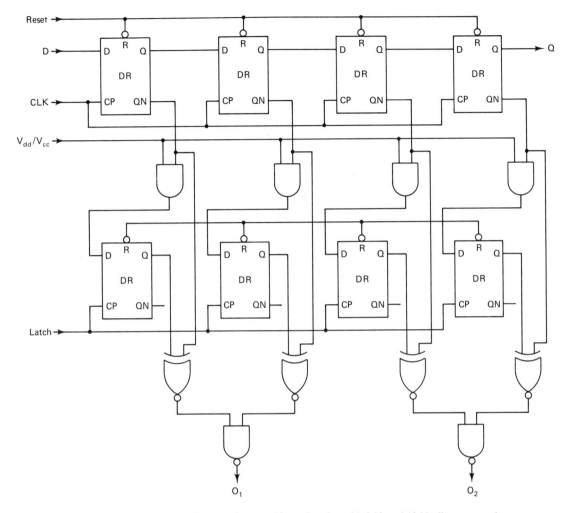

Figure 4.7 Four-bit slice sometimes used in conjunction with 2-bit and 10-bit slice as example.

By contrast, the fan-out of most CMOS devices for reasonable switching speeds is about 4 when driving into a like component (i.e., a transistor inside the array as opposed to one of the peripheral buffer transistors). The reason is the larger "on" resistance of CMOS transistors (compared to that of bipolar devices), which together with the load capacitance forms the *RC* time constant of both charge and discharge of the gate capacitances being driven.

The fan-in is similarly limited by the resultant *RC* time constants. In this case, the RC time constant of concern is that between the gate being driven (fanned into, if you will) and what it subsequently drives. For a NAND gate, the series resistance in the high-to-low propagation path is equal to the number of inputs times the "on" resistance of each NFET. Because of this difference, high fan-in gates in TTL must be changed into lower fan-in gates via Boolean algebra and DeMorgan's theorems.

Sec. 4.8 Technology Conversion

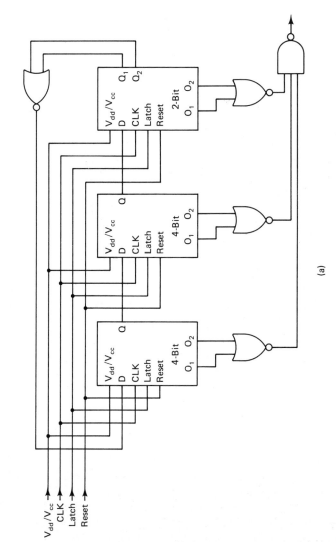

Figure 4.8 Ten-bit slice formed from 2-bit slice and 4-bit slice: (a) diagram showing nonuniform nesting; (b) composite block diagram.

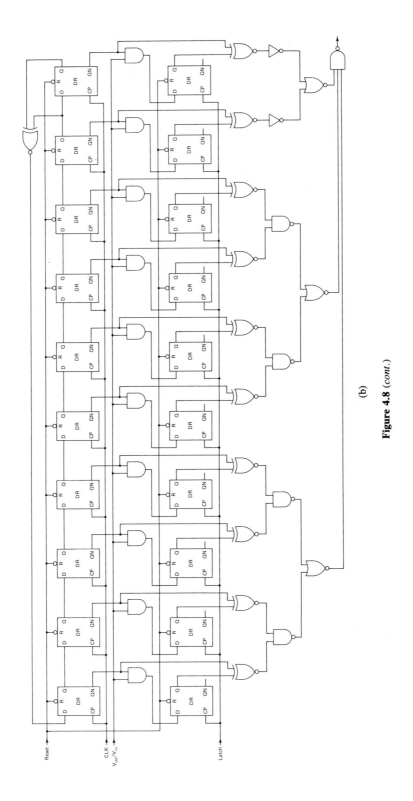

Figure 4.8 (*cont.*)

Some TTL gates will drive very high loads, such as 30 to 50 mA, and are used as clock and bus drivers. These must be converted to CMOS structures employing either paralleled array devices or more likely, peripheral buffer transistors. In some gate arrays it is necessary to use one or more stages of internal drivers to drive the buffer devices. The additional time delays of the buffer drivers must be figured into the propagation delay calculations when doing an initial feasibility analysis.

Another major difference between CMOS and TTL is the inability to use the wire AND in the former. In CMOS there are only two types of circuits that can utilize a wired configuration. These are circuits using three-state devices and circuits using "transmission gate" devices. Neither of these can be configured to a wire AND, however. Both are discussed in Chapter 9.

Yet a third major difference between CMOS and TTL is that most flip-flops, registers, and some latches (as well as other functional elements) require both phases of the clock. (This implies the use of structures called "transmission gates," which most gate array vendors use, or "floatable drivers," sometimes called "clocked inverters," which most gate array vendors do *not* use. It is possible to build such devices without transmission gates, however.)

The other phase of the clock may be in the same state as that of the original clock phase for a time equal to the propagation delay of the inverter that produced it. (This time will vary with temperature, voltage, and signal polarity and must be considered in the feasibilty analysis.)

The clock circuits so designed must use low fan-outs if circuit speed is an issue. The reason is that the times spent in the same state (high or low) of the two clock phases must be minimized.

All of the above must be taken into account. A subtlety is that the conversions must be done in such a way that the propagation delays of the resultant paths track over temperature, voltage, and wafer processing variations. Because the factors determining such variations are multiplicative rather than additive, dummy loads or extended paths may have to be introduced. For example, the ratio of CMOS propagation delays at 85°C to those at −55°C is about 1.7. A signal with a propagation delay between two points of three-fourth of a clock period at −55°C will wind up in the next clock cycle at 85°C. One with a propagation delay of one-half a clock period will not skip a clock cycle. If the two signals interact with each other, the results will be different at different temperatures.

4.8.2 Conversion of TTL to ISL/STL

Because there is no standard family of ISL/STL parts as there is for CMOS and TTL, the reader may not be familiar with the conversion to the former. For purposes of conversion, the salient difference is that ISL/STL is a collector fan-out logic and the extensive use of the wire AND to create logic functions in ISL/STL.

As with CMOS, the maximum fan-out of an internal ISL transistor is four in the gate arrays offered by the major vendors of gate arrays in this technology. Unlike CMOS, in which a fan-out of 4 is a rule of thumb, the fan-out-of-4 limitation in ISL is due to the four *usable* Schottky diodes on the outputs of each ISL transistor. It is, hence, a difficult limitation. (Buffer devices internal to the various ISL arrays can be used to increase the fan-out, but there are only a limited number of such devices on a given gate array.) The word "usable" is used because generally five such diode outputs are in place in the different arrays, but only four of the five can be employed. The five give more flexibility with regard to routing.

Unlike CMOS, the number of outputs that can be wire AND'd, and hence the fan-in of ISL, can be quite large. Values of 10 are common for ISL gate arrays. Because the fan-in is accomplished via the wire AND function, which, of course, uses no transistors, the reader can appreciate that ISL is very powerful in such applications.

An example of conversion from TTL to ISL is given in Chapter 10.

4.9 IF YOU MAKE A MISTAKE . . .

Obviously, no one purposely makes a mistake. However, sometimes they do occur even when using the newer CAD tools. The reasons are several. First, there are generally more modes in a system than can possibly be simulated within a reasonable cost and time. Murphy's law says that the ones the designer thinks are trivial or not important are those that do not work and in which the program is most interested.

Second, the increased densities of the available chips exacerbates the problem.

Third, on occasion, changes are required to the design for ancillary reasons. A typical reason is that the competition came out with a superior widget while your gate array was in the design stages, and it is now necessary to revamp the design of the system and the gate array.

Fourth, on occasion, even gate array designers make mistakes.

Following are the tasks involved:

- Find the problem.
- Determine a fix for the problem. This involves proving that the fix works using simulation or some other tool.
- Figure out how to insert the fix into the layout.
- Check to make sure that something which was right previously has not been made wrong by the fix.
- Remake the mask(s) and a new set of parts.

The reader can correctly imagine that there is no standard time and cost associated with the tasks involved. The problems of finding missing lines in a real-life example were mentioned earlier in this section. The costs and times for a new set of masks can vary between a few hundred to two thousand dollars *per mask* and take two to several weeks. The larger figure applies to the cost of masks for wafer "steppers" (Chapter 11), used by many if not most gate array firms working at feature sizes of 1.5 to 3 μm. The costs and times of the other tasks are even less firm. They can range from nil to almost the costs of doing the entire design all over again.

Adding or changing interconnect lines or functions to the layout may be trivial or may entail relaying out the entire chip. A great deal depends on where in the layout the changes must be made.

Beyond a certain point, however, it simply does not pay to keep trying to patch up a layout if there are a great many changes. It is better to relayout the chip. Where that crossover point is depends on how well laid out the remaining portions of the chip are and how critical the parts to be changed are with regard to propagation times and other parameters.

CAD tools are essential for the more complex chips. Too often, however, such tools are used in place of *thinking*. It is there that they are misused. It is sometimes *too easy* to let a computer churn out a huge simulation which may not check the parameters of most interest.

4.10 SUMMARY

In this chapter the flow of gate array design was described together with the various permutations that are possible. The object is to lay the groundwork into which discussions in later chapters can fit. A distinction was made between designing with macros and designing with transistors. Reasons for both methods were given. Trade-offs between breadboarding and logic simulation were described. The important subject of interaction with the gate array vendor was discussed. The question of what happens if a mistake is made was answered as well as it can be without a specific case to talk about. Technology conversion details were given.

REFERENCES

1. Cole, B., "How gate arrays are keeping ahead," *Electronics,* Sept. 23, 1985, pp. 48–52.
2. Mullin, M., "High-density CMOS gate arrays," Electronic Design, Mar. 13, 1986, pp. 86–98.

Chapter 5

Feasibility, Circuit Analysis, and Partitioning

5.0 INTRODUCTION

The purpose of this chapter is to discuss feasibility, circuit analysis, and partitioning. These topics are interrelated. The focus is primarily on these topics as they relate to individual gate arrays. There is also a brief discussion of partitioning from a systems viewpoint (i.e., how one partitions a large system, such as a computer, into a number of gate arrays and other chips). An important rule known as Rent's rule will be introduced. Work at Honeywell and Arizona State University aimed at minimizing the number of gate arrays will be briefly discussed. Finally, some of the cost and time drivers of gate arrays will be mentioned.

5.1 FEASIBILITY

The first step in doing a gate array design is determining whether a gate array can be built within the available time, economic, and technical constraints of a given program. Often some trading off among these parameters is possible. For example, if a gate array is to be incorporated in a product that is already in production, the parameters of the marketplace, and hence the value of the gate array, may be rigidly defined for the product *as it exists*. However, if the gate array enables the product to be much more competitive and/or longer lived, the economics of use of the gate array may change. The gate array may do this either by incorporating the added features on the gate array itself or by replacing enough circuitry that additional functions can be provided in the space thus freed up.

Although there are many places in a product's life where a program to develop a gate array chip can begin, three fairly distinct points can generally be distinguished. The first is a concept for a circuit that does not yet exist either in a schematic or on a breadboard. The second is a circuit that has at least been designed to the schematic level and which may or may not have been breadboarded. The third is a product that has been in production for a period of time. A subset of the third option is a standard IC which is being removed from the marketplace by a semiconductor vendor and which it is desired to replace by a gate array.

The first option is generally the easiest to work with. The very firm technical constraints imposed by an existing product (third option) may make it very difficult to design a gate array chip. This is especially true when the gate array is to replace a standard IC being withdrawn from the marketplace, as happens occasionally. For example, in the case of the gate array die, *fixing the pin-outs before the chip is designed* imposes restrictions on the interior of the chip which may be severe. (Fixing the pin-outs of the gate array die is necessary to avoid relaying out the PCB to accommodate the gate array.)

This is not to say that gate arrays should not be developed for circuits already designed. Indeed, the initial experiences of many, if not most companies with gate arrays is exactly that. (The reason is that companies often prefer to have a "fallback" position in place before proceeding with a new methodology.) After using gate arrays to replace existing circuits a few times, engineers and management begin automatically to consider their use in new designs. They then begin to realize the advantages that can accrue from designing a custom part "up front" (i.e., when the system is being initially designed).

This section deals primarily with technical feasibility. The economic and time aspects of feasibility are much more changeable as vendors enter and leave the marketplace and as CAD tools and their amortized usage costs change. Cost and time drivers are discussed briefly at the end of the section.

Technical feasibility generally involves several topics. These are partitioning; circuit propagation delays; I/O interfaces, including drive requirements of the outputs; power dissipation and cooling; regularity of the circuit; technology choices; temperature and voltage effects; and questions of whether the circuit will fit on a given gate array die of a size that will permit the desired package to be used. The reader may want to review Chapter 2 for information that is technology dependent.

The discussion below assumes that the designer of the circuit to be implemented on a gate array (representing the "program") is not the same as the designer of the gate array. This situation is often the case. Small companies that lack in-house gate array design capability often make use of the design expertise existing at either a selected vendor or at an independent gate array design house. Larger companies often have their own gate array design staffs which take circuits designed by other groups and convert them to gate arrays. If the gate array designer and the "program" are one and the same, the interface is, of course, considerably simplified.

5.2 TECHNICAL FEASIBILITY

For circuits of any substantial complexity, feasibility analysis often has two phases. Distinctions between the two phases are not hard and fast. Sometimes one moderately long phase (which could be construed as a combination of the two phases) is used. Everything depends on the characteristics of the individual circuit, product, and company.

The first phase is a quick look lasting anywhere from a few hours to perhaps two days to see if it is worth proceeding further. During this phase, circuits are weeded out that have too many pin-outs, too high a frequency requirement (but see below), too much memory, or which do not meet either the economic or time constraints of the program or both. Precision analog circuitry embedded in the middle of the circuit may also cause the circuit to be rejected as a gate array candidate.

An experienced gate array designer can generally tell very rapidly whether it is worthwhile proceeding further and whether redesign of the circuit would make it a more viable candidate for a gate array. He or she can also suggest alternatives worthy of further study as a result of this brief look. For example, partitioning the circuit differently or using a different word structure if memory devices are to be used may greatly ease the task of gate array implementation. For complex circuits which operate close to the limits of frequency of the technology involved, this preliminary look will indicate where further study is needed. Two parameters that *are* very difficult to estimate sometimes are the number of array cells that must be used for *routing* and the adequacy of the routing channels.

For simple circuits not operating close to the frequency and cell count limits (see below), the simple look in the first phase of feasibility analysis may be all that is required. The reader must keep in mind a fundamental tenet: *Every circuit is different in terms of economic, available time, and technical factors; therefore, use the information in this book as guides but not as gospel. Results depend heavily on the experience level of the personnel implementing the gate array as well as the tools available.*

Some circuits simply cannot be put on gate arrays *unless* the circuit is changed. A prime example in the author's experience was a circuit that had embedded in the middle a very large number of precision differential comparators. Although some companies offer gate arrays with some analog functions, the need to have large capacitors coupled with too many outputs made it unrealistic to put all of the circuitry on one gate array.

The second phase can last anywhere from a few days to two weeks or more. This second phase, if it is done, might better be called analysis and partitioning. For a complex circuit operating close to the frequency and density limits of the technology selected, some analysis and partitioning will be done to indicate that the circuit *can* be implemented on a gate array. At that point, a period of time may elapse while the program weighs the gate array approach against other approaches.

When dealing with a circuit for which there is a lot of pressure with regard to time and schedule, there is often a great temptation to forgo all or much of the "up front" analysis and partitioning. When dealing with simple circuits *and* experienced gate array designers, this is sometimes possible. However, in general, the time not spent in adequately studying the circuit will make its implementation on a gate array both considerably more risky and more time consuming.

If the decision is made to build a gate array, further study and analysis may be done to evolve the final partitioning. For example, if converting to a CMOS gate array from a circuit designed in some form of TTL, such as ALS or LSTTL, logic gates with greater than four inputs will generally have to be either paralleled or broken into a series of gates with smaller numbers of inputs. The additional delays introduced in this process (including that of the "ripple back" effect of paralleling) will have to be accounted for in the feasibility calculations.

Once it has been established that the additional delays in the critical paths will still allow the specification to be met, the circuit may be put aside while the aforementioned review by the program takes place. At the completion of this review, the circuit may then be systematically redesigned to accommodate the greater-than-four fan-in gates in the example.

Because of the large number of arrays available in the various technologies (with more becoming available all the time) [1] rarely can a circuit that has passed the initial feasibility examination by an experienced gate array designer *not* be implemented on a gate array because of technical factors. However, the closer the circuit operates to the limits of the technology involved, the greater the amount of analysis that must be done.

During the second phase, logic simulations may be done to verify the correctness of the logic (if not already done in some other fashion). The logic simulator will also enable the testability of the circuit to be determined. (This is a case where some of the steps enumerated in Chapter 4 run together for simple circuits.) Alternatively, this may be done manually by tracing through the logic paths from input to output or by using a PCB (printed circuit board) if the circuit has been breadboarded.

Table 5.1 is a list of factors to consider in doing the initial feasibility analysis. Some of these factors are self-explanatory. Others will be discussed in detail below. Although the topics are discussed under the heading of the first phase, they apply equally well to the second phase, which is a more in-depth look than the first phase.

First phase—the quick look. The first time a gate array designer looks at a candidate circuit, several questions are going through his or her mind simultaneously. The first is probably a consideration of the operating clock frequency. However, the experienced gate array designer does not necessarily reject CMOS, for example, as the technology of choice simply because the *stated* (by the program) clock frequency is too high for CMOS (see Section 5.2.7).

5.2.1 Technology Choices

The first look at feasibility provides insight into the technology trade-offs. If the *entire* circuit must be run at ECL speeds, that technology obviously is the only choice. Because the number of functions that can be implemented on an ECL chip is considerably less than the number than can be implemented on an HS CMOS or ISL/STL

TABLE 5.1 ASSESSING FEASIBILITY

INFORMATION NEEDED

Maximum clock/data transfer frequencies; through how many levels of logic?
Critical timing paths
Noncritical timing paths—how loose?
Loads that have to be driven, including expected PCB runs
Available input drivers
Restrictions on pin-outs
Prioritized list of technical, economic, and timing factors
Any analog circuitry needed
Are there especially difficult parts that can be put outside the gate array chip?
Max./min. temperature
Voltages

ANALYSIS AND PARTITIONING:
THINGS TO LOOK FOR AND CONSIDER

High frequency clocks: often are divided down into manageable frequencies
Take slices out of a memory so that parts of it can be put on a common chip with random logic
Can two similar circuits be made into one composite design with mode select options?—decreases development costs
What is roughly the ratio of "global" path lengths (long chip runs) to local path lengths (< 2 cells distant)? Will selected array accommodate this?
Where should function cell (macro) paralleling be used, if at all?
High-fan-in and high-fan-out gates
Loads that must be driven
Number of I/Os
What are the prioritized reasons for using a gate array—speed, power, cost, etc. . . .
Where are the clocks? What do they drive? Must they be regenerated locally? How will they be distributed?
Are there one or two ICs that if left off the gate array can make the gate array chip feasible? Result might be one or two standard ICs plus one gate array IC replacing, say, 150 standard ICs.
If two or more PCBs are being combined onto a few gate array chips, partitioning across PCB boundaries can be done because now all PCBs are on chips on the same board
Any A/Ds or DACs or other analog circuit embedded?
What percent of array cells will be used?
Where are the critical timing paths? What possibilities exist for races and spikes? Do certain paths' timings have to track?
Can certain signals be multiplexed to save pin count?
What package type is required? Will the die selected fit it?
What cooling is available? What is maximum power dissipation? What are ambient temperatures (max./min.)? What is maximum junction temperature?
What voltages are available?
Must certain package pins be predetermined signals? (affects layout)
Any unusual EMC (electromagnetic compatibility) or other environmental conditions?

gate array, two or more ECL chips might have to be used and the increased power will have to be tolerated. If the entire circuit is not running at ECL speeds, some combination of HS CMOS and ISL/STL gate arrays (if more than one gate array is to be built) might be more profitably used. It may also be possible to implement the circuit on a GaAs gate array.

Factors that often strongly influence the technology and vendor choice are the vendors and technologies used for previous gate array chips by a given company. This is particularly true if a company has developed an in-house capability to do some or all portions of the gate array development cycle. As noted elsewhere in this book, there is very little standardization among vendors with regard to tools, background layer structures, and to a certain extent, even terminology. Therefore, one tends to use the vendors and technologies previously used to avoid developing design aids, macros, and so on, for another vendor's gate array. This is not to say that design aids *must* be developed by the user for a given gate array. Many gate array vendors supply quite complete design aid packages. However, there is a very strong tendency to develop one's own set of aids to accommodate the particular combination of resources and needs of the user organization once a company has designed several gate arrays with the same vendor family of gate arrays.

The broad frequency ranges into which the various technologies fall are given in Table 5.2. The reader must bear in mind that the numbers in Table 5.2 are strictly there to put the user "in the ballpark." Moreover, improvements in technology are occurring at a relatively fast pace. The reader should check with the appropriate vendors to determine if the parameters have changed significantly. The numbers in

TABLE 5.2 ROUGH PARAMETERS OF GATE ARRAY TECHNOLOGIES

Technology	Maximum number of equivalent two input gates	Maximum clock frequency at 5 V and 85°C (MHz)	Power per gate
Metal-gate CMOS	550	1	>5 μW/MHz
HS CMOS 5-μm one-layer metal	3,000	25	5 μW/MHz
HS CMOS 3-μm two-layer	>5,000	>40	20 μW/MHz
HS CMOS 2-μm two-layer	>14,000 two-input	>50	20 μW/MHz
ECL core	5,000	>300 internal	>4–6 W/chip
GaAs (SDFL)	>2,000	>1000	4–6 W/chip

(Use information in this table with *Caution!*)

Table 5.2 represent a random sampling and should not be taken as being extremely precise. The methods for producing such numbers differ from vendor to vendor and are not even remotely standardized.

The reader should see Chapter 2 for additional comparisons among technologies. In particular, Fig. 2.31, which shows a comparison of drive capabilities versus technologies, will be of particular interest.

5.2.2 Technology Selection Based On Logic Function

Although this factor rarely plays a dominant part in technology selection, the reader should be made aware that some technologies are better at some logic functions than others. For example, ISL/STL is a wire AND technology. More important, the fan-in of the wire AND is much higher than the fan-in of CMOS and ECL AND gates. (The latter are not wire AND.) Fan-ins for the current popular ISL and STL AND gates are 10 and 8, respectively, as contrasted to the normal 4 maximum for CMOS. Therefore, for example, if the circuit to be implemented on a gate array is dominated by or has a very large number of high fan-in gates, the designer may select ISL/STL over other technologies.

ECL tends to be an OR/NOR gate technology. The OR/NOR functions are obtained by paralleling transistors which are driven by the same current source. ANDing is obtained in ECL by means of *series gating*, which requires a separate reference voltage (normally internally generated) for each level of series gating. The number of such levels, and hence the fan-in for AND/NAND, is normally 2 to 3. The AND function in ECL is also obtained by "collector dotting," explained in Section 10.2.4. The OR function in ECL can also be obtained by emitter dotting (wiring emitters together).

CMOS is inherently neither an AND/NAND nor an OR/NOR technology. A given vendor's gate array may favor one or the other, depending on the "on" resistances of the PFETs versus those of the NFETs. The "on" resistances are set by several factors, including sizes of the transistors and effective channel length. In this case, favoring a NAND gate means that the propagation time of the high-to-low transition is less than the low-to-high transition of the NOR gate and *VV*. These are the respective transitions used in detection for the two elements.

GaAs in its FET permutations (the dominant permutation by far) of BFL, SDFL, and E/D (enhancement/depletion; see Chapter 2) is a NOR gate technology.

5.2.3 Size of the Array—Cell or Transistor Count

After selection of the technology and consideration of the gate arrays available from suitable vendors in that technology, the gate array designer probably next determines the size of the array and hence the package size. (The actual order of consideration of the factors will be set in all likelihood by the priority established by the program. Moreover, many of them will be considered in parallel.)

The terms *cell count* is used for both ECL and CMOS arrays. The term *transistor count* is used for ISL/STL arrays even though for the latter, one transistor does not necessarily make one logic gate (see Chapter 10). For simplicity, the term "cell count" will be used to denote both terms. Too often the term "gate count" is used for both terms. This can be very misleading. Also, the size of the array is frequently couched in terms of the number of "equivalent two-input gates." The two-input gates referred to are either NAND or NOR (but are not EXNORs, for example). Normally the number of cells required to make a two-input NAND is the same as that required to make a two-input NOR.

Generally, ECL array vendors include the functionality capability of the I/O structures in the equivalent gate count. For example, because of the differential "tree" structure inherent in ECL (Chapter 10), the interface macro from a given pad to the array core will be at least a NOR/OR gate.

When determining cell count for the particular technology selected, the astute gate array designer often does a certain amount of informal partitioning. The reason is that the next item to be determined is the number of I/Os that must be accommodated. There may be other reasons as well.

First, part of the circuit design may not be frozen. If time constraints outweigh space constraints, a program may elect to build on a gate array only those portions of the design which are currently firm.

Second, there may be a "kernal" of the design which is common to other circuits in the system. It is to the designer's advantage to know what the cell (and I/O) count is of that particular piece by itself so that macros (hard or soft) can be designed to accommodate it.

As a rule of thumb, one should not use more than 80% of the cells in an array without first consulting the array vendor. The author has had gate arrays successfully autorouted which used up to 94% of the active devices, and this number is not uncommon. However, a great deal depends on the goodness of the placements of the macros and the degree of regularity of the circuit. It may also depend on the "sequentialness" of the circuit in conjunction with the type of cell used. See the GE study mentioned in Chapter 3.

Determining cell count. An example of cell counts of common logic functions is given in Table 5.3. Table 5.3 is a list of the cell counts for CMOS with three-input cells. Cell counts for arrays with other numbers of cells will scale proportionately. However, there may be differences due to the types of cells used. For example, the transmission gate function used by most CMOS vendors to implement flip-flops and latches wastes one-third of a three-input cell (see Chapter 9).

The number of FETs actually utilized may be estimated. There are programs in existence at some gate array vendors, such as LSI Logic, which automatically calculate the number of array cells used. From this, the number of FETs can be calculated if desired. (Normally, the cell count number is sufficient).

Cell calculator. Universal Semiconductor offers a slide rule for calculating the cell cell counts. The rule is peculiar to their own array and is based on the 7400 TTL family of standard parts. The user simply places the index over the TTL part desired and reads the cell count required to make that part on a gate array. An example of the slide rule is shown in Fig. 5.1.

TABLE 5.3 CELL COUNTS FOR THREE-INPUT-CELL CMOS

Gate	Number of cells required[a]
Two- to three-input (NAND or NOR)	1
Four-input (NAND or NOR)	2
Two- to three-input (AND OR INV) (AOI)	2
Four-input (AND OR INV)	3
XOR	2
D flip-flop	4–6
D flip-flop with preset and clear	6
Buffer for each output	≥2
Buffer for each input	≥1
One-bit dynamic shift register with clear	2
4-to-1 MUX	3
JK flip-flop	6
3-to-8 binary decoder	10
16-to-1 MUX	≥40
Four-bit adder with carry out	28
Counter: 4-bit, binary, up/down, synchronous count, asynchronous clear, parallel load, cascadable	38

[a] Add 30 to 40 cells for miscellaneous buffer inverters when using the 1,000-, 1440-, or 1960-cell arrays

Other frequent users of gate arrays have simple programs that query the user as to the number of each type of standard part and/or accept manual entry of such parts. They then quickly give the user a cell count in one or more gate array families.

5.2.4 Laying Out on a Larger Chip Initially

A nice feature of using a gate array that is a member of a family of gate arrays is that a given design can initially be laid out on a larger member of the family and then later put on a smaller member to decrease both die and package cost. The process of transferral *may* be relatively easy if the design is stored on an IGS (Interactive Graphics System) *unless* there is a problem with the interconnects. Also, the area occupied by the layout should fit the next lowest chip size in the family without re-layout (and rechecking).

As an example, consider the layout in Fig. 5.2. This design, which is on a 2000-cell CMOS array, seemingly is a good candidate for putting on the next-smaller-size chip (the 1440-cell array) in the family *except* that the interconnects needed for high speed all come out on the top and bottom. If the smaller chip were used, some of these interconnects would have to come out the sides (and hence go through a significant number of polysilicon underpasses) of this single-layer metal gate array family. This is unacceptable not only because of the speed degradation of the signals affected but also because those signals would no longer track the others with regard to delay times. For this application, that was unacceptable. For other applications, it might have been acceptable.

If the program has specified a given package type and size, this specification will limit the size of the die that can be accommodated. However, the number of

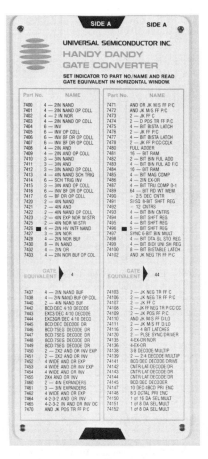

Figure 5.1 Cell calculator slide rule. (Courtesy of Universal Semiconductor, Inc.)

cells that exist on a given size of die is largely determined by the feature sizes (width and length) of the transistors, the number of bonding pads, and their associated buffers.

For example, a well-known 12,000-transistor CMOS array with 116 bonding pads and with 5-μm feature sizes is roughly 300 mils on a side. A 3-μm-feature-size array with roughly 16,000 transistors and 156 bonding pads in the same technology has dimensions that are somewhat the same. (Because the numbers of transistors per array cell in the two arrays are different, it is more useful to compare them on a transistor than on a cell basis.)

As a rule of thumb, the internal cavity size of the package must be 20 to 40 mils greater on each side to accommodate the die. However, there are often great differences in the cavity sizes of packages that have the same external dimensions, number of pins, and pin spacings. For example, an AMP 24-pin (100-mil center-to-center spacings of package pins) DIP which is 600 mils wide and about 1.2 in. long has an internal cavity which is 375 mils on each side. By contrast, another member of the same family with exactly the same external dimensions as above has an internal cavity size which is far smaller.

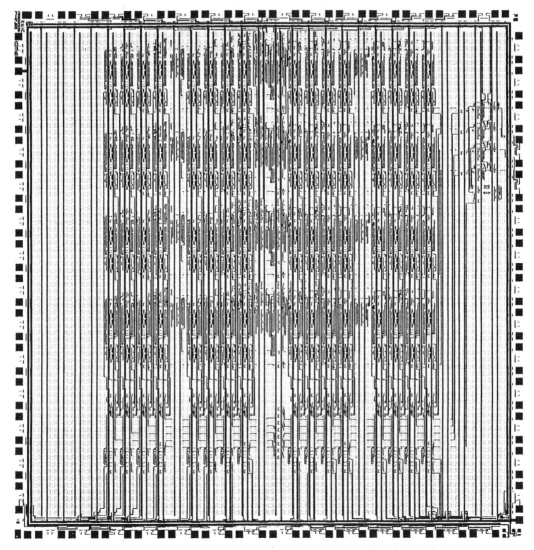

Figure 5.2 Layout for chip showing how much space is left over. It was put on the larger chip because of the pin-outs. Note that every I/O on the top and bottom is used. (Courtesy of Sanders Associates, Inc. and International Microcircuits, Inc.)

5.2.5 Pin-Outs

Determination of the pin-outs of the die and package (i.e., which I/Os go and come from what chip pads and package pins should be left to the discretion of the chip designer if possible. If this is not done, the chip layout is complicated, sometimes unduly so. This is particularly true when the parameters of signal frequency and/or circuit density (number of functions implemented on the gate array) are being pushed to their limits. One of the factors to be determined during the feasibility analysis is what pins, if any, *must* be predetermined and what the possible effects are on the design.

If a given package type must be used, this must also be taken into consideration, from three standpoints. First is, of course, that the gate array selected will have a die size small enough to fit in the cavity (see Section 5.2.4). The second is that the package so selected enables the I/Os to/from the die bonding pads to be routed to the package pins without crossing each other.

Suppose that all 16 pads on the top edge of a gate array die are used. (This might occur in a single-level metal gate array where the highest speeds occur by running the metal interconnects vertically instead of using the slower polysilicon underpasses, which route the signals horizontally and under the power and ground buses.)

If this die is put in a 40-pin DIP which has 10-pin connections on each of the four sides of the cavity, some of the wires will have to cross portions of the die inside the package to get to the pin connections on the side of the package cavity. (Package cavities are generally square with equal numbers of pins on each side.) This crossing the die is highly undesirable because of the danger of shorts. In this case, a better choice would be a 64-pin package, such as a chip carrier, a flat pack, or even a DIP. See the discussion in Chapter 11 for more information.

The third point that must be sometimes considered is the differential time delays of the package pins. This topic is important when, for example, there are a lot of high-speed data bus lines that must track each other in propagation delay. The propagation delays between the longest and shortest pins of a 64-pin DIP, for example, differ by a factor of about 1.6.

Pin-outs-limited circuits. A fairly common problem in partitioning is what to do when the number of pins of the circuit exceeds that of the package into which the die will go. One possible solution is to use a die that will fit in the package which has more array cells on it than did the die first considered. This would occur if the feature size of the second die is smaller than that of the first die (i.e., a die with 2-μm feature sizes has more array cells, in general, than does one with 3-μm feature sizes). The increased functionality may allow the designer to repartition the system in such a way that fewer I/Os are required.

A second way is to consider MUXing of signals so that a given pin can be shared among several outputs which normally would have their own separate bonding pads and package pins. The multiplex is controlled via signals generated internally perhaps in relation to external stimulii. The same caveats mentioned under bidirectional pins (below) hold here also. Obviously, the data rates of the pins being MUXed must be low enough that the combined data rate does not exceed that of the technology being used.

Still another way of overcoming pin-out limitations is to make chip pads bidirectional so that they can be used as both inputs and outputs. This generally involves making use of three-state circuitry inside the chip. The control for the circuitry inside the chip generally requires one or more pins to be used for bidirectional control. This means that it is often not useful to try to save pins by this method unless at least several pins are made bidirectional. If four control lines were input to the chip for controlling bidirectional characteristics, the 16 possible states of the four lines might be sufficient to save the four pins used plus others. A similar statement holds for the same lines used to control MUXing.

An example of implementing bidirectionals is given in Chapter 9. A problem with this method, of course, is the added complexity of the circuitry outside the chip to which the bidirectional pins are attached. Such circuitry must recognize and deal with the signals to and from the bidirectional pins.

5.2.6 Gate Array Families

As noted in Chapter 1, most vendors offer "families" of gate arrays. If a given design will not fit on one family member because of limitations of the numbers of either cells or pads, it will often fit on another member of that family. This gives the user considerable flexibility. Moreover, for a second application, which is perhaps unrelated to the original application, a user may wish to use a portion of a given gate array design. It is normally a very small matter to transfer a portion of a design to another smaller gate array chip in the same (and sometimes even in a different) family of a given gate array vendor.

Table 5.4 lists the sizes and numbers of pads of members of a popular 2-μm, double layer metal family (the LSI Logic 7000 series). Note that two permutations

TABLE 5.4 LSI LOGIC ARRAY SPECIFICATIONS

Device Number	Gate Count	% Max. AUR[a]	Max. I/O Pads	Assigned VDD Pads	Assigned VSS Pads	Assigned VSS2 Pads	Max. Pads
7000 Series (P-Version Standard Pad Pitch) Silicon gate, oxide isolated, 2.0 micron, double-layer metal							
LL7080	800	90	44	2	4	2	52
LL7140	1443	90	58	2	4	2	66
LL7220	2224	85	70	2	4	2	78
LL7320	3192	85	80	4	8	4	96
LL7420	4242	85	98	4	8	4	114
LL7640	6072	80	122	4	8	4	138
LL7840	8370	80	150	4	8	4	166
LL71000	10013	70	158	4	8	4	174
7000 Series (C-Version Tight Pad Pitch) Silicon gate, oxide isolated, 2.0 micron, double-layer metal							
LL7080	880	90	60	2	4	2	68
LL7140	1443	90	78	2	4	2	86
LL7220	2224	85	98	2	4	2	106
LL7320	3192	85	112	4	8	4	128
LL7420	4242	85	134	4	8	4	150
LL7640	6072	80	170	4	8	4	186
LL7840	8370	80	206	4	8	4	222
LL71000	10013	70	216	4	8	4	232

[a]Maximum Array Usage Recommended

Source: CMOS Macrocell Manual, LSI LOGIC Corp., July, 1985. Used with permission.

of this family are offered. They differ by the number of pads offered on each size chip. Also listed is the maximum array utilization recommended. (As always, the reader should consult the vendor for the most current information.)

5.2.7 Exluding Circuit Elements

A factor sometimes overlooked in feasibility analysis is the effect of excluding certain elements from the gate array chip to make it more doable. For example, the clock frequency may be 175 MHz, but it may be divided down to, say 35 MHz almost immediately upon entering the circuit. By not trying to put the handful of ICs that use the >35 MHz frequency on the gate array chip, the entire rest of the circuit may possibly be put on a HS CMOS gate array instead of being forced to go to an ECL gate array. The ECL gate array might not have enough array cells to accommodate the circuit or might use too much power. (HS CMOS gate arrays *will* run faster, sometimes considerably faster, than 35 MHz, with some qualifications. This example is used simply to provide a clear-cut case.)

5.2.8 Power Dissipation of the Package

As more and more elements are put on a chip this becomes of concern even for CMOS. (It has always been of concern for ECL and to a lesser extent, to ISL/STL.) CMOS has the advantage that power is used *internally* only when the devices are switching. Because in a typical circuit only 10 to 30% of the gates are switching at any one time, CMOS has an inherent advantage of 3 to 10 over other technologies, all else being equal.

The power dissipation of output buffers of CMOS which are continually sinking and/or sourcing conduction current (as opposed to displacement current of a capacitive load) is not decreased by the percentage of gates that are switching. Such cases occur when CMOS is interfacing to one of the bipolar technologies, such as TTL. (The power dissipation of the output buffers of CMOS is decreased by the percentage of gates that are switching when CMOS is interfacing to CMOS.)

In cases where most of the power is dissipated in the buffers, there will be little difference among the technologies insofar as power dissipation is concerned.

The power dissipation of the package and the ambient temperature, T_a, must be such as to maintain the required junction temperature, T_j, of the chip. A typical calculation is

$$T_j = T_a + P_d \times O_{ja} \qquad (5.2)$$

where P_d is the power to be dissipated and O_{ja} is the thermal impedance of the package from junction to ambient. For military-grade circuits, T_j can be as high as 175° C but is preferably much, much less. Commercial circuits use 150° C maximum.

Equation (5.2) is an equation developed during the era of SSI (small-scale integration) when chip temperatures were uniform across the chip. The user must be much more careful when using bipolar technologies or even high speed (40 MHz), high (>50%)-duty-factor CMOS. Indeed, most ECL gate array manufacturers require that the chip power be equalized as much as possible in all four quadrants of the chip.

Calculation of power dissipation. Because of the number of approximations and guesstimates that have to be made, power dissipation calculations sometimes have to be regarded as highly approximate. This is especially true of CMOS for which it is extremely difficult to guess what percentage of gates use switching at a given time. Some gate array vendors, such as AMCC, have CAD programs which extract the current used from each of the macros in the design and give a printout of the current at each voltage. From the information, the power can be calculated quite precisely. (If ECL is interfaced to T^2L, there will be current at the T^2L operating voltage as well as at the ECL operating voltage.)

As suggested above, the calculation of power dissipation involves calculation of the power dissipated by the internal array structure and adding it to the power dissipated by the buffers.

Rough calculation of power dissipation of the buffers. An approximation to the power dissipated by the buffers is V_{ol} (the output low voltage) times the current being sunk, I_{ol}, times the number of such buffers. If CMOS is driving CMOS, this figure should be multiplied by the guesstimated number of gates switching at any one time. This assumes that the product is greater than or equal to the product of V_{oh} (output high voltage) and I_{oh} (output high current), which normally it is. Otherwise, either the total number of buffers can be multiplied by the output high power (V_{oh} times I_{oh}) and used in place of the number above. Alternatively, an estimate can be made of the respective numbers of buffers that will be in each state for time periods longer than the thermal time constants of the package and each such number multiplied by the proper power product, and the results added together. The latter method is cumbersome and would not be used except in unusual circumstances in which, for example, the chip buffers were driving control circuitry.

The power dissipation of buffer outputs which are switching large capacitive loads can be estimated as $CVVf$, where C is the capacitance of the load, VV the square of the voltage, and f the switching frequency.

The power dissipation of the internal array structure will now be discussed for representatives of the separate technologies.

Rough calculation of power dissipation of the array.

CMOS. The power dissipation of CMOS driving CMOS can be calculated as in Section 2.7.2. The total capacitance has to be estimated. Manufacturers normally supply the gate capacitance. This is typically in the range 0.2 to 0.3 pF per CMOS gate pair (one NFET and one PFET connected together) for CMOS with feature sizes in the range 3 to 5 μm. To this must be added a guesstimate of the capacitances of the wiring channels as well as the output capacitances of the logic devices themselves. The latter in CMOS comes from the source/drains of the transistors. The resultant capacitance depends on several factors, including the number of source/drains (and hence the logic configuration), the capacitance of the source/drains and other contributing capacitances (see Chapter 6 for the Spice model of a MOSFET), and the logic state. (The logic state is important because the capacitances involved in the source/drains are either depletion or diffusion, depending on the state and in either case are voltage variable.)

In lieu of going through such a calculation (which is really one of the outputs that a circuit simulation program can provide much more easily), a common method is simply to assume that CMOS nodal capacitances are 1 pF. This is reasonable, although sometimes a little conservative. Depletion capacitances of CMOS source/drains with feature sizes in the range 2 to 5 μm are typically in the range 0.3 to 0.45 pF. Diffusion capacitances are often negligible. A three-input NAND gate in its worst-case logic state will have three depletion capacitances in parallel. An inverter, however, will have only one. Therefore, 1 pF is probably a good estimate.

ISL. An example of the power calculation of ISL is given in Table 5.5. The values calculated include not only the internal gates but also the Schottky buffers driving inside the chip (see Chapters 3 and 10), the transceivers, and the input and output buffers.

Capacitances of interconnect. Three parameters that are difficult to assess are the number, lengths, and capacitances of the interconnect lines. IBM, LSI Logic, Inc., and other companies have programs that extract the first two parameters from the layout itself. The third is calculated from the first two knowing the characteristics of the interconnect. However, until a layout is actually done, the effects of the time delays due to layout cannot be totally analyzed.

A variety of algorithms exist for estimating the capacitance of the interconnects for purposes of calculating power dissipation and speed degradations before layout is done. Sometimes some percentage (such as 25%) of the gate capacitance is assumed. The number may be higher for single-level metal arrays in which poly-

TABLE 5.5 EXAMPLE OF POWER CALCULATION

Generally: Maximum power (mW) = 0.25 × Number of ISL gates
+ 0.25 × Number of Schottky buffers
+ 12 × Number of transceivers[a]
+ 8 × Number of output buffers[b]
+ 5 × Number of input buffers[c]
+ 0.5 × Load current (mA)[d]

Specifically:
216 = (0.25) (863 ISL gates)
  6 = (0.25) (22 Schottky buffers)
 24 = (12)   (2 transceivers)[a]
136 = (8)   (17 output buffers)[b]
 85 = (5)   (17 input buffers)[c]
 76 = (0.5) (8 mA maximum I_{ol})(19 maximum low outputs)
543 Maximum power (mW)

[a]Transceivers include TTS, TOC, TEOC.
[b]Output buffers include AP, TS, OC, EOC, IOD.
[c]Input buffers include INB, IOCD, EOCD.
[d]Load current = Maximum I_{ol} for selected temperature range × total number of output buffers and transceivers that can be at a low output simultaneously.
Source: Signetics, Inc.

silicon with higher capacitance is used, and it may be lower for two-layer metal arrays. Other algorithms use fan-out of each gate to calculate the capacitive load on that gate.

An alternative employed by some companies is to include in each macro's time-delay specification the time delay of a given amount of interconnect length. (This partly explains why some gate array vendors' claims for maximum frequency of operation are so much different from those of other vendors with a similar technology. The less conservative vendors ignore the effects of interconnect delays, as well as delays due to voltage, temperature, and processing variations.)

5.2.9 Temperature Effects

An important consideration in feasibility analysis is that of temperature effects. These are technology dependent.

CMOS temperature effects. MOS transistors, of which CMOS is a subset, exhibit a very pronounced temperature dependence of propagation delay. This is due in large measure to the temperature dependence of the so-called "mobility" parameter of solid-state devices. Mobility has the units of cm/s per V/cm (more often seen as $cm^2/V\text{-}s$). Hence mobility is the rate of change of carrier (hole or electron) velocity with change in electric field strength. The carriers of interest for this discussion are those which exist in the conduction band (as opposed to the valence band) of the semiconductor.

The reader will recall from Chapter 2 that there are four factors that influence the speed of a MOS device. Of the four, the one that temperature affects is mobility. Because most of the transistor action of a MOSFET takes place within an infinitesimal layer of silicon at the junction between the gate oxide and the silicon substrate, the mobility of interest is the surface mobility.

The surface mobility (hereinafter called simply the "mobility") varies with a number of parameters other than temperature. These include doping level and thickness of the layer in which the mobility is being measured. Michael Splinter of Rockwell published an interesting curve for the latter parameter, which showed that at thicknesses of less than about 2600 Å (0.26 μm), the mobility of the electrons is actually *less* than that of the holes. See Appendix 2A.2. This is intended to give the reader a "feeling" for scaling effects and also to make the reader aware that the electron mobility is not always greater than that of the hole mobility.

The user, of course, cannot control the doping level and thickness of the silicon. The user can only control the temperature or accommodate the differences that exist in device characteristics with variation in temperature.

The mobility of silicon decreases with increase in temperature. This causes the propagation delay to decrease monotonically with increase in temperature. Table 5.6 gives rules of thumb for this delay decrease with temperature increase. The reason that the mobility decreases with increase in temperature is that the mean free path of the carriers decreases. This, in turn, is due to increased thermal agitation of the crystalline lattice with increase in temperature.

TABLE 5.6 RULES OF THUMB FOR INCREASE IN PROPAGATION DELAY WITH INCREASE IN TEMPERATURE OF MOS DEVICES

Temperature (°C)	Delay ratioed to $-55°C$
-55	1
20 (room temp.)	1.4
85	1.7
125	2

Although these numbers are fairly widely accepted and correllate reasonably well with a number of independent sources, the reader should use these *only as estimates. Actual ratios can vary from these numbers by more than plus and minus 20%.* However, this is still better than the 40% variation of propagation delays generally quoted by semiconductor manufacturers for the wafer-to-wafer variations. (The ratios listed above will be in *addition* to the wafer-to-wafer variations.)

The conduction factor K in the equations for the MOS transistor in Chapter 2 depends on temperature as shown in Table 5.7. These numbers are reasonably close to the values given by many manufacturers for the variations in propagation delay versus temperature given in Table 5.6. The values are in terms of a normalized conduction parameter K', which is normalized to the ratio of the effective width to the effective length of the FET.

TABLE 5.7 VARIATION OF CONDUCTION FACTOR WITH TEMPERATURE

T (°C)	Normalized K (μAV^2)
-55	1.60
0	1.14
$+25$	1.00
$+85$	0.76
$+125$	0.65
$K(T) = [(T + 273)298]^{-3/2}$	

Source: Design Manual HC Series: CMOS Gate Arrays, California Devices, Inc., 1982; p. 151 (temperature variation of conduction factor).

ISL/STL temperature effects. The variation in propagation delay with temperature for ISL is shown in Fig. 5.3. Compare to Fig. 2.7 for STL.

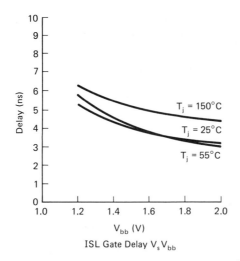

ISL Gate Delay $V_s V_{bb}$

Figure 5.3 Propagation delay versus base current of ISL for various temperatures. (Courtesy of Signetics.)

ECL temperature effects. ECL comes in both temperature-compensated and non-temperature-compensated forms. A difference between these forms is that the former requires more circuitry to provide the compensation.

The longer the path, the greater is the effect of temperature. Temperature variations can be extremely significant. A problem is that the factors listed above multiply the *total* of the propagation delays of the *active* elements in a path.

Suppose that a system is operating synchronously with several competing paths of unequal time delays and that skipping clock cycles cannot be tolerated. The shorter paths will suffer less change than the longer paths. The result may be that some paths wind up in a following clock cycle, whereas others do not as temperature increases. This, of course, would destroy the synchronism of the system. Pipelining techniques can sometimes be used judiciously provided that the system will tolerate the additional number of clock cycles. (Note that this is not the same as dropping clock cycles as temperature increases.)

For example, suppose that a CMOS system is operating at a clock period of 40 ns; suppose further that the delay through two critical paths is 32 ns in one case and 25 ns in the other case at room temperature. It is also assumed that both signals are generated at the same time by the same element.

As the temperature is increased to 85°C, both signals will be in the same clock period (32 × 1.2 = 38.4 and 25 × 1.2 = 30). The first signal will be only marginally there because of realistic rise time and perhaps jitter problems, however.

At 125°C, the first signal will definitely be in a different clock cycle than the second signal (32 × 1.4 = 44.8, while 25 × 1.4 = 35). The result is that the first signal has skipped a clock cycle due to temperature changes.

The values of 1.2 and 1.4 used in the calculation above were obtained by ratioing the delays at 85 and 125°C respective to room temperature instead of to −55°C. (Thus, 1.7/1.4 = 1.2.)

While on the subject, it should be noted that the variation in propagation delays of polysilicon and metal with temperature are negligible.

5.2.10 Analysis of Propagation Delays

Unless the highest clock frequency is less than about 10% of the maximum frequency of the technology involved (and sometimes even then), analysis of the propagation delays of critical paths is done. The closer the system operates to the limits of the technology involved, the more extensive generally is the analysis. The reader must be made aware, however, that propagation delays due to wafer-to-wafer variations can be 40% or more, as noted earlier. Therefore, any analysis will have to consider this.

There are several ways to do propagation delay analysis. These range from a simple paper-and-pencil calculation to use of a circuit simulation program, such as SPICE or TRACAP (see Chapter 6). In the first phase of feasibility analysis alluded to above, in which the object is to grossly determine feasibility, a simple paper-and-pencil calculation may be done to determine how much margin exists. During the actual analysis and partitioning that follows, one of the simulation programs may be used.

Alternatively, a *logic* (as opposed to a circuit) simulation program, such as Tegas or Texsim, may be used to do propagation delay analysis. In this case, the logic simulation is run with propagation delay parameters appropriate for the expected worst-case values to be seen for temperature, voltage, clock skew if present, and additional delays due to wafer processing. The logic simulation output is then studied to see what problems, if any, have occurred and what margin exists.

When using worst-case values for logic simulations, the user has to be careful to not make unrealistic cases. For example, if CMOS is running off a battery, the slower speed due to a higher temperature will be somewhat offset by the higher voltage available from the battery at the higher temperature. The higher voltage will cause CMOS to run faster. (Consult the battery manufacturer for voltage versus temperature values.)

Also, for parameters, such as wafer variations and clock skew, which do not depend on one another, one can usually reasonably expect that the worst parameters will not all occur at the same time. If the events can be regarded as statistically independent, a cumulative effect of the square root of the sum of the squares approximation can often be usefully used.

Importance of using edge delays of signals. When propagation delays are of concern, it is important to use the edge delays of logic structures. Edge delays are the propagation delays of the low-to-high and high-to-low transitions as opposed to the average propagation delays of these structures. This is particularly true when CMOS and ISL technologies are being analyzed. CMOS logic gates have the property that either the PFETs or the NFETs (or both in the case of AOIs and OAIs) are in parallel. If the NFETs have the same "on" resistance as that of the PFETs, there will be substantial differences between transition times from low to high and high to low. Moreover, the transition times will depend greatly on the amount of capacitive load, and hence fan-out, being driven. Although bipolar devices also have this dependence of transition time on fan-out, it is not nearly as severe as it is in the case of the MOS technologies (CMOS in this discussion). The reason is that the resistance value in the RC time constant in the case of CMOS transistors in the array is much larger than in the case of bipolar devices. (The RC time constant determines how rapidly the capacitance of the succeeding transistor gate can be charged or discharged.)

Dependence of CMOS logic gate propagation times on states of the input signals. Consider the three-input CMOS NAND gate whose FET diagram is given in Chapter 2. (See Chapter 9 for more information on the conversion of common logic functions into FET diagrams.) Because there are three PFETs in parallel, the low-to-high propagation time (to charge the capacitance of the succeeding gate) will depend on how many of the three PFETs are turned on at the time the transition from low to high begins. (Each FET, when on, behaves essentially as a resistor of value typically in the range of a few to several thousand ohms.) In the case of a NOR gate, the opposite situation occurs; that is, the high-to-low propagation time will depend on how many of the input signals are high at the time the switching transition begins.

Table 5.8 gives the "on" resistances for a popular 5-μm CMOS family. "On" resistances depend on W/L and other parameters and can be considerably different for other families even at the same gate length from those in Table 5.8. For example, the buffer and array FET on resistances for a popular 3-μm family are 100 and 400 Ω, respectively. See Chapter 2 for a discussion. Curves showing the high-to-low and low-to-high propagation times of the signals are given in Chapter 2.

The reader is cautioned that the foregoing representation of the FETs as lumped resistors is a simplification. The simplification is, however, very useful for making the point. If more precision is desired, a circuit simulation program, such as SPICE or TRACAP, should be used.

The practical problem associated with this property is the potential for race conditions that exists. Too frequently, only the maximum propagation delays of the signals are considered in analysis. When there are two or more parallel paths, the minimum propagation delays must also be considered. This is particularly true in the case of synchronously clocked systems.

TABLE 5.8 IMI HS CMOS TRANSISTOR CHARACTERISTICS

	Nominal (Ω) "On" resistance	
	5 V	10 V
Array P	5000	3000
Array N	7000	6500
Big N	800	600
Bigg N	600	400–500
Big P	600	400–500

CMOS delays are sometimes stated in terms of the average of the high-to-low and low-to-high delay values. Such numbers can be extremely misleading and are essentially worthless unless the values in the two directions are essentially the same. Average values should never be used for analysis.

Paper-and-pencil methods for critical path analysis. Instead of using logic or circuit simulations to determine whether a given circuit is capable of running at the required clock speeds, a paper-and-pencil analysis is often sufficient. There are several ways in which such analysis can be done. One method is shown in Fig. 5.4. In this method, each of the delay elements including any wired OR or wired AND gate is represented as a block. If layout paths, such as the elements labeled "poly A," "poly B," and so on, in Fig. 5.4, are known, these are included. Otherwise, an estimate is sometimes made for their effects.

Each block contains a statement of the element, its fan-out, and its propagation delay for the edge (high to low or VV) of interest of the signal. Setup and hold times must be included for any memory elements in the paths. By adding up all of the propagation delays, the overall propagation delay can be determined very rapidly.

The block diagram shown in Fig. 5.4 is useful for other purposes as well. It can be used by test personnel to ensure that the critical path delays are met. (A logic simulation is not as explicit.) If a schematic entry system is available, the diagram can be prepared on that.

Another use for Fig. 5.4 is to spot race conditions by showing the paths which are in parallel and which later affect the same element.

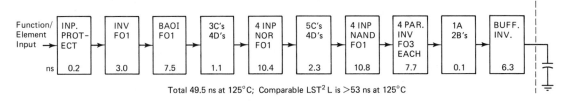

Figure 5.4 Paper-and-pencil method of delay analysis. A's, B's, C's, and D's refer to poly A's, poly B's, poly C's, and poly D's, respectively. <0.1-ns time delays (one poly A, for example) are ignored. 4∥ INV refers to paralleled inverters (in this case, four). FO fan-out.

Sophisticated propagation delay models. [2, 3, 4, 5] Penfield et. al. [6] and Mead et. al. [7] have studied the problem of MOS circuits with multiple fan outs driving paths that are unequal in length. Models have been developed. When the driving point resistance (which would be the sum of the on resistances of three PFETs when a three-input CMOS NOR gate is driving the lines from low to high, for example) is not negligible with respect to the resistance of that being driven, then there is no closed form solution [5]. However, Penfield has been able to bound the solution.

Using the values of on resistance in the above table in conjunction with the resistances of poly underpasses for a typical 5-μm process, there is no closed-form solution when more than about seven poly underpasses being driven by an inverter are involved. A three-input gate in the same family driving from its high impedance (i.e., low to high for a NOR gate) can tolerate about 24 such underpasses before the closed form solution breaks down. In the opposite transition, however, there is no closed-form solution for greater than about one or two polys. (But note that these are not general statements because of the aforementioned variation of on resistances among gate array families.)

5.2.11 Freedoms and Restrictions of Gate Arrays

The author, whose involvement with gate arrays goes back many years, has seen the pendulum of interest in gate arrays swing from "prove that they are useful" to "gate arrays can do it all." The truth lies somewhere in between, although nowadays there is a far greater range of possibilities not only in gate arrays but in other forms of custom VLSI as well. The purpose of this section is simply to indicate a few of the subtleties in the use of gate arrays that were not appropriate for other chapters.

Adding small extra functions. One of the important but subtle differences between designing with gate arrays and designing with discrete (i.e., standard) parts is that small extra functions can often be added to the chip at essentially zero additional cost. A small function could, for example, be a flip-flop or two organized as a 2-bit counter. If the rule of thumb about not using more than 80% of cells in the gate array is followed, there will generally be extra cells available on the chip. Moreover, in MOS there will often also be more bonding pads on the chip than are actually used in the design. The IMI 2000-cell (3000 gate) chip alluded to earlier, for example, has 116 such bonding pads. The LSI Logic LL71000 (10,000 gate) chip has 216 bonding pads.

Putting independent circuits on the same chip. The author called this concept "share-a-chip" in internal memos before he became aware of the important Mead–Conway work [8, Chap. 2] in which the same thing was done. Each design is totally invisible to the other user's designs except from the standpoint of possibly requiring a larger package than would otherwise be necessary due to the larger die size. The author is currently using this method with a GaAs gate array whose NRE is quite high.

The advantages of putting independent circuits on the same chip are that circuits whose economics are too low by themselves to permit being put on a gate array can sometimes be "piggybacked" on an array quite economically. NRE costs are highly nonlinear with array size. The advantage to the circuit being piggybacked is that there is a larger production run of the part.

The disadvantages are the increased costs of testing and possibly increased package size and die size. However, for many gate array projects, NRE, not die or package and test cost, is the dominant cost factor. Each combination has to be weighed on its own merits.

The other big problem is that of ensuring that both (or all, if more than two designs are being done on the same chip) designs are complete by a given date which does not hold up either (or any other) design on the chip. (The masks, of course, are composites of the individual designs.)

As an example, consider the IMI 2000-cell (three inputs per cell) gate array. As noted, this has 116 bonding pads for I/Os and can accommodate a variety of independent circuits, if desired. The die itself will fit in a 40-pin (Kyocera) DIP or larger package.

There are many different ways in which this method can be used. In one scenerio, the 2000 cells are divided into eight groups *averaging* 200 cells (300 gates) each. (The current 6000 to 10,000 gate arrays would have a correspondingly higher number allocated to each design.) This allocation would fit in the rule of thumb about using no more than roughly 80% of the 2000 cells in the array. (To accommodate different requirements, some might be larger than others.) This is the scenerio that might be used for an in-class exercise, for example. Obviously, the number of groups and the number of cells and bonding pads allocated to each group could be varied.

Each group would be allocated 11 bonding pads. (Power and ground connections would not have to be included in the pad allocation to each group because these would be common to the chip.) Each such design would go in the 40-pin package (or larger) mentioned above. Note that only the bonding pads for each individual circuit have to be bonded to the package pins because *each circuit is totally independent of the others* (ie., a 116-pin package does *not* have to be used simply because there are 116 pads which have connections to them).

In a second scenerio, a relatively small circuit utilizing 200 cells might be put on the same chip with a larger circuit using 1400 cells. Bonding pads allocated to the smaller circuit might far outweigh those allocated to the larger circuits, or the reverse might be true. This is the scenario in which a circuit whose economics do not permit it to be put on a gate array by itself is piggybacked with another circuit to share costs. Figure 5.5 shows how the chip might be typically blocked out.

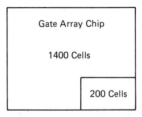

Figure 5.5 Block diagram of layout showing a circuit that uses 200 cells on the same chip as one that uses 1400 cells. (These numbers are in reality 125% of the values shown, in line with the rule of thumb that only 80% of the cells in an array are used.)

If the economics of the circuit being piggybacked later improve, it might then be feasible to separate the two designs. This can be done relatively simply on an interactive graphics system. The smaller chip might go on the 200-cell member of the IMI family and the larger circuit might go on the 1440-cell member of the IMI family. Smaller packages could then be used for each circuit. Testing routines for the two circuits would remain roughly the same except for the difference in package types and pinouts.

How much circuitry can 200 cells replace? The answer is roughly 9 to 20 MSI/SSI ICs. Replacing this number of standard parts might make the difference between using an expensive and unreliable multilayer PCB and a simple, inexpensive double-sided PCB.

If the processor of the gate array wafers has a computerized wafer test system, some improvement in yield may occur. Even though die yield decreases with increase in die size, the fact that only (using the numbers from this example) one-tenth of the chip has to work to provide a useful part that means that the effective die area will be one-tenth that of a 200-cell chip with only one design on it. Hence the yield will be considerably higher than that of the 2000-cell chip with only one design on it.

With a uniform distribution of defects across the wafer, the yield of 200 cell parts will be seven times that of the 1400 cell parts. The yield of the latter would be slightly lower than that of a 1400-cell chip all by itself. Whether it would be any higher than a 200-cell chip with only one design on it is moot. Note that for the family of gate arrays used, the die cost will not be higher because of die size. The reason is that a 2000-cell gate array would have to be used regardless of whether the 200-cell chip is piggybacked on it or not simply because there is no family member larger than 1750 (1400/80%) cells and less than 2000 cells. Moreover, the die cost may actually be lower because of the additional number of chips bought to satisfy the 200-cell requirement. The die cost will, however, be slightly higher than for the 1440-cell die because there are fewer dies per wafer of the 1960-cell gate array.

It must be stated that when more than a few simple additional functions, such as the 2-bit flip-flop counter mentioned above, are put on the chip, the design cost of the chip will increase, although not necessarily in direct proportion to that added. In summary, a few small functions can be added inexpensively, but the cost increases nonlinearly with the amount of additional circuitry added.

Use of inverters and transistor inputs. Along the same lines as the above but worthy of treatment by itself is the notion of the use of inverters and transistor inputs. A distinction is made between the two because sometimes only the latter is used (for purposes of equalizing loading). This notion applies especially to CMOS gate arrays which have more cells than do ECL gate arrays.

Inverters and transistor inputs can be used for a wide variety of functions. An important class of applications is their use as dummy loads to provide delays that track over temperature and voltage. For example, in CMOS, by "hanging" a dummy transistor input or an inverter, paralleled if necessary to achieve the desired loading, on the output of a gate, that gate can be made to have the same loading *which tracks over temperature* as that of a similar gate which actually uses the added fan-out. This technique is part of a patent that has been filed.

As a second example, suppose that one has a 16-bit bus that must be routed across the chip, and suppose further that all 16 members of the bus must track in delay to within a fraction of a nanosecond. If the user is dealing with a single-layer metal gate array, some of the bus members will be forced to go through more polysilicon underpasses than others. A "ballpark" number for a popular 5-μm process is that going through 16 poly underpasses is equivalent in delay to roughly the delay of one inverter *at room temperature* (i.e., 16 polys constitute a fan-out of about 1 for that particular gate array). (Consult the gate array manufacturer of the gate array being used at the time for the number for a given gate array.)

Therefore, when running a signal bus horizontally (i.e., in the direction perpendicular to the direction of the distributed power and ground buses), one works out a scheme before layout as to how many underpasses will be gone through before an inverter is used to reconstitute the signal. A typical scheme might be that for every eight such underpasses, one inverter will be used. Although some signals in the bus could be routed with fewer delays, extra inverters (poly underpasses) will be inserted to equalize the delays.

It must be remembered that while the delays of inverters vary with temperature, the delays of polysilicon underpasses are *largely temperature invariant*. Therefore, in groups of signals that must track over temperature, the same numbers of poly underpasses and the same number of inverters are used in each path. The designer *cannot* substitute two (to retain the signal polarity) inverters for a group of poly underpass delays in one leg but not in others, for example.

Elimination of unused functions. If a logic designer is using standard parts on a PCB and needs to add just one more inverter or NAND gate, for example, he or she must often add another IC package to the PCB. Although there are generally unused functions existing in IC packages already on the PCB which might be used, these are often in the wrong places geographically on the PCB. It is sometimes easier to add a chip than to try to "snake" (i.e., fit) an additional line on the PCB from one of these geographically unsuitable other IC packages. Note that *every* part of *all* of the IC packages *must be powered*.

The statement is often made (including in this book) that "only the parts of the gate array actually used are powered." This is strictly true only when design using transistors is done. When macros (see below) are used, the result is the inclusion of some excess circuitry.

An important difference between the gate array and design using standard parts is that in the gate array approach, *the unused functions can be eliminated if desired*. Even if one is using macros for gate array design, the unused portions can be eradicated. On an IGS, the elimination is generally relatively simple. The problem is in making sure that the remaining lines are reconnected properly. This adds to the line-checking problem, however.

For each individual circuit, a judgment must be made as to the need for (in terms of power savings, circuit density, and speed increase) and value of eliminating unused portions of macros. Bear in mind that some vendors will not allow modification of their macros under any circumstances. If the design is being done by the user organization, there is generally more flexibility.

Some vendors have computer programs that automatically eliminate unused portions of macros. An example is the Gate-Gobbler program of LSI Logic, Inc. mentioned earlier.

5.2.12 Retrofitting Existing Parts

On occasion, a user wants to try to replace an existing standard part with a gate array. The reasons are typically one or more of the following.

- The standard part is being withdrawn from the manufacturer's product line, and no other standard parts are suitable.
- A second source is needed for the standard part.
- The withdrawal of the standard part occurs late in the life cycle of the user's product. At this point, there is neither a great deal of money (because the user's product is selling at a cost that reflects both the effects of competition and the learning curve) to pay for the retrofit nor as much incentive as there would be if the product were early in the life cycle.
- The replacement gate array has to meet the "form, fit, and function" exactly of the standard part being withdrawn. This may be difficult to do from the standpoint of the pin-outs alone of the gate array.

The author's experience is generally that replacing a standard part with a gate array does not work very well for several reasons. First, the standard part being withdrawn from the market is often difficult to make. Gate arrays are generally not processed in unusual ways with processes designed to yield, for example, a very high breakdown voltage at the expense of other parameters. (Telmos does make a high-voltage breakdown gate array.)

Second, the standard part being withdrawn may be a selected version of a part that is generally of much lower quality. A case that the author reviewed involved a program's desire to replace a precision op amp. (This happened to be a potential replacement using a semicustom linear chip.) The problem was that the precision op amp was selected from a population of probably a million or more parts. The ones that tested best were simply given a different number and sold at a higher price.

Unlike the standard part, the semicustom part could not be produced in the millions. Even if it could, there was no market within the company or elsewhere for the less precise versions from which those meeting the specifications were to be selected. More correctly, there *was* a market but it could not be addressed (because of the lack of distribution and sales channels) by either the semicustom parts maker or the user organization. This is often the case.

5.3 PARTITIONING

In the course of doing a feasibility analysis, partitioning may or may not be done. It certainly will be done if the circuit is too large to fit on the gate array to be used. If no gate array has been selected, trade-offs will be done using different trial partitionings during feasibility to see if the circuit will even grossly fit on a given array and to bracket cost and time numbers for development of such a gate array and to compare these with the costs and times required using a different combination of gate arrays and standard ICs. On the other hand, if the circuit will fit entirely on a single gate array with the desired technical and economic parameters, partitioning may not be done at all.

Partitioning a circuit to permit its implementation on one or more gate arrays is a difficult process to describe. Probably no two people partition in exactly the same way. Partitioning involves a fair amount of trial and error based on the characteristics of the gate array being worked with. The gate array or the gate array family being worked with will have often been dictated by nontechnical factors. It may be that a given company has been using vendor XY's gate array family and requires that all designs be done in that family. It may be that economic constraints prohibit a gate array which has a high up-front NRE costs. Similarly, a program operating under tight time constraints may require a success on the first pass. This will prohibit an iteration of the design and implies the use of sophisticated CAD tools.

Very frequently, package size and type, allowable heat dissipation, and use or nonuse of a multilayered PCB will be dictated by the packaging engineers. Frequency of operation and allowable power supply voltages may also have been set. Within such a framework, the gate array designer still has some latitude in which to operate.

Very frequently, partitioning done during feasibility will be redone when the decision is made to proceed with implementing the circuit on a gate array. During feasibility pin and cell counts (number of cells required to implement the circuit) will have been established. (Of course, if a different gate array is used, these will change.) In any case, the focus is on several factors.

Suppose that it has been determined that a given circuit is too large to fit on a desired gate array the size of which has been dictated by other factors. The only choice here is to exclude parts of the circuit. When faced with this situation, the clever designer will try to retain for implementation on the gate array those parts of the circuit which are expected to have high usage. It may be that a "kernel" of the circuit is used in many different places in the company's products and systems. Logic designers tend to build designs from that which they have used successfully before. Portions of designs are thus often common.

Putting these common parts on a gate array, even if it means that extra outputs or a little extra circuitry must be used, will often make the gate array suitable in many applications. This, in turn, will lower the amortized cost of each part. (Recall that the supposition is that the original circuit is too big and that only a portion of it is to be implemented. Therefore, there will probably be space for a small amount of extra circuitry to make it a common part.)

A second option that might be available is to design the gate array on two gate arrays of a given family for prototyping the system. If the development time of the system itself is to be rather long, it may be that the gate array vendor will have a gate array large enough to accommodate the entire circuit by the time the system prototypes have been through evaluation and test. The additional cost and time required to put the gate array design on a single chip of the newer array may be minimal. The cost and time will depend on several factors. These are as follows:

1. The percentage of the original circuit that had to be excluded or put on two chips.
2. The gate array vendor's methods of operation. If the gate array vendor does mask design from a logic simulation program, such as Tegas, it is a reasonably simple matter to combine the files of the two chips. Knowing that both parts of the circuit function correctly by themselves is a big advantage.

The increased density of the newer and larger gate array may come about either because of a die size increase (with all of the internal cell configurations remaining the same except for being larger in number) or an increase in density due to a straightforward lithographic shrink (from, say, 5-μm feature sizes to 4-μm feature sizes). By treating the two gate array designs as two big macros, the gate array vendor may be able to combine them quite readily on an IGS. If this is to be done, the form factors of the two circuits should be made such that they can later readily be combined in this fashion. Among other things, this means that the I/Os for interconnect among the two pieces as originally laid out for the two initial gate arrays will be properly positioned at the circuit boundries and in rough apposition to one another.

Consider now a different scenario. Suppose that the circuit under consideration has a *large* block of memory or, alternatively, a microprocessor embedded in the middle of it. Current gate arrays are too small to enable either device to be implemented on the gate array and still have room for anything else. (This situation is changing, however, as newer arrays become available. See the discussion in Chapter 1.)

What often complicates the problem is that there are very wide buses going in and out of these devices. The number of I/O pins of a gate array which tries to provide the SSI/MSI "glue logic" (i.e., the nonmemory/ nonmicroprocessor logic) will often be prohibitively large. The partitioner of the circuit in this case may elect to do one or more of the following. He or she may:

1. Convert the memory segments and possibly the microprocessor segments to a bit-slice architecture. Instead of dealing with 16-bit wide buses, the user might then have to deal, for example, with only 4-bit-wide buses. This will simplify the I/O considerations.
2. Try to change the sequence of operation (if the sponsoring program permits it) in such a way that elements (the microprocessor and memory) which cannot be put on a gate array talk directly to one another without ancillary glue logic in between. The object is to provide an architecture in which the logic to be implemented on the gate array exists without the microprocessor and memory in the middle of it.

5.3.1 Incorporating Two Similar Designs into One Circuit on One Gate Array

There is another often overlooked possibility which affects both partitioning and possibly feasibility analysis. To "get the most mileage" out of the gate array design, the designer may make use of the ability to customize the chip to add functions which are useful in some applications of the chip but not in others. In effect, two separate chip designs are incorporated into *one* circuit. This is useful especially when the two designs differ by, say, less than 10 or 15%. (This technique is *not* the same as that mentioned earlier, in which two or more separate and independent designs are incorporated into different parts of the chip. In the latter case, each such circuit has its own portion of the chip. In the case discussed in this section, two circuits are molded into one circuit and one chip design.) One or more control pins determine which of the circuit functions will be used in a given application.

A benefit of incorporating two somewhat similar designs into one design is that it increases the number of parts to be bought. Even for small quantitity buys, the increased NRE to put the additional circuitry on the same chip will often be far less than the amortized costs to develop and build the two designs on two separate gate arrays. The same comments made under the discussion of "share-a-chip" apply here also.

Even with all these variables factored in, the two-circuits-on-one-chip approach is often very attractive from a cost standpoint. This is particularly true when one of the two circuits is a small circuit in a product or program which cannot afford the residual NRE if implemented on a gate array by itself.

5.3.2 Partitioning a Circuit into Multiple Gate Arrays and across Printed Circuit Board Boundaries

A factor sometimes overlooked in partitioning a circuit to be put on several gate array chips is the effect of including elements from more than one PCB on a given gate array. It is often advantageous *NOT* to have the system partitioned into gate arrays in the same way that it was partitioned into PCBs. Consideration of this factor will often greatly simplify pin-outs. When doing partitioning, the gate array designer looks across PCB boundaries to choose the circuit elements that are to go on a given gate array. By taking elements from different PCBs to go on a given gate array, a much better partitioning is sometimes effected. This is sometimes also true when a single gate array is to be used to replace only a small fraction of a total system.

For example, suppose that there are five PCBs in a system and that two gate arrays will be used to replace 85% of the circuitry. By using a partitioning which is different from that used to create the PCBs, significant savings in pin-outs may result.

One reason is that some circuit designs "evolve" over a period of time. They often contain redundant elements or elements that represent, in effect, patches to the system. These patches may have been put in the most readily available portion of the group of PCBs. Careful analysis of the system will elicit such design areas, and careful partitioning can often "clean them up."

The gate array designer doing such analysis, however, must beware that some circuits function only because certain patches slow the signals down advertently or inadvertently in such a way as to prevent race conditions. Redesign of the circuit is invariably preferable to trying to replicate the time delays of the patches.

This is an example of what is meant by the author's statement earlier that time spent "up front" in analysis of the circuit to be implemented often pays off handsomely later. An aid to picking up such problems is logic simulation using as nearly as possible the actual time delays of the chips *and* the PCB itself.

A second reason that the partitioning for the gate array may be different from that of the system of PCBs is that most PCBs are designed to accommodate standard data book chips. It may be that there is a very wide bus that goes across two or more of the PCBs. By including this on one of the gate array chips, a significant reduction in the number of I/Os can be achieved.

5.4 DISTRIBUTION OF CLOCK AND OTHER HIGH-FAN-OUT SIGNALS

Distribution of high-fan-out signals can pose significant problems to the system partitioner and layout designer or layout CAD program. In the latter case, the need to route such signals may cause the routing program to route other signals in deleterious paths. The problem is exacerbated when such signals must track each other in relative propagation delay times. An example of the latter occurs when a clock signal is divided down into a clock distribution "tree" with several paths that must have identical characteristics.

The floor plan technique mentioned in Chapter 4 is helpful in dealing with high-fan-out signals and their drivers. Examples of high-fan-out signals include clock and reset lines of flip-flops and other circuits.

A key requirement of high-fan-out drivers is that they have low source and sink impedances to minimize the times required to charge and discharge the capacitances of the lines being driven. The term "buffer" is sometimes used to denote high-drive-capability (in the preceding sense) drivers.

Generically, there are two ways in which high-fan-out signals can be distributed across the chip. Most configurations are combinations of the two.

5.4.1 Single-Driver Approach

The first is to have a single driver capable of driving the entire fan-out from a single place on the chip. Typically, this place is along the periphery, to make use of one or more of the I/O buffers. Alternatively, some gate array background structures have high-drive-capability circuits built into the middle portions of the array. A benefit of this approach is that circuitry aimed at minimizing the skew between clock and its inverse (see below) can be rather extensive because it is common to all lines that are driven.

5.4.2 Problems with the Single-Driver Approach

A problem with this approach is that the large currents that flow tend to be localized into a relatively small number of lines. Because all or a very large percentage of clock or reset lines will often switch at the same time in a circuit, very high displacement and conduction currents can occur. Because more current is flowing on a given line (than if the same current were flowing in several lines on the chip), the voltage drops will be larger on the smaller number of lines.

A second problem is that the inductive couplings among the high-current and other lines on the chip will be greater because of the larger fields generated by the larger amounts of current.

5.4.3 Tree-Driver Approach

The second approach is to use the tree structure mentioned above. In this structure a moderate-sized driver drives a few to several "tree drivers," which are all in parallel with one another. Chip line lengths to and from the drivers are made as similar as possible. Often this will be accomplished by putting each of the tree drivers at one end of a column of array cells.

Flip-flops to be driven by the clock and reset lines would be symmetrically distributed as close as possible to the dirvers in each of the columns. Alternatively, a given column of array cells might have placed on it a number of groupings of macro combinations. Each grouping might consist of the tree drivers required to drive that particular group of, say, flip-flops plus the flip-flops themselves. [This configuration would violate the "rule" about making the lines *to* the tree drivers as short as possible but would enable the output lines from the tree drivers to be made as short as possible. Moreover, the "violation" would occur on what is probably the low-current side of the tree drivers (i.e., the inputs).]

5.4.4 Typical Mistakes

Two typical mistakes are often made during design of high-fan-out signal drivers. The first is to forget that a buffer (driver) inverter is still an inverter and to get the inverse of the desired signal. This mistake often occurs when the basic circuit is changed during the design (a very bad happening but one that is sometimes necessary) in such a way that the fan-outs of clock and reset lines are increased still further.

The second is to not make all the paths have the same number of inversions to enable relay tracking. In a complex tree structure, this is not as simple as it might sound.

5.4.5 CMOS Drivers

When CMOS flip-flops and latches are driven, the problem is often more complicated. The reason is that such devices are often made of transmission gates (TG's—see Chapter 9). These require both the clock and its inverse. The fan-out to both the clock and its inverse is 4 (*each*). The fan-out of a reset signal is typically 2 per flip-flop. This means that a group of, say, 20 flip-flops would require a fan-out of 80 on both clock and clock inverse and a fan-out of 40 on the reset line.

Normally, an even number of inversions is required in the drivers to keep the polarities of the signals the same as those of the original clock or reset line. One benefit of the TG approach is that a given clock signal can sometimes be the inverse of the actual clock. This means that only one such *series* inversion needs to be present in the tree drivers. The polarity of clock and its inverse are simply switched at the flip-flops. *Caution:* This technique *cannot* be used when one edge of the clock or reset pulse (but *not* the other edge) is carefully controlled or synchronized to some other signal.

Some CMOS gate array vendors have driver circuitry built into the flip-flop macros so that only one clock line needs to be input. This is an example of *local regeneration* of the clock signal. A problem with this approach is that it is sometimes slower than other approaches. Moreover, it does not lend itself to circuitry required to minimize the skew times between the clock and its inverse. The reason is the need to keep such regeneration circuitry minimal because it is so distributed.

EXAMPLE

As an example, the IMI 71960 chip has 70 rows of three input array cells. If a given six- (normally only five are needed for a flip-flop with reset, but it is easier to talk about six) cell flip-flop is spread across two columns, then the depth of each flip flop will be three rows.

It will be seen in Chapter 9 that a triple parallel inverter takes only one cell in IMI's 71960 structure. Hence two such triple-paralleled inverters would occupy only two such array cell positions. Each would be capable of driving a fan-out of 12 with low propagation delay. It will also be shown in Chapter 9 that a master-slave D flip-flop has a fan-out of four to both the clock and its inverse. Therefore, if one of the triple-parallel inverters is used to drive the clock line and the other (positioned on the second of the two columns) is used to drive the clock inverse line, the two taken together can drive three such flip-flops.

Hence, a given grouping made up of two triple-parallel inverters and three master-slave D flip-flops will occupy a space of two columns by 10 rows in this example. This means that the 70 rows of the IMI 71960 chip can accommodate 70/10 or about seven such groupings or a total of 21 such flip-flops and their drivers on the two columns used. (Actually as noted there will be at least 21 cells left over.)

It is true that one disadvantage of this combined tree-and single-driver approach is that the upper parts of the lines see more current and hence suffer more loss than do the lower parts of the line, as noted in the discussion above.

5.5 SYSTEM PARTITIONING— RENT'S RULE

Up until now the discussion has been concerned with a single individual gate array. (The prior section discussed building a gate array from elements on different PCBs; this topic relates to partitioning the entire system.) For maximum effectiveness, however, gate arrays should be designed taking into consideration the overall system. Therefore, the greater problem is the partitioning of the *system* into gate arrays and the resultant simulation of such a system. A special-purpose piece of hardware that addresses the latter problem is the Zycad LE-1002 series discussed in Chapter 8. This series permits simulation and fault analysis of very large (up to 1,500,000 equivalent gates) circuits. (They can also be used to perform rapid simulations on much smaller circuits.) This is discussed briefly in a later chapter.

The problem of partitioning large systems into a combination of gate arrays and other elements is addressed briefly in this section. Many of the same techniques to be described below can also be used to help determine the wiring channel requirements in the design of the basic gate array chip itself (i.e., in the design of the background layers). The reader can appreciate that because a gate array is meant to accommodate an almost infinite variety of circuits as opposed to just one, the determination of the numbers of wiring channels and the pin-outs involves a certain amount of empirical judgment (otherwise known as pure guesswork).

The objects of this brief section are to make the reader aware of such efforts and to point him or her to some of the pertinent literature. The problem addressed is basically that of partitioning a large group of circuitry in the form of blocks into a series of modules taking into consideration the pin-out requirements of each block. (Other permutations of the same techniques consider also the wiring channel requirements. However, for partitioning into gate arrays the wiring channels are external to the gate array itself by virtue of being on the PCB on which the gate array is mounted.)

A block corresponds to what has been termed a macro up to this point in this book. It may be as simple as a primitive element, such as a logic gate, or as complex as the entire chip. A module is a container of such blocks. In this case, a module represents a gate array. Other names for a module used in the literature are "circuit pack" and "cluster."

The reader will appreciate that the same techniques can also be used to partition systems into PCBS (and indeed, this was their original use). Equally important, Schmidt [8] has shown that the techniques can be applied to the placement problem as well.

Logic partitioning of systems (large groups of logic) generally is aimed at one or more of the following goals:

- Minimization of numbers of clusters
- Minimization of circuit delays
- Minimization of interconnects among the clusters
- Obtaining equal-size clusters

A fair amount of work has been done on semiempirical techniques using what has become known as *Rent's rule*. This rule basically states that the number of pin-outs (total of inputs and outputs) P on a module (gate array or very large macro for purposes of this book) which is made up of G blocks of random logic is

$$P = k_1 G^{k_2} \tag{5.3}$$

where k_1 and k_2 are constant; k_1 is the average number of pins per block. As formulated by Schmidt, k_1 can be made to cancel out of the expression; however, it is not necessary for this to occur in order for the relation to be useful. The value of k_2 is typically in the range 0.5 to 0.67.

As Schmidt has pointed out, although attempts have been made to justify Rent's rule formally, the greatest justification is the very large amount of empirical data gathered from "several companies, a wide range of circuit sizes, and several technologies."

5.5.1 Applying Rent's Rule to System Partitioning into Gate Arrays

Landman and Russo [9] applied Rent's rule to the problem of determining the average number of pins per module and the average number of blocks per module. Their work utilized four circuit designs (which they call block graphs) of computers from three different design groups within IBM. The equivalent number of NOR circuits ranged from 671 to 12,700. Thus they obtained a reasonably good cross section of design methodologies.

Their work showed that relation between the number of blocks and the number of pins has two distinct regions. The number of modules determines into which region a given circuit will fall.

For large numbers of modules (the region of interest to readers of this book), k_2 was shown to be in the range 0.57 to 0.75.

As a simple application of Rent's rule, consider a circuit made up entirely of two-input, one-output gates. k_1 would then be 3. Suppose that all logic is broken into clusters composed of 80% of 2000 (two inputs/cell) gate arrays. Then $G = 1600$. Therefore, from Rent's rule, the number of pads that are likely to be used for each such gate array would be in the range 67 to 252. Most such current gate arrays have on the order of 100 or so pads in that size.

Rent's rule points out the need to go to larger macros (so that G becomes smaller) as the number of gates in the array becomes larger.

Peltzer [10] has applied Rent's rule to proving that the total number of connections per circuit decreases as the number or circuits increases. Although Peltzer's work is aimed at wafer-scale integration, there is a direct analog to that of the system partitioning problem.

5.6 WORK AIMED AT MINIMIZING THE NUMBER OF GATE ARRAYS REQUIRED BY A SYSTEM

Palesko and Akers [11] have developed algorithms to do some of the partitioning automatically. Their object is to minimize the number of gate arrays and the number of interconnects. They observe that in a typical case studied by them, a 1500-gate, 140-pin gate array has an average pin utilization of more than 90% but an internal gate utilization of only 60%. Therefore, although their object is to minimize the numbers of gate arrays to be used, they recognize that the number of interconnects among the gate arrays must also be minimized.

They divide the logic into clusters. Each cluster is initially composed of a seed element (logic function). The undistributed logic (that not used for seeds) is added to the seed elements in such a way that the clusters are "grown." The clusters in this case represent the gate arrays to be formed from the partitioning. The growing process continues until the clusters run out of gates or run out of pins. Iterative procedures are then used to effect optimization. The user can then interact with the program if desired to effect final partitioning.

The object of their work is minimize the number of gate arrays of a given size required to implement a given system. Results of tests on logic circuits ranging in size from 14,000 to 60,000 gates show that the procedure partitions logic to an average of at least 24.8% fewer gate arrays than do other commonly used algorithms.

5.7 COST AND TIME DRIVERS IN GATE ARRAYS

Although this is a book on design, a small amount of information will be given on cost and time drivers in gate arrays. The object is to try to pinpoint areas where some of the larger times and costs are incurred.

It would be somewhat impractical to try to give absolute cost and time numbers themselves. Gate arrays are a fast-moving, highly competitive technology. More competitors are entering the marketplace quite regularly. Newer tools which produce faster turnaround, greater chance of first-pass success, and sometimes lower costs after their development costs have been amortized are constantly being developed.

Newer, faster, more dense gate arrays which are more amenable to use with the better CAD tools are coming on stream. The addition of a significant new CAD tool or the entry into the market of a superaggressive competitor can radically change the pricing policies of a given group of companies. Also, no two circuits are alike in terms of technical, economic, and time parameters. For all of these reasons any absolute numbers given would be quite transitory.

Because gate array design is quite labor intensive, increased development times imply increased development costs. There are exceptions. Costs to operate expensive machines, such as VLSI testers and high-speed computers, also get into the picture. Computer costs for a high-speed [7.5 MIPS (million instructions per CPU seconds)] machine, such as an Amdahl V-8, are typically $7500 per hour. A given design might use one to three hours or more of such time, depending on the complexity of the circuit. Also, gate array wafer costs are sometimes much higher initially than those of standard parts.

5.7.1 Time Drivers

The time required to convert a *verified* logic circuit into a gate array varies widely with vendor, CAD tools available, experience of personnel doing the design (especially if done in-house), circuit complexity, completeness of specification, and testability of the circuit. The latter is the point most often overlooked by neophyte gate array developers but is a "sleeper" that can negate an otherwise fine effort. Too often, testability is addressed *after* the circuit is designed partly because testing occurs later in the process cycle and partly because there is often pressure to get to the point where the PG tape can be shipped to the mask vendor.

Much of the often quoted 12 to 20 weeks for obtaining a gate array of any degree of sophistication is queue time as opposed to actual working time. This is particularly true of the time allocated to getting the mask(s) made. Although times of two weeks from receipt of PG tape to "working plates" [the actual mask(s)] is often quoted, the author has seen cases where more than seven weeks elapsed. The actual working time for masking and etching the wafer (exclusive of checking, which will be discussed below) is less than 14 hours even for the larger chips (305 mils on a side).

5.7.2 Time to Study Computer Output

In this computerized age, a factor often forgotten or ignored is the time required for human beings to study the output of what the computer does. Although many mundane tasks have (thankfully) been relegated to the computer, human intelligence must still be used on occasion. The salient example that comes to mind is that of studying printouts from simulations which exhibit the timing characteristics of the signals.

If a given circuit is "pushing" (operating close to the limit of) one or more technology dependent parameters, more time will have to be spent analyzing under what set of conditions the circuit failed to operate properly in order to determine the *performance envelope* of the circuit. The performance envelope is the set of conditions, such as temperature, voltage, clock rate and clock skew, and processing variations, under which the simulations say the circuit operates.

In such cases, significant amounts of time can be taken answering the "Why did it fail under set of conditions number one but run perfectly well under set of conditions number two?" and "Does it run if I change the circuit design slightly?" questions.

5.7.3 Regularity of the Circuit

The more regular the circuit is, the easier it will be to lay out generally. In any layout, whether regular or not, a high percentage (of total available cells) of cell utilization in conjunction with a large number of signal paths will increase layout time and costs. This is especially true when there is a large number of paths that must be routed over a distance of more than a few cells in length. This is generally true even when using current autorouting programs. The higher the cell utilization, the greater will be the percentage of lines that must be hand routed. The percentage will be smaller for some circuits than for other circuits with the same cell utilization, due to the differences among the circuits.

5.7.4 Adequacy of the Macro Library [12]

An important subject that must be addressed during the feasibility analysis is the adequacy of the macro libraries of the vendors under consideration. This can greatly affect the time and costs to do the design. As noted earlier, when time and cost outweigh performance, design with macros is preferable to design using transistors. It is both faster and less risky.

However, unless a given vendor's library contains roughly 100 (the writer's number but generally corroborated) separate macro cells *exclusive of permutations,* the chances are fairly high that some design with transistors will be needed. Permutations include, for example, the same function which is paralleled to increase drive capability, the same function that has a layout which is the reverse of its mate or which has a different form factor, and so on.

The chances of having to use transistor design are especially great when the frequency limits of the technology involved are being "pushed." In this case, the increased size of macros containing functions not being used for that specific design can often not be tolerated. Repeated use of such elements forces often critical chip line lengths to be longer than necessary. Ironically, even in this computer age, a human mind can sometimes find ways to "snake a line" in an area of the chip already crowded with lines better and faster than can a computer. This statement is generally not true for a multiplicity of such lines.

5.8 SUMMARY

The first step in a gate array design is assessing the feasibility. In this chapter the important parameters of propagation delays, temperature and power effects, pin-outs, drive requirements, and so on, were discussed. If analysis indicates a "go" situation, partitioning is probably the next step, although some or much of this may have been done in the feasibility phase. System partitioning was also discussed. Note was made of important work at Honeywell and Arizona State University. Use of Rent's rule was mentioned.

REFERENCES

1. Cole, B., "How gate arrays are keeping ahead," *Electronics,* Sept. 23, 1985, pp. 48–52.
2. Carter, D., and D. Guise, "Effects of interconnections on submicron chip performance," *VLSI Design,* Jan. 1984, pp. 63–68.
3. Carter, D., and D. Guise, "Analysis of signal propagation delays and chip level performance due to on-chip interconnections," International Conf. on Computer Design: VLSI in Computers, Port Chester, NY, Nov. 1983.
4. Wakeman, L., "Transmission line effects influence high speed CMOS," *EDN,* June 14, 1984, pp. 171–77.
5. Sinha, A., J. Cooper, and H. Levenstein, "Speed limitations due to interconnect time constants in VLSI integrated circuits," *IEEE Electron Devices Letters,* April 1982, pp. 90–92.
6. Rubenstein, J., P. Penfield, and M. Horowitz, "Signal delay in RC tree networks," *IEEE Trans. on Computer-Aided Design,* July 1983, pp. 202–11.
7. Lin, T., and C. Mead, "Signal delay in general RC networks with application to timing simulation of digital integrated circuits," 1984 Conf. on Advanced Research in VLSI, MIT, Cambridge, MA, Jan. 24, 1984.
8. Schmidt, D. C., "Circuit pack parameter estimation using Rent's rule," *IEEE Trans. CAD Integrated Circuits and Systems,* Oct. 1982, pp. 186–92.

9. Landman B. S., and R. L. Russo, "On a pin vs. block relationship for partitions of logic graphs," *IEEE Trans. Computers,* Dec., 1971, pp. 1469–79.
10. Peltzer, D., "Wafer scale integration: the limits of VLSI?", *VLSI Design,* Sept. 1983, pp. 43–47.
11. Palesko, C., and L. Akers, "Logic partitioning for minimizing gate arrays," *IEEE Trans. CAD Integrated Circuits and Systems;* April 1983, pp. 117–20.
12. Gabay, J., "Unlocking the mysteries of CAD/CAE," *Electronic Products,* Mar. 3, 1986, pp. 48–53.

Chapter 6

CAD for Gate Arrays

6.0 INTRODUCTION

It is not true that CAD (computer-aided design) tools *must* be used for gate array development. Relatively small, inexpensive gate arrays can be designed (laid out in this context) using only a pencil and a simple vendor-supplied layout sheet. Pasties (mentioned in Section 4.2.2) can be used to speed this process and to make it more reliable. Nevertheless, almost every gate array designed today makes use of CAD in one form or another; and this trend is accelerating at a very rapid pace. Moreover, the place of smaller gate arrays is being taken by PALs (programmable array logics) in some cases.

Chapter 4 gave the overall flow of gate array design and showed how the various CAD tools interrelate to one another in the gate array implementation process. The purpose of this chapter is to describe in detail three important CAD programs that are pertinent to the gate array methodology. The three are schematic entry, circuit simulation, and logic simulation. Also included is a brief mention of layout verification tools.

The weighting of each topic is roughly proportional to a combination of the extent of use, the useful (for gate array design) options available to the user, and the author's estimate of the degree of explanation required. Some of the tools to be described in this and subsequent chapters have many options that are useful for other than gate arrays. These other options will not be described.

Subsequent chapters will describe additional CAD tools. As in previous chapters, an overview is first given followed by a more detailed discussion of each tool.

The intent of this chapter is to illustrate the capabilities of the different programs. It is *not* to try to teach the reader how to use them. Nevertheless, it is instructive and sometimes necessary to show some of the detailed syntaxes in order for the reader to appreciate some of the subtleties of the programs.

A few printouts of actual results are included to give the reader a feeling for the pertinent programs. The most important parts of such printouts are explained in the text.

The reader will appreciate that one or more books could be devoted to each program in each category. Therefore, it is necessary to judiciously filter the information available. The method of presentation is first to describe the CAD classifications (e.g., logic simulators) and then to focus in on a particular member or sometimes members of that class.

At the time of this writing, CAD tools for VLSI, including those for gate arrays, are in a state of dramatic change. The number of vendors and their program offerings in this area rivals the number of new gate arrays being offered by the gate array manufacturers. By even very conservative counts, there are well over 100 such programs.

The large number of programs gives the potential user a very significant problem of selecting those which are most suitable for his or her needs. Compounding the problem is the cost in both dollars and time of the effort required to understand a given program to the point where a true judgment can be made of its usefulness *in the given user's environment.*

6.1 CAD TOOLS MOST NEEDED AND USED

CAD tools most needed by and used for gate array development are listed below. Many of these are also used for other purposes independently of the gate array methodology. For example, logic simulators can be used to verify logic regardless of whether the logic is to be implemented on a PCB (printed circuit board), gate array chip, or full custom chip.

- Schematic entry (includes netlist extraction and nested module expansion programs)
- Testability analysis programs; see Chapter 7
- Logic and behavioral simulators [1], [2]
- Auto place and route (P&R)
- Manual P&R using "correct by construction" techniques
- Circuit simulation
- Test vector generation and analysis (may be part of logic simulation)
- Layout checking
    Connectivity
    Design rule verification
    Layout extraction and comparison to original schematic

- PG (pattern generator) generation (converts layout to reticle used to make mask)

6.2 SIMULATORS

If coding of the data base with the network interconnects cannot be done with schematic entry, it will have to be manually encoded for simulation (logic verification), testability analysis, and test vector generation and analysis to take place. In this chapter, coding of the simulator program with the netlist and logic simulation of the resultant circuit are covered. In Chapter 7, testability analysis, and test vector generation and analysis are covered.

Although the normal procedure would be to do a testability analysis on the circuit after encoding and before extensive logic simulation, the subjects of testability analysis and test vector generation and analysis are so interrelated that they need to be put together under the topic of "design for testability."

6.2.1 Overview

Several types of simulators are used in VLSI and PCB design. These are circuit, logic, register transfer, behavioral or functional, and mixed mode or multilevel. The last is a combination of two of the others. An example of a multilevel simulator is the Helix simulator of Silvar-Lisco. This simulator will run a simulation on a circuit that is partly described behaviorally and partly (or wholly) described at the gate or other macro level. This permits a circuit to be simulated to verify the architecture, for example, before effort is spent designing the internal structure of all the blocks.

Behavioral/functional simulators are sometimes called hierarchical simulators (although the latter word is becoming badly abused as a popular buzzword). They permit a circuit to be described in terms of the I/Os and/or opcodes of the instruction set if the device is a microprocessor. Register transfer simulators also allow this feature. However, register transfer simulators synthesize the logic described by the opcodes or other inputs from registers. An example of a popular register transfer-level simulator is the TML simulator of Terradyne, which is part of their LASAR simulation system.

6.2.2 Logic Simulators versus Circuit Simulators

The distinction between these two topics can often be "fuzzy." The purpose of this section is to contrast the two to avoid this problem at the outset.

The purpose of logic simulators for purposes of this book is to verify that the logic is correct before a gate array chip is built. Logic simulators simulate gates and functional elements such a flip-flops and registers to predict the output of a circuit statically (truth table) or dynamically as a function of time.

By contrast, a circuit simulator, such as SPICE or TRACAP, uses the sizes and characteristics of the active (transistors, diodes, etc.) and passive (underpasses, resistors, capacitors, etc.) devices. It, too, gives outputs at given nodes as a function of time. These outputs can be used to derive the maximum and minimum propagation times as a function of voltage, temperature, fan-out (load capacitance, source and sink current, etc.) and other parameters. *The maximum, minimum, and nominal propagation times used in logic simulators often come from circuit simulators.*

Both logic and circuit simulators can plot node signals as a function of a selected time interval in response to a set of time-varying input signals. The signals from a circuit simulator display the actual rise and fall times of the nodes under consideration. By contrast, a logic simulator neither uses nor displays rise and fall times. All transitions of most logic simulators are regarded as instantaneous.

The plot using a logic simulator, such as Tegas, can show not only the high and low transitions and the transitions to and from high impedance (three-state) but also the points in time where potential race and hazard conditions exist. These are shown as letter symbols, and the user must use ancillary information to determine if a true race or hazard condition does indeed exist. Warning messages about spikes being produced and the times at which they are produced are also given by the logic simulator. A circuit simulator under the same conditions will show the actual time-varying waveform.

6.2.3 Typical Logic Simulators

A list of typical logic simulators is given in Table 6.1. Some of these, such as Rockwell's Simstran and its companion Logtrans programs, use Boolean expressions to represent the gates. For example, a two-input NOR gate with inputs A and B would be denoted by (A + B)−, where the − represents the inversion. Others, such as Tegas, use either Boolean expressions or primitives (gates, etc.) to denote the functions.

TABLE 6.1 LOGIC SIMULATORS

Simulator name	Source
Tegas5	GE/CALMA, Austin, Tex.
Texsim	GE/CALMA, Austin, Tex.
Logcap	Phoenix Data Systems, Albany, N.Y.
Simstran/Logtrans (in-house)	North American Rockwell, Anaheim, Calif.
Logsim	ERADCOM/RCA, Camden, N.J.
HI-LO2	GenRad, Concord, Mass.; also DEC and Cirrus
HI-LO3	
Lasar 5 and 6	Terradyne, Boston, Mass.
SILOS	SILOS, Incline Village, Nev.
Logis	ISD
Cadat	HHB-Softron, Mahwah, N.J.

Some of these, such as Texsim B and HI-LO 2, are "mixed-mode" (see later in this chapter) simulators. A comparison between HI-LO 2 and Cadat is given in Table 6.2. *Caution:* Because of the rapid change and intense development efforts

TABLE 6.2 HI-LO 2, Cadat Comparison

	HI-LO 2	Cadat
Primary use (to date)	PCB	IC
Number of logic states modeled	4	12
Number of primitives supported	15	90
Number of TTL devices in libraries	1000+	300
Automatic test generation	Yes	No
Fault collapsing	No	Yes
Functional modelling language	Yes	Yes (user can only write "C" program)
Timing assertion support	Yes	No
Interactive control	Yes	No
Checkpoint restart	Yes	No
Vectors of signals	Yes	No
Bidirectional gate modeling	Yes	Yes

Courtesy of CAE, Inc. Features are subject to change. Consult vendors for most current information.

in this highly competitive environment, the reader should *always* consult the manufacturers for up-to-date information. Helix from Silvar-Lisco is also a mixed-mode simulator but tends to be used primarily as a higher level simulator.

6.3 CIRCUIT SIMULATION

The differences between logic simulation and circuit simulation were discussed in Section 6.2.2. The purpose of this section is to give the reader a feeling for one of the more widely used circuit simulators, the SPICE circuit simulator. This originated at the University of California at Berkeley and has been incorporated into a huge variety of CAD programs and tools, including some of the current workstations, such as Mentor and Daisy. It is also used in the Dracula layout verification system of ECAD (see below).

The SPICE circuit simulator can be used to model the parameters of many kinds of active and passive circuits, including those of pure custom LSI. The current version is 2G.6.

6.3.1 Uses and Problems of Circuit Simulation

In gate array work, circuit simulation is used primarily to characterize macros. Usually, this is done by the gate array vendor because the vendor is the one who knows best how to "tweak" the individual parameters of the model to make the results agree with what is seen over a period of time. Occasionally, a user organization that designs its own macros will use SPICE for such characterization.

Because SPICE parameters involve process information that is generally highly proprietary to a given vendor, however, such parameters are often very hard to get. Unless a user is tightly "coupled" to a given gate array vendor, it is generally not privy to such information.

Even when the proper "tweak factors" are not available from the gate array vendor, running a circuit simulator is sometimes useful to find out whether a given trial circuit implementation has merit. This can be done by a comparison of the *relative* propagation delays of the old and new implementations.

The simulator will be illustrated using parameters for CMOS. SPICE also has models for bipolar. Because as noted, SPICE parameters are proprietary, those given in this section do not represent the process of an actual gate array vendor. Nevertheless, they are indicative to some degree of actual processes in the current range of feature sizes. *Do not use them for design.*

The syntax used below is a small subset of the possibilities available with SPICE.

6.3.2 Coding and Running the SPICE Simulator

The purpose of this section is not to teach the reader how to use SPICE but rather to indicate some of the parts of a circuit simulation program and how they can be used. The reader should see Chapter 2 for background information.

The SPICE FET model is given in Fig. 6.1. The meaning of the parameters is given in Table 6.3.

The reader will note that some of the 42 parameters listed are interrelated. The parameters used in this section are listed in Fig. 6.2.

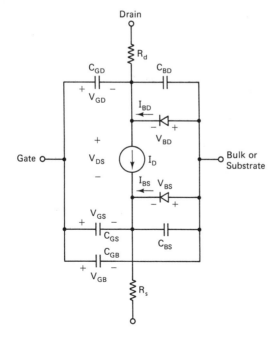

Figure 6.1 SPICE FET model.

TABLE 6.3 PARAMETERS OF SPICE FET MODEL

	Name	Parameter	Units	Default
1	LEVEL	Model index	—	1
2	VTO	Zero-bias threshold voltage	V	0.0
3	KP	Transconductance parameter	A/V^2	2.0E-5
4	GAMMA	Bulk threshold parameter	$V^{0.5}$	0.0
5	PHI	Surface potential	V	0.6
6	LAMBDA	Channel-length modulation (MOS1 and MOS2 only)	1V	0.0
7	RD	Drain ohmic resistance	Ω	0.0
8	RS	Source ohmic resistance	Ω	0.0
9	CBD	Zero-bias B-D junction capacitance	F	0.0
10	CBS	Zero-bias B-S junction capacitance	F	0.0
11	IS	Bulk junction saturation current	A	1.0E-14
12	PB	Bulk junction potential	V	0.8
13	CGSO	Gate–source overlap capacitance per meter channel width	F/m	0.0
14	CGDO	Gate–drain overlap capacitance per meter channel width	F/m	0.0
15	CGBO	Gate–bulk overlap capacitance per meter channel length	F/m	
16	RSH	Drain and source diffusion sheet resistance	Ω/square	0.0
17	CJ	Zero-bias bulk junction bottom capacitance per square meter of junction area	F/m^2	0.0
18	EJ	Bulk junction bottom grading coefficient	—	0.5
19	CJSW	Zero-bias bulk junction sidewall capacitance per meter of junction perimeter	F/m	0.0
20	MJSW	Bulk junction sidewall grading coefficient	—	0.33
21	JS	Bulk junction saturation current per square meter of junction area	A/m^2	
22	TOX	Oxide thickness	M	1.0E-7
23	NSUB	Substrate doping	$1/cm^3$	0.0
24	NSS	Surface state density	$1/cm^2$	0.0
25	NFS	Fast surface state density	$1/cm^2$	0.0
26	Tpg	Type of gate material: +1 opposite of substrate −1 same as substrate 0 Al gate	—	1.0
27	XJ	Metallurgical junction depth	meter	0.0
28	LD	Lateral diffusion	meter	0.0
29	UO	Surface mobility	cm^2/V-S	600
30	UCRIT	Critical field for mobility degradation (MOS2 only)	V/cm	1.0E4
31	UEXP	Critical field exponent in mobility degradation (MOS2 only)	—	0.0
32	UTRA	Transverse field coef (mobility) (MOS2 only)	—	0.0
33	VMAX	Maximum drift velocity of carriers	m/s	0.0

	Name	Parameter	Units	Default
34	NEFF	Total channel charge (fixed and mobile) coefficient (MOS2 only)	—	1.0
35	KF	Flicker noise coefficient	—	0.0
36	AF	Flicker noise exponent	—	1.0
37	FC	Coefficient for forward-bias depletion capacitance formula	—	0.5
38	ETA	Static feedback (MOS3 only)	—	0.0
39	DELTA	Width effect on threshold voltage (MOS3 only)	—	
40	THETA	Mobility modulation (MOS3 only)	1/V	0.0
41	KAPPA	Saturation field factor (MOS3 only)	—	1.0

Source: A. Vladimirescu, "The Simulation of MOS Integrated Circuits Using SPICE2," ERL Memo No. ERL-M, Electronics Research Laboratory, University of California, Berkeley, Jan. 1980.

```
*************************   SPICE 2G.6   *************************

ONE INVERTER DRIVING THREE INVERTERS AS LOADS

****      MOSFET MODEL PARAMETERS           TEMPERATURE = 27.000 DEG C

******************************************************************

            N.X1      P.X1      N.X2      P.X2      N.X3      P.X3      N.X4

TYPE        NMOS      PMOS      NMOS      PMOS      NMOS      PMOS      NMOS

LEVEL       2.000     2.000     2.000     2.000     2.000     2.000     2.000

VTO         0.800    -0.950     0.800    -0.950     0.800    -0.950     0.800

KP          9.50D-06  4.75D-06  9.50D-06  4.75D-06  9.50D-06  4.75D-06  9.50D-06

GAMMA       1.700     0.900     1.700     0.900     1.700     0.900     1.700

PHI         0.740     0.640     0.740     0.640     0.740     0.640     0.740

LAMBDA      2.57D-05  5.79D+00  2.57D-05  5.79D+00  2.57D-05  5.79D+00  2.57D-05

RD          1.25D-02  9.00D-02  1.25D-02  9.00D-02  1.25D-02  9.00D-02  1.25D-02

RS          1.25D-02  9.00D-02  1.25D-02  9.00D-02  1.25D-02  9.00D-02  1.25D-02

CBD         9.17D-11  3.76D-11  9.17D-11  3.76D-11  9.17D-11  3.76D-11  9.17D-11

PB          0.950     0.850     0.950     0.850     0.950     0.850     0.950
```

Figure 6.2 Parameters used.

CGSO	4.20D-10	3.36D-10	4.20D-10	3.36D-10	4.20D-10	3.36D-10	4.20D-10
CGDO	4.20D-10	3.36D-10	4.20D-10	3.36D-10	4.20D-10	3.36D-10	4.20D-10
CJ	4.13D-04	1.94D-04	4.13D-04	1.94D-04	4.13D-04	1.94D-04	4.13D-04
TOX	7.15D-08	7.15D-08	7.15D-08	7.15D-08	7.15D-08	7.15D-08	7.15D-08
NSUB	1.95D+16	3.87D+15	1.95D+16	3.87D+15	1.95D+16	3.87D+15	1.95D+16
TPG	1.000	1.000	1.000	1.000	1.000	1.000	1.000
LD	4.00D-07	6.50D-07	4.00D-07	6.50D-07	4.00D-07	6.50D-07	4.00D-07
UCRIT	3.33D+04	3.33D+04	3.33D+04	3.33D+04	3.33D+04	3.33D+04	3.33D+04

Figure 6.2 (*cont.*)

Numbering the nodes. The user's first task is to number the nodes of the circuit. In the case of MOS transistors, such as CMOS, this includes the sources, drains, and gates, and if back gate bias is used, the substrate as well. (See Chapter 2 for a definition of these elements.) There will be one node number for the interconnection of, for example, the output of a CMOS device (drain of NFET connected to the drain of a PFET) and the gate of that which is being driven by the PFET/NFET combination (see Fig. 6.3a. A second example is given in Fig. 6.3b.).

Coding the program (writing the syntax). An example of coding a very simple circuit is given in Fig. 6.4. It basically consists of several parts. The first part is a header that identifies the program. The remaining parts follow in order and give the program the following information:

1. Input voltages and signals. See below.
2. Temperature (will default to 27°C if value is not input).
3. Parameters of the FET models (will default to internal values of values that are not input). A "level" statement is embedded in the MOS model. For MOS, SPICE has three levels, which vary in sophistication, which it can run. These differ in formulation of the I–V characteristic.
4. Parameters and hookup (node connections) of the individual FETs that make up the circuits and subcircuits, if used. (A subcircuit corresponds to the nested module mentioned in Chapter 4.) Each FET is given an occurrence name, such as M1. This is followed by four numbers, which are, respectively, the node connections of the drain, gate, source, and body. This is followed by the model name to be used for that individual FET. On most CMOS chips, gate array or non-gate array, there will be two models, one for the NFETs, the other for the PFETs. The next values are the sizes of the individual FETs: length, L; width, W; and the areas of the drain and source, AD and AS. Defaults are available. AS will default to AD, for example. If it were desired to know the current or the power of a given FET, an I or a P would be input here also.

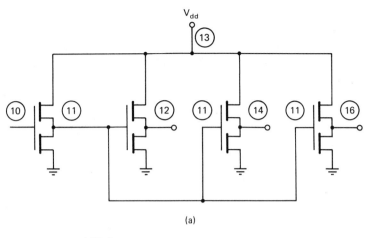

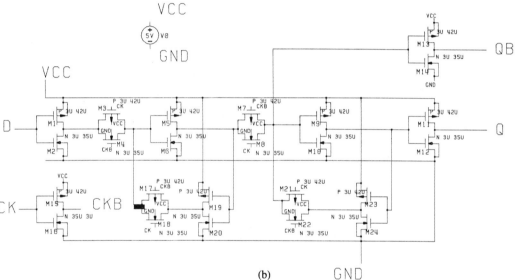

Figure 6.3 (a) Node numbering of circuit (nodes are encircled), (b) another example. (Courtesy of Mentor Graphics, Inc.)

```
************************** SPICE 2G.6 **************************

ONE INVERTER DRIVING THREE INVERTERS AS LOADS

****      INPUT LISTING                TEMPERATURE = 27.000 DEG C

*****************************************************************
*
* THIS IS INV3.IN WHICH CONNECTS THE OUTPUT TO 3 LOADS EQUAL TO THE ORIGINAL
```

Figure 6.4 Coding of simple CMOS circuit in SPICE.

```
** CIRCUIT
*
*V(1) 1 0
*
VDD 13 0 5
VIN 10 0  PULSE(0 5  0NS 0NS 0NS 15NS 30NS)
*
*
*
.SUBCKT   INV1 1 2 3
*              NODES ARE INP OUTP VDD

M1 2 1 3 3 P L=4U W=36U AD=13.8P
M2 2 1 0 0 N L=4U W=36U AD=13.8P
*
**************************************************************************
*TYPICAL VALUES
.MODEL N NMOS (VTO=0.8 GAMMA=1.7 PHI=0.74 LAMBDA=2.57E-5 UCRIT=3.33E4
+TOX=715E-10 NSUB=1.95E16 RD=1.25E-2 RS=1.25E-2 CBD=9.175E-11 CGSO=4.2E-10
+CGDO=4.2E-10 PB=0.95 LD=0.4E-6 KP=9.50E-6 LEVEL=2
.MODEL P PMOS (VTO=-0.95 GAMMA=0.9 PHI=0.64 LAMBDA=5.7865-5 UCRIT=3.33E4
+TOX=715E-10 NSUB=3.87E15 RD=9E-2 RS=9E-2 CBD=3.76E-11 CGSO=3.36E-10
+CGDO=3.36E-10 PB=0.85 LD=0.65E-6 KP=4.75E-6 LEVEL=2)
**************************************************************************

.ENDS INV1
*
**
* OUTER CIRCUIT
X1 10 11 13 INV1
X2 11 12 13 INV1
X3 11 14 13 INV1
X4 11 16 13 INV1
*           NOTE: ORDER HAS TO BE SAME AS IN SUBCKT DEFTN
*
*
*
***
*
.WIDTH IN=80 OUT=80
*
*
*.DC VIN 1 5 .1
*.PRINT DC V(1) V(2)
*.PLOT DC V(1) V(2)
*
*
.TRAN 1NS 30NS
.PRINT TRAN  V(10) V(11) V(12) V(14) V(16)
.PLOT TRAN   V(10) V(11) V(12) V(14) V(16)
*
*
.END CIRC
```

Figure 6.4 (*cont.*)

For gate arrays, all of the internal array NFETs will generally have the same W and L values. A similar statement holds for the PFETs. The buffer FETs will have different W values from either the array NFETs or PFETs. There may also be special-purpose FETs on the chip (such as those specifically placed to drive the buffers from the internal array FETs). These will have other values of W. [The length is generally that of the minimum feature size (e.g., 2 μm).]

5. Makeup of the outer circuits from the subcircuits if the latter are used. In this simple example, the same subcircuit, INV1A, is used four different times, as shown in Fig. 6.3. The output of the first inverter fans out to three other replications. Note that the drains of the PFETs in all subcircuits connect to V_{dd}, node 13. Node 11, which is the output of the first subcircuit, is the gate input of the other three. Note that the node numbers of the outer circuit do not have to correspond to those of the inner circuit with regard to node number. *They must correspond, however, with respect to order of inputs* (i.e., the order stated in the comment of the subcircuit definition, in this example).

   Note that combinations of FETs and subcircuits can be used to make up the outer circuit. This would be an example of nonuniform nesting mentioned elsewhere. Note also that the nesting can be multiple levels deep (i.e., a subcircuit can call another subcircuit which is inside it, etc.). Doing this, however, can increase computer CPU time under some conditions.

6. Type of analysis. DC, AC, TRAN (for transient), etc. The numbers that follow are the step, stop, and, if desired, start and maximum times. The timing increment defaults to 1 if no value is given.

7. Output instructions. What signals should be printed and/or plotted. Figures 6.5 and 6.6 give examples. The instructions are given in Fig. 6.4. The V stands for voltage, (I for current could also have been specified) and the numbers in parentheses are the node numbers.

8. Minor formatting instructions, such as the width of paper in terms of numbers of columns.

Comments form an important part of the documentation of the program and can be placed anywhere within the program. They are preceded by an asterisk.

Input signals and voltages. The first step in coding the program is to define the input signals, including the power supply voltages. In the usual case of CMOS operating with just one voltage on the entire chip, there will be a single supply voltage, V_{dd}. Ground is denoted by node 0. Hence, the term for the power supply voltage is VDD 13 0 5, where 13 is the node (in this example) where power is attached to the chip or subcircuit.

In the case of a gate array, the entire V_{dd} bus would be labeled node 13 (for the example used, the node number is at the discretion of the user). The 5.0 is the value of the power supply voltage for the given run. Separate runs could be made with V_{dd} at other values, such as 4.5 V and 5.5 V. Any source such as that of a logic gate, for example, which connects to V_{dd} will have the node number 26 in this example.

```
****************** SPICE 2G.6 ******************

ONE INVERTER DRIVING THREE INVERTERS AS LOADS

****    OPERATING POINT INFORMATION    TEMPERATURE = 27.000 DEG C
*************************************************************

**** MOSFETS

             M1.X1      M2.X1      M1.X2      M2.X2      M1.X3      M2.X3      M1.X4
MODEL        P.X1       N.X1       P.X2       N.X2       P.X3       N.X3       P.X4
ID        -6.93E-12   1.93E-12  -1.93E-12   6.93E-12  -1.93E-12   6.93E-12  -1.93E-12
VGS          -5.000      0.000      0.000      5.000      0.000      5.000      0.000
VDS           0.000      5.000     -5.000      0.000     -5.000      0.000     -5.000
VBS           0.000      0.000      0.000      0.000      0.000      0.000      0.000

             M2.X4
MODEL        N.X4
ID         6.93E-12
VGS          5.000
VDS          0.000
VBS          0.000
```

Figure 6.5 Tabular output of SPICE.

*************************** SPICE 2G.6 ***************************

ONE INVERTER DRIVING THREE INVERTERS AS LOADS

****   TRANSIENT ANALYSIS   TEMPERATURE = 27.000 DEG C

****

TIME	V(10)	V(11)	V(12)	V(14)	V(16)
0.000E+00	0.000E+00	5.000E+00	1.545E-08	1.545E-08	1.545E-08
1.000E-09	5.000E+00	5.025E+00	-5.788E-02	-5.788E-02	-5.788E-02
2.000E-09	5.000E+00	3.939E+00	1.758E-01	1.758E-01	1.758E-01
3.000E-09	5.000E+00	3.609E+00	3.981E+00	3.981E+00	3.981E+00
4.000E-09	5.000E+00	2.526E+00	4.819E+00	4.819E+00	4.819E+00
5.000E-09	5.000E+00	1.451E+00	4.879E+00	4.879E+00	4.879E+00
6.000E-09	5.000E+00	7.110E-01	4.952E+00	4.952E+00	4.952E+00
7.000E-09	5.000E+00	3.022E-01	4.976E+00	4.976E+00	4.976E+00
9.000E-09	5.000E+00	5.759E-02	4.989E+00	4.989E+00	4.989E+00
1.000E-08	5.000E+00	2.525E-02	4.995E+00	4.995E+00	4.995E+00
1.100E-08	5.000E+00	1.123E-02	4.998E+00	4.998E+00	4.998E+00
1.200E-08	5.000E+00	5.040E-03	4.999E+00	4.999E+00	4.999E+00
1.300E-08	5.000E+00	2.201E-03	5.000E+00	5.000E+00	5.000E+00
1.400E-08	5.000E+00	9.756E-04	5.000E+00	5.000E+00	5.000E+00
1.500E-08	5.000E+00	4.374E-04	5.000E+00	5.000E+00	5.000E+00
1.600E-08	5.000E+00	1.881E-04	5.000E+00	5.000E+00	5.000E+00
1.700E-08	0.000E+00	4.771E+00	6.497E-01	6.497E-01	6.497E-01
1.800E-08	0.000E+00	4.928E+00	8.885E-03	8.885E-03	8.885E-03
1.900E-08	0.000E+00	4.964E+00	2.566E-03	2.566E-03	2.566E-03
2.000E-08	0.000E+00	4.980E+00	1.223E-03	1.223E-03	1.223E-03
2.100E-08	0.000E+00	4.988E+00	8.056E-04	8.056E-04	8.056E-04
2.200E-08	0.000E+00	4.993E+00	4.850E-04	4.850E-04	4.850E-04
2.300E-08	0.000E+00	4.996E+00	2.863E-04	2.863E-04	2.863E-04
2.400E-08	0.000E+00	4.997E+00	1.707E-04	1.707E-04	1.707E-04
2.500E-08	0.000E+00	4.998E+00	1.036E-04	1.036E-04	1.036E-04
2.600E-08	0.000E+00	4.999E+00	6.307E-05	6.307E-05	6.307E-05
2.700E-08	0.000E+00	4.999E+00	3.930E-05	3.930E-05	3.930E-05
2.800E-08	0.000E+00	5.000E+00	2.425E-05	2.425E-05	2.425E-05
2.900E-08	0.000E+00	5.000E+00	1.485E-05	1.485E-05	1.485E-05
3.000E-08	0.000E+00	5.000E+00	9.169E-06	9.169E-06	9.169E-06

*************************** SPICE 2G.6 ***************************

Figure 6.5 (*cont.*)

```
ONE INVERTER DRIVING THREE INVERTERS AS LOADS

****         TRANSIENT ANALYSIS              TEMPERATURE =   27.000 DEG C

****************************************************************

LEGEND:

*:  V(10)
+:  V(11)
=:  V(12)
$:  V(14)
O:  V(16)

    TIME       V(10)

*+)---------   0.000D+00      2.000D+00      4.000D+00      6.000D+00      8.000D+00
               - - - - - - - - - - - - - - - - - - - - - - - - - - - - - - - - - -
=$O)---------  -2.000D+00     0.000D+00      2.000D+00      4.000D+00      6.000D+00
               - - - - - - - - - - - - - - - - - - - - - - - - - - - - - - - - - -
0.000D+00   0.000D+00  *            X                .         +         .                  .
1.000D-09   5.000D+00  .            X                .         X         .                  .
2.000D-09   5.000D+00  .           .X                +         *         .                  .
3.000D-09   5.000D+00  .            .         +      .         *         X                  .
4.000D-09   5.000D+00  .            .   +           .          *         .         X        .
5.000D-09   5.000D+00  .      +     .                .         *         .         X        .
6.000D-09   5.000D+00  .   +        .                .         *         .         X        .
7.000D-09   5.000D+00  . +          .                .         *         .         X        .
8.000D-09   5.000D+00  .+           .                .         *         .         X        .
9.000D-09   5.000D+00  +            .                .         *         .         X        .
1.000D-08   5.000D+00  +            .                .         *         .         X        .
1.100D-08   5.000D+00  +            .                .         *         .         X        .
1.200D-08   5.000D+00  +            .                .         *         .         X        .
1.300D-08   5.000D+00  +            .                .         *         .         X        .
1.400D-08   5.000D+00  +            .                .         *         .         X        .
1.500D-08   5.000D+00  +            .                .         *         .         X        .
1.600D-08   5.000D+00  +            .                .         *         .         X        .
1.700D-08   0.000D+00  *            .   X            .         +         .                  .
1.800D-08   0.000D+00  *            X                .         +         .                  .
1.900D-08   0.000D+00  *            X                .         +         .                  .
2.000D-08   0.000D+00  *            X                .         +         .                  .
2.100D-08   0.000D+00  *            X                .         +         .                  .
2.200D-08   0.000D+00  *            X                .         +         .                  .
2.300D-08   0.000D+00  *            X                .         +         .                  .
2.400D-08   0.000D+00  *            X                .         +         .                  .
2.500D-08   0.000D+00  *            X                .         +         .                  .
2.600D-08   0.000D+00  *            X                .         +         .                  .
2.700D-08   0.000D+00  *            X                .         +         .                  .
2.800D-08   0.000D+00  *            X                .         +         .                  .
2.900D-08   0.000D+00  *            X                .         +         .                  .
3.000D-08   0.000D+00  *            X                .         +         .                  .
               - - - - - - - - - - - - - - - - - - - - - - - - - - - - - - - - - -
                                             (a)
```

Figure 6.6 (a) Plot of voltages versus time; (b), (c) examples of SPICE outputs from Mentor workstation. (Courtesy of Mentor Graphics, Inc.)

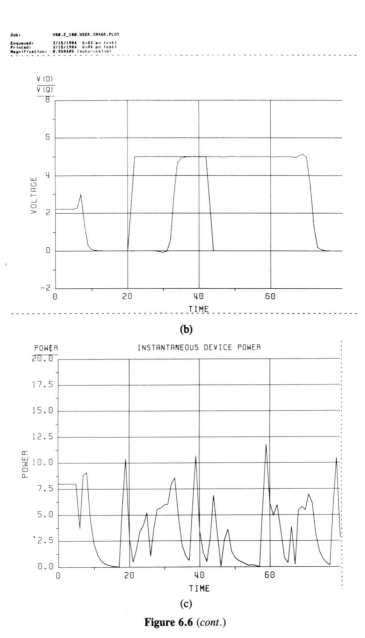

Figure 6.6 (*cont.*)

Signal Inputs. Time-variable input signals, such as V_{in} (for input signal), are denoted in the form

VIN 10 0 PULSE (05 ONS ONS 15NS 3ONS)

where 10 and 0 (ground = V_{ss}) are nodes between which the signal is input. The values in parentheses are, respectively, the initial and pulsed values of voltage (0 and 5 in this case) delay, rise, fall times (all zero in the example), pulse width, and pulse period (15 and 3ons, respectively).

Models of active and passive devices. The next step is to give the models of the active and passive devices being used. There will be one or more models for the NFETs and one or more for the PFETs (plus one for each type of diode, Schottky, zener, regular, etc., if bipolar or other process were being used). There is also one for each *separate* type of resistor (one whose Ω/square differ, not one whose resistor value differs).

In short, wherever a *process* relating to a device type is different from other devices of the same class (such as transistor or diode), a model must be specified. Simply using a different *size* device does not constitute a process change. For example, in the coding below, the W/L (see Chapter 2) of the PFETs is a variable. However, the process is the same for all the PFETS; hence only one model is specified for the PFETs. A similar statement holds for the NFETs.

Default values. As noted earlier, the FET model in SPICE contains 42 parameters. Some of these are intrinsic properties of silicon (or the material being used). Others change only rarely. Once the labor of getting a valid model has been done, the model can be used repeatedly changing only the W/L of the FETs as noted. If desired, any of the parameters can easily be changed by stating the parameter followed by an equal sign and the value. A parameter not specified defaults to that embedded in Spice.

The user must be careful not to conflict one of the parameters so defined with one of the default parameters that is derived from other values.

Operation. SPICE can be run in either batch or time-sharing mode. In the time-sharing mode, a good rule of thumb is that the circuits to be simulated should not contain more than about 32 or so FETs.

A final word. SPICE is a "fun" program to run. It provides impressive (but sometimes depressive—misspelling intended) printouts and plots.

A problem is that users sometimes put too much faith in its results without adequately questioning the models that were input to it or the suitability of the default parameters for the particular application at hand—GIGO (garbage in, garbage out).

The same can be said for any of the computer programs. However, with other programs, such as logic simulation programs, the user can much more readily spot-check the results. It takes more skill (or rather perhaps different skills) to spot-check the results of a circuit simulator, such as SPICE.

For example, knowing the propagation delays and the truth tables of logic functions, any designer can verify the timing relations of a logic simulation. However, few can readily take into consideration oxide thicknesses, transconductances, threshold voltages, and so on, to check the adequacy of a SPICE output. Indeed, it is precisely the difficulty of making such analysis that makes SPICE the valuable tool that it is *properly used with proper models*.

6.4 CHECKING THE LAYOUT OF THE MASK VERSUS THE SCHEMATIC

For cases in which manual intervention is used to do layout, it is very useful to have a method of checking the layout of the mask against the circuit that exists on the schematic and which was simulated, and so on. Such cases include complete manual placement and routing as well as cases where the auto P&R program does only a percentage of the complete placement and routing task.

Three programs that do this are the MTRACE program of Rockwell, the MASKAP program of Phoenix Data Systems, and the LVS portion of the layout verification system of ECAD. The MTRACE program compares the layout to inputs from the Simstran/Logtrans simulation program. The LVS (which appropriately stands for "layout versus schematic") compares to a layout entered via schematic entry. Both of these programs were originally designed for use with pure custom design.

6.5 SCHEMATIC ENTRY

6.5.1 Overview

Schematic entry refers to designing a logic or other circuit on a computer-driven video terminal. Logic symbols representing gates, flip-flops, and other functional elements are called from the computer's memory and interconnected using a "puck" with buttons to delineate specific actions or a joystick. Moving the puck across a tablet moves crosshairs on the graphics terminal.

The crosshairs can be used to select menu commands, such as "insert (or delete) line" and "move component on screen to position specified by pushing one of the buttons on the puck." Alternatively, the crosshairs can be used to select component symbols from the components library and to place them in position on the screen. There are generally submenus and sub-submenus callable by the crosshair being positioned on a command. The menu commands are often located along the sides or bottom of the display. These concepts apply also to workstations, which are discussed in Section 8.2. Sometimes prompt commands and menus occur on an ancillary video terminal as well and the use types in a few characters on occasion.

The logic structures so created can be stored with a user-selectable symbol as macros and used (replicated) to form larger circuits. The schematics used in Section 4.7 were created using a schematic entry system.

The data base created by this process can be used in several ways. First, it can be used to run a thermal or pen plotter or other hard-copy device to create copies of the resultant schematic for checking, documentation, and other purposes. As with logic simulation, this function is totally independent of whether the circuit will be implemented on a PCB (printed circuit board), gate array chip, or other means.

Of particular interest to the VLSI designer, however, is the usage of the schematic entry data base in logic simulation, testability and fault analysis, and auto

P&R (placement and route) programs. A major problem in designing VLSI chips is the need to transfer larger amounts of data in a completely error-free fashion. By deriving both the logic simulation and P&R from the same data base, the possibilities of errors are greatly reduced.

Equally importantly, the schematic so drawn represents that which will be simulated and later built. The test vectors derived from the logic simulation and/or the automatic test vector generator part of the simulator will test the part. Thus the entire package of schematics, simulation, chip layout, and test of parts can potentially be derived from the same data base.

Of considerable importance also is that any changes made to the schematic are reflected in that used for simulation and analysis.

At the time of this writing, however, only a handful of such integrated systems exist. There is a great deal of intense activity in this field toward making integrated packages containing the foregoing items because of the resultant large payoffs.

Nearly every major manufacturer of IGSs (interactive graphics systems) offers schematic entry in one form or another. In addition, schematic entry programs are available from vendors such as Silvar-Lisco and CGIS (the owner of Tegas described earlier). They are also a standard feature on workstations.

Other terms used in place of schematic entry are schematic capture, symbolic entry, or sometimes symbolic layout. These are not to be confused with the symbolic layout methodologies of Rockwell (Chapter 8), AMI (the SLIC system), Bell Labs, Applicon's CASL (color assisted symbolic layout), and so on. The latter refer to creating the layout of a pure custom IC by using symbols to represent the layers of the circuit. For example, an X would represent the metal contact to a P^+ diffusion such as a MOSFET source. In schematic entry, the symbols being encoded are not alphanumerics and they represent a higher level than a portion of a transistor.

Figure 6.7 shows a picture of a typical schematic entry system. In this case it is a workstation. More will be said about these in Chapter 8. As noted above, it consists of a graphics terminal with a "puck" for data entry. The puck is moved over the tablet to move a crosshairs on the screen of the graphics terminal. Buttons on the puck enable the user to designate selection of menu items and placement of components and lines.

The meaning of the buttons, a listing of commands, and other prompts are sometimes displayed on a separate alphanumeric terminal located to the right of the graphics terminal. On occasion, the user may have the option (or be required to) type in a few simple characters. However, most of the work is done with the puck. (See Chapter 8 for a list of input devices other than the puck.)

This section describes the operation of a schematic entry system from a major CAD manufacturer, Silvar-Lisco. This is representative of other such systems. On occasion, to illustrate a point, the author may deviate from the Silvar-Lisco methodology. This section represents only a small portion of the capabilities of the Silvar-Lisco software package.

Figure 6.8 is a photo of the screen of the graphics terminal. The screen is broken into five separate sections in the Silvar-Lisco method. To the left of the drawing area is a list of the components or macros (functional elements, such as D flip-flops, logic gates, etc., sometimes called modules) that the user can select to create the

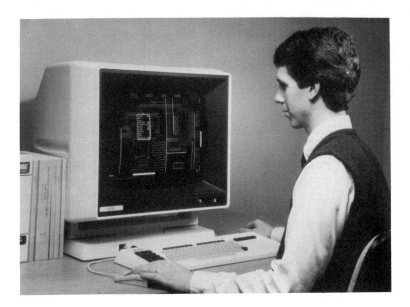

Figure 6.7 Schematic entry system. In this case it is a workstation. (Courtesy of Mentor Graphics, Inc.)

schematic. (The list can be scrolled vertically using the puck to bring in other macros not shown.) In this example, some of the elements are named M7EF730 AND M7EF731 (D latches), M7EF372 (OR gate), and M7EF360 (four-input OR/ NOR gate).

To the right of the drawing space is a list of the functions (such as zoom in or out, select a component, connect a line, etc.) that can be performed. In the example, the puck has been used to select the menu item "plot" in order to plot or copy what is on the screen. This is indicated in reverse video in the upper right side of the picture/ screen.

The reader will note that in each case there is a main category (e.g., CONNECTION) followed by a submenu (e.g., enter, delete, style). The submenu item is selected after the main menu item, also with a simple push of one of the puck buttons. The entire process takes about 1 second.

The reader will also note menu items that enable the user to add text to the lines and symbols (very important, as will be described later); add to or delete from the menu; copy, delete, rotate, and move components; and save and quit commands. Not shown is a command that enables a group of components to be moved together. There are also commands to build other symbols. (See the "free line" command.)

More complex functions may cause a submenu to be called up. If "copy group" (not shown) is selected, for example, the user will then be prompted to use a given one of the four buttons (e.g., number 3) on the puck to define the group of components to be moved. This command applies to components that the user has previously placed on the drawing part of the screen.

The third section is the drawing space itself, hereinafter called simply "the drawing." Standard sizes of drawings (A, B, C, and D), as well as user-selectable

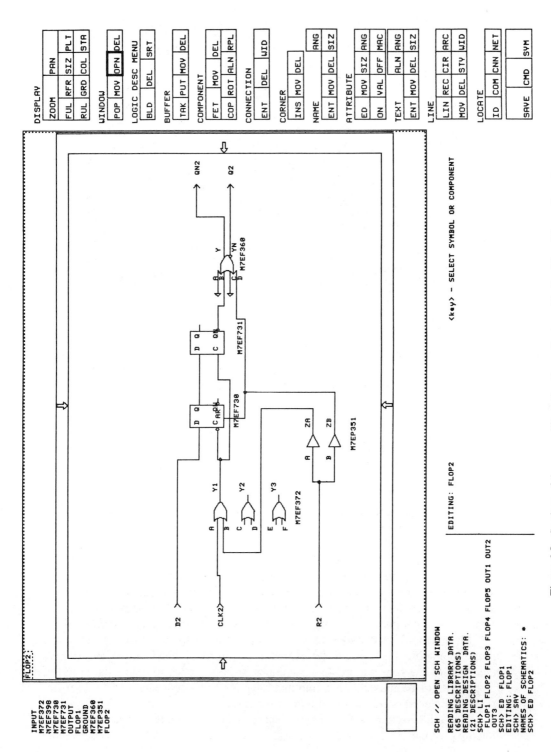

Figure 6.8 Sections of the screen illustrating the sections used in the Silvar-Lisco schematic entry system. (Courtesy of Silvar-Lisco, Inc.)

sizes, are offered. The fourth section is a small diagram in the lower left of the screen showing how much of the drawing is being viewed by that which is represented on the screen at a given time. It also shows where on the drawing that portion represented by the screen view is. In this example, the drawing (size A) is filled. See Figures 6.8.

The last section is a cue line on the bottom of the screen which prompts the user for more information and enables text to be entered for naming signals and nodes. In this case the system is asking the user to tell it the scale on which the user wants the plot to be done.

Components (macros) not on the drawing can be placed there from the menu by placing the crosshairs over "component" and then over "fetch," moving to the list of macros on the left, placing the crosshairs over the desired macro, and then using the crosshairs to designate the desired location of that particular macro on the drawing. The appropriate buttons on the puck are used to effect the selection and transfer. Figure 6.8 shows how different menu items have been placed from the menu to the drawing.

Connecting lines to the various components is done with much the same sequence except that the list of components is not used. The "connection, enter" commands are used. Buttons on the puck are used to effect the connection. The program is capable of making a trial interconnect between two components whose terminals have been selected. Other examples of schematic capture are given in Figures 8.16(a), (b), (c).

6.5.2 Naming of Signals and Nodes

It is desirable for the user to name the signals and nodes of the schematic with names that are meaningful to him or her whenever possible. The reason is that if this is not done by the user, the system will produce such names itself. The names so produced will mean little to the user, in general. Although a list can be checked to identify such names in the system, this is a tedious process. Names are entered by selecting the appropriate menu item and typing in the desired text.

6.5.3 Netlist Extraction and Simulator Interface

The data base created by schematic entry must still be formatted. This is done in two (or more, depending on the needs of the designer) separate programs. The first, called NLE or net list extractor, formats the file so that a subsequent file can code it properly for Tegas, Helix, or other simulators or for other functions, such as auto P&R. The second program is the interface program itself, such as ITegas (interface to Tegas), IHelix, and so on.

6.5.4 Creation of New Library Elements

Functional elements not in the library of components can be created from the existing elements by simply naming a schematic created as above as a component and giving it both a name and a graphical designation (typically a rectangle). The representation also shows the I/Os and the reference point of component.

Drawing segments of common shapes, such as arcs, are part of the menu. These can be used to designate components created with special symbols. They are then added to the library using another command on the menu.

The process above can be repeated to create "nested macros." The utility of nesting is explained in Section 6.7.

6.6 TEGAS LOGIC SIMULATOR

One of the more widely used logic simulators is TEGAS (Test Generation and Simulation), owned by General Electric's Calma Division. Tegas is representative of logic simulators. It is available on CDC (Control Data Corporation) Cybernet's timesharing services and can also be licensed to other companies as well by GE/Calma. It runs on most of the popular mainframe and minicomputers including Digital VAXs, Amdahl 470s, Prime, and IBM machines. It is also resident on the Tegas workstation.

There are over 100 installations of Tegas worldwide. It is part of the CAD systems of many gate array vendors, such as those of GE/Intersil, AMCC, Harris, Signetics, UTMC/Mostek (the Highland system), and LSI Logic, Inc. (the LDS-II system). (In the latter, the extended version of Tegas called Texsim is used.)

Tegas has evolved over several years into successive versions. The current version is Tegas5 (or TegasV). The version of Tegas5 offered by Control Data Corp. on their timesharing network is called CC-TDL (Cyber Control—Tegas Design Language). An offshoot of Tegas5 called Texsim is oriented specifically toward MOS devices, as is another offshoot called Texmos. Tegas, Inc., also offers a schematic entry package (TGate), a workstation, and display generator. Important CAD programs, such as the schematic entry package from Silvar-Lisco, interface to Tegas. Many workstations, such as those from Mentor and Daisy, interface to it (as well as to other simulators). There is also a behavioral version called Texsim B.

One of the advantages of Tegas is that unlike many simulation programs, it is quite complete in the sense of offering subprograms that do automatic test generation, testability analysis of the circuit, analysis of the usefulness of a given set of test vectors, and fault analysis of the circuit.

Tegas is a logic simulator as opposed to mixed-mode simulators such as HELIX and HI-LO II.

This chapter will concentrate on Tegas5/CC-TDL, hereinafter called simply Tegas.

6.6.1 Purpose of This Section

The purpose of this section is not to teach the reader how to use Tegas but rather to indicate the usefulness of simulation programs such as Tegas. Any simulation program has so many options that entire books could be devoted to them. The treatment below is intended to highlight those aspects which the author has found to be particularly useful.

6.6.2 Uses of Tegas

The name Tegas, which stands for "test generation and simulation," describes its main functions. These are:

- Verification of the static and dynamic characteristics of a logic circuit
- Test pattern (test vector) development
- Fault analysis and grading
- Testability analysis

This section focuses on the makeup of Tegas and on its use in logic verification. Subsequent chapters deal with the other topics.

6.6.3 Features of Tegas

In what follows, the terms *propagation delay, time delay,* and *simple delay* will all be used to denote the high-to-low and low-to-high propagation times of a functional element unless otherwise noted. Tegas denotes low-to-high and high-to-low propagation times as rise and fall delays. This leads some users incorrectly to assume that the rise and fall *times* of the circuit elements are being used.

In addition to the main functions, Tegas has several other very useful features. These are:

Connectivity. Tegas will give a list of outputs that are unconnected. The designer can then peruse this list and determine which of these, if any, should have been connected. For example, when using a D flip-flop, the designer may have used only the Q output. The Q_B (inverse of Q) output may not have been used. Tegas will include the Q_B output in the list of unconnected signals.

Testability. The Tegas program COPTR, described in Chapter 7, allows the testability of a circuit to be determined as soon as the network has been compiled.

Quality of test vectors. One of the simulation runs, called DFS, is an inexpensive way to place an upper bound on the quality of a given set of test patterns. A list of the nodes not exercised by that set is also printed out. This enables the user to determine rapidly where additional work in generating test patterns is required. Sadly, DFS has been largely replaced by the new mode 4.

A more expensive-to-run program, called mode 5 (and its sister, mode 4), gives a detailed analysis of the quality of a given test vector set (See Chapter 7).

Automatic test vector generation (ATG). Tegas will generate test vectors for the circuit if desired. See Chapter 7.

Signal names. This topic may seem trivial but in reality is not. Two different signals given the same name will provide very interesting results! (They will be wired together.) Tegas prints out a list of the signal names used. (Hence the user can rapidly select an unused signal name without fear. This topic is especially important when modifying the existing coding of a circuit. It also has importance because it is to the user's advantage to keep the signal names as short as possible, especially when multiple levels of "nesting" (see below) are used. This limits the number of combinations of alphanumeric characters available.

Signal locations. Tegas also prints out a list of all of the line numbers where each signal appears. This permits the designer to rapidly check the location of each signal.

Restart. A beneficial property of Tegas is that the simulation can be stopped at a given point, analyzed, and then restarted from the point where simulation stopped. This property enables long simulation runs to be made in pieces. Initially, the user may simply want to get an indication that the circuit functions without regard to making sure of speed or all values or combinations of parameters. Later runs, which may depend on the earlier values, can be made without repeating the earlier portion of the simulation. This saves computer cost.

Partitioning. Large circuits can be partitioned into smaller portions. The smaller portions can be simulated individually. The simulation outputs from one subcircuit can be saved and used as inputs to other subcircuits.

Partitioning must be done in such a way that control signals from the subcircuits to which the data are being passed are not dependent on the data being passed. As a simple example, it would be impractical to partition in such a way that the clock signal for a D flip-flop in one subcircuit was generated in another subcircuit and dependent on the Q output of that flip-flop in the first subcircuit.

Printed circuit board (PCBs). Tegas can be used equally well with PCBs as well as custom LSI. When used to predict the characteristics of PCBs, a mode (mode 3) of operation in which combinations of minimum and maximum propagation delays of the circuit elements is sometimes used. This mode is almost never used to simulate custom LSI because unlike the elements on a PCB, the time delays of the elements on a single chip all track each other. They all scale up and down the same way with regard to temperature, voltage, and wafer-processing variations.

Control files. An important feature of Tegas is the ability to use control files. These are separate files that can be prepared by the user and called by the Tegas simulation program. They are analogous to macros in that the files can be used for many different circuits. They typically are written and called in three different areas. These are (1) modification of the time delays of the elements due to temperature, voltage, and fan-out; (2) specification of the stimuli to the circuit (input signal changes and test vectors); and (3) (sometimes) specification of the signals to be displayed and the format for displaying such signals. Different combinations of parameters can be selected from libraries and used to form a composite control file.

Activity analysis of each node.

Fan-out extraction from netlist. TEGAS can be made to extract the fan-out of each gate in the circuit from the netlist and calculate the propagation delay of that gate under those circumstances. The user can program equations representing such delays based on particular vendors and particular gate arrays into Tegas.

This feature is particularly important when working with MOS technologies because these are highly dependent on temperature, voltage, and fan-outs. The feature is contained in the OPDELAYS command used in the simulation part of Tegas.

Nesting. Nesting can be done up to 31 levels. Nesting is the process of building larger elements from smaller elements. For example, gates are used to make latches (innermost nesting level), which in turn are used to make D flip-flops (next innermost level of nesting), which in turn are used to make counters (next level of nesting), which are used to make pulse generators (next level of nesting). The outermost level of nesting calls or references the next innermost level of nesting via the USE command (see below), which in turn calls the next innermost level of nesting, and so on.

Fields. There are four separate fields in the module call in Tegas5. If any one of the fields is different between two separate modules, Tegas5 treats them as being different from one another. This greatly aids in giving meaning to modules and hence helps significantly in the debugging process. Unused and default fields are denoted simply by slashes (/).

6.6.4 Makeup of Tegas5

Tegas5 consists of two physically separate programs. These are the preprocessor and the simulator. The former consists of three subsystems: a linking loader, a compiler, and a file management system. Its purpose is to translate the user's network into a form that is appropriate for the simulator.

The simulator produces all the useful output from Tegas, including testability analysis (COPTR) and automatic test generation (ATG). By specifying the appropriate mode in the simulator program, a nominal timing analysis (mode 2), a worst-case timing analysis (mode 3), or one of several types of fault generation and analysis runs (mode 5) can be made. There are also two limited-performance but very inexpensive modes. Mode 1 is used for logic simulation, and mode 4 is used for fault simulation.

Another program, DFS (detectable fault simulation), allows the user, rapidly and inexpensively, to place an upper bound on the value of a given set of test vectors.

Various display options are available. These can be incorporated in the simulation runs or written as a separate display program.

6.6.5 Important Interfaces to and from Tegas

Because Tegas is so widely used, important interfaces have been developed for it. To create the net list, Silvar-Lisco, Mentor, Zycad, Daisy (via AMCC), and other companies have developed interfaces from their schematic entry programs to Tegas. This means that the same data base used to create the schematic is also used for the coding, which in turn greatly reduces the possibilities for error.

A second important interface, in this case *from* Tegas, is to major VLSI testers, notably the Fairchild Sentry series.

Yet a third major interface, again *from* Tegas, is to automatic placement and routing programs. The LDS-II system of LSI Logic, Inc., has such an interface, for example. (In this case, it comes from the associated Texsim program. The current version of LDS, LDSIII, uses a simulator developed in-house.)

The ZYCAD accelerator (Chapter 8) also has an interface in Tegas.

6.6.6 Libraries in Tegas

If coding of the preprocessor is to be done manually (as opposed to coming from schematic entry), simplification can be achieved using elements from one or more of the libraries in Tegas. Such use will also result in faster turnaround and fewer errors. The libraries are of three major types. The first type is that built into Tegas itself. It is called the master module library and contains both primitives and standard parts. The second type is that created by the user using one of the libraries of the former. It is called the user module library. Less often used than either of the other two is the PLA module library, which contains the modules defined by the user using Boolean equations. It is used less often because it is much more expensive to run than either of the others. Nevertheless, it is available for those functions that are more easily described in terms of Boolean equations.

Libraries of standard parts. To make it easy to model circuits employing standard parts, Tegas has a number of libraries that contain both the pin-outs and the time delays of the standard parts. Because the pin-outs are the same, the user can use the data books containing the standard parts to do the coding of the network if this has not already been done by a schematic entry program. (Alternatively, the user can find the part in the appropriate Tegas library.) The current list of libraries includes one with roughly 400 TTL, LSTTL, and STTL parts, another with the 4000 series of CMOS parts, and another with the 10k series of ECL parts. The 8080B and 2900 series of microprocessors are also included. More parts are being added.

A benefit to the user of the libraries of standard parts is that the coding of the network is considerably simplified. Without the libraries, the user would have to manually input the time delays of each standard part to the appropriate primitives.

A second benefit of the libraries is that the user can use them to make modules (see below) of his or her own which like those in the standard libraries can be used repeatedly in different circuits. The time delays associated with these modules can be either those from the standard parts or can be permuted by the user. For example, there is no totem-pole output EX-NOR in the LSTTL family of standard parts. Such a function can be created, however, from the combination of an LS7486 EX-OR followed by an LS7404 inverter. This part will behave exactly the same as a combination of the two elements (the EX-OR and the inverter) and will have propagation delay values which are the sum of those for the elements. Alternatively, the user can assign such delay values.

Tegas library of primitives. Tegas also has a library of primitive elements. This library consists of gates, flip-flops, MUXs (multiplexers), priority encoders, three-state gates, one-shots (monostable multivibrators), and primary inputs and outputs. On a higher level there are NXN multipliers, PLAs (programmable logic arrays), ALUs, RAMs, counters, registers, and ROMs. There are also a clock element and elements that enable detection of certain fault phenomena. The default time delay is unity for all primitives.

User-defined libraries and modules. A module is a functional element, such as a MUX (multiplexer), flip-flop, or register that exists in the master module library of Tegas5 or is user created and stored on a user library.

A module in Tegas corresponds to a macro; that is, it is a predefined interconnect among circuit elements that can be as large or as small as the user desires. As with any macro, it can be used over and over in different circuits, can be permuted to form new macros, can be replicated on a given circuit, and so on. Those gate array users who do their own designs generally develop libraries of macros, some or all of which may be supplied by the gate array vendor. The information in the modules is primarily interconnect and propagation-delay-time information. The elements in the user-defined libraries may be developed from the library of standard parts, from the library of primative elements, or from a combination of the two.

The user libraries are generally called directories—not to be confused with the directories exisiting on the computer being used. The latter will be called computer directories and subdirectories.

6.6.7 Comparative Analysis Using Subdirectories

By setting up computer subdirectories, the user can easily create module libraries for each vendor being worked with. Each computer subdirectory corresponds to one vendor or to the gate array family of one vendor.

The technique is to copy the macros from one subdirectory into another and then to change the time delays to correspond to those of the new vendor or the new gate array family of the same vendor. This is a one-time effort. A given gate array vendor may even (for a fee) supply the appropriate time-delay information.

Many, if not most, vendors retain and use the same library when building a new gate array family. A few macros may be added or deleted, but the vast bulk of the libraries generally remain the same with regard to function. *What does change* are the propagation delays of the macros. However, the functionality of the macros in going from, for example, a 3-μm family to a 2-μm family is generally the same.

A major benefit of this technique (or some permutation of it) is that it enables the user to rapidly answer questions about what kind of performance improvement would be obtained by using a different gate array family. A given circuit that has been coded using the propagation delay values of one family does not have to be recoded to be run with the delay values of the second family if it is run in the subdirectory of the second gate array family. By running simulation programs in the second subdirectory, none of the circuit has to be recoded. There is no conflict with the Tegas directory name (in the Tegas module itself, *not* the Tegas master library and *not* the VAX directory, of course) being the same but implying different delay values as long as operation is carried out in different subdirectories. The technique works on VAX series of computers and presumably works on other computers as well.

An alternative to the use of subdirectories of macros for each vendor or vendor's family is the use of the appropriate OPDELAYS commands for each vendor's offerings. Comparison is made by causing the OPDELAYS commands to force the appropriate time delays to be used in each case.

6.6.8 Examples to Be Used

To provide the reader with a measure of understanding of Tegas, the example shown in Chapter 4 will be used. Logic designers will recognize this as a 2-bit slice consisting of two D flip-flops which are synchronously clocked. The Q outputs of each of the two D flips-flops are input to both a register (another D flip-flop) and to an EXNOR. A comparison can thus be made between a particular pattern of the counter which has been stored and the instantaneous pattern of the counter. When the latter has reached a given state, the output of the NAND gate goes low to indicate detection. This is a useful circuit that has the additional benefit of being expandable to show some other features.

The circuit is long enough to permit meaningful points to be illustrated, but not so long as to confuse (hopefully) the reader (or use up half the publisher's stock of paper with printouts).

The 2-bit-slice example will then be nested with replications of it and with a 4-bit slice (also made from it) to form the 10-bit slice also shown in Chapter 4.

6.6.9 Nonuniform Nesting

Note that the 10-bit slice contains the 2-bit slice directly (i.e., one level down) and also two levels down by virtue of being inside the 4-bit slice. This is what the author terms *nonuniform nesting*. Not all simulators, including those of some of the workstations, can handle nonuniform nesting. It is, hence, one of the features that distinguishes one simulator from another.

The reader can readily appreciate the utility and desirability of this property even from the simple example given above. The coding would be more complicated even for the simple example above if the user had to develop from "scratch" a special circuit for the 4-bit slice.

6.7 GENERAL FLOW OF LOGIC SIMULATION

The preprocessor program is created and used as an input to the simulation program, as noted above. The preprocessor file tells Tegas (1) what modules are being used, (2) in what locations (directories) they exist, (3) what their interconnects are if they are made from primitives, and (4) what their delays are if not already specified in the library definitions.

The general flow of the process was given in Chapter 4. It will be summarized briefly here.

After the preprocessor program is run and debugged, simulation programs that use it are debugged and run. For large circuits (greater than 5000 gates), portions are often run to decrease both personnel and computer costs. This is especially true if the circuit being simulated contains elements that are repeated throughout the circuit. In the initial debugging phases of the simulation modes, very simple changes are put in. Normally, the testability of the circuit is determined using the COPTR program before extensive simulation runs are made. If COPTR indicates that needed nodes cannot be tested adequately, modification of the circuit design is made to ensure testability.

When the testability of the design is adequate, more extensive simulation runs are made. Generally, several runs are made with mode 2 after the simulation has been debugged. Debugging removes signal changes that are too fast and/or which violate setup and hold times of the elements. What seems to be too fast a signal change may in reality be a race condition or other design deficiency of the network. In this case also the circuit is redesigned to overcome the deficiency.

The next step is to determine what faults can be detected in the nodes of the circuit. This is done using the "stuck at (1 or 0)" methodology. Fault simulation is then done using mode 5 or possibly mode 4. The same test patterns developed for the mode 2 simulations can be used with mode 5. Alternatively, TEGAS offers the user the choice of the ATG mentioned above. Because fault simulation is relatively expensive, it is important to simulate only those faults that will be encountered in actual operation of the circuit. After chip layout, the circuit will generally be resimulated using the propagation delays of the actual network which have been extracted from the layout and input to the simulator. The delays are input using the TCF (Tegas circuit file).

6.7.1 Network Encoding of the Preprocessor

Constructing the preprocessor program involves telling Tegas what functional elements are connected to what other elements. In effect, the net list must be given to Tegas. This can be done in several ways. Probably the most common way currently is to manually figure out the interconnects and then to type them in on a computer terminal.

Increasingly, output files from schematic entry programs are being used to input the net list. This has a dual advantage. First is that the net list encoded does indeed represent what is on the logic diagram. An additional advantage occurs if the same file (properly translated by *software,* not by error-prone human beings) is used in the place and route (P&R) program. The internal interconnects and propagation delays of previously defined modules do not have to be encoded.

The most important parts of the preprocessor are inputs, outputs, definition of which modules are to be used, what their propagation delay values are, and how they are interconnected. (If the intent of this section were to teach the reader how to use Tegas, a number of other topics would have to be addressed as well. Such teaching is beyond the scope of this book, however.) In what follows, the four sections outlined above are highlighted.

The coding of the preprocessor is as shown in Fig. 6.9. The input and output portions of the coding are self-explanatory. The two remaining sections of interest are the USE and DEFINE sections.

The USE section has four main purposes. They are to

1. List the names of modules that will be used. For example

   USE SN74LS74///MASTER
   (MASTER is the name of the library
   containing standard parts)

   EXNOR///CHIPLSI
   (EXNOR is a user-created module, and
   it exists in a directory called CHIPLSI)

2. Permit assignment of a name to each such module by which the module can be referenced. This name is called the *type name*. It is generally a shortened version of the module name. (Still a third name, the occurrence name, will be used in the define section.) For example

   USE

   DF = SN74LS74///MASTER

   EXN = EXNOR///CHIPLSI

3. Specify the number of inputs and outputs of those primitives that require such specification. For example, primitive type 6 is a multi-input NAND gate. If it is to be used as a three-input NAND gate with type name ND3 and as a four-input NAND gate with type name ND4, it would be listed in the USE section as

   USE

   ND3 = NAND(3,1),

   ND4 = NAND(4,1)

   where NAND is the primitive name, and the first digit inside each set of parentheses denotes the number of inputs. (The second digit denotes the number of outputs, in this case 1. Some more complex modules will have multiple output possibilities.)

4. Permit specification of the propagation delay values for that module. Alternatively, a delay name can be used. This has the advantage that the delay values of all modules with a given delay name can be changed simply by respecifying the value for that one delay name.

   Alternatively, the values of delay corresponding to a given delay name can also be specified in a separate delays section.

```
$ TY 2BIT.TDL
"2bit.tdl from 4BIT.TDL
COMPILE;
DIR DEMO;
OPTIONS XREF,REPL;
MODULE 2BIT;

"*******************************************************************"
"                                                                   "
"             K E Y    P O I N T S                                  "
"                                                                   "
"   - APPLIES TO PRINTED CIRCUIT BOARDS AS WELL AS GATE ARRAYS      "
"                                                                   "
"   - IS TRANSPORTABLE AMONG GATE ARRAY VENDORS                     "
"                                                                   "
"*******************************************************************"
"                                                                   "
"         C O D I N G    O F    T H E    C I R C U I T              "
"                                                                   "
INPUTS
        RESET,CLK,LATCH,DIN;

OUTPUTS
        O1,O2,Q01,Q02;


"         C A L L   M A C R O S   F R O M   L I B R A R Y          "
USE
    DF = SN74LS74///MASTER, "EXAMPLE OF STANDARD IC FROM MASTER LIBRARY"

    EXN = EXNOR///CHIPLSI,  "EXAMPLE OF USER DEFINED MODULE. NOTE THAT IT"
                            "EXISTS IN A USER LIBRARY DIFFERENT FROM THE "
                            "CURRENT USER LIBRARY"

    2A= SN74LS08///MASTER,  "USE NAME IS 2A; MODULE NAME"
                            "IS SN74LS7408///MASTER"

    IV = SN74LS04///MASTER; '


"      D E F I N E   I N T E R C O N N E C T S   A M O N G   M A C R O S"
DEFINE
        "CODE THE COUNTER FLIP-FLOPS - NOTE THAT EACH SN74LS74 CONTAINS"
        "TWO D FLIP-FLOPS AND THAT THE OUTPUT OF THE FIRST, DF1(5), IS "
        "INPUT TO THE SECOND"

     DF1(Q1=5,Q1B=6,Q2=9,Q2B=8) =DF(DIN=2,DF1(5)=12,CLK=3,CLK=11,
        RESET=1,RESET=13,NC/1/=4,NC/1/=10);
```

Figure 6.9 Coding of the preprocessor of the 2-bit slice.

```
        "THIS 2 BIT SLICE WAS MADE FROM THE 4 BIT SLIDE BY DELETING "
        "(OR PUTTING IN QUOTES) THE EQUATIONS IN QUOTES BELOW. (NAND"
        "GATES WERE CHANGED TO INVERTERS"

        "DF2(Q3=5,Q3B=6,Q4=9,Q4B=8) = DF(Q2=2,DF2(5)=12,CLK=3,CLK=11,"
        "    RESET=1,RESET=13,NC/1/=4,NC/1/=10);"

            "CODE AND GATES AT REGISTER INPUTS - NOTE THAT SIGNAL NAME HAS"
            "BEEN MADE EQUAL TO ELEMENT OCCURRENCE NAME; THIS CAN BE DONE"
            "BECAUSE THERE IS ONLY ONE OUTPUT FROM THE ELEMENT"

         A1=2A(DF1(6),NC/1/); "ONE INPUT COMES FROM Q1B; THE OTHER IS TIED HIGH"
         A2=2A(DF1(8),NC/1/);
        "A3=2A(DF2(6),NC/1/);"
        "A4=2A(DF2(8),NC/1/);"

"THE AND GATES ARE AN EXAMPLE OF AN ELEMENT THAT EXISTS IN THE PARTICULAR"
"MODULE (=MACRO) WHICH IS NOT USED IN THIS PARTICULAR APPLICATION. IN CMOS"
"THESE LOAD THE QB OUTPUTS OF THE D FLIP-FLOPS LESS THAN THE D INPUTS OF THE"
"REGISTERS."

            "CODE THE REGISTERS USING SN74LS74 FLIP-FLOPS"
         DR1(Q40=5,Q41=9) = DF(A1=2,A2=12,LATCH=3,LATCH=11,RESET=4,
             RESET=10,NC/1/=1,NC/1/=13);
        "DR2(Q42=5,Q43=9) = DF(A3=2,A4=12,LATCH=3,LATCH=11,RESET=4,"
        "    RESET=10,NC/1/=1,NC/1/=13);"

         "CODE THE EXNORS"

         X1=EXN(DF1(6), DR1(5));
         X2=EXN(DF1(8), DR1(9));
        "X3=EXN(DF2(6),DR2(5));"
        "X4=EXN(DF2(8),DR2(9));"

            "CODE THE INVERTERS AT THE EXNOR OUTPUTS"

         O1=IV(X1);
         O2=IV(X2);

           Q01=(DF1(5));
           Q02=(DF1(9));
          "Q03=(DF2(5));"
          "Q04=(DF2(9));"

END MOD;
END COMPILE;
FILEMGR;
END FILEMGR;
LOAD 2BIT///DEMO;
END;
```

Figure 6.9 (*cont.*)

Two examples that show the two cases are

USE

ND3 = NAND (3,1)///(6,4,8,4,2,6),

ND4 = NAND (4,1)///(DELAYZ),

The six digits in the delay portion of ND3 represent, respectively, the nominal, minimum, and maximum propagation delays low to high and high to low, respectively.

The DEFINE section of the preprocessor tells Tegas what the interconnects among the elements are. In the example of the 2-bit slice, one term in the coding will be that of the first D flip-flop. Its (user-selected) OCCURRENCE name will be DF1. The net list equations are of the general form

occurrence name(outputs) = type name(inputs)

"Explicit notation" will be used to denote what signals connect with what pins. (There is also an implicit notation.) The form of explicit notation is

signal name = pin name or number

The pin name or number will come from the definition of the module in the appropriate library. If standard parts are being used, the pin numbers from the data books can be used because the Tegas libraries are keyed to them. In explicit notation, the order of the inputs or outputs is immaterial (as long as the outputs are on the left side of the equation and the inputs are on the right side). A sample preprocessor output is shown in Fig. 6.10.

```
*****************************
ENTERING COMPILER.
*****************************
    1.  DIR DEMO;
    2.  OPTIONS XREF,RPL;
    3.  MODULE 2BIT;
    4.
    5.
    6.  '******************************************************************'
    7.
    8.  '                K E Y     P O I N T S                              '
    9.
   10.  '     - APPLIES TO PRINTED CIRCUIT BOARDS AS WELL AS GATE ARRAYS    '
   11.
   12.  '     - IS TRANSPORTABLE AMONG GATE ARRAY VENDORS                   '
   13.
   14.
   15.  '******************************************************************'
   16.
   17.  '          C O D I N G    O F    T H E    C I R C U I T             '
```

Figure 6.10 Pertinent parts of the preprocessor output.

```
18.
19.   INPUTS
20.        RESET,CLK,LATCH,DIN;
21.
22.   OUTPUTS
23.        O1,O2,Q01,Q02;
24.
25.
26.
27.
28.   '    C A L L    M A C R O S    F R O M    L I B R A R Y              '
29.
30.   USE
31.       DF = SN74LS74///MASTER, 'EXAMPLE OF STANDARD IC FROM MASTER LIBRARY'
32.
33.       EXN = EXNOR///CHIPLSI,  'EXAMPLE OF USER DEFINED MODULE. NOTE THAT IT'
34.                               'EXISTS IN A USER LIBRARY DIFFERENT FROM THE '
35.                               'CURRENT USER LIBRARY'
36.
37.
38.       2A= SN74LS08///MASTER, 'USE NAME IS 2A; MODULE NAME'
39.                              'IS SN74LS7408///MASTER'
40.
41.       IV = SN74LS04///MASTER;
42.
43.
44.   '    D E F I N E    I N T E R C O N N E C T S    A M O N G    M A C R O S'
45.
46.   DEFINE
47.        'CODE THE COUNTER FLIP-FLOPS - NOTE THAT EACH SN74LS74 CONTAINS'
48.        'TWO D FLIP-FLOPS AND THAT THE OUTPUT OF THE FIRST, DF1(5), IS '
49.        'INPUT TO THE SECOND'
50.
51.       DF1(Q1=5,Q1B=6,Q2=9,Q2B=8) = DF(DIN=2,DF1(5)=12,CLK=3,CLK=11,
52.           RESET=1,RESET=13,NC/1/=4,NC/1/=10);
53.
54.        'THIS 2 BIT SLICE WAS MADE FROM THE 4 BIT SLICE BY DELETING '
55.        '(OR PUTTING IN QUOTES) THE EQUATIONS IN QUOTES BELOW. (NAND'
56.        'GATES WERE CHANGED TO INVERTERS'
57.
58.       'DF2(Q3=5,Q3B=6,Q4=9,Q4B=8) = DF(Q2=2,DF2(5)=12,CLK=3,CLK=11,'
59.     '    RESET=1,RESET=13,NC/1/=4,NC/1/=10);'
60.
61.        'CODE AND GATES AT REGISTER INPUTS - NOTE THAT SIGNAL NAME HAS'
62.        'BEEN MADE EQUAL TO ELEMENT OCCURRENCE NAME; THIS CAN BE DONE'
```

Figure 6.10 (*cont.*)

```
63.            'BECAUSE THERE IS ONLY ONE OUTPUT FROM THE ELEMENT'
64.
65.        A1=2A(DF1(6),NC/1/); 'ONE INPUT COMES FROM Q1B; THE OTHER IS TIED HIGH'
66.        A2=2A(DF1(8),NC/1/);
67.        'A3=2A(DF2(6),NC/1/);'
68.        'A4=2A(DF2(8),NC/1/);'
69.
70.    'THE AND GATES ARE AN EXAMPLE OF AN ELEMENT THAT EXISTS IN THE PARTICULAR'
71.    'MODULE (=MACRO) WHICH IS NOT USED IN THIS PARTICULAR APPLICATION. IN CMOS'
72.    'THESE LOAD THE QB OUTPUTS OF THE D FLIP-FLOPS LESS THAN THE D INPUTS OF THE'
73.    'REGISTERS.'
74.
75.
76.
77.        'CODE THE REGISTERS USING SN74LS74 FLIP-FLOPS'
78.        DR1(Q40=5,Q41=9) = DF(A1=2,A2=12,LATCH=3,LATCH=11,RESET=4,
79.            RESET=10,NC/1/=1,NC/1/=13);
80.        'DR2(Q42=5,Q43=9) = DF(A3=2,A4=12,LATCH=3,LATCH=11,RESET=4,'
81.        '   RESET=10,NC/1/=1,NC/1/13);'
82.
83.
84.        'CODE THE EXNORS'
85.
86.        X1=EXN(DF1(6), DR1(5));
87.        X2=EXN(DF1(8), DR1(9));
88.        'X3=EXN(DF2(6),DR2(5));'
89.        'X4=EXN(DF2(8),DR2(9));'
90.
91.
92.
93.
94.        'CODE THE INVERTERS AT THE EXNOR OUTPUTS'
95.
96.        O1=IV(X1);
97.        O2=IV(X2);
98.
99.
100.       Q01=(DF1(5));
101.       Q02=(DF1(9));
102.       'Q03=(DF2(5));'
103.       'Q04=(DF2(9));'
104.
105.
106.   END MOD;
```

Figure 6.10 (*cont.*)

```
OCCURENCE INPUT UNCONNECTED,    A1(4)
OCCURENCE INPUT UNCONNECTED,    A1(5)
OCCURENCE INPUT UNCONNECTED,    A1(9)
    .
    .
    .
OCCURENCE OUTPUT UNCONNECTED,   A1(6)
OCCURENCE OUTPUT UNCONNECTED,   A1(8)
OCCURENCE OUTPUT UNCONNECTED,   A1(11)
    .
    .
    .
OCCURENCE INPUT UNCONNECTED,    A1(12)
OCCURENCE INPUT UNCONNECTED,    A1(13)
    .
    .
    .
OCCURENCE OUTPUT UNCONNECTED,   O1(8)
OCCURENCE OUTPUT UNCONNECTED,   O1(10)
    .
    .
    .
OCCURENCE INPUT UNCONNECTED,    O2(11)
OCCURENCE INPUT UNCONNECTED,    O2(13)
OCCURENCE OUTPUT UNCONNECTED,   O2(4)
OCCURENCE OUTPUT UNCONNECTED,   O2(6)
OCCURENCE OUTPUT UNCONNECTED,   O2(8)
OCCURENCE OUTPUT UNCONNECTED,   O2(10)
...ID  ERRMSG MOD
       MODULE NOT FOUND IN DIRECTORY,

REPLACED AS 2BIT/GATE/1/DEMO ,

TDL SYMBOLIC CROSS-REFERENCE,

    SYMBOL              TYPE            REFERENCES

    2A                  MODULE          38    65    66
    A1                  OCCURENCE       65    78
    A2                  OCCURENCE       66    78
    CLK                 MODULEIN        20    51
     .
     .
     .
    LATCH               MODULEIN        20    78
    O1                  PIN/OCCR,       23    96
    O2                  PIN/OCCR,       23    97
    Q1                  SIGNAL          51
```

Figure 6.10 (*cont.*)

```
    .
    .
    .
Q01             MODULEOUT       23      100
Q02             MODULEOUT       23      101
RESET           MODULEIN        20      52      78      79
X1              OCCURENCE       86      96
X2              OCCURENCE       87      97
```

```
    1.  END COMPILE;
*****************************
LEAVING COMPILER.
*****************************
    1.  FILEMGR;
*****************************
ENTERING FILE MANAGER.
*****************************
    1.  END FILEMGR;
*****************************
LEAVING FILE MANAGER.
```

Figure 6.10 (*cont.*)

6.7.2 Simulation (Figs. 6.11, 6.12)

Now that the module has been coded, it is necessary to simulate it. As in the case of the preprocessor, the object is not to try to teach the reader how to use Tegas but rather to make him or her aware of the possibilities with these techniques by showing how some of the more important operations are performed. The operations selected are those which relate most strongly to the logic designer's normal mode of operation—putting signals in and observing the resultant outputs.

```
NOTES TO READER: THE SYNTAX WHICH FOLLOWS HAS BEEN SLIGHTLY (SOMETIMES)
PERMUTED TO ILLUSTRATE A GIVEN FEATURE. SOME COMMANDS HAVE BEEN ELIMINATED TO
REDUCE CLUTTER ON THE FIGURES. DO NOT USE FOR DESIGN. THE TOP AND BOTTOM
DELIMITERS OF THE DESCRIPTION WHICH PRECEDES THE COMMAND ARE THE %%% AND
(((((( RESPECTIVELY. QUOTATION MARKS SIMPLY MEAN THAT THE COMMAND IS NOT
BEING USED IN THE GIVEN RUN. (NOT ALL OPTIONS WILL BE USED EVERY TIME.)

%%%%%%%%%%%%%%%%%%%%%%%%%%%%%%%%%%%%%%%%%%%%%%%%%%%%%%%%%%%%%%%%%%%%
    DESCRIBE THE MODE OF OPERATION. MODES 1 AND 2 ARE USED FOR LOGIC
    SIMULATION. MODE 1 IS A SIMPLIFIED VERSION OF MODE 2. MODES 4 AND 5 ARE USED
    FOR FAULT SIMULATION. MODE 4 RUNS ABOUT 3 TIMES FASTER THAN MODE 5.

((((((((((((((((((((((((((((((((((((((((((((((((((((((((((((((((((((
    MODE 2;
    LOAD+STATS;

%%%%%%%%%%%%%%%%%%%%%%%%%%%%%%%%%%%%%%%%%%%%%%%%%%%%%%%%%%%%%%%%%%%%
```

Figure 6.11 Simulation input.

```
BRING IN STATEMENTS FROM A CONTROL FILE CALLED OPDELAYS_3U. THESE MIGHT
REPRESENT THE DELAY VALUES FOR A GIVEN VENDOR'S 3 MICRON PROCESS (AS OPPOSED
TO ITS 2U PROCESS) FOR EXAMPLE.

(((((((((((((((((((((((((((((((((((((((((((((((((((((((((((((((((((((

"CONTROL OPDELAYS_3U $"

%%%%%%%%%%%%%%%%%%%%%%%%%%%%%%%%%%%%%%%%%%%%%%%%%%%%%%%%%%%%%%%%%%%%%

INPUT SIGNALS TO THE CIRCUIT WHICH COME FROM A PRIOR RUN. THE
PRIOR RUN MAY REPRESENT TEST VECTORS WHICH HAVE BEEN GENERATED BY THE
AUTOMATIC TEST VECTOR GENERATOR (ATG) OR MAY COME FROM SIMULATION OF A
CIRCUIT  WHICH IS CONNECTED TO THE CIRCUIT BEING PRESENTLY SIMULATED

(((((((((((((((((((((((((((((((((((((((((((((((((((((((((((((((((((((

"FILEINPUT;"

%%%%%%%%%%%%%%%%%%%%%%%%%%%%%%%%%%%%%%%%%%%%%%%%%%%%%%%%%%%%%%%%%%%%%

   DETERMINE WHAT IS TO BE SAVED
3 OPTIONS ARE SHOWN

(((((((((((((((((((((((((((((((((((((((((((((((((((((((((((((((((((((

"SAVE TYPES 4,9;" " primary inputs and outputs"
SAVE ALL;
"ADDSAVE + TYPES PI + REPORT" "adds signals to what has already been saved"


%%%%%%%%%%%%%%%%%%%%%%%%%%%%%%%%%%%%%%%%%%%%%%%%%%%%%%%%%%%%%%%%%%%%%

TELL SYSTEM TO MONITOR ACTIVITY OF SIGNALS

(((((((((((((((((((((((((((((((((((((((((((((((((((((((((((((((((((((

ACTIVITY OSCDET;
SIMSETUP$


%%%%%%%%%%%%%%%%%%%%%%%%%%%%%%%%%%%%%%%%%%%%%%%%%%%%%%%%%%%%%%%%%%%%%

ILLUSTRATION OF CHANGE (DENOTED BY C) COMMAND FORM OF INPUT STIMULI]

(((((((((((((((((((((((((((((((((((((((((((((((((((((((((((((((((((((

C RESET, LATCH 0 0;  CHANGE SIGNALS RESET AND LATCH TO 0 AT TIME ZERO
C RESET 1 100;
C LATCH 1 1500;
C CLK 0 0,20000,400; CHANGE SIGNAL CLK TO 0 AT TIME 0 AND REPEAT EVERY 400
                    TIME UNITS UNTIL TIME 20000

C CLK 1 300,20000,400;
C DIN 1 200,20000,800;
C DIN 0 600,20000,800;

%%%%%%%%%%%%%%%%%%%%%%%%%%%%%%%%%%%%%%%%%%%%%%%%%%%%%%%%%%%%%%%%%%%%%
(((((((((((((((((((((((((((((((((((((((((((((((((((((((((((((((((((((

SIMULATE ;
ACTIVITY OSCDET + REPORT ALL;

$
```

Figure 6.11 (*cont.*)

```
RESULTS OF RUNNING THE MODE 2 SIMULATION DESCRIBED IN PREVIOUS FIGURE
THE LINES LABELED TCC ARE COMMANDS IN THE COMMAND FILE (MOST OF WHICH WERE NOT
SHOWN PREVIOUSLY)
%%%%%%%%%%%%%%%%%%%%%%%%%%%%%%%%%%%%%%%%%%%%%%%%%%%%%%%%%%%%%%%%%%%%%%
   LOADED MODULE  2BIT
   END OF SIMULATION SCHEDULED FOR TIME    140000.
   TCC   76. SIMULATE ;

   MODE 2 SIMULATION COMMENCED.
%%%%%%%%%%%%%%%%%%%%%%%%%%%%%%%%%%%%%%%%%%%%%%%%%%%%%%%%%%%%%%%%%%%%%%

   TIME WHEN ALL X's (UNDEFINED STATES) GO AWAY IN THE SIMULATION IS 200 TIME
   UNITS (UNITS ARE SCALED TO REPRESENT THE PROPER VALUES)

(((((((((((((((((((((((((((((((((((((((((((((((((((((((((((((((((((((((

   NET IS FIRST INITIALIZED AT TIME      200

%%%%%%%%%%%%%%%%%%%%%%%%%%%%%%%%%%%%%%%%%%%%%%%%%%%%%%%%%%%%%%%%%%%%%%

   SPIKE WARNINGS; SPECIFICS OF WHEN AND WHAT CAUSED THE SPIKE ARE GIVEN

(((((((((((((((((((((((((((((((((((((((((((((((((((((((((((((((((((((((

   ...WA  DELSCH SPK
         DUE TO AN INPUT CHANGE AT TIME      1525
         A SPIKE WILL BE PRODUCED AT TIME    1533
         ON X1/EX1(O1)
   ...WA  DELSCH SPK
         DUE TO AN INPUT CHANGE AT TIME      1535
         A SPIKE WILL BE PRODUCED AT TIME    1536
         ON X1(XNO)

   TCC   77. ACTIVITY OSCDET + REPORT ALL;

%%%%%%%%%%%%%%%%%%%%%%%%%%%%%%%%%%%%%%%%%%%%%%%%%%%%%%%%%%%%%%%%%%%%%%

   ACTIVITY OF EACH NODE

(((((((((((((((((((((((((((((((((((((((((((((((((((((((((((((((((((((((

   ***** OSCILLATION CONTROL STATISTICS ***** TIME =     140000 *****

      -- TRANSITIONS --    SIGNAL
      ALLOWED  OCCURRED    NAME

                     2    (RESET)
                   101    (CLK)
                     2    (LATCH)
                    50    (DIN)
                    51    Q1
                    51    Q1B
                    50    Q2
                    50    Q2B
                    51    A1/Y1.3
                    50    A2/Y1.3
                     2    Q40
                     1    Q41
                    52    X1/EX1(O1)
                    52    X1(XNO)
                    50    X2/EX1(O1)
                    50    X2(XNO)
                    52    O1/Y1.2
                    50    O2/Y1.2

   TCC   84. DISPLAY SIGNALS  DIN, CLK, Q01, Q02;
                                              PAGE  1.  1
```

Figure 6.12 Printouts showing some of the pertinent parts of the simulation output. Included are the simulation parameters and the display results.

TIME	(DIN)	(CLK)	(Q01)	(Q02)
0	XXX	0	XXX	XXX
25	XXX	0	0	0
200	1	0	0	0
300	1	1	0	0
313	1	1	1	0
400	1	0	1	0
600	0	0	1	0
700	0	1	1	0
713	0	1	1	1
725	0	1	0	1
800	0	0	0	1
1000	1	0	0	1
1100	1	1	0	1
1113	1	1	1	1
1125	1	1	1	0
1200	1	0	1	0
1400	0	0	1	0
1500	0	1	1	0
1513	0	1	1	1
1525	0	1	0	1
1600	0	0	0	1
1800	1	0	0	1
1900	1	1	0	1
1913	1	1	1	1
1925	1	1	1	0
2000	1	0	1	0
2200	0	0	1	0
2300	0	1	1	0
2313	0	1	1	1
2325	0	1	0	1
2400	0	0	0	1
2600	1	0	0	1
2700	1	1	0	1
2713	1	1	1	1
2725	1	1	1	0
2800	1	0	1	0
3000	0	0	1	0
3100	0	1	1	0
3113	0	1	1	1
3125	0	1	0	1

Figure 6.12 (*cont.*)

The file created by the preprocessor can be used by the simulator for several purposes. These are:

1. Static and dynamic verification of digital logic design, including race, hazard, and spike analysis.
2. Interactive design analysis. This refers to answering questions of the type: "What happens if a given change to the design is made?" It is true that a good logic designer can often see what will happen under nominal conditions when a simple design change is made without the aid of a simulator program. However, without the aid of a simulator or extensive and expensive tests on a breadboard, it is virtually impossible to determine what will happen under extremes of parameter variations.

Any designer who has had to put a complex PCB in a temperature chamber, acquire the necessary high-temperature probes because use of long wires to bring signals out from the temperature chamber could not be tolerated, and then when something appears to be acting anomously wonder whether it is the test fixture itself or the PCB under test will understand the value of simulation.

To raise the temperature to 125°C from 25°C, for example, requires none of the above "hassle." It involves simply using one statement in the Tegas simulation program:

OPDELAYS ALL 141P

where 1.41 is the ratio of the delays (roughly) at 125°C to those at room temperature (25°C) for MOS devices. It will be seen that similar relatively simple changes apply to other parameter variations.

However, this is not intended to imply that logic simulation is without its problems or is infinitely faster than building and testing PCBs. It is generally superior to the latter once a user has become facile with it, but it *does* take time.

A major difference between building and testing PCBs and doing logic simulation is that in the case of the latter, the parts to be used are always available (in computer memory) if they exist at all. Any designer who has had to "scrounge" a standard part or a piece of test equipment for a PCB knows the frustrations of such tasks. By the same token, certain functions do not work well in the various simulators. Some of these are being improved. Currently, very few of the simulators do a good job of modeling CMOS transmission gates, for example.

3. Assessment of the effectiveness of diagnostic tests. Primarily, this refers to determining how effective a given set of input test patterns is in terms of exercising the nodes of a gate array or other circuit in such a way that the differences can be seen at the outputs.
4. Generation of test vectors noted earlier.
5. Investigation of the parameter tolerance of a logic design. For example, the user may have to use a clock that has a ±10% variation in the "skew" (duty factor, i.e., ratio of the time of the positive half-cycle to that of the negative half-cycle).

If the user were to do this with a breadboard, he or she would have to acquire a fairly large number of sample of clock generators, find those that exhibit the maximum skew, and then make repeated tests to determine the effects of such skew, hoping that all calibrations of equipment remained the same throughout the tests. By contrast, the user of a simulation program would simply make successive runs with the clock skew at the different values.

One approach is not necessarily less costly than the other. A high-speed computer can cost from $3000 to $7500 per CPU hour (these are actual numbers from two prominent gate array vendors), depending on the speed (MIPS) and other parameters. A long simulation run of a complex circuit can take a substantial amount of CPU time. By the same token, however, the loaded value of a person's time can also be substantial.

6. Determination of what percentage of the nodes can be exercised under any circumstances (DFS, mentioned earlier).
7. Generation of production (go/no go) and diagnostic tests.
8. Generation of tester interface files. Although this interface is no longer supported by GE/Calma, the interface programs are available to interface to the following testers: (a) GenRad 1795 PCB tester, (b) GenRad model 16 IC tester, (c) Fairchild Sentry V through VIII IC testers, and (d) Terradyne J125/J325 IC testers.

Basic parameters of simulation

Timing Increments. All timing in Tegas is in unit increments which may or may not represent different actual times for the different technologies. For example, in ECL, where propagation delays in nanoseconds are known to two decimal places, 100 time units might represent 1 ns. By contrast, in CMOS only 10 time units might be used because propagation delays are generally no more precise than one decimal place. Because Tegas is event driven, activity does not occur at (and the user is not charged for) the smaller time increments unless the user has made it occur by specifying clock or other signal changes at those increments.

Strengths Attribute. Although Tegas presently lacks this feature, it is important enough to warrant explanation. The strengths attribute is intended to characterize the ability of a node to charge or discharge that to which it is connected. There are typically three strengths that are specified. These are high impedance (high Z), resistive, and driving. A node that has a *high Z strength* can be neither charged nor discharged. A three-state buffer output is one example of such a node.

A *resistive strength* means that charge can be added to or removed from a node at a limited rate. A node connected to a power supply through a resistor is an example of a node with a resistive strength.

A *driving strength* means that charge can be added to or removed from a node at an unlimited rate. A node connected directly to a power supply rail or to ground is an example of a node that has the driving strength attribute.

Strengths versus States. The literature on simulators contains phrases such as "three-strength simulator" coupled with "nine-state simulator." One can justifiably wonder how a binary signal can take on so many states. There are two types of answers to this question.

The first has to do with the definition of a state. Some simulator vendors make a distinction between logic levels and logic states. In this terminology the number of logic states is the product of the number of logic levels and the number of strengths that the simulator is able to model. Thus a simulator, such as that of Mentor, which is advertised as a "nine-state simulator," incorporates three strengths and three logic levels. The three strengths are those noted above. The three logic levels are 1, 0, and unknown.

By contrast, Tegas5, which possesses no strengths attribute, nevertheless includes, for example, the high-impedance state in its list of states and is described by GE/CALMA as a seven-state simulator. The seven states are 1, 0, unknown initial condition, high-impedance state, upward and downward transitions within the "ambiguity region," transition to or from a high-impedance state, and potential response to a race condition.

Race, hazard, and spikes and signal cancellations. A *race condition* occurs when two signals that enter the same gate travel parallel paths. If the signal that is supposed to enable the gate arrives later than the signal coming to the other input, the result will be a false indication of timing and probably a foreshortening of the output pule. Race conditions are detectable in Tegas by observation of the signal values themselves. These can be put in offset form for easier viewing.

A *hazard condition* occurs when the signals entering a gate are changing state simultaneously in opposite directions or in such a way that the output may generate a spike.

When an input signal is passed into a series string consisting of three similar elements, three cases can be distinguished. These are the input signal width is greater than and less than the low-to-high propagation time (cases 1 and 2, respectively) and less than the high-to-low propagation time (case 3). *For the condition in which the propagation times are as shown in the example* (*but not necessarily for other conditions*), the following observations can be made.

- A foreshortening results from the skew (difference between the two times), even in case 1.
- A spike is generated in case 2.
- The signal is canceled in case 3.
- The signal in case 2 applied to the input of another element of the same type will produce signal cancellation at the output.

Tegas detects these conditions and prints out a list of the times at which they occur, the names of the signals involved, and the type (spike or signal cancellation) of condition. An example of such a printout is given in Fig. 6.12.

Modification of macro delays—the OPDELAYS command. The OPDELAYS command enables the user to modify the propagation delays encoded in the preprocessor for a given simulation run. This saves computer time because the preprocessor does not have to be reencoded with the new delay values (corresponding to perhaps another vendor's gate array or to a different temperature). It is also more reliable because in the process of reencoding, errors could be made.

All of the permutations of the commands may be applied to either all of the circuit elements or only to selected elements, such as all two-input NAND gates. The modifications may take one *or more* of several forms (i.e., the command can be successively applied). The important ones are:

1. A percentage multiplier. This is useful for temperature dependence.
2. A multiplier that is dependent on fan-out. Tegas extracts the fan-out from the preprocessor file and multiplies it by this factor.

3. An additive or subtractive constant.
4. Replacement of the existing delay by a new delay which is *totally independent* of the former delay. This is an extremely important property because it avoids the need to examine the existing delay values of each macro and then figure out how to modify them algebraically to the values desired.

Forms 2 and 4 are useful for inputting the equations of propagation delay versus fan-out.

The command can also be used to cause the values built into the macros being used to be printed out. An example of the simplicity of the command to enable runs at different temperatures to be made was given earlier. As a second example, suppose that it is desired to change the values of propagation delay versus fan-out for three-input NAND gates ND3 based on new information from the gate array vendor or from new circuit simulation results. The equations for the delays in the example are

$$L\text{-}H = 4 \times \text{fan-out} + 10 \text{ (low to high)}$$
$$H\text{-}L = 3 \times \text{fan-out} + 20 \text{ (high to low)}$$
(6.1)

(Also, it would probably be necessary to change other delay values, but putting these in would make for too long an example.)

The syntax would be (comments in parentheses are not part of the Tegas commands)

OPDELAYS TYPES ND3 + RISE 10

OPDELAYS TYPES ND3 + RISE A400PF

OPDELAYS TYPES ND3 + FALL 20

OPDELAYS TYPES ND3 + FALL A300PF

The effort involved in modifying such time delays can be considerably reduced by using copies of the aforementioned control files which have been edited to incorporate the new values.

Logic and timing simulation [3]. Logic simulation takes place in one of three modes noted earlier in the chapter, but no distinction will be made in this section among the modes. The logic and timing simulation modes are 1, 2, and 3. Modes 1 and 2 utilize nominal propagation times in a four-state simulation to provide a measure of how a given circuit will typically perform. The four states are 0, 1, X (undefined), and Z (high impedance).

Mode 1 is useful for logic verification of complex networks requiring long simulations where computer time is significant. It is also useful in doing initial debugging of the logic. It obtains high speed by eliminating certain time-consuming activity monitoring functions. Mode 2 enables the user to code many activity-monitoring and other commands which increase computer cost but which give the user a great deal of information on what is happening inside the network. Mode 3 is a worst-case timing simulator. Each signal in mode 3 can take on one of seven states. These were described under the topic "Strengths versus States."

Input stimuli. There are four different ways in which input stimuli can be applied to the circuit to be simulated. Three of the ways involve the CHANGE, FILEINPUT, and TESTPATT (for TESTPATTern) commands. These are common to all the modes of simulation. Both the CHANGE and the TESTPATT commands have permutations. The fourth way is to use a primitive element, appropriately called CLOCK, in the coding of the preprocessor. Using this primitive element has the advantage of significantly reducing the memory storage requirements of the program. The user thus has a great deal of flexibility in applying and developing input stimuli.

Each of the four methods will now be discussed briefly. The reader will want to refer to this section when going through the detailed example that follows this section.

The CHANGE Command (Abbreviated C in the Printouts). Except in the case of a step function, two or more lines are required for each change command. Normally, the first line tells when the signal or signals change to a 1 state, and the second tells when they change to a 0 state. If the signal is to remain at a low or a high state for a period of time and then periodically change state, an additional line is used to denote this. See signals EXN604 and CLK in Fig. 6.12. The change command may be repeated indefinitely for asymmetrical signals or for signals that are periodic part time, then aperiodic, and then, for example, periodic again. The Tristate level is also valid in the change command.

Example 1

> change I1 I2 Z 10,30,35,52 67$

[Read: Change signals I1 and I2 to signal level Z (the high impedance level of a three-state device) at times 10, 20, 30, 35, 52, and 67 time units. The time units could be interpreted as nanoseconds, microseconds, 1/10 ns, 10 ns, or whatever the user wishes as a scaling factor. The time units DO have to be integer, however.]

Example 2

> C I3 I4 I5 1 50,2000,100$

(Read: Change signals I3, I4, I5 to signal level 1 at times beginning at time 50 and ending at time 2000 with a time increment of 100.)

Example 2 is incomplete. Tegas has been told when signals I3, I4, and I5 go to a 1 but not when they go to a 0. Specifying when the signal goes low is a slight bit "tricky." It is in this area that a lot of mistakes are made by first-time users. If a 50% duty cycle (i.e., a square wave) is desired, the second change command might read

> C I3 I4 I5 0 100,2000, 100 $

The TESTPATT Command. The TESTPATT command can be more useful to the logic designer than the change command once he or she learns to use it. The reason is that it is more akin to the way most digital logic designers think and also more like what is actually displayed on a logic analyzer, such as a Biomation. Also, it is sometimes superior to the change command in terms of minimization of computer storage requirements.

The TESTPATT command involves a series of lines of code. Each line of code represents a *parallel input* of values (1's, 0's, z's, or even x's). Each line is input to the circuit at a predetermined time. The times of input can be absolute (e.g., at 100, 250, 350 ns) or relative to (delayed from) the immediately preceding test pattern by a stipulated amount. The command can be used in any of the simulation modes. An example of its format is

TESTPATT

(*Comment:* This calls all of the TESTPATTerns in a particular group. The command can also be used more than once, if desired, in a given simulation run but would normally not be so used.)

111111	25
10X 00	60

In this example, two TESTPATTerns (111111 and 10X-00) have been defined to occur at times 25 and 60. The second TESTPATTern contains the usual 1's, 0's, and undefined or don't-care states. The blank space means that the value of that test vector has not changed from the previous test vector and is a convenience to the user.

The meaning of each of the signal levels is given by the IVECTOR command, which appears prior to the TESTPATT command. For example, if the user were using the signals a, b, d, p, q, and m, the format would be

IVECTOR a,b,c,d,p,q,m

Where the order of the signals defines the signals in the test vector (TESTPATTern).

Clocks. Clocks can be treated like any other input signal using the change command. Alternatively, there is a clock primitive (element 53) which has cycle times specified by the delays given. The nominal, minimum, and maximum delays (in the delay statement of the preprocessor) are used to specify the number of time units at 0, the number of time units at 1, and the rise/fall times, respectively. Hence asymmetrical pulses are readily created. The change command (in the simulator part of Tegas) is used to start these clock elements. More than one clock signal can be applied to the system at once; as in real life, these clock signals do not have to be synchronized to each other.

The FILEINPUT Command. The user can specify what signals are to be saved by Tegas. Type 9 signals, which are primary outputs, can be saved and used as inputs to submodules to which the given module connects.

An important use of the FILEINPUT command is to enable vectors generated by the ATG to be used in simulations. A second important use is to enable signals from a simulation that has been run over a given time interval to be input to the circuit to continue the simulation over a successive time interval without repeating the simulation over the initial time interval.

Control files. Control files contain alphanumeric data stored in records. Each such control file can contain up to 25 separate records. The amount of data that each record can contain is limited by the default characteristics of the computer being used, but is generally more than adequate for even large circuits.

Control files are often used to call libraries of OPDELAYS commands mentioned previously to accommodate different-feature-size macros or those from different vendors.

Examples. Examples of printouts using some of the above are given in Fig. 6.12 for the 2-bit slice. The reader should try to correlate the change commands with the resultant outputs shown in the display command.

A signal tracing "trick." This section is included to show the reader what can be done with simulators once the user is facile with them. Suppose that a given input is not supposed to be affecting a given signal internal to the chip but appears to be so doing based on the simulation results. It would be handy to have a probe by which one could follow the signal. (Bear in mind that even a small chip of 800 gates may have 2200 or more, mostly internal signals. Exhibiting a small portion of these may yield 100 pages or more of printout, depending on the length of the simulation and number of signals exhibited.) Following a signal and its effects through this morass can become rather difficult (but is probably no more difficult than tracing a signal on a complex PCB with a scope probe or logic analyzer).

One way of implementing such a probe is to purposely insert either z or x values for the input in question at the appropriate time in the test pattern when the phenomenon is occurring. Although the remainder of the simulation run will be wasted, the user should be able to see how the z or x propagates (assuming that there are normally no z's or x's in the signals of interest). The technique is somewhat akin to tracing radioactive dye through the human body.

An aid in doing the tracing aside from the use of z or x state at the input is to use the computer's "find" command (which every computer has) to scan the simulation listing starting at roughly the point in time where the effect is expected to occur.

6.8 SUMMARY

The first part of this chapter described CAD tools most often needed and used. This was followed by a description of the differences between logic and circuit simulators as well as logic versus hierarchical simulators. Circuit simulation and its uses were then described. An overview of logic simulation was given followed by detailed description of use of the widely used Tegas simulator. The purpose is not to teach the reader how to use Tegas but rather to show what such a simulator can do to make the gate array design process more error free.

REFERENCES

1. Hiserote, J., J. Morris, and R. Hunter, "Semicustom IC simulation on CAE workstations," *VLSI Systems Design,* Dec. 1985, pp. 94–100.
2. Odryna, P., "Simulator merges speed of logic verifiers with circuit level accuracy," *Electronic Design,* Nov. 28, 1985, pp. 97–106.
3. Gabay, J., "Unlocking the mysteries of CAD/CAE," *Electronic Products,* Mar. 3, 1986, pp. 48–59.

Chapter 7

Design for Testability and Test Pattern Development

7.0 INTRODUCTION

An extremely important part of gate array design is that concerned with making the gate array testable. [1, 2, 3, 4, 5] Although probers, such as the Wentworth, allow probing of lines on the chip down to about 2 to 3 μm, such probing destroys the line generally and in any case is not a substitute for careful design. Therefore, it is essential that methods to test the gate array be incorporated into the design from the start, not as pieces to the design added on just before the design goes to the mask shop.

This philosophy is the opposite of many logic designers whose work is implemented on printed circuit boards. In the latter, it is much easier to probe the individual signal paths; and such probing does not destroy the path. Too often, this allows testability aspects to be addressed after the printed circuit board has been designed and is being or has been built. In circuitry implemented in gate arrays, this philosophy must change.

Aside from the pressures to get a design finished in a certain period of time, a lot of chip logic designers regard design as glamorous and challenging. Testing is regarded too frequently as a dull, boring, and routine task which does not exercise the designer's creative processes. Although this might or might not have been true for circuits implemented in printed circuit boards, it definitely is not true for design of VLSI chips. Configuring a design to enable maximum (or at least the required degree of) testability while minimizing the loading of the circuits inside the chip and the additional chip area required requires highly creative design and thinking.

A distinction needs to be made between probing signal lines on the chip and probing test points brought out to pads on the chip. Unlike the former, probing pads does not normally make them unusable. There is no problem in bonding wires to 4-mil-on-a-side bonding pads that have been probed.

The discussion that follows is couched in terms of the Tegas5 program owned by GE/Calma. It is used for the same reasons mentioned earlier: it is the most widely used logic simulator in the world; it is representative of other widely used simulators, and the subprograms involved in its use form a fairly complete package. An outline of its makeup was given in Chapter 6. The flow of its use will be similar to those of other simulators with the same capabilities.

The flow discussed in Section 7.1 is not the only flow that could be followed. Indeed, with a complex and multifaceted set of tools, such as Tegas5, HI-LO, Cadat, Lasar, or any of the others, one would expect many permutations to occur to meet individual needs. Nevertheless, the flow which follows has several salient features.

First, it is intended to make maximum use of the "smarts" of the designer. One of the drawbacks of high-powered CAD tools is that there is a temptation to substitute computer time for thinking the problem through. Computers can generate reams of impressive-looking printout which, through no fault of theirs, lack fundamental information needed to prosecute the design.

Second, it is intended to minimize computer time. The reasons are twofold. First, every computer the writer has worked on (among them IBM 4043s, VAX 11/785s, DEC20s, Amdahl .V6s and .V8s) can be swamped by numbers of people running CAD programs. This slows everyone down and makes every person less efficient. Also, when a given computer becomes too loaded, the cost of additional computers and additional software licenses must be considered. (Other options, such as using a hardware accelerator such as Zycad and buying workstations, are discussed elsewhere.)

7.1 OVERVIEW OF THE FLOW

Design for testability involves several tasks. The first is a determination of exactly what is to be tested. VLSI chips generally have many modes in which they can be operated. Even in the case of custom VLSI (of which gate arrays are a part), not all of these are exercised in system operation. Because of the need to minimize test time on VLSI testers, it is essential to avoid testing unused modes unless such testing sheds light on those modes which are used. VLSI testers have typical operating costs of a few hundred to several hundred dollars per hour. (The actual costs depend on how the $500,000 to $2 million cost of the machine is amortized, as well as the loaded labor and maintenance costs.)

The second task is that of determining what tests are necessary to verify the design and what tests are necessary for routine (production) tests. Also, the user may want to incorporate some tests to isolate problems if the tools involved in the design and fabrication of the gate array do not use the same data base.

The designer is then in a position to start the design of the circuitry to make the selected tests. Such design will take place in parallel with technology conversion.

Estimates will be made of the effects of the added circuitry for testing on speed, chip size, and usage of I/Os.

Logic or possibly circuit simulation may be used to verify the loading effects of the added circuitry. A program such as COPTR may be used to determine the observability, predictability, and testability of the nodes in the circuit. The circuit design will be modified as required. Normally, the logic is statically (functionally) verified via simple simulations or other means before and after COPTR is used just to make sure that the circuit does what it is supposed to do. The process above will then normally be iterated a number of times. COPTR is inexpensive to run, unlike many other programs. Moreover, it can be run any time there is a compiled netlist (TDL file) available. From Chapter 6 the reader will note that such a file is the input to all the simulator modes of Tegas5 and comes from compiling the result of either schematic entry or manual coding of the netlist.

Testability analysis programs other than COPTR are:

- SCOAP, developed by Sandia. Modifications of this program are used by UTMC in their Highland set of CAD tools for gate arrays and by Daisy in their workstations.
- Camelot, from Cirrus Computers (which developed the highly regarded HI-LO simulator).
- ARCOP, from MATRA design systems.
- COP, from Texas Instruments, an outgrowth of SCOAP.
- TMEAS, from Bell Telephone Laboratories.
- Testscreen, from Sperry Univac.
- Functional Level Testability Analyzer, from NEC.

Most of these work in much the same general fashion as COPTR in terms of analyzing the controllability and observability of circuit nodes.

The testability analysis program provides a means to check on what nodes have no connection to them. An output node that has nothing connected to it cannot be observed. (In COPTR this shows up as a -1 on the printout.) Similarly, an input node to which nothing is connected cannot be controlled. The designer can thus check at this early stage in the design when the testability analysis program is run to make sure that all nodes that are hanging should be hanging.

The output from the preprocessor in Tegas5 also lists the nodes that are not connected. (The preprocessor contains the netlist of the circuit. It is run before any of the simulation or other subprograms of Tegas5, including COPTR, are run.) Normally, any nonconnected items on this list that should be connected are connected. Otherwise, a single connection which has not been made that should have been will generate a list of nodes going toward the outputs of the circuit that are uncontrollable (because there is no connection to a circuit input). A corresponding statement holds for the outputs and the observability criterion. It should be noted that all of the foregoing activity takes place in the absence of any stimulus (what will later be called test vectors or TVs) to the circuit.

When the designer is satisfied that the circuit is testable within the constraints determined above and that it performs statically as it should, more extensive (and more expensive) *dynamic* (i.e., at speed) simulations can be run. *One of the most expensive parts of making a simulation is the cost of person power to study the results.* This is why in the author's flow delineated herein, such efforts are minimized until the designer is quite sure that the circuit will not have to be modified (to, for example, make it more testable or to hook up a path that should have been connected). It is also why thought *must* be given to *what* is to be simulated and *how* it will be simulated and displayed.

The circuit will then also be simulated with different combinations of speed parameters which take into account variations in propagation delays of the circuit elements with temperature, voltage, fan-out (usually done during all simulations of CMOS circuits), wafer-processing variations, clock skew, and so on. The extent of these latter simulations depends on the needs (financial, time, and degree of risk) of the program and how closely the circuit is "pushing" the technology involved with regard to speed.

When the designer is satisfied that the circuit performs statically and dynamically correctly and that it is testable, the quality of the stimuli can be tested and generation of additional stimuli to exercise more of the chip can take place. These tasks have been lumped into the categories of "fault analysis" and test vector generation. "Quality of the stimuli" refers to analysis of the ability of the inputs used to simulate the circuit with regard to their abilities to exhibit "faults" (see Section 7.4.1) in the network.

There is no standard definition of fault analysis of which the author is aware. In this book it refers to the general process of determining the quality of a set of test vectors. This also implies analyzing the faults on a circuit.

It is appropriate to treat these subjects in the same chapter as design for testability because the two go hand in glove. Moreover, on (generally rare) occasions it may be desirable to make a circuit change to decrease the number of test patterns involved. For example, the circuit may have been designed at the outset without level-sensitive scan design techniques in order to minimize chip area and number of stage delays. However, suppose that test patterns developed exceed those allowed (typically 10,000) by the gate array vendor. A trade-off analysis may then show that incorporation of such circuitry at least in selected portions of the chip is more cost and time effective than trying to determine even a semioptimal set of test patterns.

The next section deals with design for testability. This is followed by discussions of fault analysis and test vector generation.

7.2 DESIGN FOR TESTABILITY

Design for testability involves two key attributes: controllability and observability. These are the output parameters of testability analysis programs.

Test patterns (vectors) must be capable of controlling nodes of interest within the circuit, and the result of such control must be capable of being observed at one of the circuit outputs.

Williams and Parker [6] break design for testability into two categories: adhoc and structured. Adhoc approaches include partitioning, degating, adding extra test points, bus architecture systems, and signature analysis. To this list, the author would add the use of multiplexers (MUXs) and demultiplexers (DEMUXs) to control and observe desired points within the circuit.

By contrast, structured approaches have the objective of being able to observe and control the state variables of a sequential machine. There are four categories of structured approaches: random access scan (a multiplexer technique), level-sensitive scan design (LSSD) and scan path, scan/set logic, and built-in logic block observation (BILBO). LSSD and scan path enable the test generation problem to be reduced to generating tests for combinatorial logic.

7.2.1 Ad hoc Techniques

For gate array design, the most useful adhoc approaches are probably the use of degating, MUXing, DEMUXing, and test points.

Degating is the process of enabling an externally generated control signal to be substituted for a signal in the system that controls a point of interest. It is normally implemented using what will later turn out to be a degenerate AOI (AND-OR-INVERT). This consists of the AND gate feeding the NOR gate, as shown in Fig. 7.1.

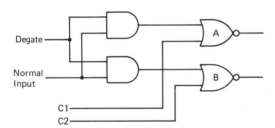

Figure 7.1 Example of degating. Holding the degating signal low enables the inputs to the sections to be controlled, and elements A and B, to come from signals C_1 and C_2. (After Williams & Parker, *Proc. IEEE*, Jan. 1983, p. 101.)

The problem with degating (as with many of the other techniques) is the additional time delay introduced by the degating circuitry. Moreover, unless degating is to be applied to only a small number of signals, the externally generated signals will have to be combined in some logical fashion to derive a larger number of control signals. This is necessary to minimize the number of package pins allocated to the degating control signals.

Such internal generation of control signals causes two problems. The first is the chip area required for such internal generation. The second is that the more complex the generation circuitry is, the greater is the liklihood that a defect in the chip will reside in the *generation circuitry itself* instead of that which the generation circuitry is testing. Similar comments apply to all the other methods.

Degating uses degenerate AOIs (see Chapter 9). These are a form of MUX. However, MUXs and DEMUXs can be used for much more. In particular, they can be used to make points either observable or controllable. Examples of such use are given in Fig. 7.2. The earlier comments about the need for additional circuitry when the point to be observed or controlled is deeply embedded in the circuit apply here also.

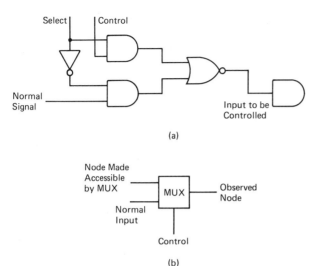

Figure 7.2 (a) MUX used to control a node; (b) MUX used to observe a node.

7.2.2 Structural Techniques

Structural techniques are somewhat more amenable to automatic computation and analysis than are adhoc techniques. They are based on usage of registers and latches which are present either in the initial design of the circuit or after the structural techniques have been designed in. The object is to make the circuit for testability purposes as much like a combinatorial circuit as possible. Testing of the latter is a well-understood and (sometimes barely) tractable problem.

Level-sensitive scan design. One of the most useful structural techniques is the LSSD (Level-Sensitive Scan Design) approach of IBM. The term "level sensitive" means that its operation is not dependent on rise, fall, or delay times of the circuit. A purpose of LSSD is to reduce the problem of testing sequential networks to one of testing combinatorial and sequential networks. It does this by adding extra elements (which typically increase the size of the chip by 4 to 20%), including hazard-free latches and multiplexers at the inputs to the flip-flops. This, of course, increases the time delays of the flip-flops and may make the technique unusable in very high speed applications.

An example of such a D flip-flop is shown in Fig. 7.3. To permit the use of LSSD techniques, the D flip-flops are connected as shown in Fig. 7.4. Note that the D flip-flops are connected in two different configurations. The first is the configuration necessary to perform the circuit function (counter, register, etc.). The second configuration is that of a serial shift register. (Of course, if the normal circuit function of the D flip-flops is that of a serial shift register, the process is simplified considerably. Normally, this will not be the case for *all* D flip-flops in the circuit.)

When the control signal TE is in its low state, the D flip-flop functions as it normally does in the particular circuit (i.e., as a counter, a register, or whatever). When TE is high, the shift register connection is invoked. By clocking the resultant

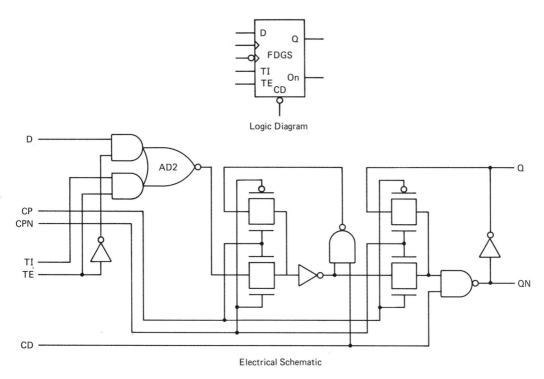

Figure 7.3 Example of D flip-flop that can be used with LSSD techniques. (Courtesy of LSI Logic, Inc.)

shift register and by keeping track of the polarity of the bits that are shifted out as a function of the clock cycle number, the contents of each D flip-flop at the time the test began can be determined. This process, of course, destroys the information that was stored in the D flip-flops at the time the test began. A possible way to prevent this loss of information is to have the output of the resultant serial shift register chain feed back to the first D flip-flop in the serial shift register chain and to make the number of clock pulses used to serially shift out the signal equal to exactly the number of stages in the resultant shift register.

The reader should note that the resultant serial shift register is the serial connection of all D flip-flops (which have scan test inputs) in the circuit regardless of their original purposes.

Signature analysis. A second major structural technique is signature analysis. This is used by CDC, National (in the SCX6260 array), and others. In this method, an on-chip linear feedback shift register generates pseudo random test patterns which are sequenced through the circuit inputs. (The chip inputs are multiplexed to permit doing this.) The outputs are compressed into signatures which are compared against signatures stored in memory external to the chip, such as ROM. The chip area increase to accommodate this technique is typically 12%.

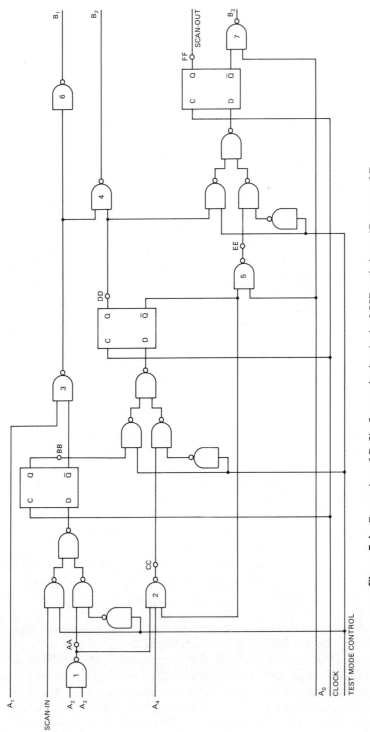

Figure 7.4 Connection of D flip-flops or latches in the LSSD technique. (Courtesy of Dr. Satish M. Thatte, Texas Instruments, Inc.; all rights reserved.)

7.2.3 Comments on Methods for Design of Testable Circuits

Problems with the additional circuitry needed to implement the methods just described have already been mentioned. A few other comments are in order. First, LSSD and its relatives configure memory elements into a serial chain as noted. A problem is that to exercise such an arrangement in a manner that permits analysis requires the storage and analysis of what can be a long serial code. This can be time consuming and expensive, depending on the circuit and the facilities available. Resnick [7] has outlined an alternative approach which is used on a gate array chip designed at CDC. The reader is referred to Resnick's paper for an explanation.

The second comment is that if loading and speeds are critical in a given path, the reader should be aware that as a rule of thumb, a transmission gate (see the discussion of CMOS macro design in Chapter 9) loads an output about $2\frac{1}{2}$ times more than does a normal gate of a transistor. For this reason, MUXs in such applications should not be made from transmission gates. (Use the normal AOI structure.)

7.3 TESTABILITY ANALYSIS USING COPTR

COPTR is an example of a class of programs designed to aid the designer in making circuits more testable. It is also used in conjuction with an ATG (automatic test generator) program to generate test vectors.

The acronym COPTR, which stands for "controllability, observability, predictability, testability report" describes quite accurately its function. The importance of controllability and observability attributes of signals was mentioned in Section 7.2. COPTR goes one step further and computes a testability parameter based on the previous two and then issues a predictability report of the testability of the circuit. By scanning the resultant report, the designer can, early in the design cycle, determine what parts of the circuit will be difficult or impossible to test. The circuit can then be modified until the desired degree of testability of the entire circuit is achieved.

Moreover, the designer can readily determine what individual signals in the circuit are untestable or will be hard to test. He or she may find that a given critical signal, for example, is not testable. If the signal is in a high-speed path, the designer may elect to modify the circuitry around the given signal (rather than that in the high-speed path itself), to try to make the signal more testable. Rerunning COPTR will give the designer instantaneous feedback of the results of such design changes on the testability of the circuit and individual signals in the circuit.

COPTR can be run as soon as the netlist is compiled. The designer does not have to wait until simulation of any kind is done. Therefore, before spending effort and CPU time on simulation, the designer knows that the circuit is testable. If simulations indicate that some of the testability of the circuit will have to be compromised or modifications made to the circuit, COPTR can be run again after such modifications are made.

```
NOTES TO READER: THE SYNTAX WHICH FOLLOWS HAS BEEN SLIGHTLY (SOMETIMES)
PERMUTED TO ILLUSTRATE A GIVEN FEATURE. SOME COMMANDS HAVE BEEN ELIMINATED TO
REDUCE CLUTTER ON THE FIGURES. DO NOT USE FOR DESIGN. THE TOP AND BOTTOM
DELIMITERS OF THE DESCRIPTION WHICH PRECEDES THE COMMAND ARE THE %%% AND
(((((( RESPECTIVELY. QUOTATION MARKS SIMPLY MEAN THAT THE COMMAND IS NOT
BEING USED IN THE GIVEN RUN. (NOT ALL OPTIONS WILL BE USED EVERY TIME.)

%%%%%%%%%%%%%%%%%%%%%%%%%%%%%%%%%%%%%%%%%%%%%%%%%%%%%%%%%%%%%%%%%%%%%
THIS SHOWS HOW THE "NET" OPTION OF THE COPTR SUBPROGRAM OF TEGAS CAN BE USED
TO TRACE AND CHECK THE FANOUTS OF THE SIGNALS AS WELL

(((((((((((((((((((((((((((((((((((((((((((((((((((((((((((((((((((((
%%%%%%%%%%%%%%%%%%%%%%%%%%%%%%%%%%%%%%%%%%%%%%%%%%%%%%%%%%%%%%%%%%%%%
REPORT THE TOTAL NUMBER OF FANINS AND FANOUTS

(((((((((((((((((((((((((((((((((((((((((((((((((((((((((((((((((((((

NUMBER OF FANINS         1391
NUMBER OF FANOUTS        1394

LOADED MODULE  STING

%%%%%%% READER: NOTE THAT THIS IS NOT THE 2BIT SLICE USED IN OTHER EXAMPLES

TCC    8.   "COPTR + REPORT <option>; OPTIONS   MAX - exhibits only problem nodes"
                                                "NET - lists fanout of each element"
TCC   10.                                       "ALL - lists all nodes and their COPTR"
TCC   11.                                              "parameters"
TCC   12.
TCC   13.   COPTR + REPORT NET;

%%%%%%%%%%%%%%%%%%%%%%%%%%%%%%%%%%%%%%%%%%%%%%%%%%%%%%%%%%%%%%%%%%%%%
THIS SHOWS THAT SIGNAL JAM0I FANS OUT (IS CONNECTED) TO 4 OTHER SIGNALS
LABELED G06/S, G06/R, G03/-001 AND G02/S. THE PRINTOUT ALSO SHOWS THAT 3
LEVELS OF LOGIC MUST BE SET TO CONTROL JAM0I TI EITHER A COMBINATORIAL ONE OR
ZERO BUT THAT 411 LEVELS OF LOGIC MUST BE SET TO OBSERVE JAM0I

(((((((((((((((((((((((((((((((((((((((((((((((((((((((((((((((((((((
```

	CONTROLLABILITY				OBSERVABILITY				
CDT	COMBINATIONAL		SEQUENTIAL					FO.	
NUM	ZERO	ONE	ZERO	ONE	COMB.	SEQ.	TEST	NUM	DEVICE NAME
26	3	3	0	0	411	0	411		JAM0I
					1011	0	1011	1	G06/S
					1110	0	1110	2	G06/R
					411	0	411	3	G03/-001
					1012	0	1012	4	G02/S

```
%%%%%%%%%%%%%%%%%%%%%%%%%%%%%%%%%%%%%%%%%%%%%%%%%%%%%%%%%%%%%%%%%%%%%
THE -1'S SHOW THAT SIGNAL L101/G IS NOT CONNECTED TO ANY OUTPUT

(((((((((((((((((((((((((((((((((((((((((((((((((((((((((((((((((((((
```

					-1	-1	-1	1	L101/G
					-1	-1	-1	2	G39/-002
					-1	-1	-1	3	G35/-001
1084	-1	16	-1	0	724	0	724		P7_53
					724	0	724	1	G40/S

Figure 7.5 (a) COPTR printout showing (a) node analysis and (b) summary of data.

```
            COPTR   STATISTICS
    TOTAL ZERO COMBINATIONAL UNCONTROLLABLE NODES       128
    TOTAL ONE COMBINATIONAL UNCONTROLLABLE NODES.       149
    TOTAL ZERO SEQUENTIAL UNCONTROLLABLE NODES...       128
    TOTAL ONE SEQUENTIAL UNCONTROLLABLE NODES....       149
    TOTAL COMBINATIONAL UNOBSERVABLE NODES.......       213
    TOTAL SEQUENTIAL UNOBSERVABLE NODES..........       213
    TOTAL TEST UNOBSERVABLE NODES................       213

    AVERAGE ZERO COMBINATIONAL CONTROLLABILITY...        95
    AVERAGE ONE COMBINATIONAL CONTROLLABILITY....       147
    AVERAGE ZERO SEQUENTIAL CONTROLLABILITY......         0
    AVERAGE ONE SEQUENTIAL CONTROLLABILITY.......         0
    AVERAGE COMBINATIONAL OBSERVABILITY..........       601
    AVERAGE SEQUENTIAL OBSERVABILITY.............         0
    AVERAGE TEST OBSERVABILITY...................       601

END OF SIMULATION RUN.
$
```

Figure 7.5 (b) (*cont.*)

Because COPTR is so inexpensive to run, runs using it can be made on modules which are to be nested within other modules before the outer modules are designed. Obviously, if the inner modules are not testable, the outer modules are not either. Two examples of the printout obtained by the use of COPTR are given in Fig. 7.5. In all columns, the greater the number, the more difficult it will be to either control or observe a given node. The circuit used is the 4-bit slice shown in Chapter 4.

The first printout shows a portion of the controllability/observability table produced by running the COPTR program. The reader will note that there are four columns under the heading "controllability," which are labeled, respectively, as combinatorial zero and one and sequential zero and one. The number under each column is the number of nodes that must be controlled to meet the stated condition. For example, consider CDT (circuit description table) number 26. The combinatorial controllability to a zero is 3 (i.e., just three nodes have to be set to control each node to that state). The sequential controllability to a 1 (fourth column in the controllability table, fifth column in the printout) is the number of clock pulses required to set its output to a 1. Because no flip-flops are involved with this node, the value is zero.

The reader will note the rows of −1's in Fig. 7.5(a). The nodes represented by these values are neither controllable nor observable and hence are not testable. The elements which these nodes connect to are not even in the circuit according to the COPTR printout. And indeed they are not. From Chapter 4 the reader will recall that some standard LSTTL parts were used to code the example. Included were quad NAND (74LS00) and NOR gates. Only one of each of the four devices was used. Therefore, COPTR properly reported that the other devices were not in the circuit. The designer would make a check to see if they *should* have been connected into the circuit, thus saving a potentially expensive rework later.

The reader will also note that Q_4 (the output from the last flip-flop in the chain) is controllable but not observable, and hence not testable. The reason is that it is an unconnected output, which again the designer would verify.

The table in Fig. 7.5(b) gives statistics of the circuit. The first two sections give the totals of the controllable and observable and uncontrollable and unobservable nodes. These numbers provide an important measure of just how difficult it will be for test patterns to be generated and an idea of how extensive the patterns must be. For example, there are 22 nodes that simply cannot be controlled. Because some of these nodes drive other internal nodes, one would expect to find that it will be impossible to generate test patterns for some of the stuck-at values (see below) of some of the nodes. This will turn out to be true.

To proceed with a discussion of the flow, it is necessary first to define a few terms.

7.4 IMPORTANT DEFINITIONS

7.4.1 Faults and the "Stuck-At" Concept

Almost all simulators model faults using the stuck-at model. In this model, the voltage of a node is fixed at either a high state (SA_1) or a low state (SA_0) or sometimes at high-impedance (SA_Z) or undefined (SA_X) states. These stuck-at values are called *faults*. El-ziq [8] has noted the limitations of this model, especially for MOS circuits (see Section 7.11).

7.4.2 Master Fault File (MFF)

The first step in fault analysis is to generate a list of all the faults in the circuit and break them into classes that are equivalent to one another. Figure 7.6 is an example of equivalent faults. Either of the inputs being stuck at 0 is equivalent (cannot be distinguished at the output) to the output node being stuck at 1.

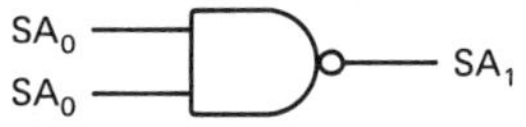

Figure 7.6 Equivalent faults.

Figure 7.7 shows a portion of the MFF for the 2-bit slice used above. The faults are collapsed into equivalent fault classes. Each class is given a number. The lead fault in each class is that which is closest to the input. The terminology used in the MFF is sometimes a trifle confusing. Consider fault 14 (fault class 14). This has two equivalent faults in it. The first is D_{in} (D input to the first flip-flop), SA_1 treated as an output (i.e., it is the output of a signal coming in on, in this case, from the outside of the chip). In effect, each signal path is treated as two separate points: the output of a driver and the input to that which is being driven. This is realistic. Either the input of a device or the output driver could be stuck high or low. A confusion is that the two points have the same node name, in this case, D_{in}.

The second fault in that class is D_{in} acting as an input to the first flip-flop. When D_{in} is SA_1, Q_1, the output of the first flip-flop, is also SA_1.

The listing in the MFF does not depend on any set of test vectors. It is strictly topology related. *All fault analysis programs will reference the numbers of faults on the MFF that are detected or potentially detected by a given set of test vectors.*

```
NOTES TO READER: THE SYNTAX WHICH FOLLOWS HAS BEEN SLIGHTLY (SOMETIMES)
PERMUTED TO ILLUSTRATE A GIVEN FEATURE. SOME COMMANDS HAVE BEEN ELIMINATED TO
REDUCE CLUTTER ON THE FIGURES. DO NOT USE FOR DESIGN. THE TOP AND BOTTOM
DELIMITERS OF THE DESCRIPTION WHICH PRECEDES THE COMMAND ARE THE %%% AND
((((((( RESPECTIVELY. QUOTATION MARKS SIMPLY MEAN THAT THE COMMAND IS NOT
BEING USED IN THE GIVEN RUN. (NOT ALL OPTIONS WILL BE USED EVERY TIME.)


%%%%%%%%%%%%%%%%%%%%%%%%%%%%%%%%%%%%%%%%%%%%%%%%%%%%%%%%%%%%%%%%%%%%%%%%
DESCRIBE THE MODE OF OPERATION.
(((((((((((((((((((((((((((((((((((((((((((((((((((((((((((((((((((((((

TCC    9.   MODE 5;
TCC   10.   LOAD;

LOADED MODULE   2BIT

%%%%%%%%%%%%%%%%%%%%%%%%%%%%%%%%%%%%%%%%%%%%%%%%%%%%%%%%%%%%%%%%%%%%%%%%
   INSTRUCT SIMULATOR TO PRINT RESULTS (PRTON) AND GENERATE MASTER FAULT FILE
(((((((((((((((((((((((((((((((((((((((((((((((((((((((((((((((((((((((

TCC   11.   PRTON;
TCC   12.   GENERATE;

%%%%%%%%%%%%%%%%%%%%%%%%%%%%%%%%%%%%%%%%%%%%%%%%%%%%%%%%%%%%%%%%%%%%%%%%
SOME RESULTS OF GENERATION OF MFF (MASTER FAULT FILE)
(((((((((((((((((((((((((((((((((((((((((((((((((((((((((((((((((((((((

DL   AUTOMATIC FAULT GENERATION SUBSYSTEM -- GENERATED FAULTS LISTING.
```

FAULT NUM	ELEMENT	SIGNAL	FAULT TYPE	STUCK VALUE
1		(RESET)	OUTPUT	S-A-1
2		(RESET)	OUTPUT	S-A-0
3		(CLK)	OUTPUT	S-A-1
4		(CLK)	OUTPUT	S-A-0
5		(LATCH)	OUTPUT	S-A-1
6		(LATCH)	OUTPUT	S-A-0
7		TIEUPTO1	OUTPUT	S-A-0
8	A2/Y1.3	Q2B	INPUT	S-A-1
9		A2/Y1.3	OUTPUT	S-A-1
	Q41	A2/Y1.3	INPUT	S-A-1
10		A2/Y1.3	OUTPUT	S-A-0
	Q41	A2/Y1.3	INPUT	S-A-0
11	A1/Y1.3	Q1B	INPUT	S-A-1
12		A1/Y1.3	OUTPUT	S-A-1
	Q40	A1/Y1.3	INPUT	S-A-1
13		A1/Y1.3	OUTPUT	S-A-0
	Q40	A1/Y1.3	INPUT	S-A-0
14		(DIN)	OUTPUT	S-A-1
	Q1	(DIN)	INPUT	S-A-1
15		(DIN)	OUTPUT	S-A-0
	Q1	(DIN)	INPUT	S-A-0
16	Q1	(CLK)	INPUT	S-A-1
17	Q1	(RESET)	INPUT	S-A-1
18	Q1	(CLK)	INPUT	S-A-0
19	Q1	TIEUPTO1	INPUT	S-A-0
20	Q1	(RESET)	INPUT	S-A-0
21		Q1	OUTPUT	S-A-1
22		Q1	OUTPUT	S-A-0
23		Q1B	OUTPUT	S-A-1
24		Q1B	OUTPUT	S-A-0
25	Q2	(CLK)	INPUT	S-A-1
26	Q2	Q1	INPUT	S-A-1
27	Q2	(RESET)	INPUT	S-A-1
28	Q2	(CLK)	INPUT	S-A-0
29	Q2	Q1	INPUT	S-A-0
30	Q2	TIEUPTO1	INPUT	S-A-0
31	Q2	(RESET)	INPUT	S-A-0
32		Q2	OUTPUT	S-A-1

Figure 7.7 Portion of MFF of 2-bit slice.

```
            (Q02)         Q2           INPUT     S-A-1
    33                    Q2           OUTPUT    S-A-0
            (Q02)         Q2           INPUT     S-A-0
    34                    Q2B          OUTPUT    S-A-1
    35                    Q2B          OUTPUT    S-A-0
    36      Q40           (LATCH)      INPUT     S-A-1
    37      Q40           (RESET)      INPUT     S-A-1
    38      Q40           (LATCH)      INPUT     S-A-0
    39      Q40           (RESET)      INPUT     S-A-0
    40      Q40           TIEUPTO1     INPUT     S-A-0
    41      Q40                        OUTPUT    S-A-1
            25            Q40          INPUT     S-A-1
    42                    Q40          OUTPUT    S-A-0
            25            Q40          INPUT     S-A-0
    43      Q41           (LATCH)      INPUT     S-A-1
    44      Q41           (RESET)      INPUT     S-A-1
    45      Q41           (LATCH)      INPUT     S-A-0
    46      Q41           (RESET)      INPUT     S-A-0
    47      Q41           TIEUPTO1     INPUT     S-A-0
    48      Q41                        OUTPUT    S-A-1
            27            Q41          INPUT     S-A-1
    49                    Q41          OUTPUT    S-A-0
            27            Q41          INPUT     S-A-0
    50      27            Q2B          INPUT     S-A-1
    51      27            Q2B          INPUT     S-A-0
    52                    27           OUTPUT    S-A-1
                          X2(XNO)      OUTPUT    S-A-0
                          O2/Y1.2      OUTPUT    S-A-1
            (O2)          O2/Y1.2      INPUT     S-A-1
    53                    27           OUTPUT    S-A-0
                          X2(XNO)      OUTPUT    S-A-1
                          O2/Y1.2      OUTPUT    S-A-0
            (O2)          O2/Y1.2      INPUT     S-A-0
    54      25            Q1B          INPUT     S-A-1
    55      25            Q1B          INPUT     S-A-0
    56                    25           OUTPUT    S-A-1
                          X1(XNO)      OUTPUT    S-A-0
                          O1/Y1.2      OUTPUT    S-A-1
            (O1)          O1/Y1.2      INPUT     S-A-1
    57                    25           OUTPUT    S-A-0
                          X1(XNO)      OUTPUT    S-A-1
                          O1/Y1.2      OUTPUT    S-A-0
            (O1)          O1/Y1.2      INPUT     S-A-0
FAULT GENERATION COMPLETE.    57 FAULTS GENERATED FROM A TOTAL OF    81.
TCC    13.   NAMELST;

LIST OF PRIMARY INPUTS, PRIMARY OUTPUTS   AND PRIMITIVE OUTPUTS

   1. (RESET)       2. (CLK)        3. (LATCH)      4. (DIN)
   5. (O1)          6. (O2)         7. (QO1)        8. (QO2)
   9. Q1           10. Q1B         11. Q2          12. Q2B
  13. A1/Y1.3      14. A1/Y2.3     15. A1/Y3.3     16. A1/Y4.3
  17. A2/Y1.3      18. A2/Y2.3     19. A2/Y3.3     20. A2/Y4.3
  21. Q40          22. DR1/Q1.6    23. Q41         24. DR1/Q2.6
  25. X1/EX1(O1)   26. X1(XNO)     27. X2/EX1(O1)  28. X2(XNO)
  29. O1/Y1.2      30. O1/Y2.2     31. O1/Y3.2     32. O1/Y4.2
  33. O1/Y5.2      34. O1/Y6.2     35. O2/Y1.2     36. O2/Y2.2
  37. O2/Y3.2      38. O2/Y4.2     39. O2/Y5.2     40. O2/Y6.2
TCC    14.   END TDL;

END OF TDL RUN.
$
```

Figure 7.7 (*cont.*)

7.4.3 Test Vector

A test vector is the spatial representation of the set of input signals in parallel. For example, 0010 might represent the input signals clock, D, reset, and clear *in that order* which are input to the circuit at a given time. It is important to note that a TV (test vector) can be created by means of one of the change commands listed in Chapter 6.

7.4.4 Detectability

A fault (a stuck at 1 or 0) is said to be detectable for a given input pattern (test vector) when that fault (and *only that fault*) inserted into the circuit produces an output which is different from the output produced when the fault is not present.

7.4.5 "Bad Machine"

The single fault inserted into the network at a given node produces what is known as a "bad machine." This is a copy of the circuit that has the given node held or "stuck" at the given value (1, 0, or Z).

7.5 FAULT ANALYSIS TOOLS

The next step in the procedure is to analyze the set of stimuli that the designer developed to simulate the circuit. The analysis is couched in terms of the numbers of faults on the MFF that can be detected by a given set of stimuli. Although it is true that an ATG (automatic test vector generator) could be simply run at this point, such a run would be wasteful of computer power. The stimuli already developed by a designer who has quite thoroughly simulated his or her circuit will generally cause a very high percentage of faults to be detected, as defined above.

Several techniques are available to analyze the "goodness" of a set of test vectors. Goodness in this context refers to the number of faults on the master fault file (MFF) that can be detected by the given set of test vectors.

7.5.1 DFS Program

One program used for analysis of the given set of test vectors is the DFS (detectable fault simulation) program. This is an inexpensive-to-run program which places an upper bound on how many faults (on the master fault file) can *potentially* be detected with that set of test vectors. Unless the number of such faults is relatively high, the user discards that test vector set or else retains it and adds to it. Faults reported on by DFS are potentially detectable (if not undetectable) because DFS assumes that every node in the network can be observed at the output. DFS also gives an indication of how many nodes (actually, the percentage of nodes on the so-called "master fault file," which is the set of collapsed faults) are being toggled with that set of test vectors as inputs. This is important because logic simulation may not have exercised the circuit in the way in which it will be used because of limitations on computer time.

An example of the output of a DFS run is shown in Fig. 7.8. The reader will note that the faults not detected were placed in another file.

```
TCC   33.  DFS;
DETECTABLE FAULT SIMULATION ACTIVATED.
SIMULATION STOPPED AT TIME    140000 FOR THE FOLLOWING REASON.
    A SCHEDULED STOP

DFS OPTION COMPLETE.  THERE WERE    35 FAULTS POSSIBLY DETECTABLE
OUT OF    57.  MAXIMUM PERCENT POSSIBLE DETECTABILITY IS  61.40.
TCC   34.   FAULTFILE 3;

THE FOLLOWING TABLE CONTAINS THE NUMBERS OF THE FAULTS ON FILE  9.
        7           8           9          11          12          14          15
       19          22          23          29          30          33          34
       40          41          47          48          50          53          54
       57
THERE ARE    22 FAULTS ON FILE  9.
TCC   35.   END;

  END OF TDL RUN.
FORTRAN STOP
$
$
$
$
```

Figure 7.8 Results of DFS run.

7.5.2 Mode 5

The next step is to see how many of the potentially detected faults were truly detected by the set of TVs. To do this, the MFF is replaced by the file containing the potentially detected faults. (Because it is somewhat expensive to run, the MFF is sometimes copied to a holding file.) This is done using mode 5 (or mode 4), which is more extensive but more costly (than DFS) to run. Both mode 4 and its more-expensive-to-run cousin, mode 5, determine which of the faults that DFS says are potentially detectable are truly detectable *with the given set of test vectors used*. A different set would give different numbers (on the MFF) of faults which are detectable.

The result is that the potentially detected faults have now been written to two files. The first is a second file containing those that are not detectable (so now there are two files containing undetectable faults). The other contains those faults that are definitely detectable.

Note that the expensive mode 5 has been run on only a fraction of the MFF (the potentially detected fault file), not on the entire MFF. This saves cost and loading of the computer.

The user can now analyze the faults that are truly undetectable and either manually or automatically generate TVs that will detect them. It may turn out that simply by changing the state of another control line and repeating the change commands or test patterns, the designer can cause a large group of faults to be detected that were not detected before. In each case, the additional stimuli are added to those that existed before to provide a complete set of test vectors for the tester to be used to test the actual gate array. (Simulation of the faults, however, is done only with the new stimuli commands as opposed to the complete set of stimuli.)

The advantage of this method is that increasingly smaller files (those containing the faults that remain undetected) are worked with each time. This saves large amounts of computer power.

Modes 4 and 5. Fault analysis using mode 4 or 5 involves putting a fault on a node of a network and then seeing if that fault can be detected at the outputs of the circuit *with the given set of test vectors* at the inputs.

Types of fault simulators. [9, 10] There are at least four different methods of fault simulation: serial, parallel, deductive, and concurrent.

In the *serial* method, one bad machine is simulated with one good machine (a circuit with no SA faults). In the *parallel* method, 31 bad machines (for a 32-bit computer, i.e., one less bad machine than there are bits in the computer word) are simulated against one good machine on each pass of the computer. (In fault simulation, the computer program automatically does the assignment of faults to the nodes on each pass.) In the *deductive* fault simulator, only the good machine is simulated, but all faults that are detectable on an internal or terminal node up to that point in the circuit are deduced simulataneously and carried along at each stage. *Concurrent* fault simulation is somewhat similar to deductive simulation in that it deals with lists of faults at each node. However, in the concurrent simulator, the lists are updated only if they are different from those of the good machine. Concurrent fault simulation requires more storage of the fault lists associated with each node than does deductive fault simulation but is favored over the latter for simulations involving high-level descriptions.

Tegas5 uses parallel fault simulation.

7.5.3 Path Sensitization

The faults listed under each fault number in the MFF are equivalent to each other. This does not say that they are detectable *without* proper sensitization of the circuit via an appropriate test pattern. This means that the test vector has to be set up so that it sets the stages of the other inputs around that being controlled to the proper values. For example, in the case of the two-input NAND gate used earlier, one of the two inputs would have to have an input signal of 1 to enable a SA_0 condition to be detected on the other. The other input would also have to have a 1 generated by the test vector (which, in general, will be many levels removed from that input). Therefore, the output remaining high would indicate that one of the three equivalent faults (one fault number) in Fig. 7.6 with the two-input NAND gate in the MFF has been detected by that pattern (of two 1's).

7.5.4 Outputs from Mode 5

There are several useful outputs from mode 5. The first is a list of the fault numbers of the MFF which the given set of test vectors will detect together with another list of those the set of vectors will not detect. An example of these outputs is shown in Fig. 7.9 for the case involving the 2-bit slice circuit used above.

The second main output from running mode 5 is a "dictionary" that lists each good vector and the change in that vector that will exhibit the stuck-at faults listed. An example of such an output is shown in Fig. 7.9. In each case, the good and bad output vectors are listed together with the time and the signal names and the number on the MFF. (The times listed are those at which the outputs actually occur, starting with the first. They are generally, but not necessarily, in the same order in which the input test vectors occur. Therefore, it is difficult to correlate the output test vectors with the input test vectors.)

The CPU time required is proportional to XN^2, where N is the number of gates or elements and X is the average number of nonequivalent faults per gate or element. The exponent 2 is for a two-state simulator. Therefore, iterations of mode 5 using small numbers of fault classes on the MFF will generally result in less CPU time being used overall.

```
NOTES TO READER: THE SYNTAX WHICH FOLLOWS HAS BEEN SLIGHTLY (SOMETIMES)
PERMUTED TO ILLUSTRATE A GIVEN FEATURE. SOME COMMANDS HAVE BEEN ELIMINATED TO
REDUCE CLUTTER ON THE FIGURES. DO NOT USE FOR DESIGN. THE TOP AND BOTTOM
DELIMITERS OF THE DESCRIPTION WHICH PRECEDES THE COMMAND ARE THE %%% AND
((((((( RESPECTIVELY. QUOTATION MARKS SIMPLY MEAN THAT THE COMMAND IS NOT
BEING USED IN THE GIVEN RUN. (NOT ALL OPTIONS WILL BE USED EVERY TIME.)


%%%%%%%%%%%%%%%%%%%%%%%%%%%%%%%%%%%%%%%%%%%%%%%%%%%%%%%%%%%%%%%%%%%%%%%%
DESCRIBE THE MODE OF OPERATION. MODES 1 AND 2 ARE USED FOR LOGIC
SIMULATION. MODE 1 IS A SIMPLIFIED VERSION OF MODE 2. MODES 4 AND 5 ARE USED
FOR FAULT SIMULATION. MODE 4 RUNS ABOUT 3 TIMES FASTER THAN MODE 5.

(((((((((((((((((((((((((((((((((((((((((((((((((((((((((((((((((((((((



TCC    12.   MODE 5;
TCC    13.   LOAD;

LOADED MODULE   2BIT

%%%%%%%%%%%%%%%%%%%%%%%%%%%%%%%%%%%%%%%%%%%%%%%%%%%%%%%%%%%%%%%%%%%%%%%%
TELL SIMULATOR THAT A MASTER FAULT FILE EXISTS AND COMMAND IT TO NOT GENERATE
A NEW ONE
(((((((((((((((((((((((((((((((((((((((((((((((((((((((((((((((((((((((

TCC    14.   NOGEN;

%%%%%%%%%%%%%%%%%%%%%%%%%%%%%%%%%%%%%%%%%%%%%%%%%%%%%%%%%%%%%%%%%%%%%%%%
TELL WHICH VECTORS ARE TO BE INCLUDED IN THE  INPUT VECTOR TEST LIST AND GIVE
THEIR RELATIVE POSITIONS SPATIALLY
(((((((((((((((((((((((((((((((((((((((((((((((((((((((((((((((((((((((

TCC    16.   IVECTOR
TCC    17.   RESET, CLK,LATCH
TCC    18.   $
THE FOLLOWING SIGNALS HAVE BEEN INCLUDED IN THE INPUT   VECTOR LIST.
(RESET)    (CLK)     (LATCH)
```

Figure 7.9 Mode 5 output.

```
TCC    19.   CONTROL SIMSET$
TCC    23+   SIMSETUP$
THE OUTPUT VECTOR WILL CONSIST OF THE FOLLOWING SIGNALS.
           1.  (O1)
           2.  (O2)
           3.  (QO1)
           4.  (QO2)

%%%%%%%%%%%%%%%%%%%%%%%%%%%%%%%%%%%%%%%%%%%%%%%%%%%%%%%%%%%%%%%%%%%%%%%
INPUT THE STIMULII
(((((((((((((((((((((((((((((((((((((((((((((((((((((((((((((((((((((((

TSS    24+   C RESET, LATCH 0 0;
TSS    25+   C RESET 1 100;
TSS    26+   C LATCH 1 1500;
TSS    27+   C CLK 0 0,20000,400;
TSS    28+   C CLK 1 300,20000,400;
TSS    29+   STOPSIM 140000
TSS    30+   $
END OF SIMULATION SCHEDULED FOR TIME     140000.
TSS    31+   END SIMSETUP$

TCC    33.   SIMULATE;

MODE 5 SIMULATION COMMENCED.
SIMULATION STOPPED AT TIME    140000 FOR THE FOLLOWING REASON.
    A SCHEDULED STOP

%%%%%%%%%%%%%%%%%%%%%%%%%%%%%%%%%%%%%%%%%%%%%%%%%%%%%%%%%%%%%%%%%%%%%%%
RESULTS OF RUNNING THE MODE 5 FAULT ANALYSIS ARE SHOWN BELOW. ON EACH PASS,
UP TO 31 FAULTS ARE SIMULATED. THE NUMBER OF DETECTED FAULTS IN EACH PASS IS
DETERMINED AS IS THE CUMULATIVE NUMBER THAT HAVE BEEN DETECTED.
(((((((((((((((((((((((((((((((((((((((((((((((((((((((((((((((((((((((

    PASS   1   FAULT STATISTICS

                              THIS  PASS                    TOTALS
                        NUMBER         PERCENT         NUMBER         PERCENT
FAULTS DETECTED          6.  OF  31.    19.35           6.  OF  31.    19.35
FAULTS NOT DETECTED     25.  OF  31.    80.65          25.  OF  31.    80.65


MODE 5 SIMULATION COMMENCED.
SIMULATION STOPPED AT TIME    140000 FOR THE FOLLOWING REASON.
    A SCHEDULED STOP

    PASS   2   FAULT STATISTICS

                              THIS  PASS                    TOTALS
                        NUMBER         PERCENT         NUMBER         PERCENT
FAULTS DETECTED         10.  OF  26.    38.46          16.  OF  57.    28.07
FAULTS NOT DETECTED     16.  OF  26.    61.54          41.  OF  57.    71.93

%%%%%%%%%%%%%%%%%%%%%%%%%%%%%%%%%%%%%%%%%%%%%%%%%%%%%%%%%%%%%%%%%%%%%%%
FILE 10 IS THE UNDETECTED FAULT FILE WHICH LISTS ALL OF THE FAULTS
ON THE MASTER FAULT FILE WHICH HAVE NOT BEEN DETECTED BY THE GIVEN SET OF
TEST VECTORS
(((((((((((((((((((((((((((((((((((((((((((((((((((((((((((((((((((((((
THE FOLLOWING TABLE CONTAINS THE NUMBERS OF THE FAULTS ON FILE 10.
         1          2          3          4          5          6          8
         9         10         11         12         13         14         15
        16         17         18         20         22         23         25
        27         28         29         31         33         34         36
        37         38         39         41         43         44         45
        46         48         50         53         54         57
THERE ARE     41 FAULTS ON FILE 10.
$
```

Figure 7.9 (*cont.*)

7.5.5 One Lack

One feature currently lacking is an automatic or semiautomatic technique for eliminating those test vectors which are doing the least. For example, one test vector might be capable of detecting 10 faults. Another might be capable of detecting only one. If that one is included in the 10 that the first vector encompasses, the second vector should be eliminated. The only way to do this currently is to manually edit the list of test vectors using the information contained in the output from mode 4 or 5. Because some gate array vendors restrict the number of TVs that can be used to, for example, 10,000, this factor is sometimes important. (Complex arrays with many modes of operation and those employing long-maximal-length feedback shift register sequence generators are two examples of circuits that require many TVs.)

7.6 SUBTLETIES

7.6.1 Faults That Cannot Be Detected

At least four classes of faults (on the MFF) exist that cannot be detected. These can be broadly classified as wired to power supply, redundant, clock, and inverter faults. Examples of each will be given. There is no standard nomenclature.

The reader will note that in order to detect a fault, two things must happen. First, the input to the node in question must be set to the opposite state of the stuck-at condition (i.e., a 1 for SA_0). Second, it must be possible to detect a difference between the effects of that node at an output under the two conditions (the stuck value and the opposite state).

Wired to power supply. Unused portions of macros are sometimes tied to one of the power supply rails. Depending on how this is done (either with a global or local option which specifies that all no connects are logical 1's or 0's or with a power or ground macro), the simulator may include the item as a fault class in the MFF.

Redundant. Paralleling is one example of a widely used technique wich produces untestable faults. It is typically used to speed up a path by lowering the driving-point resistance. Examples of its use in CMOS are given in Chapter 9. Another example is given in Fig. 7.10(a). To detect a SA_0 on the input to, say gate B, a 1 must be placed on its input and a 0 must be placed on the input to A. This will then allow the C output to be determined as the inverse of the B output. The reader will note that if the blocks labeled logic A and logic B have the same truth table, there is no way for the input to be configured so that it will simultaneously produce a 0 on the A output and a 1 on the B input.

A method of overcoming this difficulty is shown in Fig. 7.10(b). The additional gates enable the proper test to occur. However, if the purpose of the paralleling was to gain speed, some of that speed at least has been lost by the addition of the extra gate levels of delay.

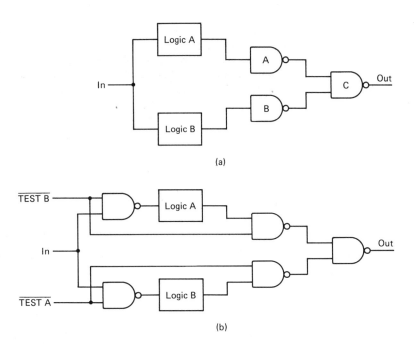

Figure 7.10 (a) Example of redundant fault; (b) its cure. (Courtesy of Signetics; all rights reserved.)

Another very common example of a redundant fault is shown in Fig. 7.11. The reader should confirm that a SA_1 at input from D to E could never be detected as a fault.

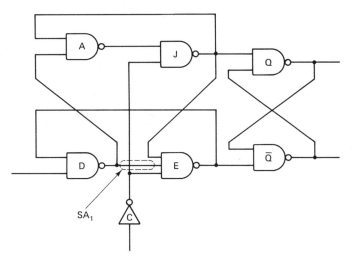

Figure 7.11 Example of a redundant fault: flip-flop path. (Courtesy of Signetics; all rights reserved.)

Sec. 7.6 Subtleties

Clock faults. Latches made from fixed-size macros (see Chapter 4) from some gate array vendors sometimes lack reset capability simply because there are not enough transistors in the cell. (Many of these of which the author is aware are being corrected in newer versions of the respective gate arrays.) Flip-flops made from these latches will come up in the "X" (undefined) state at the start of simulation. They will stay that way for the entire simulation for stuck-at faults on the clock lines because there is no way to clock in initializing values. Therefore, *for simulation purposes, all flip-flops should be resettable*. Unfortunately, as seen, this is not always possible. Such faults are generally nonfaulted on the MFF.

Inverter-type faults. This is really a subset of the redundant faults class. An example is shown in Fig. 7.12. In ECL, where both a signal and its inverse are available and where the logic swing is very low, it is common to drive long paths with both phases of the signal as shown. The reader will note that there is no way for a SA_0 on the lower input to the receiver to be detected. The reason is that D always follows directly the input because of the SA_0.

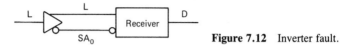

Figure 7.12 Inverter fault.

Faults that cannot be detected can give a designer "fits." Unless precautions are taken, they show up as undetected faults (of the MFF) during fault grading. The designer then has the choice of either gambling that all the undetected faults cannot be detected or laboriously verifying that such is the case by correlating the undetected faults with the schematic. If the latter is done, the faults are removed from the MFF by means of a nonfault procedure. Alternatively, their percentage is added to the existing fault coverage.

A more common method is to fault to only the macro level but not to allow the system to generate faults for the nodes inside the macro. This eliminates many faults of all classes because faults internal to a macro are not listed in the MFF. (Recall that Tegas relates all faults to the MFF.)

Before being allowed in the library, the layout of the macro and its coding are thoroughly checked to minimize errors due to not faulting within the macro. Actual faults that occur within the macro are likely but not guaranteed to appear at the macro outputs for a given set of test vectors. The larger the macro, the less likely is the occurrence of such faults. The reader can see that thought must be given to how a macro will be modeled fault-wise when coding the macro for inclusion in the macro library.

Arguments exist as to the minimum fault coverage needed. In theory, if the faults that cannot be faulted are removed from the MFF, it should be possible to get 100% fault coverage. The problem with this approach is the amount of computer and person time required. Figure 7.13 shows that *exclusive* of no faulting as described above, it is very difficult to achieve greater than 95% fault coverage.

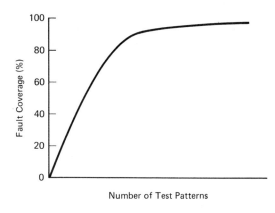

Figure 7.13 Relation between fault coverage and test set size. (Courtesy of Dr. Satish M. Thatte, Texas Instruments, Inc. All rights reserved.)

7.6.2 Decreasing the CPU Time of Fault Grading

Fault analysis is very CPU intensive. Some methods for reducing the CPU times for fault coverage have been alluded to earlier: forcing the analysis and test vector generation programs to run on only the undetected and potentially detected faults until satisfactory coverage is obtained, faulting to only the level of the macro, no faulting nodes with undetectable faults, and so on. There are others. Concurrent fault simulation is one method. A hardware accelerator is another. Zycad, for example, has a module that does fault analysis with the huge decrease in CPU time characteristic of its logic simulator accelerators.

Another entirely different approach is to use partitioning. This will decrease the costs by allowing smaller numbers of faults to be simulated at once. (See below.) The partitioning is often that inherent in the system. A difference between the partitioning used in the system and that needed for fault analysis is that in the latter, there must be I/Os of the overall circuit. By contrast, partitioning that utilizes macros nested multiple levels deep may not have the necessary I/Os to enable faults to be detected at the circuit outputs.

CPU times required. See Appendix C for typical CPU times taken for the various subprograms of the Tegas simulator and a particular gate array design. The treatment here follows that of Thatte [11] The CPU time for fault simulation can be given as

$$XTN^2 \quad \text{(for a two-state simulator)} \qquad (7.1)$$

where T is the time in CPU seconds to simulate one circuit cycle, N is the number of test vectors, and X is number of nonequivalent faults per gate.

The CPU time T required for test generation is

$$T = AN^X \qquad (7.2)$$

where A is a constant that depends on the simulator used.

If the system is partitioned into K unique partitions each of which contains M gates with X nonequivalent faults per gate, the test generation time T_m of a single such partition is

$$T_m = AM^X = A\left(\frac{N}{K}\right)^X \tag{7.3}$$

Therefore, the test vector generation time for all partitions is

$$KT_m = KA\left(\frac{N}{K}\right)^X$$
$$= \frac{T}{K^{X-1}} \tag{7.4}$$

This shows that the test generation time is reduced by the number of partitions raised to a power that depends on the average number of nonequivalent faults per gate. For example, if there are three nonequivalent faults per gate, the time required to do automatic test generation is reduced by the square of the number of such partitions.

In real life, of course, the partitions will almost never be the same size. Nevertheless, the decrease in simulation times due to partitioning is substantial.

Another important benefit of partitioning is that if a partition is modified, test generation needs to be done only for that partition.

7.7 EFFECTS OF FAULT COVERAGE ON YIELD

Williams at IBM [12], Wadsack at Bell Labs [13], and others have postulated that there is a relation between the fault coverage, F, the process yield, Y, and the field rejection rate, D. The two models and results for a 95% fault coverage and yield of 0.15 are

$$D = (1 - F)(1 - Y) \quad \text{Wadsack model}$$
$$= 0.04 \tag{7.5}$$
$$= 1 - Y^{1-F} \quad \text{Williams model} \tag{7.6}$$
$$= 0.08$$

These are important considerations because the costs of a failure in the field are three to four (and sometimes uncountable when human life and company reputations are at stake) orders of magnitude greater than the costs of a failure in the design phase. It gives reason to do more "up front." A benefit of the models above is that they enable semiquantitative values to be arrived at in terms of risk versus testing and CPU costs.

7.8 TEST VECTOR GENERATION AND USE

As noted earlier, a test vector is a set of 1's, 0's, and possibly X's (don't cares), and Z's (high-impedance states) which are applied in parallel to the circuit as an input stimuus at a given time. The user specifies the meaning of the input test pattern (which digit represents which input) via an IVECTOR command in the fault simulation. The user also specifies the absolute or relative (to the immediately preceding vector) times to apply such test patterns.

Instead of manually deriving test vectors, the user can use some form of ATG to generate them automatically. The vectors so generated are added to the stimuli provided by the schematic designer. Iterations are normally done.

7.8.1 Automatic Test Vector (Pattern) Generator

To find suitable test vectors, GE/Calma introduced a "new" automatic test pattern generator to be used in conjunction with COPTR. (The new ATG replaces the older Pathsen and Heurtest programs used in Tegas3, 4, and 5 for those readers familiar with Tegas.)

The new ATG, henceforth called simply ATG, makes use of path sensitization via the so-called "D algorithm" to develop test vectors for the circuit automatically. It also makes use of the COPTR analysis.

The reader will observe that reset is generally high except for occasional pulsings low to reset the circuit, CLK repetitiously goes high and low, and that D_{in} and Latch also behave as one would reasonably expect the circuit to behave. (However, bear in mind that the object is to generate test vectors that will exhibit the stuck-at faults in the network, not to operate the circuit in one of its modes of operation.)

The D algorithm.* One of the most widely used algorithms for test pattern generation is the *D algorithm* developed by Roth and others at IBM. In this algorithm, D and D_b are used as values. Typically, D is 1 and D_b is 0. Under fault conditions, the reverse is true. The propagation cube for a two-input AND gate with inputs a and b and output c is

a	b	c
D	1	D
1	D	D
D_b	1	D_b
1	D_b	D_b

The test cube against which this will be measured is

1	1	D
0	X	D_b
X	0	D_b

To provoke a fault, D is assigned for a SA_0 (stuck at 0) fault, and D_b is assigned for a SA_1 fault. The fault is then propagated forward toward a primary output

---

*The following development of the D algorithm is an abridgement of class notes of Satish M. Thatte of TI [11]. Used with permission.

along a sensitized path. It may also be propagated backward toward a primary input to generate the proper input condition. The combination of the forward and backward drives will generate a test for a stuck-at fault if one exists. If no test can be found, there is redundancy in the logic.

7.9 ITERATIVE EXHAUSTIVE TEST PATTERN GENERATION

There are a variety of problems with the present approach to automatic test vector generation. These include the requirements of large amounts of computer time and the dependence on the *single* stuck-at model as a criterion.

The purpose of this section is to make the reader aware of a technique that may help overcome some of the inadequacies of the present conventional approach. The technique is called *iterative exhaustive test pattern generation*. A paper [14] reviews the work to date.

Iterative exhaustive test pattern generation schemes generate the set of all possible inputs feeding a given output. Therefore, patterns are provided for any *combination* of hard faults. This is unlike the single stuck-at criterion used presently. Tang and Chen make the statement that "test patterns can be generated quite easily . . . the process and its fault coverage are no longer dependent directly on the fault model assumed . . . ."

A major problem with current iterative exhaustive test pattern generation schemes is arriving at a method to handle multiple outputs simultaneously. This is the subject addressed by Tang and Chen.

One of the keys to effective utilization of iterative exhaustive test pattern generation is partitioning to limit the size of each input subset associated with an output. This subject has been addressed by McCluskey and Nesbat [15].

The iterative exhaustive test pattern generation approach can be used in conjunction with other approaches, including the pseudorandom-pattern testing approach.

7.10 TESTER INTERFACE [16], [17]

Although GE/Calma no longer supports the tester interface, it is still widely used. Many companies have adapted it to their own use. Key elements in the tester interface are time sets, strobe times, tester cycle times, and the increased loadings of the tester itself.

Strobe times refer to the times at which the tester will sample the designated outputs of the circuit. If the output being sampled is there and of the proper polarity at the strobe time, that part of the test is passed by the circuit. Otherwise, additional software tells the tester whether to continue the test or to reject the part.

Tester cycle time is simply the repetition rate of the tester. A Fairchild Sentry series 21, for example, has a repetition rate of 20 MHz with options to 40 MHz.

Time sets. Time sets can be one of the more confusing aspects of the tester interface. The basic idea of a time set is that a given signal in one of the allowed time sets can occur (go high or low or to a high-impedance state) *only* at a given position with respect to the edge of that time set. The signal does not have to change at all; but if it does, it must change at the fixed position with respect to the time-set edge. The fixed position is specified by the user. The tester interface program reads the SAVE file generated by the logic simulation to determine whether the signal changes. Use of the tester interface program is thus restricted to circuits which in general are synchronous. (Asynchronous signals, such as resets, can readily be implemented, however.) (See Figure 11.17.)

7.11 SUITABILITY OF THE "STUCK-AT" MODEL

The stuck-at model, although while used almost universally, does have some drawbacks. First is that the stuck-at faults are always applied one at a time. This is necessary to make the problem reasonably tractable, although some work has been done with handling multiple stuck at faults. (A multiple stuck-at fault might involve one of the inputs to a two-input NAND gate SA_1 and the other SA_0, for example. A second example would be both inputs SA_0 or both SA_1. A third example would be stuck-at faults placed on multiple elements in the chain during the same pass.)

A second major drawback to the stuck-at model is that it really does not represent the majority of the fault types that occur in CMOS devices. According to Y. M. El-ziq [8], gate-to-drain and gate-to-source short circuits are the most prevalent failure mechanisms in MOS devices. The next most prevalent failure mechanisms are drain and source contacts being open. Stuck-at 1 and 0 faults are the least likely to occur.

Williams and Parker [18], [19] point out that there are a number of faults that could change a combinatorial circuit into a sequential circuit and that the use of single stuck-at faults does not cover the bridging problems.

7.12 SUMMARY

Design for testability plays a key role in gate array design. Unlike a printed circuit board, one cannot fix a gate array with a wire-wrap gun and an X-Acto knife. Therefore, testability must be built in and test vectors must be generated to ensure that the chip will be adequately tested. Common techniques, both ad hoc and structural, were delineated. Important testability analysis programs such as COPTR and SCOAP were noted. Analysis of an example circuit using the former program were given.

Because of the large amount of computer resources required for such activities, methods were given that help alleviate some of the CPU requirements. The important questions of how much testing is enough were addressed taking into consideration the effects on field failures. A new method which is quite different from those previously used, called "iterative exhaustive test pattern generation," was mentioned. Finally, the adequacy of the widely used stuck-at model was discussed.

REFERENCES

1. Singer, P., "Test software development," *Semiconductor International*, Sept. 1985, pp. 76–81.
2. McCluskey, E., "Testing semi-custom logic," *Semiconductor International*, Sept. 1985, pp. 118–23.
3. Archambeau, E., "Testability analysis techniques: a critical survey," *VLSI Systems Design*, Dec. 1985, pp. 46–52.
4. Radhakrishnan, D., and R. Sharma, "Easily testable CMOS cellular arrays for VLSI," IEEE International Conference on Computer Design: VLSI in computers," *IEEE publication 85CH2223-6*, Oct. 7–10, 1985.
5. Dervisoglu, B., "VLSI self-testing using exhaustive bit patterns," IEEE International Conference on Computer Design: VLSI in computers," *IEEE publication 85CH2223-6*, Oct. 7–10, 1985.
6. Williams, T. and K. Parker, "Design for testability—a survey," *IEEE Trans. on Computers*, Jan. 1982, pp. 2–15.
7. Resnick, D., "Testability and maintainability with a new 6K gate array," *VLSI DESIGN*, March/April 1983, pp. 34–38.
8. El-ziq, Y., "Classifying, testing, nd eliminating VLSI MOS failures," *VLSI DESIGN*, Sept. 1983, pp. 30–35.
9. Son, K., "Fault simulation with the parallel value list algorithm," *VLSI Systems Design*, Dec. 1985, pp. 36–45.
10. Waicukauski, J., et. al., "Fault simulation for structured VLSI," *VLSI Systems Design*, Dec. 1985, pp. 20–31.
11. Thatte, S.M., "VLSI Design for Testability," Seminar, Dedham, MA, April 27, 1984.
12. Williams, T., "Design for testability," NATO Advanced Study Institute on Computer Design Aids for VLSI Circuits, 1980.
13. Wadsack, R., "VLSI: How much fault coverage is enough?" International Test Conference, Cherry Hill, NJ, Oct. 1981.
14. Tang, D., and C. Chen, "Iterative exhaustive pattern generation for logic testing," *IBM Jl. Research and Development*, Mar. 1984, pp. 212–19.
15. McCluskey, E., and S. Nesbat, "Design for autonomous test," *IEEE Trans. on Computers*, Nov. 1981, pp. 866–75.
16. Huber, J., "Bridging the gap between CAE design and testing," *Electronic Products*, Mar. 17, 1986, pp. 53–57.
17. McLeod, J., "How the PC is changing testing," *Electronics*, Mar. 24, 1986, pp. 31–38.
18. Williams, T., "Design for testability: what's the motivation?," *VLSI DESIGN*, Oct. 1983, pp. 21–23.
19. Williams, T., and K. Parker, "Design for testability—a survey," *Proceedings of the IEEE*, Jan. 1983, pp. 98–112.

Chapter 8

Workstations and Other Tools

8.0 INTRODUCTION

Chapters 6 and 7 dealt with software tools for CAD. It is appropriate now to discuss some of the hardware tools that are used. It is assumed that the reader is already familiar with basic alphanumeric terminals, such as the DEC VT100 and 200 series, and is at least vaguely aware of the existence of major mini and mainframe computers, such as the VAX 11/785 and 8800 of DEC, the Data General MV20000 series, the Apollo Domain series, the IBM 3084, and the Amdahl 470 V-8. These are common machines that are used for many purposes, both technical and nontechnical. Therefore, the discussion below is aimed at hardware tools which are especially useful for the design of gate arrays. These are workstations and IGSs (interactive graphics systems) and hard-copy output devices. Again the emphasis in the latter section is on those uncommon peripherals which are most useful for the gate array implementation process.

Brief mention and discussion of special-purpose processors are included as examples of items of that type. One processor is the Zycad unit, which does simulations and fault analysis up to 1000 times faster than on conventional computers. Other special-purpose processors are the Megalogician of Daisy Systems Corp. and Realfast of Valid Logic. Both of the latter companies are workstation vendors.

Also included in this chapter are several other sections. The first is a discussion of both manual placement and routing and automatic placement and routing (auto P & R). It is appropriate to include this in this chapter because the reader needs to understand IGSs before getting into auto P&R. Exercises are included on manual P&R to enable the reader to get a feeling for the problems involved.

Another topic covered is that of designing remotely. It is appropriate to include this topic here after the reader has gained an understanding of the flow of the gate array design process and the hardware and software tools involved in it. Other sections cover hard-copy output devices and two nongate array layout techniques that may find use in gate arrays at some future time.

The first topic covered is that of IGSs (interactive graphics systems).

8.1 OVERVIEW OF INTERACTIVE GRAPHICS SYSTEMS

IGSs play several important roles in design of the gate array masks. Because workstations have taken over some of the tasks which were previously the exclusive domain of the IGS, one of the major uses of IGSs in current gate array work is in design of the basic background layers of the gate array itself. This involves creation of the design of the transistors, underpasses, bonding pads, power and ground buses, and other elements of the basic structure. This task is different from customization of the gate array structure so designed. The greater power of the IGS over that of a workstation is well suited for this task. Because this book is aimed at the user who customizes a gate array whose structure has already been designed, the latter task will not be discussed further.

A second major use of IGSs is in the design and layout of the macros themselves. This, too, is done by the gate array vendor; but it may also be done by users who design their own macros.

A third major use of IGSs is in the placement of critical macros. Such placements are used to "seed" the P&R program and are done when the user desires certain macros to be placed in certain positions in the array (for example, close to an I/O pad).

A fourth major use is in the routing of certain critical paths in the layout.

Finally, "rip-up and retry," which is done when an autorouter does less than 100% routing, is also done on an IGS.

Most of the above tasks can also be done on high-end workstations.

Although schematic entry can also be done on IGSs, few commercially available programs exist to translate the schematic entry data base into coding for simulation programs.

IGSs can also be used for many purposes other than design of integrated circuits. They are widely used in the architectural and mechanical engineering fields.

8.1.1 Definition of an IGS

The definition of what constitutes an IGS has become muddied by the introduction of many intelligent terminals, workstations, and engineering design stations. Prior to such introduction, there were essentially three major purveyors of IGSs in the United States: Calma, with its GDSI and II models; Applicon, with its 860 (and now 4245/75) models; and Computervision systems. Today, workstations, from companies such as Daisy (the Gatemaster), Mentor, Valid, CAE (Tektronix), and Metheus, can perform many of the functions in the domain previously solely occupied by the three major IGS vendors.

Moreover, intelligent terminals, such as the Sanders Vistographic and the Tektronix 4014, can perform similar functions when running programs on a large minicomputer, such as a Data General MV10000 or MV 20000 or a DEC VAX 11/785 or 8600. Compounding the definition problem is the fact that the newer Applicon 4275 and the Calma GDS II both use VAX computers as their "engines." Lastly, there is the DEC MICROVAX II, which is a standalone workstation, but which can be tied to a larger VAX such as a 750 or 785.

Although workstations and intelligent terminals can perform many of the foregoing tasks, for purposes of this book, an IGS will be defined as one of the systems noted above from one of the three major U.S. manufacturers (or their equivalent European and Japanese counterparts).

8.1.2 Properties of an IGS

An IGS so defined has several properties. First, it has as an integral part a large minicomputer such as one of those mentioned in Section 8.1.1. This is important because it relates to both the size of the chip that can be worked on at one time and to the speed of execution of the commands. (It can be argued that a workstation with only a single user will have faster response times than a system from one of the big three or equivalent with three to six users independently executing big programs.)

Second, the large minicomputer permits PG (pattern generator) tapes to be made on the IGS *without* having to transfer to a larger machine. Making such tapes requires a great deal of computing power in order to "fracture" each of the lines, contacts, and even the mask identification numbers and possibly copyright symbol into rectangles. The fracturing has to be done in a format suitable for the pattern generator at the maskmaker. The ability to make PG tapes or *large* chips is a feature which distinguishes IGSs from current workstations.

Third, the large (in this case, 32-bit) minicomputer permits sufficient points to be addressed that the characteristic grid spacings of the gate arrays being used can be accommodated. For example, suppose that transistors with 2-μm feature sizes exist on the gate array being worked with and that it is desired to have a resolution which is 1/10 the minimum feature size. This would imply that 5 resolution units are needed per micrometer. For a chip that is 300 mils (equal to about 7620 μm) on a side (not an uncommon size), 38,100 resolution units would be needed.

A 16-bit computer would permit 2^{16} = 65,536 points to be addressed on each axis. However, the normal procedure is for the address space to be broken into quadrants. The address space in each dimension of the 16-bit machine per quadrant, unless the center were offset, would be 32768. This would be sufficient for smaller chips but would be marginal or unusable for larger chips. Moreover, there are other factors that make 32-bit computers preferable to 16-bit machines. One of these is that it is necessary to make "fiducial" marks on the pattern generator tape to allow the recticle to be aligned. These fiducial marks are specified by the mask maker to be given distances outside the reticle of the chip. To do this requires that an address space considerably larger than that required to accommodate the chip alone be used. Therefore, although 16-bit machines can be used, 32-bit machines are preferable.

Another property of IGSs which distinguishes them from workstations is the much larger amount of disk storage associated with them. Whereas a workstation might have one hard disk drive with perhaps (currently) on the order of 40 to 70 megabytes of storage, an IGS will generally have one or more drives with removable disk packs. Each such pack is typically capable of storing 300 megabytes.

8.1.3 Mask Layout

Before beginning, it is essential to understand clearly that what is being laid out are the generally one or two metallization layers of the wafer and, hence, die. (TI used three on one of their STL arrays and IBM has used up to four metallization layers. These are currently exceptions, however.) The metallization layers are the interconnects among the active and passive components on the chip. The active components include the transistors, diodes, and macros (function cells) made from the transistors. The passive components include underpasses (poly blocks) and resistors, which exist in virtually every ECL macro.

A gate array layer has thousands of contacts to which the metal lines (to be defined by the metallization masks) can go. For example, the IMI (International Microcircuits, Inc.) 1960-cell chip has about 71,000 such internal contact points. Most of these are redundant to enable easy interconnect and routing. For example, the source contact for each array NFET transistor is available on both sides of the V_{ss} (ground) bus.

8.1.4 Functions of IGSs

To appreciate how IGSs can be used, it is necessary to indicate what they are and what some of their features are.

For purposes of this discussion, an IGS is basically a computer-driven display which allows the operator to:

- Call one or more background layers of the gate array from memory.
- Create macros by interconnecting the transistors, underpasses, and so on. The macros may, of course, represent I/Os as well as internal cells.
- Call previously created function (i.e., macro) cells from memory.
- Modify existing macros to form new macros.
- Interconnect them by adding lines according to the gate array layout to form larger cells or even the whole chip.

The foregoing methods all involve "drawing on the tube" using one of the methods outlined in an earlier chapter. IGSs also enable the operator to:

- Store the results of the above manipulations in memory.

- Prepare a tape (the PG or pattern generator tape) from which the masks can be made.
- Enter coordinates of lines into the computer data base by tracing each line with a crosshair and telling the computer where the start, stop, and corner turning points of each line are. (This is the process of "digitizing." Not all IGSs offer this method. It can be used in place of drawing "on the tube.")

Ancillary functions include driving various hard-copy output devices, such as photo, thermal, and pen plotters.

Figure 8.1 gives an overview of a typical IGS. A typical system consists of one to four terminals connected to the aforementioned large minicomputer. Figure 8.2 shows an operator at a typical IGS. Associated with the terminal are one or more tablets and a cursor of some type. Moving the cursor on the tablet and perhaps using one or more keys on the keyboard enables the operator to select macros, add and delete lines, change the magnification of the image on the screen, select a different portion of the layout or chip to be viewed, place and replicate macros, and so on. Subsequent figures show these functions.

Figure 8.1 Overview of typical IGS. (Courtesy of Sanders Associates, Inc.)

Figure 8.3 is a picture of the screen of the IGS. One macro has been placed on the background layer. The reader will recognize the background layer by noting the bonding pads (small squares on the periphery) and the power and ground distribution buses (pairs of vertical lines). The macro has been placed in the upper left corner. The single dots to the left of each pair of power and ground buses denote sites for

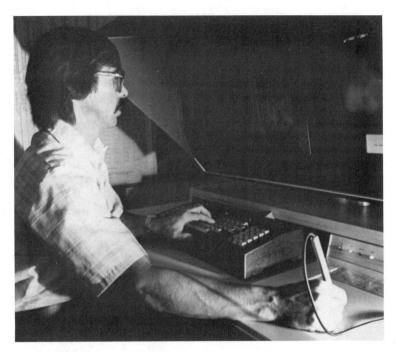

Figure 8.2 Operator at typical IGS. (Courtesy of Sanders Associates, Inc.)

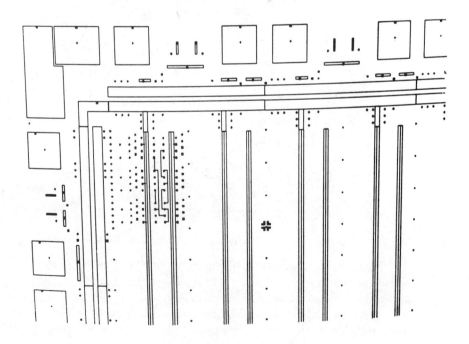

Figure 8.3 Screen of IGS. (Courtesy of Sanders Associates, Inc.)

different macros. Each macro has a reference point which when aligned with the reference point of a particular cell correctly positions that macro on the array. The cursor position denoted by the maltese cross is at the reference point for another cell of the array. See below.

Figure 8.4 shows the same array with the same macro replicated twice more. The operator in this case has chosen to leave one blank column (denoted by a pair of vertical power and ground buses) between the two replications and the original replication. One of the replications uses the reference point of the cell where the cursor in Fig. 8.3 was placed.

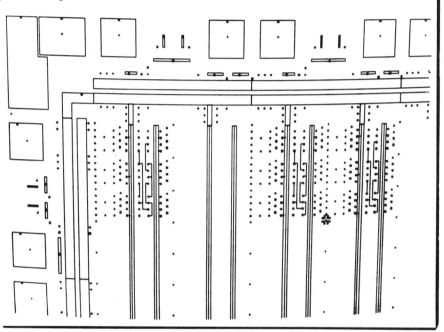

Figure 8.4 Screen of IGS with macro replicated twice more. (Courtesy of Sanders Associates, Inc.)

Note that the reference points of the array cells show where macros *can* be placed. They do not show where they will be placed. The latter decision is up to the operator or the placement program. Note also that in the case of cells with nonuniform sizes, more than one array cell may be required for a given macro.

Figure 8.5 shows a close-up view of one of the macros just placed. Figures 8.6 and 8.7 show lines that have been added and deleted by operator action.

Figure 8.8 shows the grid that may be placed on array face to enable line drawing and sizing. More commonly, either "pen lock" in either the x or y direction or "channelizing" is used to ensure straight horizontal or vertical lines. In the latter, sometimes called "snapback," the array is divided into channels. The computer will force the line being drawn into the nearest existing channel even if the operator's hand wobbles slightly. This helps avoid violation of line-spacing rules.

Unfortunately, it does not avoid and may even promote lines being connected together which were not meant to be connected together. If the operator is careless

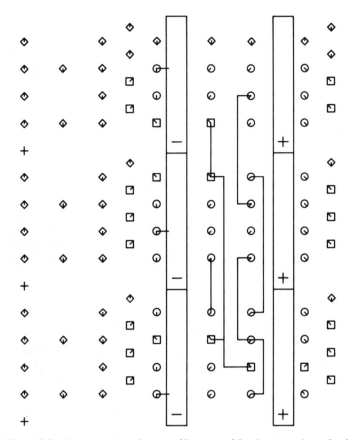

Figure 8.5 Close-up view of macro. (Courtesy of Sanders Associates, Inc.)

just for an instant and allows the cursor to go too far in the *original* direction, it will interconnect to the next point or line. (Changing direction of the line from x to y or vice versa requires operator action and is, therefore, generally not a problem.)

8.1.5 Digitizing

Digitizing is still used by many gate array vendors. In this method, the layout of the chip or portions of it is done on a medium, such as Dupont's Chronoflex, which does not change dimensions with temperature and humidity. The lines drawn on the Chronoflex may simply be the interconnects among the macros or (increasingly less likely these days), it may be the entire chip. The macros themselves may be designed in a similar fashion.

Digitizing involves tracing the lines with a crosshairs connected to the IGS computer. The position of the crosshairs is known exactly. The operator denotes changes in direction of the lines being traced by means of buttons on the crosshairs.

Figure 8.9 shows a close-up of the crosshairs over a line. Figure 8.10 shows a picture of an operator using such a digitizer. Note that a display and keyboard to the right of the operator enable the operator to monitor the progress of what is happening

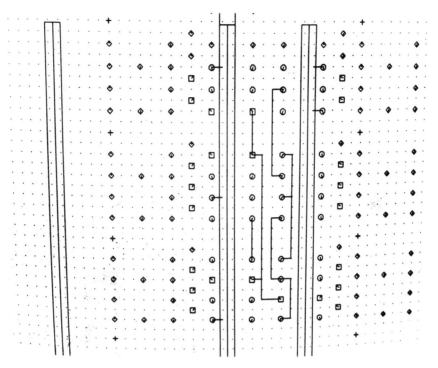

Figure 8.6 Addition of line to circuit. (Courtesy of Sanders Associates, Inc.)

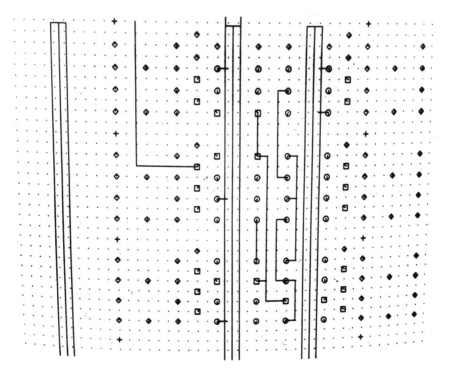

Figure 8.7 Deletion of line from circuit. (Courtesy of Sanders Associates, Inc.)

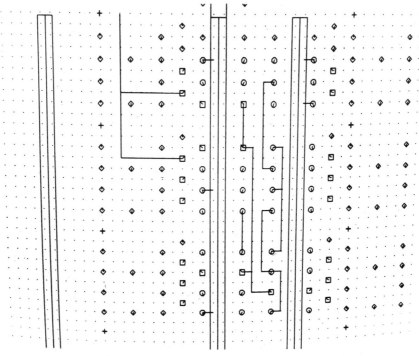

Figure 8.8 Grid place on screen. (Courtesy of Sanders Associates, Inc.)

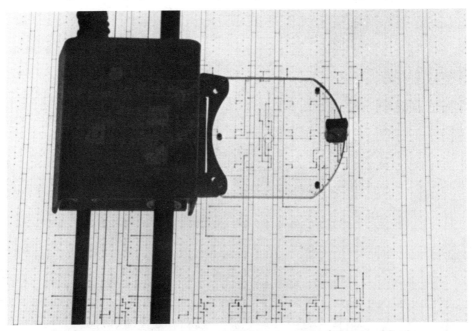

Figure 8.9 Close-up of crosshairs of digitizer over a line. (Courtesy of Sanders Associates, Inc.)

Figure 8.10 Operator using digitizer. (Courtesy of Sanders Associates, Inc.)

and to perform other functions, such as calling and placing macros. The small boxes in the upper right corner are readouts of the display of the position of the crosshairs. The size of the layout being traced is typically 200× (200 times the size of the chip itself).

The main advantage of digitizing is that it enables an expensive resource, the IGS, to be shared among a larger number of users. The disadvantage is that it has many possibilities for error. The original layout person doing layout on the tube may make a mistake. The same layout person doing layout on Chronoflex might make the same mistake. However, even if no mistake is made at this stage, the person doing the digitizing may make one or more mistakes. Hence much more time consuming checking must be done when digitizing is being used.

8.1.6 Example of a Coordinate List

When large arrays are to be designed, the floor plan mentioned in Chapter 4 is used to obtain the placements and perhaps some of the interconnects on at least a gross basis. These placements must be put into a format suitable for the IGS operator to use. Some companies have semiautomatic programs to effect this translation. A common method for those lacking such a program is to develop a list of coordinates corresponding to where each macro is to be placed. An example of such a coordinate list is given in Fig. 8.11. (The circuit name and other identifying details have been

Semiconductor Engineering Dept.
Calma Log Sheet

Page __1__ of __4__

Job Title (Code): __A_____ S/A P.T. Number _____ P.W. Number _____
Background (Code): __AT 7196∅X_____ Layer __7_____

L.S. Number	Calma Code	Layer	Grid Coord.	DIG.	GPL	C.P.	VSATK	Rework Yes	Rework No	Comp Date	Init.
1	∧TTCLBIT	20	3820, 14080	✓	✓	✓				3/26/80	/st.
			5820, 14080								
			3820, 12080								
			5820, 12080								
			3820, 10080								
			5820, 10080								
			3820, 8080								
			5820, 8080								
1	∧TTCRBIT	20	8320, 14080								
			10320, 14080								
			8320, 12080								
			10320, 12080								
			8320, 10080								
			10320, 10080								
			8320, 8080								
			10320, 8080								
1	∧TTC.A=B	20	7860, 12660								
			7860, 10660								
			7860, 8660								
			7860, 6660								
1	∧TTCI.UV1	20	7860, 14180								
			7860, 12180								
			7860, 10180								
			7860, 8180	↓	↓	↓				↓	
1	∧TTCTSB2	20	3820, 3240	✓	✓	✓				3/20/80	/st

Figure 8.11 Coordinate list of macros to be placed.

blanked out to protect the proprietary nature of the circuit.) This figure shows that macro labeled -TTCLBIT will be replicated eight times. The coordinate location of the first replication will be 3860, 14180, for example. The first coordinate represents the x axis (column coordinate), the second the y axis (row coordinate). Figure 8.11 shows that the macro labeled TTCRBIT is to be replicated eight times, with the first replication starting at 8360, 14180. Hence it is placed in the same row as the first replication of the previous macro but in a different column.

The coordinates given in Fig. 8.11 are those suitable for use by the IGS operator. He or she simply types in the coordinates of the replications after calling each of the macros in turn. (Alternatively, the cursor of the IGS could be used to position the macros using the readouts of the cursor positions existing at both terminal (and digitizer, if used).

The designer of the floor plan often prefers to use row and column numbers rather than coordinates of the IGS. A representation of an array with representative row and column numbers labeled from the lower left corner (first quadrant) is shown in Fig. 8.12. A translation chart for the array is shown in Fig. 8.13. The different demarcations denote the boundaries of the different members of the gate array family being used. In this case, six family members with array sizes of 200, 360, 640, 1000, 1440, and 1960 array cells are listed. For example, the 640-cell member has 16 columns and 40 rows. The internal coordinates that correspond to column 16 and to row 40 are 8360 and 8580, respectively.

The same coordinates apply also to the other arrays (the 1000, 1440, and 1960) that contain column 16 and row 40. This makes it easy to change from one array to another in the same family if the layout will fit or if it is desired to expand and add to circuitry which exists, for example, on a 640-cell array. In this case, the additional circuitry can go (if desired) in an area outside column 16 and row 40, respectively. The original circuitry can go on the first 16 columns and 40 rows probably without relayout. This is simply an example of a working system. The coordinates will vary with the user and the vendor.

It is useful to note one other feature of IGSs. From Fig. 8.11 the reader will note that the "layer" on which the layout is to be done is specified by the designer. The layer in this case is a separate FILE which exists in the IGS which will contain the interconnects for that application. The layer specified in this case is *not* a layer of the gate array chip.

It is very useful to lay out different classes of interconnects on different files or "layers" of the IGS. For example, all of the internal interconnects among the macros might be on one such file. A second file might contain the macros so placed on the background layer of the gate array. The interconnects among the macros might exist on yet another file. If two-layer metal is used, another file would be used for the second layer. The I/Os would be on yet another file, and so on.

The great advantage of this method is the ease by which the structures of interest can be exhibited without being cluttered with the lines of a class not needed for the particular investigation or checking taking place at the time. For example, a plot of all the interconnects on first-layer metal can be readily obtained. Examination of such a plot may reveal places where considerable improvements could be made. If such a plot were viewed while cluttered with the lines of the second layer of metal, for example, the desired features might be much harder to detect. Alternatively, more than one file may be viewed and plotted together.

When the designer is satisfied that the design is correct, all the files corresponding to a given metal layer are merged to form *one* file for that metal layer. This file is then translated into a PG tape for submission to the mask maker.

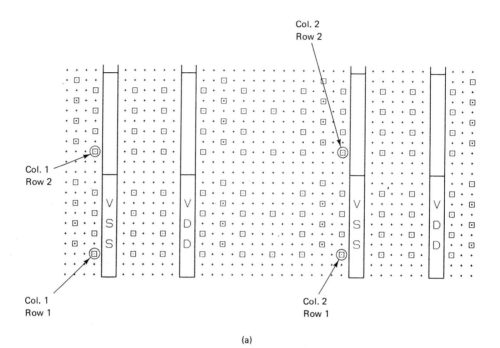

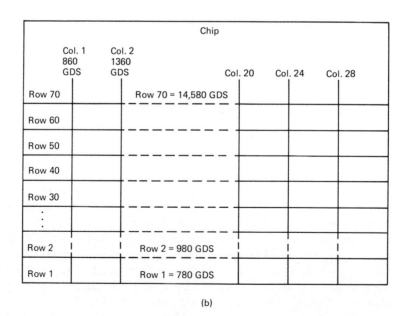

Figure 8.12 (a) Reference points of cells; (b) row column numbers of array.

Col.	X Coord.		Row	Y Coord.		Row	Y Coord.
1	860		1	780		36	7,780
2	1,360		2	980		37	7,980
3	1,860		3	1,180		38	8,180
4	2,360		4	1,380	(640)	39	8,380
5	2,860		5	1,580	Gate	40	8,580
6	3,360		6	1,780		41	8,780
7	3,860	(200)	7	1,980		42	8,980
8	4,360	Gate	8	2,180		43	9,180
9	4,860		9	2,380		44	9,380
10	5,360		10	2,580		45	9,580
11	5,860	(360)	11	2,780		46	9,780
12	6,360	Gate	12	2,980		47	9,980
13	6,860		13	3,180		48	10,180
14	7,360		14	3,380	(1000)	49	10,380
15	7,860	(640)	15	3,580	Gate	50	10,580
16	8,360	Gate	16	3,780		51	10,780
17	8,860		17	3,980		52	10,980
18	9,360		18	4,180		53	11,180
19	9,860	(1000)	19	4,380		54	11,380
20	10,360	Gate	20	4,580		55	11,580
21	10,860		21	4,780		56	11,780
22	11,360		22	4,980		57	11,980
23	11,860	(1440)	23	5,180		58	12,180
24	12,360	Gate	24	5,380	(1440)	59	12,380
25	12,860		25	5,580	Gate	60	12,580
26	13,360		26	5,780		61	12,780
27	13,860	(1960)	27	5,980		62	12,980
28	14,360	Gate	28	6,180		63	13,180
			29	6,380		64	13,380
			30	6,580		65	13,580
			31	6,780		66	13,780
			32	6,980		67	13,980
			33	7,180		68	14,180
			34	7,380	(1960)	69	14,380
			35	7,580	Gate	70	14,580

Crosshatch Direction = 15,320

Last Crosshatch in "Y" Direction = 15,340

Figure 8.13 Translation chart for array. Numbers in parentheses are the number of internal array cells on a chip. 20 GDS units = 10 μm.

8.2 WORKSTATIONS

The word "workstation" is almost as popular a current buzzword as is the word "hierarchical." Hence, it is necessary to say something about them. (The real reason is that they are a very important tool for the design of gate arrays.)

A major benefit of CAD workstations is that they tie together in *one place* all or most of the functions necessary to do a gate array design. This enables a logic design to be done, simulated, corrected, and resimulated until the designer is satisfied all on one terminal and from one data base. Because these steps are done at one terminal, the design process is greatly speeded up. If the simulation shows that a change is necessary, the designer can quickly revert to the schematic entry program to input the change, quickly revert to the simulation program to determine the effect of the change, and so on. This is far simpler than performing these functions on separate terminals (and sometimes even separate computers) using possibly even different data bases which must be converted to one another.

The speed of response enables, and hence promotes, trying different circuit configurations. This is especially important in gate array design, where ancillary functions can often be added at essentially zero cost and time added to the design cycle. The reason is that there are generally unused gate array cells (see Chapter 1) which are available to implement small additional circuit functions.

8.2.1 Problem of Definition

The purpose of claiming that the term "workstation" is a buzzword is to convey the impression that there is no standard definition of a workstation. This is true. Many vendors are coming out with machines designed "from the ground up" as workstations. Other vendors are revamping product lines to incorporate what are also termed workstations. Hence it is extremely difficult to arrive at a definition of a workstation. Nevertheless, it is possible to loosely divide workstations into two general categories. The two categories can best be described as single-user and multiuser.

8.2.2 Two Categories

The initial workstations, such as those from Avera, Daisy, and Mentor, were self-contained computer systems that existed in single-user, single-terminal configurations. This is the first category. This will be called the *CAD workstation approach* for want of a better term.

The second category is exemplified by products from DEC and Valid, which have less computer power at each terminal. Instead, a number of terminals are clustered around a host computer, which may itself be linked to an even larger computer. This will be called the *mainframe approach* for want of better terminology.

An example of the latter is the DEC workstation, MICROVAX II which has the computing power of about a VAX/730 and is suitable for schematic capture and small simulations. A number of MICROVAX IIs can be linked to a VAX 11/750, 11/780, or 11/785, which does simulation and other tasks that require substantial processor power. By unloading the 11/785 from the asynchronous interrupts of operator interactions typical of graphics entry and coding, the 11/785 can be used much more effectively.

The two categories make a general discussion of workstation quite difficult. Nevertheless, in either case, a workstation can be defined as a computer system that can be used for nearly the complete interactive design of a product. It differs from an interactive graphics system in that it is capable of doing more than just graphics. The key words here are "nearly complete," "interactive," and "design of a product." These help distinguish the workstation from other general-purpose computer systems.

The term "design of a product" implies the intent to use the machine in a specific technological area; and indeed, most such machines are bought initially for such a purpose.

Although a large number of systems could fit into the foregoing definition, a CAD workstation for purposes of this book will be further defined as a *stand-alone* computer system possessing the properties of the preceding paragraph, which, in addition, has the *software* necessary to enable all or most of the gate array design process to be done. The stand-alone part of the definition effectively eliminates the second category from the discussion temporarily.

8.2.3 Parts of a CAD Workstation

A CAD workstation generally consists of four major parts. The first is a computer with a relatively large RAM (random access memory). The second is a set of software that covers a gamut of functions from simulation to layout. The third is a high-resolution video terminal with some form of cursor movement device. The fourth is one or more mass storage devices. Typically, there is both a floppy disk drive of 1 megabyte or more capacity plus a hard disk drive with 40 megabytes or more of memory. There may also be streaming tape drives to back up the other two.

Types of terminals. A computer terminal is often characterized as being either a graphics terminal or an alphanumeric terminal. A *graphics terminal* is one on which graphics figures (rectangles, circles, and arcs, but more pertinently for CAD workstation, logic symbols and background layers of gate arrays) can be drawn. An *alphanumeric terminal*, by contrast, is one normally used for input/output of alphanumeric data. A prime example of the latter is the Digital Equipment Corp. VT-100 and 200 series of terminals.

The terminals used for CAD workstation are both alphanumeric and graphics terminals. Primarily, the software is what determines whether a terminal will be graphics or alphanumeric. However, generally, the CRT for a graphics terminal needs both higher resolution and a higher bandwidth to accommodate fast redraw times than does that for an alphanumeric. The CRTs for the Mentor IDEA 1000 and the Daisy Gatemaster are 800 by 1024 and 1080 by 830 pixels, respectively, as of this writing. This resolution is needed to provide discernible figures when a lot of logic is on the screen of the CRT in the schematic entry mode as well as for numerous other purposes.

Software. The key to any computerized design system is the software. This is even more true of a CAD workstation. A major reason for making a relatively sizable investment of $50,000 to $150,000 is to enable a complete design to be done in one place.

Two classes of software come with a CAD workstation. The first class encompasses the executive and utility programs required to make the machine perform. The second is the application specific software, which enables the specific design of the product.

The latter software falls into three categories: schematic entry, logic simulation and test, and layout of the gate array. Not all vendors offer all parts. Some include circuit simulation as well. The same comments made in an earlier chapter regarding the benefits of operating off a common data base apply here also.

In the case of logic design, whether for PCBs or for gate arrays, this would be schematic entry and logic simulation as a bare minimum. Most vendors also offer testability analysis of one type or another.

In addition to the software used for logic design, workstations used for gate array design have software that is peculiar to a number of gate array vendors. Most gate array vendors have recognized current trends and supply information to various workstation vendors. Hence a given workstation may have software from several to many gate array vendors which is resident on one of its mass storage devices.

The software from each such gate array vendor generally encompasses the background layers, library of macrocells, simulation information (propagation delay versus fan-out, for example), the information required for the CAD workstation's automatic placement and routing program, if any, and sometimes the "hooks" into the gate array vendor's mainframe computer system. The latter is needed to enable macros to be designed by the user *if* the gate array vendor allows it.

Other software typically offered is a circuit simulator, such as SPICE, and that required to make the PG (pattern generator) tape for the masks. The type and quality of this software is a key distinguishing feature among CAD workstations.

8.2.4 Interaction with Gate Array Vendors

The user must also be greatly concerned about the legal and technical arrangements whereby this software is maintained and updated. The latter is true of any computer.

License to use the software of a gate array vendor obtained by buying a CAD workstation does not necessarily imply that the gate array house will necessarily build parts for the user. The gate array vendor may have additional restrictions with regard to minimum quantities, minimum notice for a design to be built, and so on. The gate array vendor may also have certain technical requirements, such as percentage of cells that can be utilized, preferences for routing high-current signals, and so on. For these reasons, it is wise to spend time with a gate array vendor whose software exists on the workstation to iron out any potential problems *before* beginning a design.

8.2.5 Networking

Although meant to be operated as stand-alone units, most CAD workstations provide for high-speed networking among stations. This means that a given design on one station can be accessed from others. Moreover, expensive peripherals, such as plotters and high-speed printers, can be shared via networking. One distinction between working on a workstation and on a terminal connected to a mainframe computer is the frequency of such interaction. The user of a workstation would access another workstation or a mainframe only rarely and then only to send over or retrieve large blocks of data.

Examples of such interaction would be to use printers and plotters, to send or retrieve large parts of a big circuit design, or perhaps to run a simulation program that was too big for the CPU of the workstation.

As gate arrays become so large that different designers will work on different parts of the design, networking will become more important.

8.2.6 Comparison with the Second Category of Workstations

It was mentioned in Section 8.2.2 that a second category of workstations can be defined (although there are so many possiblities in the latter category that discussion is difficult). The advantage of the latter is that less expensive terminals can be distributed throughout a company or even accessed thousands of miles away using leased phone lines with throughput rates of 9600 baud or more. (The author has personally used both arrangements, but the 9600 baud is probably the minimum rate that is acceptable for transferral of data.) A second advantage is that one is virtually guaranteed that the software on the smaller machines is upward compatible with that of the larger machine in cases such as that mentioned in Section 8.2.2 involving the VAX 11/730s connected to VAX 11/750s connected to a VAX 11/780.

By contrast, a major advantage of the CAD workstation over the mainframe computer approach is that a given user of a CAD workstation never "swamps out" everyone else on the system by running a program so extensive and/or with such a high priority that all other users are left waiting. On a CAD workstation of the first type, there *are* no other users. A slight disadvantage of the CAD workstation compared to the mainframe computer approach is that of promulgating software updates.

8.2.7 Benefits of Large Amounts of RAM

A benefit of having a large amount of RAM is that the responses on the screen to various commands are much faster than if the disk has to be accessed repeatedly. This is particularly important when operations, such as schematic entry, which require the repainting of a great deal of detail on the screen are being done. Zooming in (i.e., magnification) and out are very common operations in schematic entry and other operations.

Repainting is especially important when "scrolling" from side to side or up and down. *Scrolling,* as the name implies, is moving the picture on the screen in one direction or another. Even the excellent CRTs used in CAD workstations cannot show more than a portion of a moderately large circuit without making some details unresolvable.

Figure 8.14 is a picture of typical CAD workstations. These are Mentor Graphics, Inc. workstations. The workstation in the foreground will be described. The others are very similar to it. To the right of the keyboard is the tablet on top of which is the "puck." The tablet rests on a cabinet containing the CPU and rigid and floppy disk drives. The computer and terminal are built by Apollo Computer. This and other members of the family are widely used in workstations offered by many vendors, including Silvar-Lisco and GE/Calma (Tegas workstation).

Figure 8.14 Typical CAD workstation. (Courtesy of Mentor Graphics, Inc.)

8.2.8 Typical Operation

The purpose of this section is to show how the flow of gate array design enunciated earlier in the book can be carried through using a CAD workstation. In a typical operation, the designer would first draw the schematic on the terminal using the resident schematic entry program. (Schematic entry was discussed in Chapter 6.) The logic schematic may be couched in terms of higher-order macros, such as combinations of gates. Testability analysis would be run if available. Logic simulation and timing analysis are then done on the circuit. The results of the logic simulation would be displayed and analyzed on the screen of the CAD workstation. There are many options to the flow of design, and these have been discussed previously.

Circuit simulations to determine transient characteristics of portions of the network might also be run. Any desired changes to the circuit itself arising from these analyses would be input to the data base via schematic entry. Additional simulations could then be run to verify the correctness of the changes.

The next step is to lay out the gate array chip itself. If the information for the automatic placement and routing program for the gate array (vendor and particular part) selected is in the CAD workstation, that program is run. The user may manually place certain macros and manually route them before activating the automatic P&R program.

The auto P&R program will typically place and route up to 95% of the macros. The remaining nets not routed will be listed in a table which appears beneath the gate array background layer on the split screen. (This is how it is done on the Daisy Gatemaster and is representative of how it is done on other workstations.)

Moving the cursor to one of these selects it for routing. The designer "hooks" onto the net (selects it with the puck). This causes the beginning and end points of the net to flash on the screen in a unique color together with a typical path. The user then makes the interconnect using the typical path or one of his or her own choosing. When the designer is satisfied with the circuit formulation, layout of the circuit would be done using auto P&R, manual "correct by construction" techniques, or a combination of the two.

Operation of a workstation is typically accomplished via a combination of a cursor movement device and the keyboard—the former, generally. The cursor is a point on the screen that can be moved by an external device. Moving the cursor enables many important functions to be done. These include "menu" and submenu selection; selection of areas (in the graphics mode) to be moved, copied, or deleted; definition of the timing simulation point in that mode; and so on. This type of operation was described under "Schematic entry" in Section 6.5.

Figure 8.15 shows the screen of the Mentor workstation. Figure 8.15 utilizes some of the multiple windowing capabilities of the Apollo Domain computer to exhibit different features simultaneously on the screen. The windows do not have to be the same size or shape.

The top part of Fig. 8.15 shows typical timing signals that have been displayed as a result of a logic simulation. The scale can easily be expanded, contracted, scrolled, and the more or fewer signals can be displayed. It is similar to operation of a logic analyzer in that sense. The topmost part is readouts in hexidecimal and binary of the signals; the lower part is a waveform display. Note the waveforms of "probes" 2 and 3, designated PR2 and PR3, respectively. These are mentioned below.

The lower left window is a view of a macro. The lower right is a view of the inside (in this case) of part of the macro showing where probes (indicated by reverse video) are to be placed for analysis. In this case they are placed at the output (PR3) and one of the inputs (PR2) of the EX-OR. The probes are exactly like the probes of a logic analyzer or oscilloscope (except that they do not fall off and twist leads of the chip).

The lowest left corner gives instructions to the computer on how to run the simulation program. In this case the computer is being told to run the simulation to time 99999 but to break it (stop) when the signal /f(3) and the point where probe 2 (PR2) is are both true.

Figure 8.16 shows the schematic of a circuit using Silvar-Lisco CASS.

More on multiwindowing. Another example of multiwindowing is given in Figs. 8.16(b) and (c). These are views of the screen of a Genisco 1000 terminal on which several schematics built with the Silvar-Lisco CASS (computer-aided schematic system) are displayed.

In Fig. 8.16(b) and (c), five schematics labeled "TST1 (circuit name)" followed by a number representing the page number of that particular sheet are shown. In this view, the sheets are not overlapped. The size of each individual sheet can be varied. The sheet with the dashed line around it is the one currently at the top of the (memory) stack and, hence, is the one being worked on. The menu to the left of the

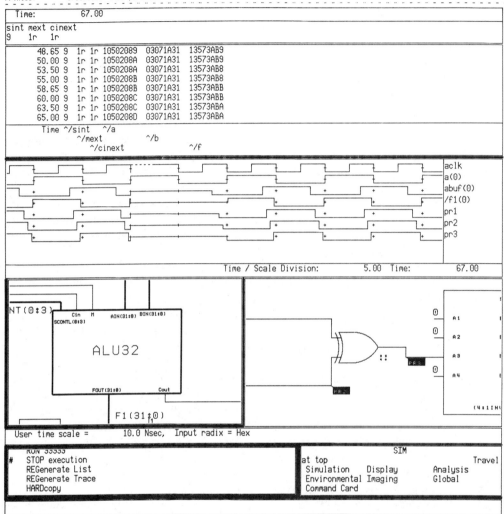

Figure 8.15 Screen of CAD workstation. (Courtesy of Mentor Graphics, Inc.)

screen contains macros which have been selected from a larger library of macros. Macros "outblt" and "output" are page and chip output connections, respectively. (A page connection, of course, means that the signal goes to another page as opposed to off the chip.)

Macro FF11 is a D-type flip-flop, and macro LOGOH is an identification block which goes in the lower right corner of each page. It is an example of a symbol which can be specially created for a given need. (The magnification used in the example is insufficient to show the printing in the logo.)

The menu on the right contains commands which are selected by placing the crosshairs (not shown) controlled by the puck on the tablet (also not shown). For ex-

318 Workstations and Other Tools Chap. 8

ample, to select one of the macros, the crosshairs would first be placed over FET (for FETch) (FET is under the word COMPONENT in the right-hand menu), would then be moved to the macro desired in the left menu, and then to the screen to make one or more placements of the components. Each placement can be adjusted before moving to the next placement of the same or different macro. (The procedure is faster than it sounds.) Appropriate buttons on the puck would be pushed each time. Connections and attributes can be entered by first placing the crosshairs under ENT (for enter) under CONNECTIONS and ATTRIBUTES, respectively, and either drawing connections or entering text.

In the lower right corner is a list of options from which the user can choose (in the mode of operation shown). For example, the user may wish to create a new schematic, edit an old schematic (in addition to those shown), and so on. In other submodes of operation, this menu tells the user what to do or what the buttons on the puck are used for in the given submode of operation.

Figure 8.16(b) shows another arrangement of the schematics in Fig. 8.16(a). In this arrangement, schematic page 7 has been made the one being worked on (top of the stack), as indicated by the dotted line around it. Page 9 has been made larger but moved farther down in the stack, as have pages 10 and 6.

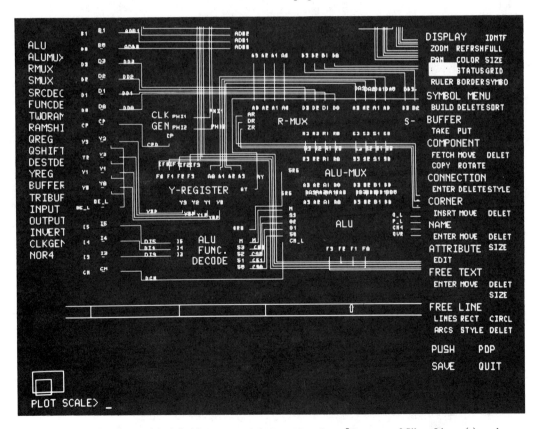

Figure 8.16 (a), (b), (c) Screens of CAD workstations. [Courtesy of Silvar Lisco, (a); and Sanders Associates, Inc., and Silvar-Lisco, (b) and (c).]

Figure 8.16 (*cont.*)

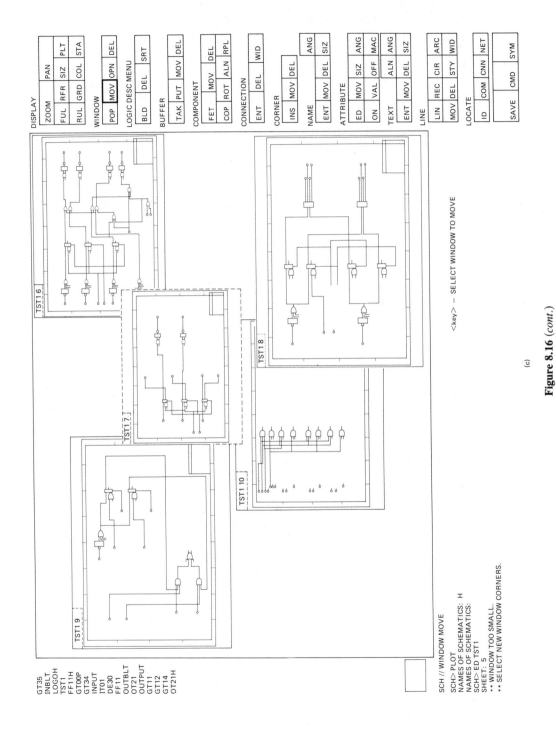

Figure 8.16 (cont.)

Being able to display multiple pages is very useful for a variety of reasons. One of these is the ability to readily check the names of signals entering and leaving pages. If these are not exactly correct, a great deal of time can be wasted (aside from the possibility of building an incorrect circuit). Among other things, most netlist extraction programs (which format the schematic capture) will give error messages if the names that are *not* used in subsequent pages go off chip via page connectors.

Copying portions of schematics or entire schematics from one window to another can be done using multiwindowing. It goes almost without saying that the entire screen can be used for one page if desired.

The purpose of the above is not to teach the reader how to use the system but rather to give the reader an inkling of how such a system works.

8.2.9 Cursor Movement Devices

The six forms of cursor movement devices in existence today are the light pen, the tablet pen, the puck, the joystick, the mouse, and the human finger. The latter (created before any of the others) is used in a variety of computers, including the HP 150. Each of these has certain advantages and disadvantages, and different users have different preferences. The most common currently is probably the puck. This has four buttons. Like the others it causes movement of a crosshairs on the screen which enables the selection of menu items, connections and so on as discussed above.

8.2.10 Menu Items

A menu is a listing of commands that can be selected by the user. Most menus are nested, meaning that submenus exist on the list of menu items. In this way the workstations are quite "user friendly." Again, see Section 6.5.

Because of the rapid improvements occurring in intelligent terminals and workstation, the reader should consult the vendor for the most up-to-date information. (The same is true of gate arrays themselves, as noted earlier.)

8.2.11 Correct by Construction

Given the complexity of the background layer, it would be very easy to make a mistake and inadvertently connect to a nearby contact point on the array. The CAD workstation prevents this by refusing to accept the route. This is what Daisy calls "correct by construction." Unless one has had to manually check the lines of a gate array layout (as the author did in the early days of gate arrays), one does not fully appreciate the benefit of this property. Needless to say, obviating the need to check the lines of a layout greatly speeds the design process and decreases the risks regardless of how such obviation is accomplished.

The CAD workstation knows what the interconnections are among the macros from the data that were entered using the schematic entry of the logic diagram. It knows where the macros that connect to the macro selected for routing are because of either the previous manual or automatic placement, which has also been entered into its data base. Other macros are routed in the same fashion. Figure 8.17(a) shows

(a)

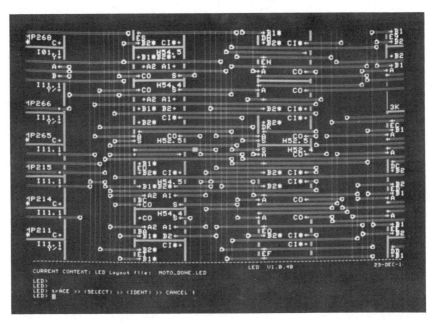

(b)

Figure 8.17 Correct by construction: (a) operator at terminal; (b) close-up of screen. (Courtesy of Daisy Systems, Inc.)

an operator using a puck to route line. The heavier lines are the second-layer metal. Figure 8.17(b) shows a close-up of the screen. Even in automated systems, manual routing is done of critical paths and in "rip-up and retry" following autorouting.

8.2.12 Hierarchical Data Base

The data bases of most workstations are "hierarchical," meaning that different levels of detail can be stored and displayed generally simultaneously. Figure 8.18 is an example of such a data base. A counter can be represented by a symbol showing simply the I/Os plus a name. It can also be represented by logic and circuit diagrams as well as simulation. Other levels are possible. When multiple replications of a macro are being used on a schematic, it is generally possible to show different levels of the hierarchy in the different replications. For example, all of the replications but two might be at the highest level (in this case, the symbol level). The remaining one might display the logic schematic and the transistor diagram of the part.

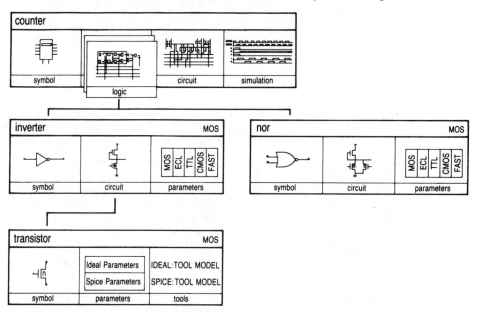

Figure 8.18 Hierarchical data base. (From S. Sapiro and R. J. Smith II, *Handbook of Design Automation,* CAE Systems, Inc., 1984. Used with permission.)

8.2.13 Electrical Trace Function

A very useful feature of workstations is the ability to highlight selected paths. This is very useful when drawing the schematic and making changes as well as for checking connectivity and other purposes. An example is shown in Fig. 8.19. Here the path from the two-input NAND gate on the left to five other gates on the right has been highlighted. The ability to highlight the different paths in different user-selectable colors adds to this ability.

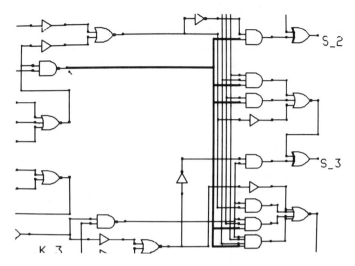

Figure 8.19 Electrical trace function. (From S. Sapiro and R. J. Smith II, *Handbook of Design Automation*, CAE Systems, Inc., 1984. Used with permission.)

8.2.14 Comparisons among Workstations

CAE has made a comparison among workstations currently offered. It is shown in Table 8.1. Because of the intense activity among workstation vendors, the features of the workstations listed are subject to rapid change. Therefore, the reader should check with all workstation vendors for the most up-to-date information at the time. The parameters listed in Table 8.1. are a starting point.

Table 8.2 gives a list of the vendors and their telephone numbers to aid the reader in getting the most current information.

8.2.15 PCs Used as Workstations [1, 2]

A number of vendors, including at least one gate array vendor, now offer CAD tools that run on personal computers. The most popular computer for such packages seems to be the IBM XT.

Universal Semiconductor, a gate array manufacturer, currently offers the UNI-CAD(TM)-1 system for about $15,000. (The price is quoted only for comparison.) The price includes Universal's library of gate array macrocells as well as a library of 7400 TTL, Intel 8000, and other standard parts, schematic capture (schematic entry) software, a net list extraction program, and a program to communicate with a VAX minicomputer. It also includes the PC with 10 megabytes of disk storage and other features. It is the same as that offered by Futurenet for non-gate-array work.

Currently, a gate array designer uses the system only for schematic capture. The data are then sent via phone link to Universal, which does the simulation and fault analysis on one of its VAXs using the Silos simulator.

A second example of a software package made to run on an IBM PC is the PSPICE program of Blum Engineering. This inexpensive ($650) package enables

TABLE 8.1 COMPARISON OF WORKSTATIONS

	CAE Systems	Daisy Systems	Mentor Graphics	Valid Logic
Schematic capture	Y	Y	Y	Y
Separate graphics editors	N	Y	Y	Y
Special purpose/ general purpose hardware	GP	SP	GP	SP
Local Tools				
Logic simulator	Y	Y	Y	Y
Functional simulator	Y	Y	Y	Y
Hardware simulator	N	Y	N	2Q'84
Timing verifier	Y	Y	Y	Y
Circuit simulator	N	Y	Y	N
Documentation package	Y	Y	Y	Y
Physical layout				
Gate array	N	Y	Y	3Q'84
Standard cell	N	Soon?	Y	N
Full custom	N	N	N	Y
Net list formatting	Y	Y	Y	Y
Waveform I/O	Y	N	N	N
Output Only	Y	Y	Y	Y
Multiple windows	Y	Y	N	N
Operating system	UNIX	Proprietary	Aegis	UNIX

Source: S. Sapiro and R. J. Smith II, *Handbook of Design Automation,* CAE Systems, Inc., 1984. Used with permission.

TABLE 8.2 MAJOR WORKSTATION VENDORS

Vendor/phone	Fast simulator
Daisy Systems Corp. (408)773-9111; local: (617)890-2666	Megalogician
Mentor (503)626-7000; local: (617)863-5776	1/4S Zycad Module
Valid (415)940-4000; local: (617)863-5333	Realfast
CAE Systems (408)745-1440; local: (617)879-5575	Zycad
Metheus (503)640-8000	
Avera (408)438-1401	
VIA (617)667-8574	

users to do SPICE modeling of up to 120 transistors, which is adequate for many uses. Speed of response is reportedly five times faster than that of SPICE running on a VAX(R)11-780.

A third example is the package offered by Personal CAD Systems, which also runs on an IBM PC. It includes schematic capture and logic simulation (as well as a PCB layout program).

8.2.16 Simulation Using Physical Modeling [3]

Physical modeling is the process of simulating a given chip by actually operating it. (One can argue semantically that the chip is not being simulated if it is actually being operated. The reader will have to bear with the semantics that currently exist for this new technique.)

The operation of the chip (or chips) is done on a hardware modeling system attached to the workstation. The signals to and from the actual chips are routed to the simulator of the workstation just as if the chips were software modules in the simulator itself.

The benefit of the technique is that it enables chips to which the gate array interfaces to be simulated in conjunction with the gate array without bogging down the gate array design effort by developing simulation models for such chips. Typically, such chips are very complex, such as microprocessors and EPROMs.

It can also be used to test the gate array itself after it has been built. The gate array is plugged into a socket. Other sockets may contain non-gate-arrays chips, such as EPROMs and microprocessors. The system of gate array and other chips is then simulated (really simulated and operated), and the results are compared to the simulation that was done with only the simulation model of the gate array chip as opposed to the actual chip itself. If there are significant discrepancies, either the chip or its model is bad.

If there are two or more gate arrays in a system, the schematic models (as opposed to the physical models) of the gate arrays are used to perform one (or more) baseline simulations. As each gate array is built, it is inserted into the socket for it and a new simulation done. If the simulation with the chip physically in place is declared valid (by analysis of the simulation), that simulation becomes the baseline. As each subsequent gate array is built, it is inserted into a socket and the simulation is updated. If any chip does not simulate correctly and the chip is declared bad, its simulation model is used to provide the signals to interface to the other chips.

Two major CAD workstation vendors offering physical modeling systems are Valid Logic Systems with its Realchip system and Daisy Systems Corp. with its PMX (physical modeling extension). Other workstation vendors with such products are Mentor and VIA Systems,

8.3 PLACEMENT AND ROUTING [4, 5, 6, 7, 8, 9, 10]

After the designer is satisfied with the design, the chip must be laid out. In Chapter 6, an overview of layout methods was given and some of the trade-offs were discussed. This section focuses on placement and routing of the macros to be used.

These macros could be those existing in a vendor library or they could be macros that have been specially created for the specific chip design.

The term *placement* refers to placement of macros on the background layers of the gate array. The term *routing* refers to routing the interconnects among the macros so placed. The routing of the interconnects among the macros and the internal routing of the macros themselves will be on one or more layers of metal. (Recall that one of the benefits of using macros is that the internal interconnects are already specified and hence do not have to be checked.)

When the layout is finished, all the lines on each of the metal layers *including the lines and contact points that are part of the background structure of the gate array being used,* will be merged into one file for *each* metal layer. Lines that are part of the background layer include the power and ground buses, for example. This file will be used to make the mask for that layer. The process will be repeated for the other layers. Masks will also be made for each of the dielectric layers which electrically isolate the metal layers from each other. The latter masks contain the "vias," which enable contact between metal layers.

Placement and routing are two related but separate tasks. Good placement makes routing much easier and enables a higher percentage of the interconnects to be routed. Conversely, bad placement makes it much more difficult to do routing.

8.3.1 Manual Routing Exercise

The purpose of this section is to give the reader a feeling for some of the problems that an automatic router or designer of auto routing algorithms faces. The exercise is also indicative of those problems encountered in doing manual routing. This is valuable for those who must complete the routing after an automatic P&R (placement and routing) program has done its work. It is also useful for those who lack access to computer-aided techniques or who need to route critical paths.

In order to participate in this exercise, the reader will have to know the grid of the IMI 70000 series of CMOS gate arrays (shown in Chapter 3) "cold" (i.e., will have to clearly understand what the connection points to the different elements are). It is helpful to remember that CMOS outputs come off drains and inputs go to transistor gates.

Two exercises are given. Exercise 1 is a "warm-up" for Exercise 2. Exercise 2 gets increasingly difficult toward the end of the routing exercise. Answers are given to both exercises. No two people route in exactly the same way; and the author does not pretend that the answers shown cannot be improved upon drastically.

A rule in both exercises is that routing cannot be done inside the boxes. The latter represent the macros. One reason is that at this point, design of macros has not yet been covered. A second reason is that a few automatic routers lack the ability to route through macros (as opposed to routing around them). This is especially true when using the gates of unused portions of the CMOS devices themselves. In the latter case, power and ground pins must be properly placed to avoid creating "phantom transistors."

The only exception to the rule above (which exists only for the purpose of this exercise) is that connections to the inputs and outputs go directly to the respective terminals, which are slightly inside the boxes.

Routing channels are denoted by dots. In a few cases, the boxes that denote the macros lie on a line of dots. In this case, it is satisfactory to route along the edge of the box. Macro designator labels have blanked portions of the grid. Reference to the original grid [Fig. 9.1(b)] should clarify any questions.

The reader should treat the macro names as being largely immaterial and not worry about whether actual macros could be fabricated in the equivalent size. (They can be.)

Benefits of the exercises. By trying the exercises, students solidify their knowledge of the grid structure as well as get a feeling for routing. This knowledge will also pay dividends when the reader comes to the design of macros.

Exercise 1

See Fig. 8.20. This exercise involves routing output A (see circle) of macro TTC8AOI#1 to four different places: input 1 of TTC8AOI#2, input 2 of TTC8AOI#3, input 1 of TRYIT.4, and input 1 of TRYIT.5. Inputs are shown on the figure.

To make it slightly more difficult (and to force the user to use the poly underpasses), three pairs of vertical lines exist as shown. (The reader should recall that the IMI 70000 series is single-layer metal.)

Hint: The reader should recall that the 70000 series, like most gate arrays, uses double-entry cells. For example, the interconnect to TTC8AOI#2 can come out the other side of the gate (input 1 to TTC8AOI#2) without incurring any additional loss in propagation delay.

Figure 8.21 is a typical solution to the problem.

Exercise 2

Figure 8.22 is a schematic of the circuit to be routed in this exercise. The knowledgable designer will readily recognize it as an oscillator, multiplier, counter, register, ... In short, it is an essentially meaningless (functionally) circuit whose sole purpose is to provide the reader with a routing problem that starts off quite straightforwardly but gets much more difficult as the number of lines which have been routed increase.

The macro layout of the circuit is given in Fig. 8.23. The reader will note that the clock and reset lines originate from the left side of the figure. The designators R and C represent the reset and clock terminals of the respective macros. Q, $\bar{Q}$ (i.e., Q inverse), and the D input to each macro are as shown. Note that the D input is on the other side of the bus bars. Again, the double-entry property of the cells must be used to avoid going significantly inside the macros. The inputs to the dual 4-to-1 MUX are as labeled. Do not forget the presence of the strobe lines, which originate as shown.

One of many possible solutions to the problem is given in Fig. 8.24.

8.3.2 Automatic Placement and Routing Programs

Current routing programs are much better than current placement programs. For this reason, manual placement followed by automatic routing is often done. It is generally much easier and much less risky and time consuming to do the placement and to

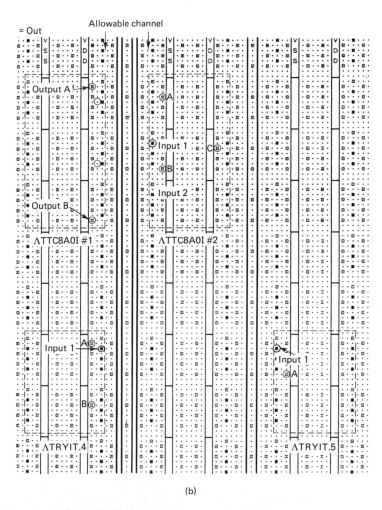

Figure 8.20 First routing exercise: trial interconnect.

check the placement of macros than it is to route and check the interconnects among the macros.

The macros are far fewer in number than are the interconnects. Moreover, they physically represent far larger blocks of chip "real estate" than do the interconnects. In the latter, it is very easy (lacking the "correct by construction" techniques mentioned in an earlier chapter) to mistakenly connect a given line of a macro to the wrong terminal of another macro. Therefore, the interconnects are both harder to do and harder to check. It is fortunate that reasonably good programs exist for this task.

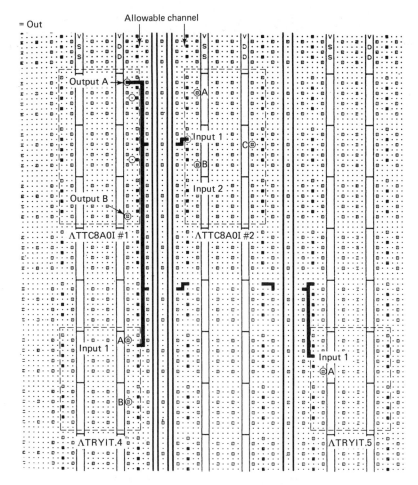

Figure 8.21 Typical solution to Exercise 1. (Readers may find many other solutions, some of which are better than that shown in one respect or another.)

Critical paths and placements. In most automatic programs, the designer can manually place critical macros before the placement program is run. Similarly, critical paths can be routed manually before the automatic router is run. One of the problems, however, is that the gate array designer almost never personally lays out the gate array and, in many cases, has no direct contact with the layout person. Moreover, because manually laying out critical paths takes longer and is subject to more risk, many, if not most, gate array vendors are reluctant to allow it. (This is one of the benefits of having in-house layout capability.)

For gate array vendors that do allow the designer to specify critical paths, the interface is often the "floor plan" mentioned in Chapter 4 with the appropriate lines marked in. Alternatively, it may be desirable for the designer to submit a block diagram of the functional elements themselves with the critical paths prioritized. *If certain signals must track each other in propagation delay, make sure that these are specified also,* even though the path delay itself may not be important or critical.

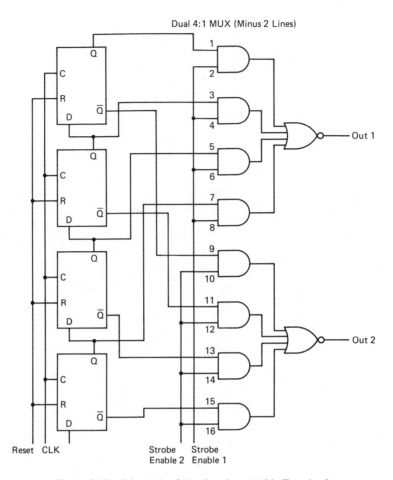

Figure 8.22 Schematic of circuit to be routed in Exercise 2.

A typical number for two-layer metal arrays is that 95% completion of the routes can generally be achieved by the automatic router if no more than 80% of the array cells are used. These numbers will vary with the gate array placement (if any) and routing programs used and especially with the circuit being implemented. Some vendors are quoting 100% automatic routability under the foregoing conditions, but this is rare. Most of the newer arrays currently available are designed to facilitate auto routing.

Routing algorithms [4], [6], [9]. Three main types of routing algorithms exist. These are called Lee, channel, and line search routers.

The *Lee* or *maze router* format is shown in Fig. 8.25. The area to be routed is broken into a grid. For gate arrays this grid naturally takes the form of the repetitive channel spacings. Each channel represents a path along which the interconnects can potentially go.

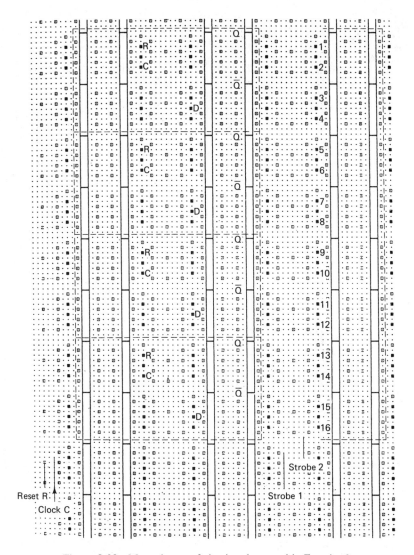

Figure 8.23 Macro layout of circuit to be routed in Exercise 2.

All the grid squares around each starting point of the line to be routed are sequentially numbered in the four major directions (up, down, left, right). Each starting point in the case of a CMOS array would generally be a drain connection, for example. The routing then proceeds by trying all of the different permutations at each point in the path. For example, when the line to be routed comes to point 3, it can go up, down, or to the right to get to point 4; when it comes to point 4, it has the same three options to get to point 5.

The Lee router is a good net router. It can find the shortest path between two points. Also, it can easily be modified to find a path that minimizes resistance and capacitance in a given path.

Sec. 8.3   Placement and Routing

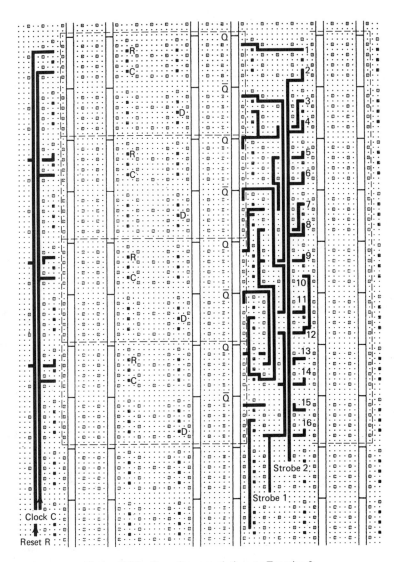

Figure 8.24 One of many solutions to Exercise 2.

There are, however, several disadvantages to the Lee router. ([5] and private communication.) First, as might be surmised, the Lee router is very slow and uses a great deal of memory space. Hightower has mentioned 4 megabytes of memory for a 1500-gate gate array. Second, it will not always find a path (100% completion of a series of routes cannot be guaranteed). Third, the paths that it does not route can be very hard to fix.

Channel routers (Fig. 8.26) require well-defined channels in which to operate. The least number of tracks required for a particular interconnect is called the *channel density*. A good channel router routes at the channel density most of the time.

Channel routers are very popular for several reasons. First, unlike a Lee router, the channel router is fast. Second, 100% routing can be guaranteed if there is enough

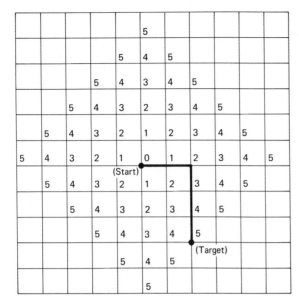

Figure 8.25 Lee router format. (From J. Werner, "Software for gate array design; who is really aiding whom?" *VLSI Design,* fourth quarter 1981. Used with permission.)

space. Third, any failures are easy to fix. The disadvantages are that the routing problem must be organized into cells because the channel router is not a good point-to-point router.

The *line search router* (Fig. 8.27), as its name implies, routes a line until it hits an obstruction. It then either turns or backs up to the last point where there was no obstruction and tries again.

The advantages of the line search router are that it is fast and no grid is needed. (This is not necessarily an advantage for gate arrays.)

Problems with the line search router are that it may not find a path even if one exists; failures are hard to fix, as in the Lee router; and it is difficult to program. However, the very major disadvantage with the line search router is that it may find a bad path.

Lee routers tend to be used to route the inside of macros by companies such as

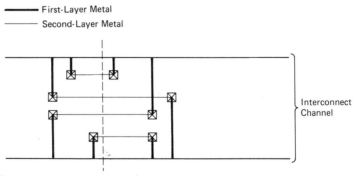

Figure 8.26 Channel router. (From J. Werner, "Software for gate array design; who is really aiding whom?" *VLSI Design,* fourth quarter 1981. Used with permission.)

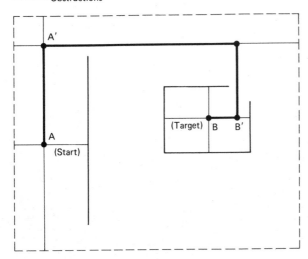

Figure 8.27 Line search router. (From J. Werner, "Software for gate array design; who is really aiding whom?" *VLSI Design*, fourth quarter 1981. Used with permission.)

TI. TI uses a channel router to route between macros. Other companies use channel routers exclusively.

As noted earlier, auto routing is seldom 100% successful. The layout person then has several alternatives. The first is to try to manually route the missing interconnects. If the paths are relatively short, this is often the best alternative. The author has often stood in awe as a layout technician at his company showed him how a given path could be routed through a seemingly incomprehensible maze of interconnects. To compound the problem, sometimes other lines had to be rerouted to enable a given interconnect. It would be extremely difficult to program the thought processes that go on simultaneously in the mind of a good layout technician as alternatives are weighed.

The need to finish a design is one of the reasons that a designer and layout person should know how to design with transistors instead of macros. If one knows the internal structure of the macro (and if the gate array vendor allows it), hard-to-route lines can sometimes be routed through macros (as oppose to treating macros as boxes that cannot be crossed).

The second alternative is to use program that performs "rip-up and retry" on congested areas, which are exactly what the name suggests.

The third alternative is to try a new placement of the macros, at least in the congested area, followed by a second or third application of the routing algorithm. This is useful if the routing algorithm is fast and is often the best alternative when there are many lines left unrouted by the first attempts.

Allocation of space in the wiring channels—Rent's rule. [11, 12] Although allocation of space, and, hence number of tracks available in the wiring channels gate array, is beyond the control of the user, it is worth at least mentioning what has come to be known as *Rent's rule*. This was briefly discussed in Section 5.3.

This rule gives the ratio of the total number of I/Os to the total number of nets (interconnects). The ratio is found to vary inversely as the fractional power $(1 - p)$ of the total interconnect count. The parameter p in the exponent is called the *Rent exponent*. This is a number that is typically between 0.5 and 0.75.

Heller et al. [11] have applied Rent's rule to the problem of determining the number of tracks required for gate arrays or other chips with a regular structure. Comparisons made with the results from actual chip designs showed agreement within 10%.

Examples. Examples of placement and routing techniques are shown in Figs. 8.28 to 8.30. In these figures, the lines which are neither horizontal nor vertical are called *air lines*. They represent the direct connection between circuit elements without regard for allowable routing channels. The benefit of air lines is that they show where congestion is highest and give an indication of what the resultant congestion would be if circuit elements were moved around.

Figure 8.28 shows how interchanging two gates (three-input NANDs and inverters in this example) which are functionally identical to one another can make routing easier. It will be noted that the inputs to the two NANDs have to cross each other to get to the proper NAND and that the outputs similarly cross. By simply reassigning (redefining) which NAND is used for which element, the problem is eased.

Similarly, by redefining the uses of the inverters in the same figure so that they have inputs and outputs which do not cross, routing is simplified in the case shown. (Actually, the overall routing picture would have to be taken into consideration;

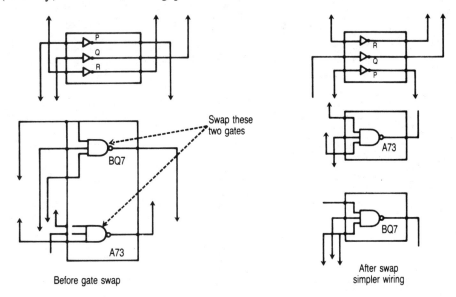

Figure 8.28 Interchanging elements makes routing easier. (From S. Sapiro and R. J. Smith II, *Handbook of Design Automation*, CAE Systems, Inc., 1984. Used with permission.)

these examples are simplified to show what can be done if other routings permit; they also show traps to avoid, where possible.)

Figure 8.29 shows that wiring congestion can be reduced simply by changing the pin assignments. For example, the input to the lower right AND gate has been taken from the lower input of the middle (vertically) AND gate on the left instead of from the upper portion of the same AND gate. This input in turn came from parts of the circuit not shown in the example [i.e., there is a fan-out of 3 to the three gates mentioned from the driver (not shown)].

The reader will recall from Chapter 3 and elsewhere that second- and later-generation CMOS gate array cells are double entry (i.e., the inputs can come to either side of the power buses). Also, the outputs can generally be taken from either side of the power buses. Therefore, the technique shown in Fig. 8.29 is very practical to implement.

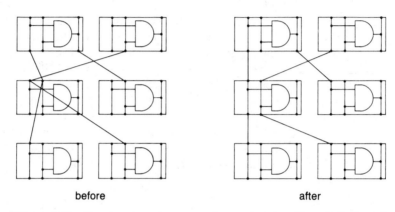

before after

Figure 8.29 Changing pin assignments reduces congestion. (From S. Sapiro and R. J. Smith II, *Handbook of Design Automation,* CAE Systems, Inc., 1984. Used with permission.)

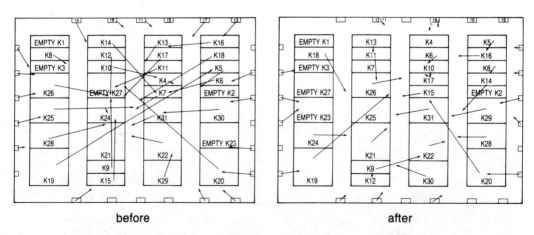

before after

Figure 8.30 Changing placements improves routability. (From S. Sapiro and R. J. Smith II, *Handbook of Design Automation,* CAE Systems, Inc., 1984. Used with permission.)

Figure 8.30 shows how changing placements of macros improves routability and emphasizes the point made earlier that good placement is necessary for a good route. The reader will note the considerable decrease in congestion that results. Although this is a hypothetical example, substantial improvements occur by replacement and sometimes by initial placement. This is one more reason for designing a floor plan (see Chapter 4) before going into formal placement and routing.

8.3.3 Three Non-gate-Array Layout Techniques

The purpose of this section is to make the reader aware of three techniques for layout of the chip. To the author's knowledge, none has yet been applied to gate arrays. There are enough similarities to the gate array methodology, however (principally in the distribution of power, ground, and terminal connections), to warrant inclusion in this book.

The techniques are useful for those groups interested in designing their own gate array background layers. If nothing else, they give alternative thought processes which may prove useful ultimately in the development of algorithms for gate array layout. The author has had personal experience with two of the methods while designing pure custom circuits using the Rockwell symbolic layout methods.

The three techniques are the Weinberger image technique, developed and used at IBM; the gate matrix technique, developed at Bell Telephone Laboratories; and the symbolic layout method, developed originally at Rockwell but also used at AMI and other companies. (It is also part of the gate matrix method of layout.)

The Weinberger image technique [13]. The Weinberger image technique is characterized by circuits, such as NANDs and NOR gates, arranged in columns. Power distribution buses are also arranged in columns across the chip. The resultant chip looks y   much like a gate array.

By applying Rent's rule (discussed in Section 5.5), Cook et al. [13] are able to show that the number of nets F entering or leaving an assemblage of containing N blocks of logic is

$$F = WN^P \qquad (8.1)$$

where P is the Rent exponent and each block has an average of W wires entering or leaving it. An important result of this calculation which may have some potential application to gate arrays is the calculation of wire channel requirements in each column. Cook et al. [13] show that for a Rent exponent of $\frac{2}{3}$, a fan-out of 3, and track utilization of $\frac{2}{3}$, the number of wiring channels required varies as shown in Fig. 8.31.

Potential Uses for the Weinberger Image Type of Analysis. It strikes the author that this type of analysis could be applied in an algorithm used before layout of a gate array using auto P&R techniques (1) to determine if the number of wiring channels is adequate (before a computer "swallowing" auto P&R run is done), (2) to determine and enable trade-offs among the number of circuits laid out in a given column (i.e., it may be advantageous, if not absolutely necessary, to use a gate array

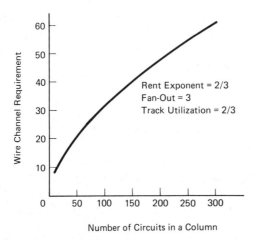

Figure 8.31 Wire channel requirements in Weinberger layout column as a function of the number of circuits in the column. (From P. Cook et al, "1 micron MOSFET VLSI technology: part III, *IEEE J. Solid State Circuits*, Apr. 1979, p. 257. Registered copyright 1979 IEEE. Used with permission.)

die containing more columns and to use fewer circuits per column), and (3) to estimate whether a larger chip in a given gate array family is needed to accommodate the wiring channels.

The gate matrix approach to layout [14], [15], [16]. The gate matrix approach to pure custom VLSI design as developed at Bell Labs uses polysilicon that runs vertically at constant pitch. Rows of *p* or *n* diffusion run horizontally. The intersection of diffusion and polysilicon (poly) forms a transistor. Poly is used for both the self-aligned gate and for interconnects among transistors. Metal interconnects can run either horizontally or vertically.

Potential Usefulness to Gate Arrays. The potential usefulness to gate arrays of the gate matrix method of layout is that it provides a potential means for designing the gate array layers themselves.

Symbolic layout. The gate matrix methodology was specifically set up to use a techinque called *symbolic layout* (not to be confused with *symbolic entry*, which generally refers to a schematic entry method). In this method, the different layers of the wafer, such as *n* and *p* diffusions, polysilicon, and metal, are represented by alphanumeric symbols. For example, the letter N might represent an *n*-type diffusion, the exclamation sign ! might represent metal, G might represent a polysilicon gate, and so on. Crossovers among these materials are also represented with alphanumerics. For example, metal crossing over polysilicon might be represented by the letter I, metal making contact to an n-type diffusion might be represented by the symbol @, and so on.

By means of this symbolic representation, pure custom circuits can be rapidly designed using a handful of relatively simple rules. For example, two @ symbols which do not connect the same *n*-type diffusion must be spaced at least one grid space apart. (The grid in this case can be the usual spacing on a terminal or printer; software will then properly interpret it in terms of the actual physical dimension of the transistors.)

Figure 8.32 is an example of a small circuit designed with symbolic layout.

```
000000000000000000000000000000000000000000000000000000000000000000000000000000000
000000000000000000000000000000000000000000000000000000000000000000000000000000000
file A30UP2.004... GATE MATRIX METHOD OF LAYOUT

                 11111111112222222222333333333344444444445555555555666666666
       01234567890123456789012345678901234567890123456789012345678901234567

  0                                                                     bb 0
  1                                                                     bb 1
  2                                                                     bb 2
  3                                                                     bb 3
  4                                                                     bb 4
  5         .         .         .         .    s11 s12          .         .  bb 5
  6                                            III III                eolt  bb 6
  7       |||||||||||||||||||||||||||||||||||||HHH|H|H|||||||||||||||||||H|||||bb 7
  8       |||||||||||||||||||||||||||||||||||||HHH|H|H|||||||||||||||||||H|||||bb 8
  9       ||||||||||||||||||||||||  XXXXXXXXXX||HHH|H|H||||||     ||||||H|||||bb 9
 10       ||||||||||||||||||||||||  ----------||HHH|H|HH|XX|| *|    X| H|vdd|bb10
 11       ||          |||||||||||| IIOOO   OOOHHHHH|H|HH---- ..|.YI----I|||||bb11
 12       XX ||X|*||     XX|||||XX I ---------  III I IIOOOOIGO|.IIOO OI     bb12
 13       -- | - .       --     -- I OOO1I OQQH|HHH|H|Y ---- ..#.I ----I     bb13
 14     IIOO HIO OII    IOO|YIIIOOII --- I -++I   I I  IOQQ||C | I  C||H     bb14
 15     I -- H - . I    I--|  I -- I CXX|X|CXXIIYHH|Y  I-+-  --| I .. II     bb15
 16db1fH|CX|H C|C|H|YIICX| |H|CX I ... I ...I   I I  I C   OOH I GOIII     bb16
 17     H .  I . - I    ...  |I .. IIGGG I GGGI III I  I.# Y|--H I ....II    bb17
 18     HIO IIIO OII    GGO  |IIGG   ... I ...  III II IOGII|XXH IIOGGO I    bb18
 19     | . II . .I     ...  |  ..   GGGII GGGIIIII I   .*     I I .... I    bb19
 20     | * II *|XH|*   ***  *  **   ... I ... IIII II  OGHHHHHH|I   ** I    bb20
 21     * | I1    1 .   ||| . ||||||||*|I *** HHHH|HH||..II|||| HHH|||H||||bb21
 22     ..|HH###HHH|###############.  ||I|||||HHHH|HHH|**|H|HHHHHHH||HHH||||bb22
 23     ##|HH###HHH||||||||#dr2t ##.* ||H|||HHHHHH|HH|||||H|HH||||||HHHH|vssbb23
 24     ##||H##JI      ##########||| | HHH|||||HHHH|HH|HHHHH|HH|**|HHHHHH||||bb24
 25     . * I..  |X|*  * ||**||||* HHII****** IIII II I|||||H  .. HH||||||||bb25
 26     | ..I..|||- .  |    ..    ..H|  ... ...IIII|HH|Y *||HH IOG I  *     bb26
 27     HH@@H##|HHO G1IY IIGG|YIIGGI|YIGGG|@@@HHHH|II I ...I  I...I ..     bb27
 281dr I ..  . I - . I  I ..| I ..  ||...|...   I II IIO OI II GOI GOI     bb28
 29 1f I CXX|* I C|C|H|YI CX| H|CXYIHHGGG|GGGIY|H|YI II...I II ..I ..I     bb29
 30     I --- | I . - I  I --  H --  HI...|...I  I II II C|H|YI C|H|* I     bb30
 31     IIOOO | IIG OIIY|1IOO |HIOO||HICXXXXC I II II  II--I II-- I-- I     bb31
 32       --- | I . -   |  -- * --- |HI-------I II III IIOOI IIOO IOO I     bb32
 33       XXX * Y *|X|* |  XX . XXX |HIOOO OOOI II III   --    -----  I     bb33
 34       ||| . *     .   ||||#|||| || -------HHHH|HHH|||XX|||| OOOIHHH|||||bb34
 35       ##### . |||##############| YIIOOO OOOHHHH|HHH||||||||||--- HHH1tti|bb35
 36       #####|###||##############|||| -------HHHH|HHH||||||||||XXX|||||||||bb36
 37       #####|###||##############||||XXXXXXXHHHH|HHH||||||||||||||||||||||bb37
 38       ...   ...                              III III                   bb38
 39       1dr1t 1dr2f                            III III                   bb39
 40          .         .         .         .     IIIIIII          .         .  bb40
 41                                              IIIIIII                   bb41
 42                                              IIIIIII                   bb42
 43                                              I1IIII1                   bb43
```

Figure 8.32 Small circuit designed with gate matrix method. (Courtesy of Sanders Associates, Inc.)

The number of FETs is purposely kept small to illustrate the method. Also, the circuit is not tightly packed (as it normally would be), for the same reason.

Figure 8.33 is the symbol table for this array. For example, polysilicon gates for NFETs are designated by the symbol G. The compiler will recognize the G and translate it to a poly mask. The width of the gate is given by the duration of the line. (The term "length" has to be reserved for the gate length, which is the grid spacing designated by one character height.). The grid on which the symbology resides can be changed simply by inputting a different scale factor to the compiler.

```
00000                         SYMBOL TABLE                            00000
                            ***************
■ N++ ISLAND          * N++ METAL CONTACT       $ N++/METAL CROSSOVER
- P+ ISLAND           X P+ METAL CONTACT        + P+/METAL CROSSOVER
I POLYSILICON         Y POLY/METAL CONTACT      H POLY/METAL CROSSOVER
I METAL               G NFET GATE               @ NFET/METAL CROSSOVER
C P+N++ METAL         O PFET GATE
  CONTACT
```

Figure 8.33 Symbol diagram of Fig. 8.32.

Figure 8.34 is the schematic of the circuit laid out in Fig. 8.32.

Potential Usefulness to Gate Arrays. Because even ordinary alphanumeric terminals, such as the very common VT-100/200 series of DEC, can be used (although with limitations on the width of the circuit), it is possible to use such terminals to design small pieces of gate arrays. The IBM PC XT and Tandy 2000 are also potentially usable. The author's group used such terminals as well as Intel MDS-240 microprocessor development terminals, which were in widespread use throughout the company for symbolic layout (not for gate matrix layout specifically).

The results were then sent to Rockwell. This illustrates what can be done. As prices of workstations drop, the approach of using such terminals will become less attractive (although substantial drops will have to occur for such prices to come close to that of a $600 VT-100 terminal linked to a minicomputer). The major limitation is the width of the circuit to the 128 columns normally found on most VT-100-type terminals. This can be improved to about 260 using an LA-120-type terminal at an increased pitch of 16 characters per inch. The latter lacks screen edit capability, however.

In currently unpublished work, the author showed that using regularly spaced metal lines as the starting point yields circuit densities which are considerably greater than are otherwise obtained. (In an in-house contest, the author's method was twice as dense as other layouts for exactly the same circuit design.)

It strikes the author that small gate array chips could possibly be laid out manually using the symbolic layout method. Each library macro would be stored in the computer's buffer memory. The screen editing functions (such as the keypad editor of the VAX series of computers) would allow the macros to be called and moved around. Interconect lines would be designated by symbols. Software to enable sideways scrolling could also be written quite readily (as it was for the MDS-240s at Sanders). This would overcome the major limitation of the alphanumerics terminals.

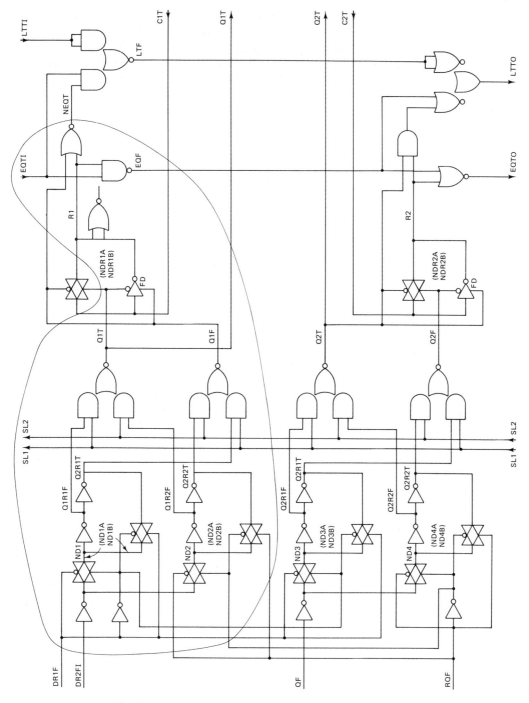

Figure 8.34 Schematic of Fig. 8.32. (Courtesy of Sanders Associates, Inc.)

Sec. 8.3 Placement and Routing

A benefit aside from the lack of investment in terminals (although there would have to be an investment in software which might outweigh all other considerations) is that the layouts so produced are very easy to read and to check. This is especially true when only the layers due to gate arrays are considered. Moreover, I/O symbols can be emplaced which would also contain information. B, for example, might stand for a bidirectional pin. (The lack of investment in terminals is based on the assumption that most companies have such terminals.)

8.4 DESIGNING REMOTELY

By the term "designing remotely" the author refers to designing a gate array using the *host computer of the gate array vendor* at a place physically separated from the host computer. It does not refer to designing a gate array at the user or other organization using CAD tools in place at the user organization and then shipping a magnetic tape to the vendor from which the mask can be made.

A typical output of the remote design is a series of netlisting, simulation, and test generation and analysis files.

Increasingly, designing remotely is being offered by major gate array vendors. LSI Logic, National Semiconductor, Motorola, and Texas Instruments are just four examples of vendors offering remoted gate array design facilities. Even distributors of electronic components, such as Scheweber, offer facilities for the remote design of gate arrays using both workstations (not part of this discussion) and remote access to the mainframe.

Two distinct possibilities exist for such remote design. In the first, a terminal at the user organization is used to access the remote host computer. In the second, a terminal at a remote branch of the gate array vendor is used. The contrast is to doing the design directly at the gate array vendor's establishment at which the host computer resides.

The purpose of this section is simply to indicate some of the real-life problems that can exist with such an arrangement as well as to indicate the positive aspects of such a method.

The author has personally designed a number of both custom and semicustom circuits using remote facilities. Because the circuits were the first designed in each case over the links from his company set up for that purpose, some of his views are based on having had to debug such links and would not necessarily be typical of a smoothly functioning link. Taken with that proviso, the author's thoughts may have some value, especially for those contemplating setting up their own links.

8.4.1 Advantages

Designing remotely has several advantages and several disadvantages. The advantages include the elimination of travel expenses if the remote station is local to the user's organization and the close proximity to the designer's resources (data books, notes, and perhaps consultation with the circuit designer if a person other than the person who designed the basic circuit being implemented on the gate array is doing the gate array design). The latter method is common in large organizations which have groups that do only gate array design.

Another major advantage to the user organization (which, however, is detrimental to the design of the chip itself; see below) is that a valuable designer is not essentially totally unavailable for other tasks and for consultation on other jobs as would be the case if he or she were physically located at the gate array vendor's establishment.

The disadvantages depend on how the remote design facility is set up and the characteristics of the designer using such facilities.

8.4.2 Problems That Can Occur

Most of the problems associated with designing remotely have to do with the status of the link and the status of the host computer. These are delineated below.

One of the most frustrating parts of designing remotely is not being sure where operational problems with the link occur. This is especially true if the designer is designing from his or her home office as opposed to a well-run field station of the gate array vendor. The latter is much more likely to be in contact with the gate array vendor's home office. Moreover, the lines themselves are likely to be better.

In the absence of such an arrangement, as for example when the user is designing from his or her home office, several maddening things can occur that are often indistinguishable from each other.

First and foremost is the telephone line itself. If it is a long-distance call, the user may have to wait for a line only to find that the connection is bad. Credit-card calls avoid the need to wait for the line but are much more expensive and no more reliable.

Once a connection is established, the line will be subject to line hits, which may or may not echo on the terminal as strange characters. The computer, of course, then sends a message to the user claiming that he or she used wrong syntax. This problem can be largely eliminated by using a packet-switching service, such as GTE's Telenet. The latter improves transmission by using error-detecting and error-correcting codes to make sure that each packet of information was received correctly. If the packet-switching entry point is located a considerable distance (100 or more miles, for example) from the designer's terminal telephone, the information to be sent can be corrupted before it reaches the "safety" of the packet-switching station.

Probably the next most frustrating problem is not knowing the status of the host computer. A property of CA tools for gate arrays is that they use a lot of computer resources. The user at a remote remote terminal may give the computer a command (to run a simulation, for example) and then wait for a considerable length of time. (The author has waited more than an hour when running on a 7.5-MIPS computer when nothing was malfunctioning, simply because it was clogged with jobs to run.)

Several possibilities exist for the delay, none of which the user can determine immediately generally. The first is that the computer is simply swamped with programs to execute. If the designer were at the vendor's facility, he or she would be able to tell that this is the case by simply checking the monitor of the computer or asking someone. When remote, it takes a long-distance call, with the problem of getting the line to get this information.

The second possibility is that the computer has been taken down because of a problem or has crashed. The remote user may not have been on the machine at the time the warning, if any, was given. Computers are sometimes taken down to correct minor problems during the lunch period of the gate array vendor. If a time-zone difference exists between the user and the host computer, this down period may correspond to prime working time for the user. (However, the same time zone difference may be to the user's advantage if he or she can get on the host computer two or three hours before those working at the gate array vendor's place of business arrive in the morning.)

The third possibility is that the controller for the remote terminals has malfunctioned or has been taken down. A fourth possibility is that remote users are given lower priority than local users on the host computer and hence are more susceptible to being swamped out.

Another problem that rears its head is that the user-organization terminals may not be of the type supported by the gate array vendor. In this case, the user may have to spend considerable effort, including generally some trial and error, figuring out how to use his or her terminals on the host.

To obtain a printout of, for example, a simulation file, high-speed links and high-speed printers are usually required. (Such printouts may number a hundred or more pages.) Seldom are these available at the user organization in a format that permits them to be accessed remotely from the host computer.

Designing from a local branch of the gate array vendor eliminates many of the aforementioned problems. The local branch will usually have installed leased lines because of the high volume of usage. These are of higher quality and allow the host to be accessed immediately instead of struggling to get the line.

Moreover, the terminals installed obviously will be supported by the host; and the user generally will be able to tell or find out easily if the system is down or simply swamped. Also, help is readily available from local applications personnel in interpreting diagnostic messages and other properties which are characteristic of the vendor's system and macro library.

In a few cases, graphics data, as opposed to the aforementioned netlisting, simulation, and test files (which are alphanumeric in form), comprise the output of a remote design. Such a case might arise when layout is done remotely, for example.

If graphics terminals are used for the transmittal of such graphics data, then high-speed lines, if not microwave links, are a must. The amount of data in a graphics package for a chip design can be huge; 1200- or even 4800-baud links simply do not suffice. (Sending a magnetic tape containing the graphics information via an overnight courier service, such as Federal Express, is sometimes a good alternative.)

The last problem that will be mentioned is a subtle one. When a designer is resident for a period of time at a gate array vendor's establishment, he or she is fully focused to the task at hand. Meetings, problems, telephone calls, and so on, that would dilute the effort if he or she were designing remotely from the user organization are not present. When the designer is designing at the vendor's establishment, there is often a great deal of visibility, with an attendant amount of pressure. The result is a very intense and very focused effort that may produce a better product in a shorter period.

8.5 SPECIAL-PURPOSE SIMULATION PROCESSORS

A property of CAD programs is that they use a lot of computer resources. When multiple users are running various CAD programs, response times decrease. The decrease for a given user obviously depends on the computer programs being run, the machine being used, the number of users running simultaneously, and the priorities assigned to each user.

When substantial slowdowns repeatedly occur, users tend to modify their modus operandi (methods of operation) accordingly to accommodate the situation and to avoid excessive waiting. Often such new ways of trying to live with an overloaded computer add risk to the design process by eliminating or minimizing steps that should neither be eliminated nor minimized.

One example of such undesirable modifications is that a change may be made to a circuit that requires a very large number of clock cycles to simulate it properly. (An example of the latter is a circuit containing a maximal-length feedback shift register sequence generator of many stages.) The designer may use manual circuit analysis and pencil and paper to decide that the change works rather than resimulate it. This is even more likely to occur if the cost of computer time is figured against a fixed project budget, and the designer believes that it is his or her error.

Although such an approach may give valid results, there is the possibility that one or more "wires" on the design were not properly connected or that the added propagation delays and delay variations over temperature were not properly accounted for in the manual analysis. Computer simulations are not foolproof either, but they do enable errors such as the above to be spotted by taking into account the global picture of the circuit operation.

A second example is when a designer minimizes the simulation runs to be made in some fashion because of the computer slowdown and the time to get results. The designer may, for example, not fully simulate the circuit over combinations of temperature, voltage, clock and clock skew (variation in duty cycle of the clock), and parameter variation due to wafer-processing variations.

A third example is that a designer does not simulate the entire circuit but rather, chooses to simulate only selective pieces of it. This is sometimes a feasible approach, especially if there are many repetitive elements and if the partitioning is such that the signals going into the replications do not depend on those generated inside the replications not simulated. Too often, however, such is not the case. The signals generated inside the replications not simulated *do* affect those that are simulated with timing or logic states different from that surmised.

Yet a fourth example involves the question of simulating an entire system, which may consist of 100,000 to 1 million gates. It can be argued that hierarchical, not gate-level, simulation would be used for this task. However, even hierarchical simulators have limitations. Even after running a hierarchical simulation, many organizations would prefer to run a final simulation at the gate level.

A determination of what comes out when, for example, a given opcode is put in to a given block will have been previously made either at the gate level or at the hierarchical level, preferably the former. (The outputs for a given opcode will have been specified at the hierarchical level; but until a gate-level simulation is done on the actual design of that block, the designer is never truly sure.)

To overcome these problems, at least three manufacturers have developed special-purpose processors, which might be called *accelerators*. The purpose of the accelerator is to do the tasks that require large amounts of computing power—logic and fault simulation.

All other functions, including coding of netlist, automatic test vector generation, and display of the results, are handled on another computer. The latter is sometimes called the *host computer*. Formatting of the circuit file into the format of the special-purpose processor and formatting of the results back into the format of the host computer or the original simulation program might also be required. These tasks would also be done on the host computer.

The host computer could be as small as a Daisy, Mentor, Valid, or CAE workstation or it could be a VAX or some other large machine. Most likely this other computer would be the host computer on which the simulation program would normally run.

The three current manufacturers are Zycad, Valid, and Daisy Systems. In addition, workstation vendors Mentor Graphics Corp. and CAE Systems, Inc. offer the Zycad product on their popular workstations. A tabulation of the vendors and their accelerator products is given in Table 8.2.

8.5.1 Zycad LE Series

Zycad offers the LE series of processors. These are designated the LE-1002, −1004, ..., −1032, respectively. Each successively numbered unit doubles both the capacity (number of primitives that can be simulated) and the throughput rate over that of the previous unit. (At the highest rates, bus contention problems may limit the highest rate to something less than doubling of the preceding, although this is not clear.) See Table 8.3.

A summary of the Zycad characteristics provided by CAE Systems, Inc., is shown in Fig. 8.35. Improvements are, of course, being implemented. Therefore, the reader should contact Zycad for the most current information.

The product of speed and capacity might be called the power of the processor. Therefore, each unit represents a quadrupling of the power of the previously numbered unit.

A few notes are in order. First, an event is a change in the output of a primitive. Some vendors use the term "evaluations per second" to rate their processors. For an event to occur, a number of evaluations must also occur. To compare to the Zycad, roughly $2\frac{1}{2}$ evaluations per event on the average should be used.

Second, a primitive is a multi-input functional structure roughly equivalent to a

TABLE 8.3 ZYCAD SIMULATION PROCESSORS

Zycad product designator	Primitives*	Speed (Events/second)
1/4 S Module	16K	0.5^6
LE-1002	64K	10^6 (is one Zycad "S" module)
LE-1004	128K	2×10^6
⋮		
LE-1032	1048K	$\geq 1 \times 10^6$

*Number that can be simultaneously simulated.

To Compare Numbers

An *event* is a change in the output of a primitive (Zycad's definition).

There are typically $2\frac{1}{2}$ to 3 evaluations per event on the average, but numbers will vary.

Example: The Valid Realfast processor is rated at 500,000 evaluations per second. This is equivalent to 200,000 events/second. Capacities must also be compared. (At the time of this writing, no other information on Realfast was available.)

Zycad Simulation Accelator

Capacity	64K-1000K gates. Plug-in expansion modules required.
Race and Hazard Analysis	None. Fancy post-processing may enable some. Allows set-up and hold tests for chosen signals only, which can pose danger of missing some glitches.
Delay	Delays < 4096 units. The way around it is to add dummy delay buffers.
	Allows each gate output to have inertial and propagational delay. Allows separate additional delay for each input of each gate to enable efficient interconnect delay modeling.
	Allows for only rise and fall delays.
Modelling	Two or more gates required to model the same, because of limited number of inputs in primitives.
MOS Circuit Modeling	Offers uni and bidirectional models to carry out switch level logic simulation.
Fault Simulation	Serial, for N faults takes N/10 to N/2 times longer than normal run.
Interactivity	Limited. Breakpoints can be set to see status of a signal on a pin.

Figure 8.35 Features of ZyCAD. (From S. Sapiro and R. J. Smith II, *Handbook of Design Automation*, CAE Systems, Inc., 1984. Used with permission.)

ZYCAD vs. Tegas Software Simulation:

Gates	Patterns	Events	Report Nodes	Tegas Time*	Zycad Time	Speedup
13K	896	8192K	1023	15K sec	6.8 sec	250
5K	1000	852K	1023	2000 sec	1.7 sec	260

*Tegas run on IBM 4341.

Figure 8.36 Comparison of IBM 4341 versus ZyCAD running Tegas. (From S. Sapiro and R. J. Smith II, *Handbook of Design Automation*, CAE Systems, Inc., 1984. Used with permission.)

Sec. 8.5 Special-Purpose Simulation Processors

macro. In general it will be more than one logic gate. This too, should be taken into consideration in doing a comparison among vendors.

Zycad's units are housed in a cabinet containing slots for 18 modules. Except for a control element that occupies the first slot and a spare power supply that occupies the last slot, all other modules are identical. The user simply buys as many modules as needed to perform the desired tasks. This is how the designations come about.

Each module is capable of processing 65,536 three-input *primitives* (not just gates) or 100,000 two-input primitives. The 15 modules plus spare module that occupy the remaining 16 slots in the cabinet can, hence, perform simulations on gate structures up to 1.5 million primitives. When the number of primitives to be simulated exceeds 100,000 (two-input primitives), the designer may want to partition the circuit so that it will make use of additional modules. (The Zycad will do such partitioning for the user, but it may not be in a way that is meaningful to the user.) Generally, such partitioning is in place anyway as part of the normal system partitioning done during initial design. For many users, a single module or two is perfectly adequate.

Formatting of the design language into the ZIF (Zycad intermediate form) code is required. This can be performed automatically by the Zycad if the Silos (from Silos, Inc; Incline Village, Nevada) simulator is used. (Zycad calls the translator ZILOS.)

The Zycad processor is potentially capable of operating with many simulation languages, such as Tegas. In the case of Tegas, it does this by a translation of the Tegas TCF (Tegas circuit file). Interfaces are in preparation for the HI-LO 2 and other simulators.

Advantages. The primary advantages of the Zycad processor are very high speed and the ability to handle very large numbers of gates. The LE-1002 module will do logic simulations 500 to 1000 times faster than a VAX-11/780. According to Zycad, the units at the high end of the series will do simulations that exceed the capacity of the VAX. Two benchmark cases provided by CAE Systems, Inc. of the Tegas simulator running on an IBM 4341 and on a Zycad processor are given in Fig. 8.36. For the two cases of 5000 and 13,000 gates, the speed-up rate is 250 and 260 times, respectively. Note that the number of report nodes is the same in both cases but that the number of events in the latter case is about an order of magnitude greater than in the 5000-gate case. The number of input patterns in both cases is roughly the same.

The very high speed means that even though the processor will simulate only one circuit at a time, moderate-sized simulations will execute so rapidly that it will appear to multiple users that the processor is doing time sharing.

Hardware interfaces. Hardware interfaces to Zycad are offered by Apollo Computer, IBM, DEC (the DR11W), and other companies.

Disadvantages. Although the Zycad unit is very attractive from the standpoints of speed and number of primitives, it does have a few disadvantages. Some of these, such as the requirement for a separate interface program, have been alluded to above. Another disadvantage, in the author's view, is the restriction to gates with no more than three inputs. While studies at IBM and elsewhere have shown that the majority of gates used in some of their (IBM's) circuits are predominately two- and three-input gates, there are nevertheless significant numbers of higher-input gates. This has been the author's experience also.

To overcome this limitation, the user must currently write a translation program to convert all gates with greater than three inputs into gates with three or fewer inputs. (The alternative is to manually recode the circuit into three or fewer input gates, but this is fraught with possibilities for error.) Sequential elements, such as D flip-flops, are coverted into one of a dozen or so other primitives in the Zycad library.

Some of the elements into which the circuit is coded may have zero delay. This would be the case, for example, in which a four-input AND gate was translated into two two-input NAND gates feeding a two-input NOR gate. The time delays of the AND gate may be ascribed to the NOR gate, and the NAND gates given zero delays.

A potentially minor restriction is that only 32 primitive types (among the 65,000 to over 1 million primitives that are being simulated at one time) can be used in a given simulation run. Because different delays can be ascribed to a given primitive *without* its counting as a separate one of the 32 allowable primitives, this is generally not a major problem. (This enables a very wide variety, much larger than the 32, of element types to be built.)

There is a certain amount of time required to get in and out of the machine. Time is also required for the host computer to translate the net list file first into three-input primitives plus, second, into the ZIF language, and third, from the ZIF language back into that of the host computer for display or other purposes. According to Zycad, for a 5000-gate circuit, this time is less than 200 milliseconds on a VAX-11/780.

It must be emphasized that the Zycad is a relatively new unit and a new concept which is still being improved. It is to be expected that the system as it evolves will have means to overcome some of the relatively minor disadvantages that exist.

8.5.2 Daisy Megalogician

Daisy has incorporated features which enable very large simulations to be done rapidly in a version of its popular workstations called the Megalogician. This unit is capable of simulating 1 million gates at a rate of 100,000 evaluations per second. It has 1 gigabyte of virtual memory. It thus overcomes one of the problems of workstations. It also voids the need to transfer data back and forth between a large computer and a special-purpose processor. Everything is done on the one system. Somewhat offsetting this is the fact that the system is a single-user system unless units are linked together via the Ethernet link provided. As with Tegas and other simulators, large simulations can be stopped and restarted without having to resimulate from time zero.

8.5.3 Economics

The interested reader should check with the pertinent manufacturers for exact prices in existence at the time. Because of the highly competitive nature of the markets in the three areas listed below, such prices are subject to marked change on short notice. The purpose of this section is simply to indicate some of the factors involved and how they interplay together. It is not to indicate a given direction to the user or potential buyer of the equipment.

The three factors involved are the cost of the special-purpose processor for a given power (speed times capacity) versus the cost of a new mainframe and a second software license for the simulator. A fourth factor might be the cost of the license of a hierarchical simulator, which would enable the entire system to be simulated also (but not at the gate level).

The user must be sure to evaluate the true power of the processor in its expected use. For example, some processors may use a lot of CPU time of the host just getting in and out of the host and also in formatting the circuit. For large (greater than 150,000 gates as an arbitrary measure) circuits, the CPU time for these two activities may be infinitesimal compared to that required to do the simulations and fault analysis. However, if the user does not plan to do simulations and fault analysis of greater than 5000 gates, the processor should be benchmarked at those circuit sizes. The CPU time saved might not be that much overall for the smaller circuits.

The user should, however, strongly take into consideration the benefits of simulating an entire system at the gate level.

Interplay among factors. An example will illustrate this. Consider the following scenarios. As noted above, *consult the pertinent manufacturers for current prices.*

Elements of the scenarios.

Scenario 1. User organization has a mainframe (VAX (R)-11/785, IBM 4341, Amdahl 470.V8, etc.) computer and the software license to run a simulation program on that mainframe. Current mainframe is swamped by simulation activity. Many persons are at low activity levels while waiting to see the results of simulations. (The cost of their time should also be considered but is so difficult to quantify and so dependent on the user organization that it will not be considered in this simple example.)

A second mainframe costs $500,000 in its "all-up" configuration (i.e., with all the necessary main memory, peripherals that cannot be shared, etc.). To this should be added a portion of the operator's time. However, this would be shared with the first machine and might be insignificant if the first machine did not fully occupy the operator's time. Also, a similar (probably smaller) cost should be allocated to that of the maintenance person assigned to the special-purpose processor. These will be neglected in the example.

A second software license of the simulation program to run on the second mainframe if the latter is purchased would cost $100,000. A facilities license (which allows an unlimited number of user machines to operate the simulation program) costs an additional $100,000.

The special-purpose processor costs $500,000 for the speed and capacity selected but provides a speed-up rate of 300 times over simulations done on the mainframe for the size of chips normally done by the user organization.

The results of this simple example show that there is an immediate savings of $100,000 and a potential savings of much more. Moreover, the users now have the ability to simulate much larger systems at the gate level if desired.

Scenario 2. Same as Scenario 1 except that the largest circuits that will ever be done by a user group are so small that the CPU time of the mainframe required to format the program for the special-purpose processor, as well as the I/O times to and from the special-purpose processor, provide only a 30% improvement in simulation time. In this case, the user group might be better off purchasing a second mainframe together with a facilities license for the software.

To evaluate some of the trade-offs, the user must answer some of the questions listed above. A key question for users that fit in the Scenario 2 category is how much *total* time is saved over running the entire simulation on the host computer in the very typical case in which there are *many* users running relatively small circuits (1000 to 5000 gates). The latter is typical of current gate array designs and is what bogs down the bulk of the current host computers in the majority of cases at gate array facilities.

Note that a hardware accelerator offers the capability of simulating the entire system *at the gate level*. A hierarchical simulator can perhaps simulate the entire system but not at the gate level without excessive CPU time.

8.6 HARD-COPY OUTPUT [17]

Until now, the discussion has centered on what computers can do. A few words are in order about the types of hard-copy outputs that can be obtained. The hard copy is used for both checking and documentation of a design. Hence it is essential that the designer be aware of a few of the terms and trade-offs. It is true that a very small number of companies use computers extensively for checking and documentation. However, even in these cases, check prints are made, and there is generally some form of hard-copy backup.

The copies to be made are generally of three types. First are the printouts from simulations and netlists. These are alphanumeric. The second is the schematic associated with the task, assuming that it was entered via schematic entry. This is both graphic and alphanumeric. The third is the class of layouts of the chip itself. This is almost entirely graphic.

8.6.1 Alphanumeric versus Graphic

The first distinction is between alphanumeric and graphics types of output devices. The former, of course, refers to those machines that will only output character streams, while the latter term refers to machines that will output shapes of various types. The shapes, although primarily graphics, may be alphanumeric as well.

8.6.2 Printers

The first category includes line printers of one type or another. The Digital Equipment Corp. LA-120 is a prime example of a moderately priced ($2000 roughly) moderate-speed unit. Higher-speed line printers are made by Plessey and others. Most users who work with computers are reasonably familiar with such machines.

Increasingly, low-priced ($350), reasonable-performance machines from companies such as Epson, Okidata (the μL 80 and 90 series), and DEC (Digital Equipment Corp.) (the LA50) are being used at alphanumeric terminals for hard-copy outputs of simulations of small circuit pieces and a host of other tasks.

Newer ink-jet plotters, which are similarly priced, just as fast, and much more quiet, are available from Hewlett-Packard.

8.6.3 Graphics Plotters

The graphics output machines are probably considerably less familiar to the average reader of this book. These fall into two main categories.

The first is the plotter category, in which each line is written by a pen of some type. Plotters are either drum or flatbed. In the former, the paper is moved back and forth in one axis by the drum beneath a set of multicolored pens which move in the other axis. The computer controls which of the pens does the actual writing.

A problem with this approach is that the moving drum actually distorts the paper slightly as it moves back and forth. The result, although acceptable for many applications, is not sufficiently precise for making overlays to lay on top of and compare to the "blowbacks" which represent the image of the reticle. (They are sufficient for many other purposes in this field, and can even be used for this purpose by sliding the print as much as an eighth of an inch on a 36-inch-wide print. However, this can be awkward and time consuming.) Examples of drum plotters are the Sanders Calcomp 960 and 1032 models.

In a flatbed plotter, the set of multicolored computer-controlled (in three axis) pens draws on a sheet of paper held down by vacuum. The result is greater precision than that of the drum plotter. Flatbed plotters are offered by Sanders Calcomp and Xyznetics.

The advantage of the drum plotter over the flatbed plotter is generally cost. The bed of the latter must be very flat (generally, it is held down with a vacuum).

Another advantage of the drum over the flatbed plotter is the smaller space requirement of the former. They can be more conveniently located closer to where the actual work is being performed if desired.

The pens themselves are of one of three types. Each type has its advantages and disadvantages. Wet ink produces masters that can be read after copies are made on the ubiquitous Bruning machines. Its disadvantage is that it very easily splotches and ruins a print after a number of hours of work. Ball point can be read somewhat after copying but has a propensity for skipping lines. Needless to say, that cannot be tolerated!

Felt pens will neither splotch nor skip, but their imprint cannot be read at all after copying. The reason for this discussion and the emphasis on copying is that one hates to mark up a print that took several hours to make. Moreover, more than one copy of either a schematic or a layout generally needs to be made to accommodate the several people normally involved in a gate array project.

8.6.4 Electro-Static Plotters

The second generic type of graphics output device is somewhat akin to a Xerox-type process in that an image of static charges is placed on a drum to which toner is applied. The toner sticks to the drum along the image lines and is then imprinted on the paper. Unlike a copier, however, the image is created not by the image of a master to be copied but rather by the output from a computer which selectively sensitizes the drum along the image lines. Small copiers that make an image of a graphics terminal screen are widely used. An example is the Versatec V-80.

The principal advantage of this technique is speed coupled with surprisingly good resolution. The Versatec 8242 will output a complex pattern roughly 42 inches on each side in a matter of a few minutes. A comparable machine from Sanders Associates Calcomp Division will do about the same. This is in sharp contrast to the six hours roughly required to pen plot an image of comparable detail. The principal disadvantage is that only one color (black) can currently be used. (Machines entering the market overcome this limitation.) For IC work where multiple layers are to be exhibited or where it is desired to make certain functions (e.g., interconnects versus I/Os versus internal wiring of macros) stand out from one another, color, while not essential, is extremely useful. Nevertheless, the ability to rapidly get a print to check is extremely helpful. Any process that takes six hours of an expensive machine's time is subject to scheduling delays as well as the time to actually do the job. The opposite is true for machines that operate rapidly. Fig. 8.37 shows a Versatec hard copy output device typically used to copy the screen of a workstation.

8.6.5 Photoplotters

Another category entirely is the photoplotter. This exposes a graphics image on Mylar film via a photographic process. A head containing perhaps 28 graphic shapes is rotated over the film, which is precisely positioned beneath it. Rotating the head selects which of the 28 shapes is to be imaged. Lines are "drawn" by repeatedly exposing a pencil-thin rectangular shape along the path defined by the computer input to the machine. When all the shapes and lines have been exposed, the Mylar sheet is developed via conventional wet chemicals used in photography.

The reader will appreciate that this is also a time-consuming process. The principal advantage of this technique is that there is no paper shrink or distortion. Mylar, while hygroscopic (absorbs water), is nevertheless entirely suitable for very fine geometries. Most of the problems with fine detail come not from the exposure mecha-

Figure 8.37 Versatec hard-copy output device. (Courtesy of Sanders Associates, Inc.)

nism but from the subsequent development of the image using wet chemical processing. Reasonable care in processing prevents this problem. The master negatives for the Chronoflex grid layout sheets shown in an earlier chapter were made using this process, with no errors.

8.7 SUMMARY

In this chapter, a number of topics were discussed that were not suitable for other chapters. These included IGSs and workstations. Comparisons among the latter were made. Placement and routing, both manual and automatic, were discussed next. The reader was given the opportunity to try two routing exercises.

Important placement and routing algorithms were given. Three currently important non-gate-array techniques which show promise for gate array work were described. The benefits and problems of designing remotely were given next.

Because of the large computer resources taken by simulation and fault analysis programs, a number of special-purpose hardware "accelerators" have come into being. These are described together with an analysis of the economics of using them in two different scenarios. Finally, hard-copy output devices were described.

REFERENCES

1. Sibbald, K., "PC CAD widens its appeal," *CAE,* Nov. 1985, p. 96.
2. "Popular IC simulator runs on IBM PC," *Electronics,* Jan. 26., 1984, p. 151.
3. E. Freeman, "Physical modeling systems let you plug VLSI chips into your workstation's logic," *EDN* Nov. 15, 1984, p. 51.
4. Werner, "Software for gate array design: who is really aiding whom?"; *VLSI Design;* fourth quarter, 1981, pp. 22–32.
5. D. Hightower, "Gate array master design considerations," IEEE 1983 Custom IC Conference, Rochester, NY, May 23, 1983.
6. A. Vincentelli and M. Santomauro, "YACR: yet another channel router," IEEE Custom IC Conference, Rochester, NY, May 25, 1983.
7. "Helping designers squeeze more on chips," *Electronics,* Dec. 2, 1985, pp. 33–38.
8. Putatunda, R., et. a., "An optimized and unique placement approach using wavefront compaction," IEEE International Conference on Computer Design: VLSI in computers, *IEEE publication 85CH2223-6,* Oct. 7–10, 1985.
9. Singer, P., "CAD tools for custom IC layout," *Semiconductor International,* Dec. 1985, pp. 56–62.
10. Dees, W.A., and M. Lorenzetti, "Channel routing in a gate array environment," IEEE International Conference on Computer Design: VLSI in computers," *IEEE publication 85CH2223-6,* Oct. 7–10, 1985.
11. W. R. Heller, W. F. Mikhail, and W. E. Donath, "Prediction of wiring space requirements for LSI," *Design Automation and Fault Tolerant Computing,* Computer Science Press, Inc., 1978, pp. 117–144.
12. B. S. Landman and R. L. Russo, "On a pin versus block relationship for partitions of logic graphs," *IEEE Trans. Computers,* C-20, 1971, pp. 1469–1479. (Has Rent's rule in it also.)
13. Cook, P., S. Schuster, J. Parrish, V. DiLonardo, and D. Freedman, "1 μm MOSFET VLSI technology: part III-logic circuit design methodology and applications," *IEEE Journal of Solid State Circuits,* April 1979, pp. 355–68.
14. Law, H., "Gate matrix layout: a practical, stylized approach to symbolic layout," *VLSI Design,* Sept. 1983, pp. 49–59.
15. Lopez, L., and Law, L., "The gate matrix method of layout," *IEEE Jl. of Solid State Circ.,* Aug. 1980, pp. 736–43.
16. Neureuther, A., "Topography simulation tools," *Solid State Technology,* Mar., 1986, pp. 71–75.
17. Kacala, J., "Hardcopy output for CAD/CAM," *CAE,* Nov. 1985, p. 36–42.

Chapter 9

Design of CMOS Macros

9.0 INTRODUCTION

The preceding chapters have dealt with the tools and technologies of gate arrays. In Chapter 4 the trade-offs between design with macros and design with transistors were discussed. Some companies prefer the term "modules" to that of macros. Others use the latter term to denote other elements.

Chapters 9 and 10 go into the design with transistors. The reasons for designing gate arrays with transistors instead of macros were given in Chapter 4. Chapter 9 is on the design of CMOS macros. Chapter 10 is on bipolar technologies, notably ECL and ISL/STL.

CMOS is by far the largest gate array technology in terms of both number of vendors and gate array offerings. Also, there are many more CMOS vendors who will currently allow the user to design his or her own macros than there are for the other technologies. Therefore, Chapters 9 and 10 are weighted accordingly.

Chapters 9 and 10 are developed under the heading of design of macros because almost without exception any transistor design will be stored in some fashion so that it can possibly be used again (possibly with modification). This statement applies to macros of any size up to and including the entire chip design. The several ways in which the design could be stored include not only on some electronic media, such as the disk of an IGS, but also in pastie form for those lacking the more expensive media.

9.0.1 Steps Involved

Macros can be as simple as an inverter or as complex as the entire chip. In the latter case, their design is quite similar to that of the chip itself.

The design of macros involves several steps. Depending on the complexity of the macro and the experience level of the designer, some of these may be skipped. However, normally it is good practice to do all of them. The reason is that once designed, a macro will be used over and over again. The possibilities for error propagation are hence very great.

There *is* the distinct case which exists in organizations that do a lot of their own designs in which a special-purpose macro is designed for a "one-of-a-kind" application. Perhaps there is the need to configure a buffer in a very special way and speed is not being "pushed." In these rare cases, some of the steps, such as circuit (as opposed to logic) simulation, might not be done, for example.

The first step is to define the functionality of the macro; that is, just what does the designer want and expect the macro to do? For complex macros, the importance of first thinking seriously about this step cannot be overemphasized. Many of the succeeding steps depend heavily on such functionality; and while it is possible to change direction partway through because of changing the specification of the macro, such "patched-up" macros are rarely as efficient as they should be and could be.

The next step is normally to define the macro in terms of its functional blocks and to compare it against the original specification. If the macro is exceedingly simple, such as a NAND gate, this step may not be necessary.

The macro may be expressed in terms of other macros using up to several levels of "nesting". (See the discussion of logic simulation in Chapter 6.) The result is often called a "soft macro."

The macro is then usually expressed in terms of the transistor diagram. Again, this step is sometimes skipped by experienced designers designing simple structures, such as NAND gates and inverters.

Coding for circuit and logic simulation is then done on the macro. If schematic entry is used for the functional diagram, the coding will often come from that. The macro is then simulated using a circuit simulator, such as SPICE or TRACAP, to determine its speed and possibly loading characteristics. It may be possible to calculate these parameters using paper and pencil in lieu of doing the circuit simulation if the macro is made up of blocks whose characteristics are given by the gate array manufacturer.

If schematic entry to input the data base for the circuit simulation is not available and the macro has been functionally simulated using high-level blocks only, *extremely* great care will have to be taken that the transistor diagram matches the functionality so simulated. The circuit simulation should tell this when compared to the logic simulation using blocks. However, because circuit simulations are often limited to a much smaller number of transistors than are used in many large macros, the results may not be readily comparable. The great need is to ensure that the mask layout of the macro produces the same results as the logic simulation, and hence the functional specification.

Using the propagation delay information from the circuit simulation, a logic simulation is then performed on the macro to verify the functionality of it using a logic simulator, such as Tegas5. For more complex macros, testability analysis, using a program such as COPTR or SCOAP from the data base used to perform the logic simulation may be done. Based on the results of this program, the circuit design may be modified to make the macro more testable. Automatic test vector generation may also be done if the macro is extremely complex.

Documentation using a form acceptable to the user organization occurs at every step. Smaller macros are rarely built in chip form and tested. However, extremely complex macros may be built on a gate array chip for test purposes.

9.0.2 CMOS Macros

The purposes of this chapter are to give the FET design of common logic structures and to show how they can be laid out in high-speed CMOS (called in the remainder of this chapter simply "CMOS" except where a distinction needs to be made among CMOS types). The structures developed in this chapter also apply to the other two forms of CMOS: metal gate CMOS and CMOS/SOS. For purposes of this chapter, high-speed CMOS is CMOS which is both oxide isolated and silicon gate. The term "high-speed CMOS," although not universal is being used increasingly by major manufacturers.

The "rules of thumb" and other generalizations, strictly speaking, apply to high-speed CMOS with a drawn 5-μm gate length and a W/L ratio of roughly 8.5 for PFETs and about half that for NFETs. However, many if not most of them also apply to smaller feature sizes as well. If in doubt, consult the array vendor. See Chapter 2 for details of CMOS itself. The reader should review Chapter 3 if he or she is not thoroughly familiar with the grids to be used in this chapter.

This chapter begins with explanations of NAND and NOR gates and then moves on to transmission gates, AND-OR-INVERTS (AOIs), flip-flops, counter stages, I/Os, and so on.

To give the reader a perspective, layouts on the HS CMOS grid structures of two different gate array vendors are utilized in the explanations. The two grid structures differ in several respects, as noted in Chapter 3. The IMI grid is entirely made up of three-input cells with transistor gates which are not broken in the middle to separate the PFETs from the NFETs and which run in a straight line to connect PFETs and NFETs which are directly adjacent to one another.

The CDI grid structure (licensed by AMI, LSI Logic, Inc., and others for their first gate arrays), by contrast, is made up of cells that contain three input subcells and two input subcells. The three input subcells are similar to those of IMI except that there is a contact to the transistor gate in the middle. The two input subcells have transistor gates which do not run in a straight line and which are routed to PFETs and NFETs which do not lie directly adjacent to one another. There are advantages and disadvantages to both methods which will become apparent in the course of the explanation. Examples of the different elements implemented on an actual HS CMOS gate array chip were shown in Chapter 3.

To further give perspective, some of the layouts are couched in terms of the background layer (as opposed to the grid structure developed at Sanders Associates, Inc.) of the IMI array. Layout sheets of the three different types were given in Chapter 3. The reader should review that material if necessary.

The author is indebted to International Microcircuits, Inc. (IMI) and to California Devices, Inc. (CDI) for permission to use layouts of their macros.

An Important Note: To enable the student to practice laying out the structures described below, permission is granted to reproduce the grid sheet labeled Fig. 9.1. This permission is strictly limited to copying only that one figure and only for the purpose of practicing layout. No other permission to copy any parts of this book is granted or implied, and reproduction of any other part or parts of the book by any means whatsoever is a violation of the copyright statutes.

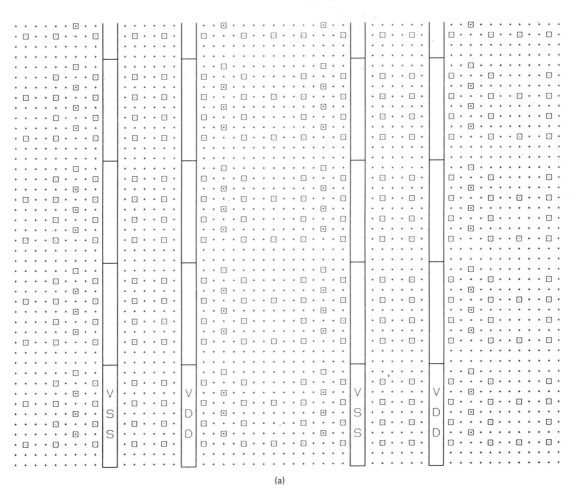

(a)

Figure 9.1 (a) Grid sheet for layout practice; (b) same grid as (a) but with smaller spacings to allow larger macros to be designed.

Sec. 9.0 Introduction

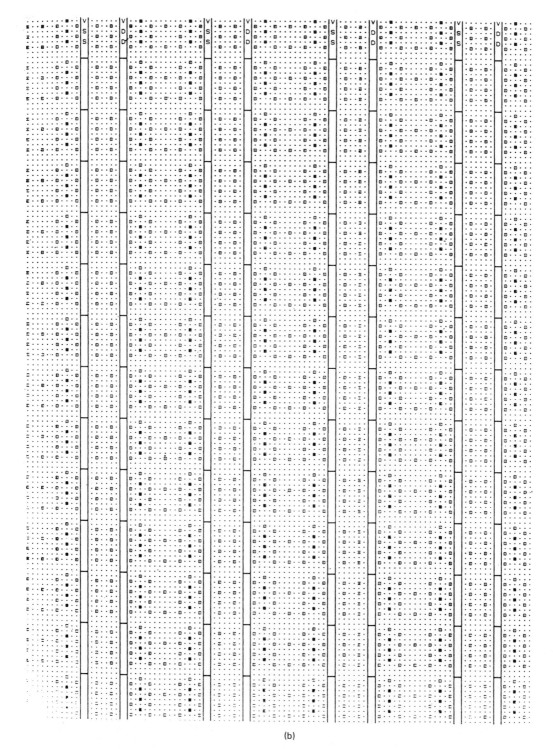

(b)

Figure 9.1 (*cont.*)

9.1 FET DIAGRAMS

Reasons for designing with transistors were given in Chapter 4. This section merely states some of the guidelines in the figures. The symbols were given in Chapter 2. The inputs to the transistors are, of course, the gates. The outputs are the drains. For the important class of structures known as transmission gates the inputs are the source/drains. Unless otherwise noted, each input goes to both a PFET and an NFET. To avoid extra lines on the diagrams, there are generally two branches of each signal.

The reader should note from Chapter 2 that a high on the input to an NFET gate turns it on and turns the associated PFET off, and vice versa for a low signal.

9.2 LAYOUT METHODS

No two people do macro layout in exactly the same way. Most of the layouts shown here have a number of acceptable permutations. Some of the permutations are shown in the figures from IMI and CDI. The more complex the device, the greater the number of acceptable permutations. (An acceptable permutation is one which compared to the original (1) takes no more space, (2) uses routing channels that are not longer, (3) has I/O points in places that are equally as good if not better, and/or (4) minimizes the number of lines to be digitized (if this is the method employed). On occasion it may not be possible to satisfy all four conditions simultaneously. The individual user needs will then dictate the permutation selected.

The writer has found that a good way to begin learning layout is first to become very familiar with existing layouts such as those shown below. Next, start permuting the layouts. Draw FET diagrams to verify what the permutation drawn actually is. This step is essential initially. Later it can be discarded for simple structures at least.

After a small amount of practice, one learns to rapidly spot certain structures embedded in a layout. For example, transmission gates on grid structures such as those of IMI and others are readily identifiable by the connection of PFETs to NFETS which are at 45 or 135 (very roughly) degrees to one another. By spotting these, one can then often spot counters and registers or sometimes EX-NORs/ORs, and so on.

NAND gates can be spotted by noting the connections to V_{dd} of the PFETs compared to those to V_{ss} of the NFETs. If the connection to V_{ss} is directly opposite one to V_{dd}, there will be a second connection to V_{dd} to effect the paralleling. NOR gates are just the opposite. Development of these skills is important also for checking of layouts even where automatic placement and routing techniques are employed (but especially where they are not). On occasion, there *is* the need to check what a computer does.

In the layouts that follow, a circle around a source/drain contact denotes an output. There generally will be a number of these because of the flexibility built into the arrays. Semicircular lines are there to indicate to the reader where the transistors are. *Note:* They are *not* actual interconnect lines.

9.3 INVERTERS

The simplest structure is the inverter. This is shown in Fig. 9.2. Figure 9.2(b) and (c) give layouts of this in the two different grid systems. (As noted above, only two different grid systems are being used even though two different permutations of one of them is being used.) The reader will note the connections to V_{ss} of the NFET, to V_{dd} of the PFET, and from the drain of the NFET to that of the PFET.

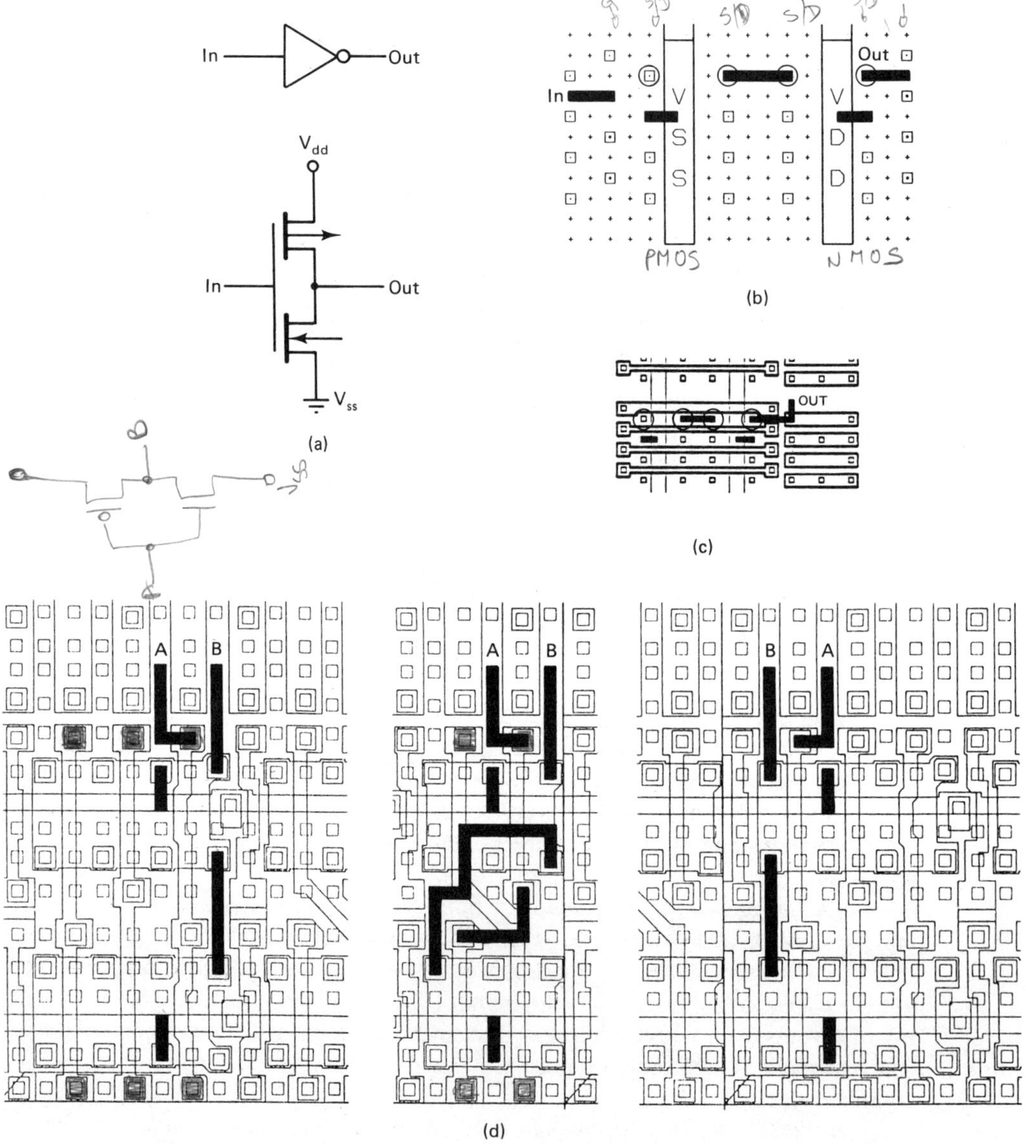

Figure 9.2 Inverter: (a) logic symbol and FET diagram; (b) layout on IMI grid (outputs exist at circled contacts); (c) layout on IMI grid; (d) versions of inverter CDI grid. [(d) Courtesy of California Devices, Inc.]

9.4 NAND AND NOR GATES

A NAND gate has one NFET in series and one PFET in parallel for each of its inputs as shown in Fig. 9.3. If the "on resistances" of the PFETs are equal to those of the NFETs, the high-to-low propagation time of a NAND gate is much longer than that of its low-to-high propagation time. Figure 9.4 dramatically illustrates this difference as a function of fan-out. Some arrays, including LSI Logic's 5000 series, have unequal "on" resistances for the two transistor types. This is done by making the W/L ratio (i.e., the size) of the NFET equal to (or sometimes greater than) that of the PFET, which in turn lowers its "on" resistance. Because the surface mobility of the electrons is much greater than that of the holes, equal W/L ratios for the two transistors will produce lower "on" resistance for the NFET.

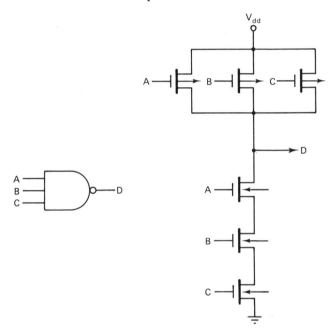

Figure 9.3 Three-input NAND gate FET diagram.

9.4.1 Benefit of Lower "On" Resistance of the NFET

A major benefit of having lower "on" resistances for the NFET is that the high-to-low propagation time of a NAND gate can now be made actually lower than its low-to-high propagation time. (Of course, as the number of inputs increases, larger and larger W/L ratios of the NFETs will be required to provide this property.) This in turn is extremely important for equal polarity detection (i.e., detection of all 1's or all 0's).

Consider, for example, a 22-stage counter for which it is desired to know when the count reaches a given value stored in a register. Decoding would be done with two-input EX-NORs which would input to a 22-input NAND gate. Using Boolean algebra to break the 22-input NAND gate into two-and three-input NAND and NOR gates shows that the signal edges of interest are the high-to-low propagation times

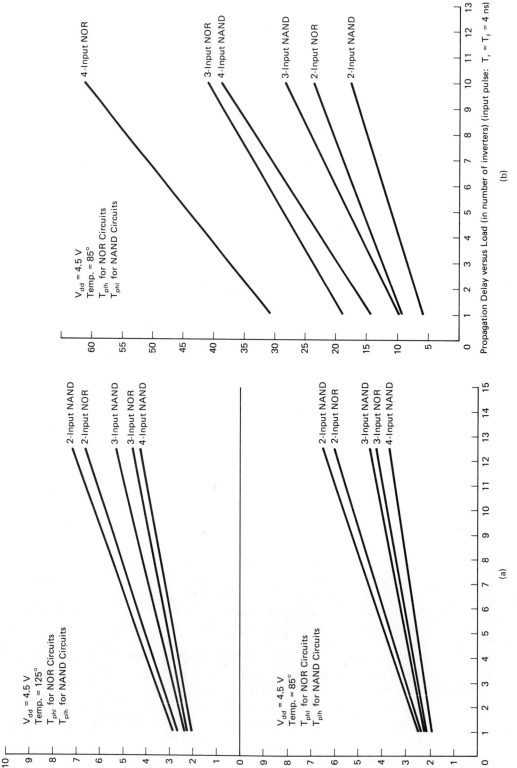

Figure 9.4 (a) Propagation delays at 85 and 125°C for NAND and NOR gates; (b) propagation delays in the direction opposite to that in part (a) for NAND and NOR gates. Note that the ratios of propagation delays can be changed by varying the relative sizes of the transistors and other parameters (see Chapter 5). Consult the manufacturer for the most current information. (Courtesy of International Microcircuits, Inc.)

for the NAND gates and the low-to-high propagation times for the NOR gates. The latter is not affected by increasing the size of the NFETs because the charge path is through the series PFET string. Nevertheless, the decrease in overall propagation time may be sufficient to avoid the need to pipeline.

9.4.2 Disadvantages

The primary disadvantages of increasing the W/L ratio of the NFETs is that both the gate capacitance and the source/drain capacitances are increased. Also, while the skew between the rise and fall times of NAND gates is decreased or eliminated, the skew between the rise and fall times of NOR gates is increased.

Layout. In the majority of grids, each gate connects to both a PFET and to an NFET. Equally important, the connection can be made from either side. Layouts of the NAND gate are as given in Fig. 9.5. A question immediately arises as to how many inputs a CMOS NAND or NOR gate can have. With rare exceptions, the answer is generally about four, possibly five. For more than about four or five inputs, the series resistance of the series NFET "string" causes too large a high-to-low propapation time for most applications. In some cases, the resultant asymmetry between rise and fall times is also intolerable.

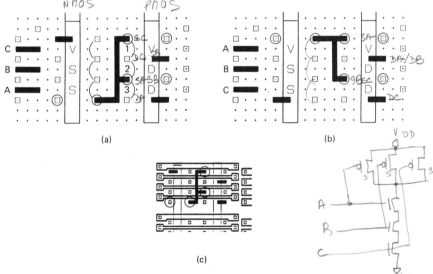

Figure 9.5 Versions of layout of three-input NAND gate: (a), (b) layout on IMI grid; (c) layout on CDI grid. The six different places from which the output can come are encircled.

A problem that many students have on initially encountering layouts is seeing both the series strings and the paralleled "layers." To aid the reader, the writer has drawn semicircular lines *which are not layout lines* to denote where the different transistors lie. The reader must fully appreciate that the source/drains are being shared.

A question that frequently arises is: If the source/drains are shared, why are the transistors not always connected together? The answer, of course, is that the transistor gate voltage determines whether or not the two adjacent transistors are connected. The reader should think of each FET used digitally as a switch controlled by the gate voltage.

Interestingly, gate isolation (Chapter 3) is the technique used by the Japanese and Honeywell in their very high density HS CMOS gate arrays. Instead of surrounding thousands of discrete cells with oxide, they use contiguous transistors and isolate adjacent transistors by means of the gate voltage.

Referring now to Fig. 9.5(b) (IMI layout) for the NAND gate, one sees that the PFETs 1 and 2 are paralleled by their common source, which is connected to V_{dd}, and by the line that connects their noncommon source/drain. However, the source/drain of 2, which is not common to 1, *is* common to 3. Thus the three PFETs are paralleled. The NFET string is much more easily seen. Start at the connection to V_{ss} of the NFET source/drain and trace through the three NFETs in series.

The top of the NFET string then connects to the bottom of the PFET layers. The connection point between NFETs and PFETs is the output. Note that although not explicitly drawn, each input goes to *both* a PFET and an NFET. In cell structures of the IMI type and of the CDI three-input subcell type, making connection to either end of the transistor gate connects *both* PFET and NFET. In cell structures of the CDI two-input cell type, one has to remember to connect the transistor gate, which is severed in the middle to connect both PFET and NFET. (However, it will be seen that this severed gate does have some distinct advantages when laying out transmission gates and other structures.)

One of the benefits of the second and later generations of CMOS (the first generation is the metal gate CMOS of Interdesign, IMI, Masterlogic, etc.) gate arrays is the multiplicity of points from which the outputs can be taken. Because each source/drain has two contacts which are electrically the same (except for a negligible amount of diffusion resistance) and because these two contacts come out on opposite sides of the appropriate power bus, interconnection is greatly simplified. (The NFETs and PFETs are clustered around the V_{ss} and V_{dd} buses, respectively.) For example, the three-input NAND gate shown in Fig. 9.5 has no fewer than six places from which the output can come. These are are denoted by circles in Fig. 9.5.

9.4.3 NOR Gates

The FET structure of a NOR gate is the opposite of that of a NAND gate, with the same number of inputs in the sense that the NFETs are paralleled and the PFETs are in series. With the background (and a little practice) on laying out NAND gates provided by the exercises above, the student should now be able to lay out NOR gates. An example of the layout of a three-input NOR gate is provided in Fig. 9.6 for those who might be a trifle unsure. The different points from which the output can be taken for the NOR gates are denoted by circles. (The circles are, of course, not part of the layout.)

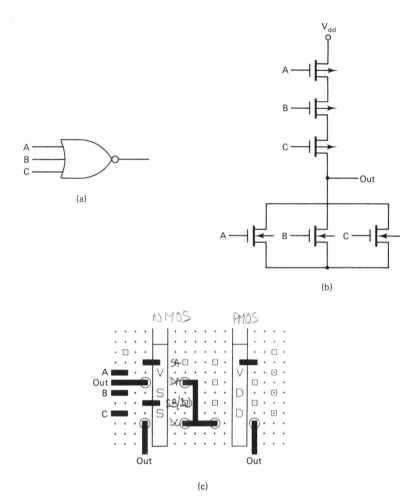

Figure 9.6 Three-input NOR gate: (a) logic symbol; (b) FET diagram; (c) layout on IMI grid. The six different places from which the output can come are encircled.

9.5 Student Exercises

Exercise 1

Lay out a three-input NOR gate on the IMI grid using connections to V_{dd} and V_{ss} which are different from those shown.

Exercise 2

Show all the different places from which the outputs can be taken.

Exercise 3

Find each of the problems with the two-input NAND gates shown in Fig. 9.7 and correct them.

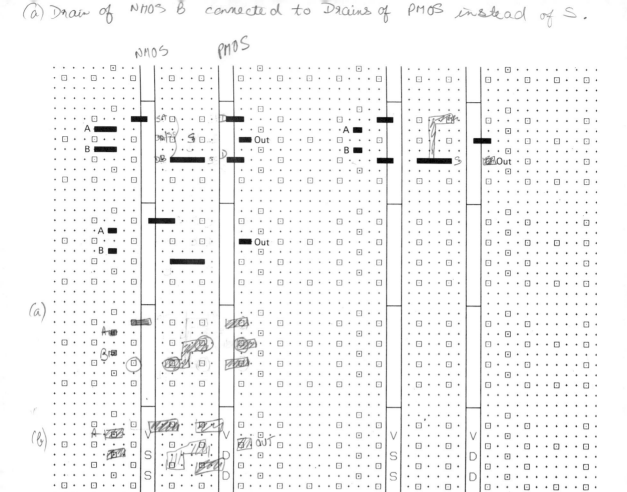

Figure 9.7 Layout for Exercise 3. Find what is wrong with the layout of each of the 2-input NAND gates and correct it.

9.6 ISOLATION OF ELEMENTS WITHIN A CELL

Consider the three-input cell of either IMI or CDI. Properly done, this type of cell can be used to build both a two input NAND or NOR gate PLUS an independent (no connection to NAND or NOR gate) inverter. The key is to use the power supply connections as isolation between the inverter and NAND or NOR. The technique is shown in Fig. 9.8 for the IMI structure. Outputs are labeled.

Exercise 1

Lay out an inverter and a NOR gate which is independent of it on the CDI grid structure.

The NAND or NOR gate output can, of course, be connected to the inverter input to create an AND or OR gate, respectively. Methods of doing this for the IMI grid are shown in Fig. 9.9. The connection to the CDI inverter from the NAND or

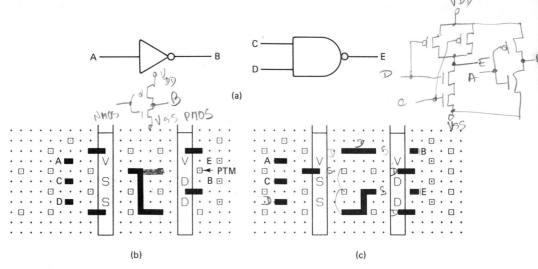

Figure 9.8 Using an array cell as both a NAND gate and as a separate inverter: (a) gate symbol; (b) wrong PT M serves as both output B and output E; (c) right outputs B and E come off separate points.

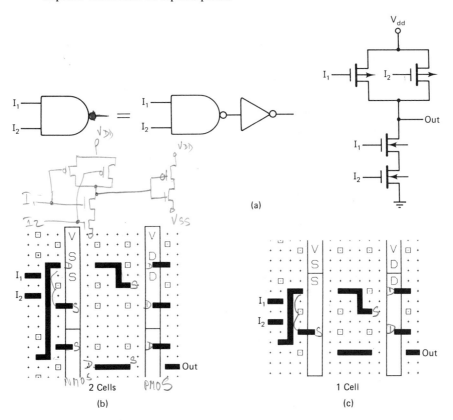

Figure 9.9 Connecting elements: making AND gate from NAND gate and inverter. The four different places from which the output can come are encircled. (a), (b), (c) are on IMI grid; (d) is on CDI grid. In (d), A, B, C are the same as I_1, I_2, and Out in (b)(c).

Sec. 9.6   Isolation of Elements Within a Cell   371

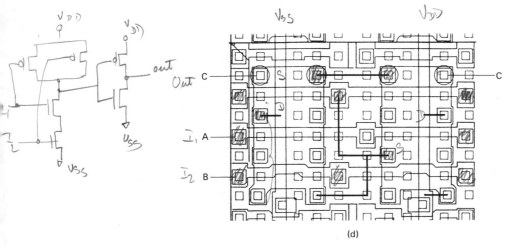

(d)

Figure 9.9 (cont.)

NOR gate can be made in a similar fashion to that done on the IMI grid. However, the CDI grid offers the interesting possibility of making the interconnect to the inverter from the NAND or NOR gate from *inside* the V_{dd} and V_{ss} buses. Figure 9.9 shows this. This saves routing channels outside the V_{dd} and V_{ss} buses for global routing.

Exercise 2

On the IMI grid, lay out an inverter with its outplut connected to the input of a NOR gate from inside the V_{dd} and V_{ss} buses.

9.7 PHANTOM TRANSISTORS

The technique of using the power supply connections as isolation between two independent logic elements can also be used to prevent what are known as "phantom transistors." These are created when a transistor gate is used to get under the V_{dd} and V_{ss} buses. Figure 9.10 shows the problem and its solution. The problem occurs when a gate is used as an underpass under the buses in single-layer metal. This might occur, for example, due to overcrowding of lines going through the array cells. In this case, the gates might have to be used.

A second place in which gates are used as underpasses is where they are used in order to equalize delays of signals. For example, the propagation delays of all 16 elements of a bus might have to track. If some of the bus signals go through a gate, all of them would. Note that the gate so used does not create an inversion: rather it simply acts as a capacitive path.

The problem is that if the rest of the array cell is used for, say, a two-input NAND or NOR gate, and if the source/drains that are shared by the FET used as throughput and the FET used for the logic gate are not tied to the rails (V_{dd} and V_{ss}), those source/drains can be swung to a voltage level by the passing through gate. The solution again is to tie the common source/drains to the rails as shown.

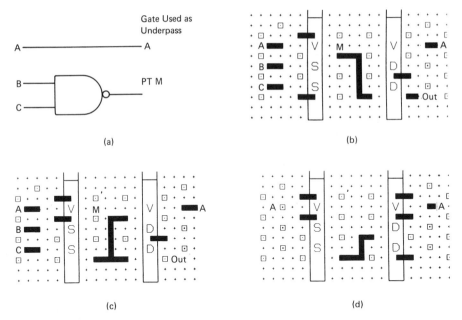

Figure 9.10 Creating phantom transistors: (a) gate symbol; (b) *wrong*: PT M goes low when A goes high, regardless of B and C; (c) *better*: n-channel transistor with A as gate is held off ($V_{DS} = 0$), but the V_{DS} of the p-channel XR is not defined; (d) *best*: both unused transistors are held off ($V_{DS} = 0$). V_{DS} is drain-to-source voltage. hand-routing technique not usually done

9.8 FOUR-INPUT GATES

A four-input gate is interesting because it introduces the reader to layout using more than one cell. As noted above, the individual array cells in most gate arrays are isolated from each other by local oxidation. Because of this, structures that occupy one cell or less are automatically isolated from each other without regard to voltages on transistor gates in nearby cells.

A way to obtain the fourth transistor pair needed for the four-input NAND or NOR gate is to borrow it from an adjacent cell on the same column using the isolation technique involving the power supply connections described above. This is shown in Fig. 9.11 using the CDI grid.

Exercise

Draw a four-input NOR gate on the IMI grid.

9.9 AND-OR-INVERT (AOI) AND OR-AND-INVERT STRUCTURES

AOIs and their cousins the OAIs (OR-AND-INVERT) are extremely useful in CMOS (and in ISL/STL). AOIs are sometimes called AND NOR gates and OAIs are sometimes called OR NAND gates.

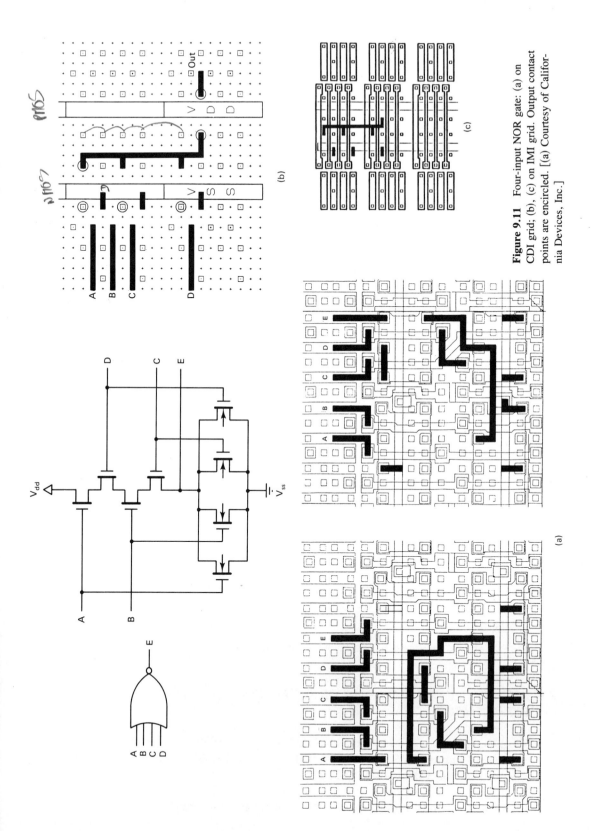

Figure 9.11 Four-input NOR gate: (a) on CDI grid; (b), (c) on IMI grid. Output contact points are encircled. [(a) Courtesy of California Devices, Inc.]

In CMOS such structures are the only structures that permit multiple levels of gating to be obtained with less than the maximum number of propagation delays. An example will illustrate the point. Consider the 2 × 2 AOI shown in Fig. 9.12.(a) If the structure were made directly as shown, there would be three levels of propagation delay through it: two for the AND gate and one for the NOR gate. By making the structure out of an AOI, the propagation times t_{phl} and t_{plh} for high-to-low and low-to-high transitions are decreased by a factor of 2 and 1.6, respectively, even in this simple example. Table 9.1 shows the direct comparison. In Table 9.1 the term "brute force" refers to making the structure out of individual NAND, INVERT, and NOR gates.

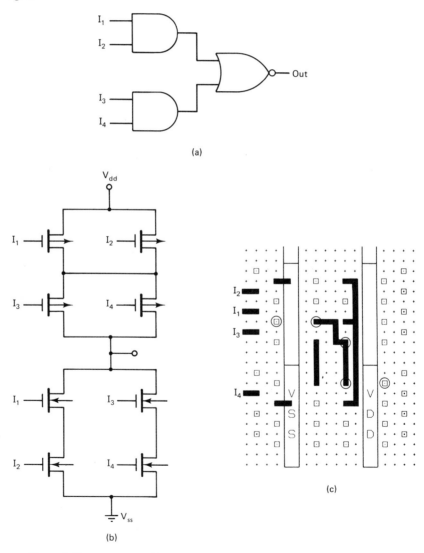

Figure 9.12 2 × 2 AOI: (a) schematic, (b) FET structure, (c) layout. Output contact points are encircled.

Sec. 9.9 And-Or-Invert (AOI) and Or-And-Invert Structures 375

9.1 COMPARISON OF PROPAGATION TIMES (ns at 85°C) for AOI AND BRUTE FORCE IMPLEMENTATIONS[a]

		\multicolumn{2}{c}{Transition}	
		High to low		Low to high
		Brute force		
NAND	t_{phl}	6	t_{plh}	2.6
INV	t_{plh}	3.5	t_{phl}	3.1
NOR	t_{phl}	2.5	t_{plh}	9.5
		12.0		15.2
		\multicolumn{2}{c}{A-O-I}	
		6.0		9.5

[a]The propagation times of the AOI are as little as one-half those of the brute-force implementation.

In Table 9.1 t_{phl} and t_{plh} are the propagation times for high-to-low and low-to-high transitions, respectively. The AOI in this case saves about 6 ns in each case over the brute-force approach. Other structures will yield similar results.

By either designing AOIs and OAIs into circuits (or by noting where AOIs could replace existing structures in already designed circuits), the user can often substantially reduce the overall propagation times of a circuit. As with any multilevel CMOS structure, however, propagation times can vary significantly depending on the number of inputs that are activated. Also in some cases, combinations of straight NANDs and NORs may have less overall propagation delay. The reason is that t_{phl} of the AOI is that of the NFET series "strings" while t_{plh} is that of the PFET series strings. It can be shown, however, that (1) a combination of only NANDs and NORs cannot replace an AOI. The two circuits perform different functions; and (2) the only way that a combination of only NANDs and NORs could have a lower propagation delay would be for the critical propagation times to be for the low-to-high transition of the NAND gate and the high-to-low transition times of the NOR gate.

Using AOIs instead of individual NANDs, NORs, and inverters also substantially reduces the number of FETs required. The symmetrical AOI shown in Fig. 9.12 requires only *Eight* FETs for its implementation. If implemented with individual NANDs, NORs, and inverters, it would require *twice* that number, or 16.

9.9.1 Symmetrical versus Asymmetrical

AOIs can be classified as either symmetrical or asymmetrical. An alternative classification scheme is degenerate versus nondegenerate, although in a sense the degenerate AOI is a subgroup (but a very important subgroup) of the asymmetrical classification. Unless otherwise stated, the comments made about the AOI apply equally well to the OAI but with opposite polarities.

Figure 9.12 is an example of a symmetrical AOI. The symmetry refers to the equal numbers of inputs to the AND gates. Figure 9.13 shows the schematic of a

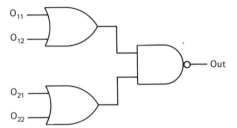

Figure 9.13 2 × 2 OAI. (Courtesy of International Microcircuits, Inc.)

symmetrical OAI. Its FET structure and layout are shown in Fig. 9.14.(a) and (b). It will be noted that the symmetry refers to equal numbers of inputs to the OR gates of the OAI.

AOIs have two properties. (Data in parentheses refers to OAIs.) First, if all inputs to at least one AND (OR) gate are high (low), the output is low (high). Second, if at least one input to each AND (OR) gate is low (high), the output is high (low). The reader can verify these properties by referring to Figs. 9.12 and 9.13.

9.9.2 An Important Property of CMOS

Unless level shifting is being done, CMOS structures must have two properties. First, if there is a low impedance to ground (required for a valid low output), there must be a high impedance to V_{dd}; if there is a low impedance to V_{dd} (required for a valid high output), there must be a high impedance to V_{ss}. Figures 9.13 to 9.14 show that the AOI and the OAI do indeed exhibit these properties.

9.9.3 Layers and Strings of FETs

It will be noted that the PFETs corresponding to the inputs to each AND gate are in "layers" (i.e., in horizontal rows) and that the number of such layers is equal to the number of AND gates. All of the PFETs in a given row are in parallel with each other. (This property is shown perhaps more clearly in the example of the asymmetrical AOI which follows, which has unequal numbers of AND gates.) The number of such PFET layers is equal to the number of inputs to the NOR gate.

A string of FETs is a series connection of two or more FETs. The NFETs in Fig. 9.12(b) are an example. However, the PFETs, in addition to being in layers, are also in series strings.

If all the inputs to at least AND gate are high, at least one of the PFET layers is an open circuit and at least one of the NFET "strings" (series combination of NFETs in this case) is a low impedance to V_{ss} (ground). Similarly, if at least one of the inputs to each AND gate is low, there is a low-impedance path to V_{dd}.

The reason for the layers of paralleled PFETs can now be seen, that is, the paralleling enables *any* one of the inputs to the AND gates to be low and still effect a low-impedance path to V_{dd}. The input that is low does not have to be in the *same* PFET string in each gate. It will also be noted that the number of NFET strings equals the number of AND gates, and that this feature allows a low on *any* input to *each* AND gate to provide a high-impedance path to ground.

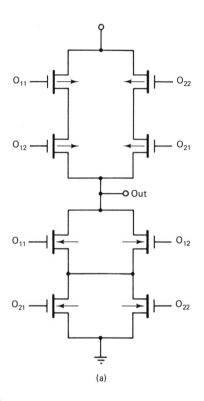

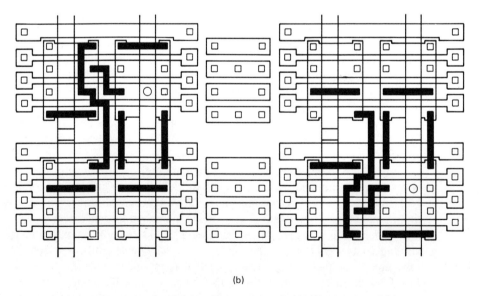

Figure 9.14 Symmetrical OR-AND-INVERT OAI: (a) FET structure (b) layout. (Courtesy of International Microcircuits, Inc.)

Asymmetrical AOIs. Figure 9.15 is an example of an asymmetrical AOI. The corresponding FET structure is given in Fig. 9.16. The reader should verify that the properties delineated above do indeed hold for the asymmetrical AOI.

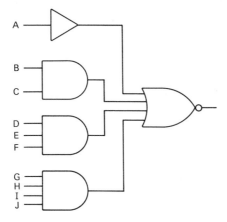

Figure 9.15 Asymmetrical AOI

9.9.4 Importance of Tying Node Capacitance to Power Supply Rails

Figure 9.16 has been drawn with the PFETs in the form of a "tree" in order to better exhibit the layering effect of the PFETs. In reality, the structure would be laid out with the PFETs in an inverted tree. The object is to tie as much source/drain capacitance as possible to the power supply rails. Taking the output from where it is shown would slow the signal considerably because of the high driving-point capacitance. When laying out asymmetrical OAIs, the NFET layer containing the most NFETs would have its common source connected to V_{ss}.

9.9.5 Degenerate versus Non-degenerate AOIs

A degenerate AOI (the author's term, but it seems to fit) is an AOI, in which one or more of the inputs to the NOR gate does not come from an AND gate but rather comes in directly from the outside of the circuit. Figure 9.17 is an example. Note that there are still two layers of PFETs (one of which has only one FET in it) and two strings of NFETs. IEEE Standard symbols are used in this diagram as an example.

9.9.6 AOIs in Complex Structures

Another reason for studying AOIs is that knowledge of their structure helps one to design more complex structures. For example, consider the EX-OR in Fig. 9.18. The reader will note that this is simply a degenerate AOI of the type shown in Fig. 9.17 with an additional NOR gate attached as shown. The NOR gate is made up of the PFETs P_1 and P_2 and the NFETs N_1 and N_2. It is left to an exercise for the reader to show that the EX-NOR circuit in Fig. 9.19 is a combination of a degenerate OAI and a NAND gate.

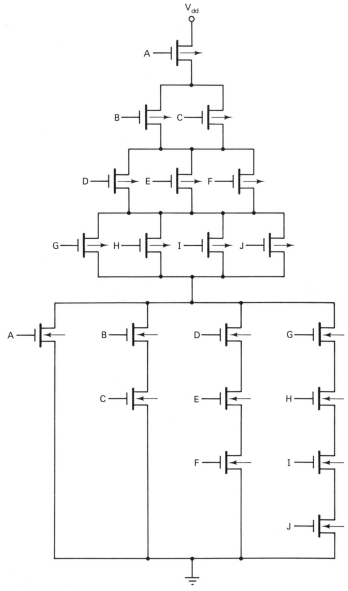

Figure 9.16 (a) FET diagram of asymmetrical AOI in Figure 9.15; (b) corresponding FET structure.

9.9.7 Chip Layouts of AOIs

In the process of teaching layout to a wide variety of students, the author has found that the layouts of AOIs are generally more difficult than those of other structures for students to comprehend initially. Part of the problem comes from the series-parallel nature of the FET structure. Another part of the problem comes from difficulty in seeing how NFETs and PFETs driven by the same gate can be arranged on the lay-

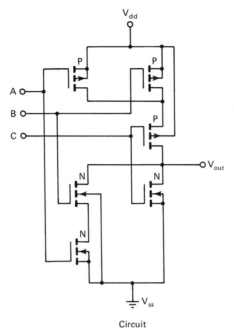

Circuit

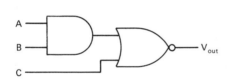

Logic Diagram

A	B	C	V_{out}
0	0	0	1
0	0	1	0
0	1	0	1
0	1	1	0
1	0	0	1
1	0	1	0
1	1	0	0
1	1	1	0

Truth Table

(a)

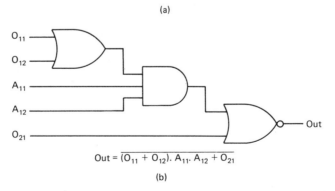

Out = $\overline{(O_{11} + O_{12}) \cdot A_{11} \cdot A_{12} + O_{21}}$

(b)

Figure 9.17 (a) Degenerate AOI. Note the different symbolism. See Chapter 2. (b) multilevel degenerate AOI; (c) layout of (b). [(a) Courtesy of Interdesign; modifications to FET symbols by the author; (b), (c) courtesy of International Microcircuits, Inc.)]

Sec. 9.9   And-Or-Invert (AOI) and Or-And-Invert Structures   381

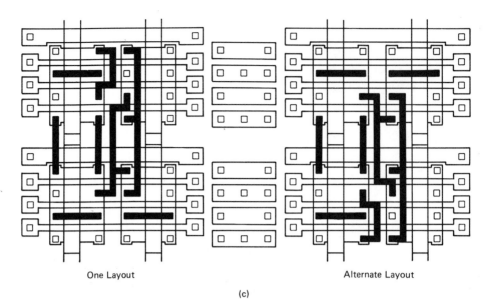

Figure 9.17 (cont.)

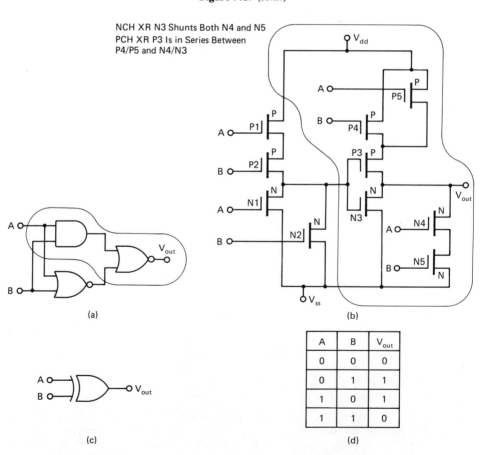

Figure 9.18 EX-OR made from degenerate AOI (degenerate AOI is encircled): (a) logic circuit; (b) circuit diagram; (c) symbol; (d) truth table. (Courtesy of Interdesign; modifications to FET symbols by the author.)

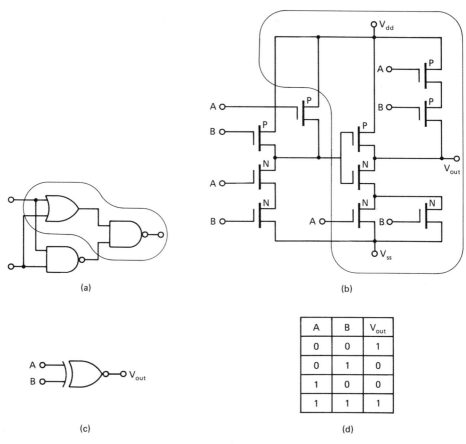

Figure 9.19 EX-NOR made from degenerate OAI (degenerate OAI is encircled): (a) logic circuit; (b) circuit diagram; (c) symbol; (d) truth table. (Courtesy of Interdesign; modifcations to FET symbols by the author.)

out. Although no two people do layout in exactly the same way, the following suggestions may be helpful.

First, draw the FET diagram. Unlike other structures where conversion from logic gate diagram is automatic, there is generally enough "trickiness" in the layout of all but the simplest AOIs to warrant this. Next, label the transistor gates that constitute the inputs to the AOI.

Now start with the top layer of source/drains of PFETs and draw lines to parallel the layers of PFETs remembering that the bottom of the top layer is the same as the top of the next layer down. The bottom of the lowest layer of PFET source/drains is the output that must be connected to the top of the NFET series strings. Using the gate identifications, lay out the series strings of NFETs.

As noted above, an important rule to remember in any layout, but especially with regard to the layout of asymmetrical AOIs, is that as many source/drains as possible should be connected to either V_{ss} or V_{dd} (whichever is appropriate) as possible.

The layout of a 3 × 3 AOI is given in Fig. 9.20. The reader should confirm the layered structure of the PFETs and the series strings of NFETs.

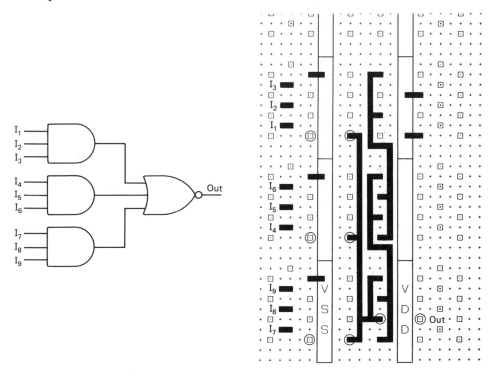

Figure 9.20 3 × 3 AOI. The purpose is to show the layered structure in the layout and also how to develop structures across cell boundaries. Output contact points are encircled. Student exercise: Find the layers in the FET diagram and correlate to those in the layout.

9.9.8 Propagation Delays of AOIs

The propagation delays of symmetrical AOIs can be found from graphs of delay versus fan-out of NAND and NOR gates. THe worst-case (maximum) high-to-low propagation time is approximately that of a NAND gate with the same number of inputs. The worst-case propagation delay in the low-to-high transition is approximately that of a NOR gate with the same number of inputs. The values are approximate because the AOI has more nodal capacitance than do the NAND and NOR gates individually, due to the larger number of FETs connected.

The situations involving asymmetrical and degenerate AOIs are more complex. Approximate values can be obtained in much the same way as above by choosing appropriate values from the graphs for NAND and NOR gates.

Minimum values of delay are sometimes of as much interest as maximum values. A signal that arrives too soon may create a race condition or cause a trigger signal to appear too soon. The reader has probably noted that the delay values of AOIs depend strongly on which combinations of FETs are on at a given time. The same is true of CMOS in general. If all inputs to a four-input NAND gate were to go low simultaneously, the output would go high much faster than if just one of them went low. The reason is that the driving-point resistance in the former case is less by a factor of 4.

Exercise 1

Find the missing line in the diagram in Fig. 9.21.

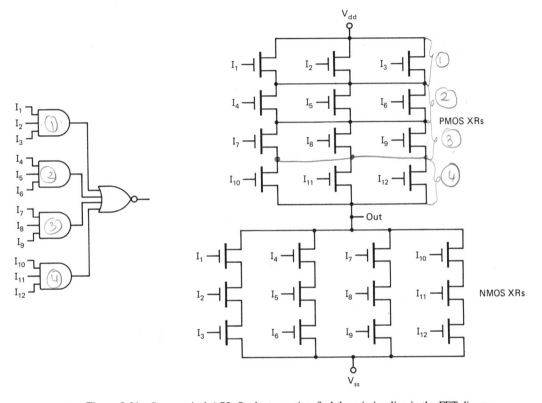

Figure 9.21 Symmetrical AOI. Student exercise: find the missing line in the FET diagram.

Exercise 2

On the grid layout sheet provided, draw the FET diagram and layout of the 4 × 2 AOI shown in Fig. 9.22.

Exercise 3

Do the same for the asymmetrical AOI shown in Fig. 9.16, being careful to place the layer containing the most PFETs closest to V_{dd}.

Exercise 4

Draw the logic diagram, FET diagram, and layout of an asymmetrical OAI which corresponds to the asymmetrical AOI in Fig. 9.16 in terms of number of inputs (to the OR gate of the AOI).

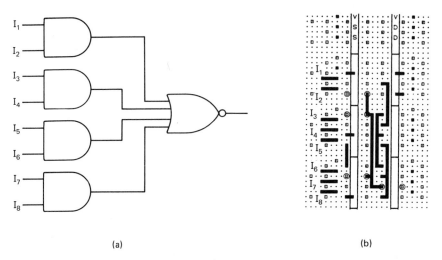

Figure 9.22 (a) 4 × 2 AOI; (b) layout. Student exercise: Draw the FET diagram and the layout of the AOI shown in this figure.

9.10 USES OF AOIs

Those readers who are digital designers are probably already familiar with many of the places where AOIs can be used. The purpose of this brief section is simply to point out a few of the perhaps less obvious places. As noted earlier, the use of AOIs saves propagation time and decreases the numbers of FETs required to implement a given function.

9.10.1 Scratch-Pad Memory

Substantial amounts of memory are not put on gate array chips. Doing so is poor usage of the chip "real estate." Memory cells are normally highly repetitive elements the design of which is optimized for that particular function. Frequently, however, it *is* both cost and space effective to have a small amount of memory on the same chip as other digital logic to avoid the need for additional chips and also to simplify I/O structures. AOIs are a useful vehicle for doing this because of the saving in number of FETs required over other static flip-flop structures and because they do not require both polarities of clock as do transmission gate structures. An example of an AOI used in this application is shown in Fig. 9.23(a).

9.10.2 Detection Circuits

AOIs can also be used as detection circuits as EX-ORs and EX-NORs. This was shown in Figs. 9.18 and 9.19.

9.10.3 Multiplexing

One of the more common uses of AOIs is as multiplexers. The 2 × 2 AOI shown in Fig. 9.12 is an example.

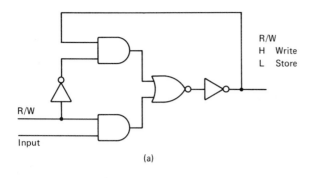

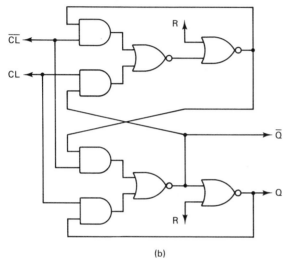

Figure 9.23 (a) Memory cell made from AOI; (b) AOI used in typical counter stage.

9.10.4 Counter Circuits

One of the perhaps lesser known uses of AOIs is in counters. Figure 9.23(b) shows an example of this. These structures can also be used in combination with transmission gates to provide synchronous circuits.

Another use of AOIs is in the decoding circuitry of Moebius (Johnson) counters. Figure 9.24 shows the degenerate AOI used in such an application.

In the BFL (buffered FET logic) form of GaAs, AOIs are also very useful. In the SDFL (Schottky diode FET logic) form of GaAs, OAIs are very useful. They are also useful in ISL/STL.

9.10.5 An Important Task That AOIs Cannot Do

Useful as they are, there is one very important task that AOIs cannot do. That task is the decoding of the all 1's or all 0's condition, which typically occurs as the output of a group of EX-ORs or EX-NORs. The reader will recall that the condition for a low on the output of an AOI is that the inputs to at least *one* AND gate must be high. Therefore, it makes no difference whether *all* the inputs to *all* the AND gates are

OCTAL COUNTER/DRIVER

The MC14022B is a four-stage Johnson octal counter with built-in code converter. High-speed operation and spike-free outputs are obtained by use of a Johnson octal counter design. The eight decoded outputs are normally low, and go high only at their appropriate octal time period. The output changes occur on the positive-going edge of the clock pulse. This part can be used in frequency division applications as well as octal counter or octal decode display applications.

- Fully Static Operation
- DC Clock Input Circuit Allows Slow Rise Times
- Carry Out Output for Cascading
- 12 MHz (typical) Operation @ V_{DD} = 10 Vdc
- Divide-by-N Counting when used with MC14001 NOR Gate
- Quiescent Current = 5.0 nA/package Typical @ 5 Vdc
- Supply Voltage Range = 3.0 Vdc to 18 Vdc
- Capable of Driving Two Low-power TTL Loads, One Low-power Schottky TTL Load or Two HTL Loads Over the Rated Temperature Range
- Pin-for-Pin Replacement for CD4022

McMOS MSI

(LOW-POWER COMPLEMENTARY MOS)

OCTAL COUNTER/DIVIDER

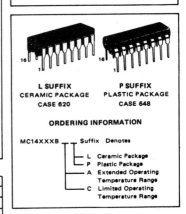

L SUFFIX	P SUFFIX
CERAMIC PACKAGE	PLASTIC PACKAGE
CASE 620	CASE 648

ORDERING INFORMATION

MC14XXXB — Suffix Denotes
- L Ceramic Package
- P Plastic Package
- A Extended Operating Temperature Range
- C Limited Operating Temperature Range

MAXIMUM RATINGS (Voltages referenced to V_{SS})

Rating	Symbol	Value	Unit
DC Supply Voltage	V_{DD}	−0.5 to +18	Vdc
Input Voltage, All Inputs	V_{in}	−0.5 to V_{DD} + 0.5	Vdc
DC Current Drain per Pin	I	10	mAdc
Operating Temperature Range — AL Device CL/CP Device	T_A	−55 to +125 −40 to +85	°C
Storage Temperature Range	T_{stg}	−65 to +150	°C

FUNCTIONAL TRUTH TABLE (Positive Logic)

CLOCK	CLOCK ENABLE	RESET	OUTPUT = n
0	X	0	n
X	1	0	n
⤒	0	0	n+1
⤓	X	0	n
1	⤓	0	n+1
X	⤒	0	n
X	X	1	Q0

X = Don't Care   If n < 4 Carry = 1, Otherwise = 0

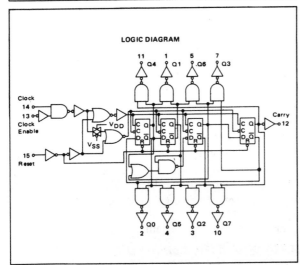

BLOCK DIAGRAM

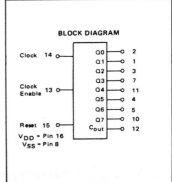

Figure 9.24 Degenerate AOI used in Moebius counter. (Courtesy of Motorola.)

high (detection of the all-1's condition); therefore, this condition cannot be detected. (Multi-input EX-ORs or EX-NORs would be used for detection in conjunction with NAND and NOR gates.)

The reader is invited to make a similar argument for the OAI.

9.11 TRANSMISSION GATES

Transmission gates (TGs) are highly useful structures. They are peculiar to CMOS, although variants of them are used in NMOS dynamic memories. Part of their usefulness stems from the fact that unlike a NAND or a NOR, they do not invert the signal when the controlling signal reverses polarity. Moreover, they can be used in dynamic CMOS circuits to facilitate holding charge on a succeeding gate and then rapidly discharging it.

Like AOIs, TGs can reduce the number of transistors required to implement a given logic function. One has to be careful to include the circuitry to obtain both phases of the clock in the cell count, however.

A third major benefit is that the polarity of the gating signals used (e.g., trigger on the high state or the low state) can be reversed without adding inversion stages by simply reversing the clock phases on the TGs.

9.11.1 Disadvantages of TGs

TGs, as implemented on most gate arrays, do have some drawbacks. These stem primarily from the fact that the transistors used to make the TG are generally considerably larger than they would be if the device were laid out on a pure custom circuit. On the latter, the W/L ratio (see Chapter 2) would be typically 1:1 or thereabouts as contrasted to the 8.5:1 typical of transistors in today's 5-μm and 20:1 in the 3-μm technology.

One of the drawbacks stated earlier to the gate array methodology is that the transistors cannot be optimally sized for a particular appilication. This is particularly true in the case of TGs. A few companies, notably Interdesign on one member of its metal gate CMOS family, have put structures made from TGs in specific parts of the gate array chip. This enables the transistors to be much better sized for the application.

Although the author has never designed with this specific chip, his experience with predefined functions in fixed positions on a gate array chip is that they are often in the wrong place or there are either too many or too few of them. (This is just one person's opinion, however.) Interdesign's placement *is* well thought out, and the number of D flip-flops available is sufficient for many uses.

The other main drawback to the use of TGs is that they require both clock phases, that is, both clock and its inverse (denoted hereafter is CLK and $\overline{\text{CLK}}$). On occasion this requires routing *two* highly buffered high-speed signals instead of one. Otherwise, the second phase must be locally generated from the first.

A related problem is that there is generally a brief period of time when both

CLK and $\overline{\text{CLK}}$ are in the same logic state because of one being derived from the other via an inversion.

Another problem is that the input loading is typically $2\frac{1}{2}$ times (rule of thumb; use with caution) that of a regular gate of a CMOS pair. This comes about because of the fact that in effect two pairs of depletion capacitances (of the source/drains) are paralleled. Each depletion capacitance is a little more than one fan-out typically (300 fF versus 280 fF, for example, for the gate of a CMOS pair in the same technology). However, the second pair of depletion capacitances is separated from the first pair (which is at the input) by an impedance equal to the parallel resistance of the two FETs. (One can also argue about the voltage variation of the depletion capacitance. However, it is a reasonable rule of thumb.)

9.11.2 Operation

A TG is made from a PFET in parallel with an NFET (both source/drains of each FET are connected to the opposite member of the other FET). The basic structure is as shown in Fig. 9.25(a). Symbols for a TG are shown in Fig. 9.25(b) and (c).

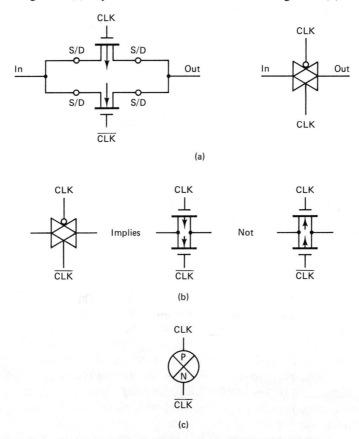

Figure 9.25 (a) Transmission gate (TG); (b) standard symbols; (c) alternative symbol.

The operation of a TG is illustrated in Fig. 9.26. Input voltage is plotted against the impedance of the device. When the input is high, the NFET will be in a relatively high impedance state even with a high on its gate because the voltage relative to its source is very low. However, the PFET will be on strongly, and hence have low impedance, because its gate-to-source voltage is strongly negative (about zero volts referenced to about V_{dd} the input value.

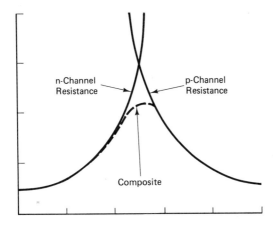

Figure 9.26 Variation of "on" resistance versus input voltage in transmission gate (noncompensated bilateral switch). (Courtesy of Interdesign.)

The opposite is true if the input voltage is low, that is, the NFET is strongly on and the PFET is in a relatively high impedance state. The TG is basically a bilateral analog switch of the "uncompensated" type. (Examples of compensated switches can be found in almost any metal gate CMOS data book of standard 4000 series parts as type numbers 4066 and 4016, respectively. These have the property of a much smaller variation of "on" resistance with variation in input signal level at the expense of greater complexity. Because of the greater number of transistors required for implementation and because the implementation on a gate array involves rerouting V_{dd} and V_{ss} buses, compensated switches are almost never implemented on a gate array.)

9.11.3 Layout

On the usual gate array layout with the PFETs on one column and the NFETs on another column, a T lays out with the transistors (one PFET and one NFET) at roughly 45 degrees with respect to one another. This is necessitated by the need to have both phases of the clock present in conjunction with the each transistor gate connecting both a PFET and an NFET.

9.11.4 Advantages in Terms of Space and Loading to a TG Pair

An interesting feature will now be noted. For grid structures of the IMI types, a *pair* of TGs can be made in the *same* space as a single TG *without* any additional loading to the clock drivers. The remaining transistors in the cell cannot be used for anything else; therefore, no additional cell space is taken up, and the loading to the clock drivers is there regardless of whether or not the second TG is implemented. Figure 9.27 shows this.

The common output of the TG *pair* still comes off the middle source/drain. The source/drain capacitance of the output *does* increase from typically 0.3 pF to about 0.5 pF. However, most of this is a depletion-layer capacitance.

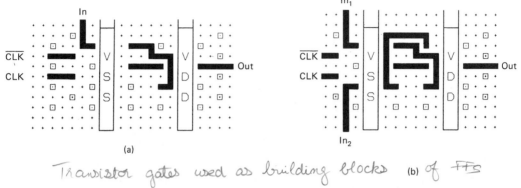

Transistor gates used as building blocks of FFs

Figure 9.27 Layout of TGs: (a) one transmission gate; (b) pair of transmission gates.

An important feature of a TG pair made in the fashion above is that when one member of the pair is on, the other is off. This feature will prove useful in several types of structures to be discussed below, including EX-ORs and D flip-flops.

A few companies, such as CDI (California Devices, Inc.), have arranged their transistor gates so that each gate goes to an NFET and to a PFET which is not directly opposite it but rather is one "row" (of PFETs) below it. These have been seen in this chapter and also Chapter 3. This permits the PFET and NFET used in the TG to be directly opposite one another, as shown in Fig. 9.28. It also means that a single TG can be made reasonably efficiently, although the gate capacitance of the one

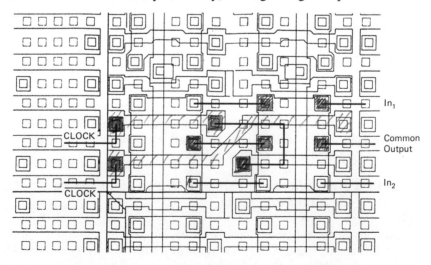

Figure 9.28 Pair of transmission gates in layout on CDI grid.

392 Design of CMOS Macros Chap. 9

gate (of the two used) which is not severed in the middle will have to be accounted for in propagation delay.

In the layout shown in Fig. 9.27(b) either gate input could be clock *if no other contacts have been defined*. However, note that defining *either* of the clock inputs *or either* of the TG pairs automatically defines *all* of the other elements in the TG. One has to be very careful in this regard. An example will illustrate the point.

Consider the EX-OR made from the TG pair shown in Fig. 9.29(a). The layouts of the TG pair and the two inverters are as shown in Fig. 9.29(b). No interconnects have been made between the inverters and the TG pair. The layout of the TG pair is the same as that shown in Fig. 9.27(b). Both inverters (for input B and control signal A) are drawn. The student should readily recognize them at this point; if not, review Section 9.3.

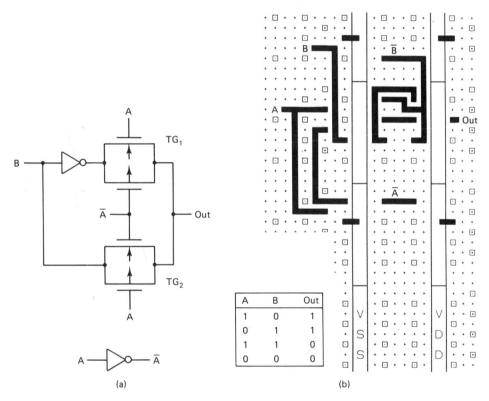

Figure 9.29 EX-OR made from transmission gate pair: (a) schematic; (b) layout with connections among TG and inverters.

At this stage of layout either signal B or its inverse could go to either of the TG pairs *or* (not and) either of the clock phases, A and $\overline{A}$, could go to either of the transistor gate inputs. Suppose that input B is drawn to the lower left source/drain of one of the TG pairs as shown. The connections of the three remaining signals ($\overline{B}$, A, and $\overline{A}$) are now fixed by this one connection. The result is shown in Fig. 9.29(b).

Sec. 9.11 Transmission Gates

It is sometimes helpful when doing a TG layout to analyze it as in the following example (which is the previous exercise). "$\overline{A}$ (A inverse) goes to the PFET whose input is B bar and to the NFET whose input is B. A goes to the NFET whose input is $\overline{B}$ and to the PFET whose input is B."

9.11.5 Operation of EX-OR Made from a TG Pair

If the reader recalls that one TG is on while the other is off, there should be little confusion in understanding the operation of the EX-OR made from TGs using the truth table supplied. It is important to note that signal A has a fan-out of 3 for the EX-OR using the rules of thumb stated above. If A is also fanning out to other elements, care must be exercised to buffer it. Otherwise, both TGs will be on momentarily and this can lead to unwanted "spikes" in the output.

Exercise

The circuit in Fig. 9.30 has a number of problems. Find and correct them. An example of the solution is given in Fig. 9.31. Note carefully the directions of the arrows in the FET diagram. (The NFETs are on the top.)

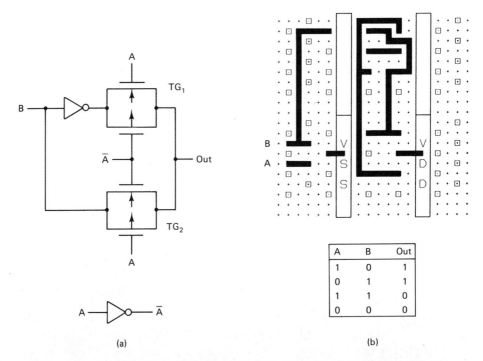

Figure 9.30 EX-OR made from transmission gate pair. Student exercise: find the problem in the layout.

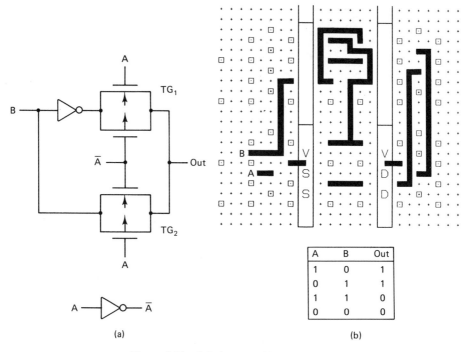

Figure 9.31 Solution to problem in Fig. 9.30.

9.11.6 TGs in Latches and Flip-Flops

TGs are used extensively in latches and flip-flops. An example of the former is shown in Fig. 9.32. This is an example of a transparent noninverting latch. Again note that when one member of the TG pair is on, the other is off. Specifically, the Q output follows the input until such time as the clock signal goes low, at which time the inner TG closes to retain the value of D at that instant. Here there is only one inversion in the direct path and reset has been provided via the NAND gate.

A D flip-flop can be made by combining two clocked latches in a master-slave arrangement as shown in Fig. 9.33. The outermost TGs in this arrangement both go "on" together and "off" together. The inner two TGs similarly operate in the same fashion except, of course, they are in opposite polarity to the outermost TGs. When the outer two are both on, the previous value is stored in the rightmost loop while the leftmost TG is admitting data. When the clock reverses again, the results are propagated to the output. The setup time is the time required for the D input to propagate through the leftmost TG, the inverter, and the reset NAND gate in the master (the left latch).

It will be noted that the reset is stored regardless of the clock phase. If the inner TGs are on, the reset is stored in the left loop; the reset is stored in the right loop if the outer TGs are on.

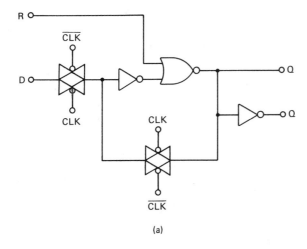

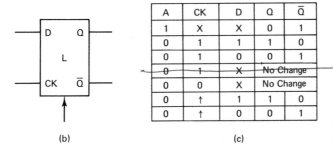

Figure 9.32 Clocked latch with reset: (Courtesy of Interdesign.)

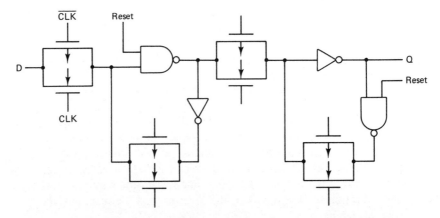

Figure 9.33 D flip-flop, which holds reset state after reset goes high regardless of clock polarities. Student exercises: (1) label remaining TG CLK and $\overline{\text{CLK}}$ signals; (2) prove above statement; (3) Does Q change state on high or low clock transition?; (4) How could you make Q change state on the clock phase, which is opposite to that in number 3?

Figure 9.34 shows a "jam" data input D flip-flop. Its operation is much the same as that of the D flip-flop described above except that the added TG permits the wired OR connection shown.

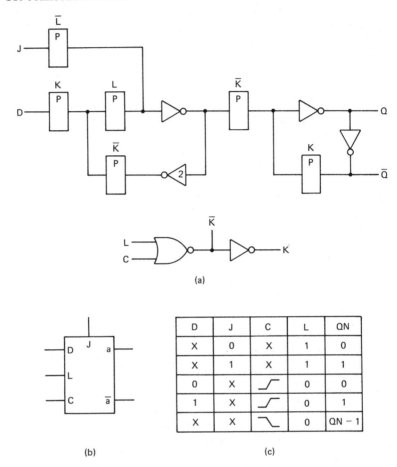

Figure 9.34 D flip-flop with jam data input and low active reset: (a) circuit schematic: (b) symbol; (c) truth table. (Courtesy of International Microcircuits, Inc.)

It is very important that the rise and fall times of the clock and clock bar (inverse) of TGs in flip-flops be very short. It is also important to have the TGs switch as much as possible at the same time. Otherwise, it is possible for spikes to occur in the output by virtue of having all TGs on at the same time, especially the two in the direct path to Q. A good rule of thumb is to limit the fan-out of the clock and clock bar drivers to 2 or less for high speed signals.

9.11.7 Multiplexers Made from TGs

Relatively high density multiplexers can be made from TGs. An example is shown in Fig. 9.35. These devices tend to be rather slow compared to those made from

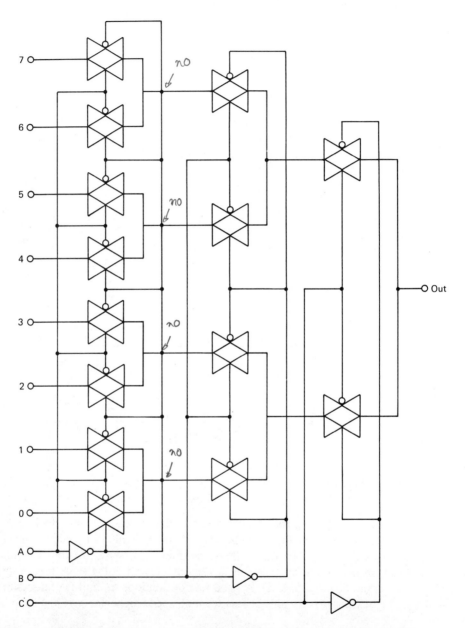

Figure 9.35 Decoder using transmission gates: 8-to-1 decode tree. (Courtesy of Interdesign.)

AOIs. Moreover, some of the density advantage is negated by the need to develop short rise and fall times of the clock phases.

9.12 FLOATABLE DRIVERS (CLOCKED INVERTERS)

A floatable driver, FD, is a structure sometimes used by manufacturers in their custom and semicustom CMOS circuitry in place of TGs. The device is basically an inverter that is clocked with both clock phases. Unlike a TG, the input comes in on the gate of a transistor.

Like the TG pair and three-state devices, it permits other outputs to be wired to its output. The FET diagram of an FD is shown in Fig. 9.36(a) and a typical layout is shown in Fig. 9.36(b).

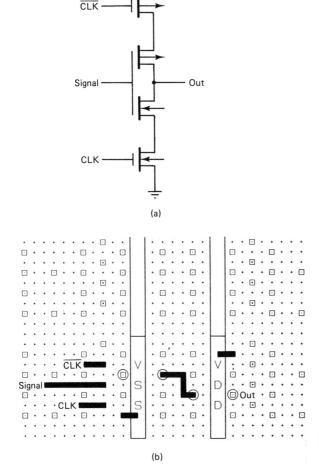

Figure 9.36 (a) Floatable driver (clocked inverter); (b) layout.

For gate array design, the FD has several disadvantages compared to a TG. First, it takes more space for a single FD than for a single TG made from a two-input cell. In a three-input cell, the space for one FD is the same as for a TG pair because the third transistor pair is wasted. Unlike a TG, a pair of FDs cannot be made in the same space and with the same loading as just one of them.

A second disadvantage is that the driving-point resistance in either direction is twice that of the TG because of the two series FETs between the output and the power supply rails.

An advantage of the structure in some applications is that unlike the TG, it is unidirectional. Also, its driving-point capacitances may be somewhat less than those of the TG.

9.13 INPUT STRUCTURES

The most common interface of CMOS (other than to CMOS itself) is to TTL in its various permutations (LSTTL, STTL, ALS, AS, etc.). The last two are, of course, the newer oxide-isolated versions of TTL. This is a defacto standard. Most MOS standard devices and even some of the newer ECL gate arrays have direct interface to TTL levels.

The TTL output low level of 0.4 V is easily below the 30% of V_{dd} or 1.5 V of CMOS for a logic low. Therefore, there is no problem for the input low.

The TTL output high level guaranteed over temperature and loading is 2.4 V even though lightly loaded devices have somewhat higher voltages. Although CMOS switches typically at 50% of V_{dd}, the guaranteed switching is 70% of V_{dd}. Therefore, inputs to CMOS from TTL must be pulled up to 70% of the V_{cc} (5.5 V maximum) of the TTL if the CMOS power supply is also that of the TTL, which it normally is. There are several ways of doing this. External pull-up resistors could, of course, be used; but these are costly in terms of space, power, and added manufacturing labor (for component insertion) and component cost. One alternative is to use chip PFETs as pull-up resistors. An example of this is shown in Fig. 9.37. The sink current capability of the TTL driver, of course, must be sufficient to accommodate the pull-up resistor.

In lieu of a fixed external pull-up resistor, a switchable pull-up could be used controlled by logic inside or outside the chip. An example of such a switchable pull-up is given in Fig. 9.38. Note that the number of FETs required is relatively small, and recall that normally there are unused FETs on the structure. Table 9.2 shows how a wide variety of pull-up resistor values can be obtained using combinations of buffer and array transistors. These combinations can be obtained in either fixed (by tying the gates of the appropriate FETs to the proper power supply rail) or in switchable pull-up form.

Switchable pull-ups are more useful than fixed pull-ups. The latter accommodate only one configuration of chips being driven from the outside world (of the chip). The reason is, of course, that a given driver can sink only so much current, and the inputs so driven by a given driver are in parallel with one another. A scheme for implementing combinations of values to accommodate various uses of the chip is

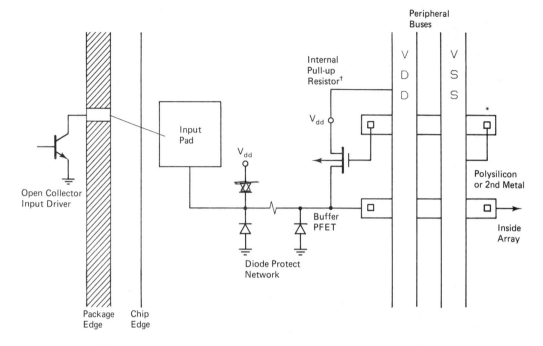

*For Bottom and Left Side Gate Can Be Tied Directly to VSS
†External Pullups Can Also Be Used and Should Be When Multiple Chips Are Connected to Same Pin

Figure 9.37 Input to CMOS from TTL.

given in Fig. 9.39. Connection point is input line. Top of string connects to V_{dd}. Consult the gate array manufacturer before using.

Major gate array manufacturers, such as IMI and CDI, use voltage-divider schemes to obtain valid input highs and lows from TTL inputs. IMI uses a pair of array FETs in a standard inverter configuration (Fig. 9.40). However, shunting the output of this inverter is the output of a buffer NFET which is also driven by the TTL input. When the TTL input is low, the PFET in the inverter is a low impedance (about 6 kΩ), while both of the two NFETs are in the off (high-impedance) state. This pulls the input high.

When the TTL input goes high to a minimum 2.4 V, all three FETs are on. Voltage V_{dd} is dropped across the voltage divider formed by "on" resistances of the three FETs. Because the W/L ratio of the array transistors is $\frac{1}{12}$ of that of the buffer devices, the resultant output signal will be about $\frac{1}{14} V_{dd}$ or about 0.4 V worst case. This is much less than the 30% V_{dd} (1.65V at 5.5V = V_{dd}). This assumes that the threshold for the PFETs is roughly the same as that of the NFETs, which it is to within generally about 0.2 to 0.3 V. Note that if the TTL device is not at its maximum fan-out, the input voltage will be greater than 2.4 V. This will cause the PFET to be at a higher impedance, and both of the NFETs to be at lower impedances, thus lowering the input signal even further and providing even more noise margin.

The 30%/70% rule of thumb for valid low and high switching applies to all

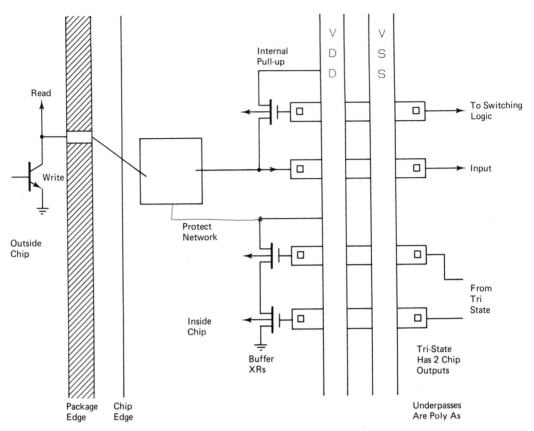

Figure 9.38 Switchable pull-up: one bonding pad used as both input and output. Student exercise: find the missing line(s).

TABLE 9.2 INPUT CIRCUITRY OBTAINING THE PULL-UP RESISTANCE REQUIRED

Method number	Pull-up resistance value desired (Ω)	Method
1	12,000	Use one big p-channel XR connected at $V_{dd}' = 5$ V
2	400	Same as method 1 but operate at $V_{dd} \sim 10$ V
3	5,000	Use one array PCH XR at $V_{dd} = 5$ V
4	2,500	Use two array PCH XRs in parallel at $V_{dd} = 5$ V
5	3,000	Use one array PCH XRs at $V_{dd} = 10$ V
6	1,000	Use three array PCH XRs in parallel at $V_{dd} = 10$ V
7	461	Parallel a big PCH XR with three parallel array PCH XRs
8	12,000	Use two array PCH XRs in series
9	8,000	Use one array PCH XR in series with three parallel array PCH XR
10	4,000	Use two series array PCH XRs in parallel with 1 PCH array XR

XR = Transistor (in this case a FET)

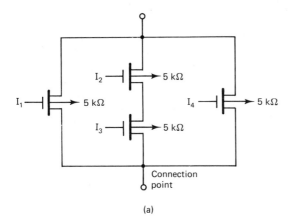

I_1	I_2	I_3	I_4	Pull-up Resistance (kΩ)
L	H	H	H	5
H	L	L	H	10
L	L	L	H	3.3
L	H	H	L	2.5
L	L	L	L	2.0
H	H	H	H	10^9

(b)

Figure 9.39 Obtaining more than one pull-up value for a given input by using logic to switch values (allows greater flexibility, but requires decoding): (a) example (this combination of PFETs replaces the single buffer PFET of Fig. 9.38); (b) values available at 5 V (other values can be obtained by changing voltage.) It probably is not worth adding SR with I4.

three types of CMOS: metal gate, silicon gate, and CMOS/SOS. Therefore, a CMOS output driving a CMOS input of even a different permutation will generally be straightforward if the displacement currents are sufficient to handle the capacitive load at the speed required.

9.14 OUTPUT STRUCTURES

Again the most common (except for CMOS to CMOS) interface is to TTL. In this case, however, the type of TTL being interfaced to is important because of the source and sink currents required of the driver. For example, each LSTTL load requires the driver to sink 360 μA and to source 20 μA. In sharp contrast, each STTL load requires 2 mA and 50 μA for sink and source, respectively. Values for FAST, AS, and ALS are in between.

The currents mentioned above are static *conduction* currents. The CMOS gate array chip must also source and sink enough *displacement* currents to switch the load capacitance in the time required. Both of these criteria must be used to determine the output structure to be used.

A common misconception is that CMOS is incapable of driving high-displacement (capacitive) or conduction-current loads. The truth is that CMOS does not drive such loads as easily as an 8-Ω ECL output, but structures can indeed be designed on most of the CMOS gate array chips to light an LED (10 mA) or to supply

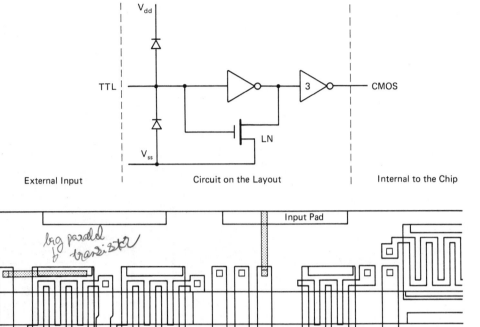

Figure 9.40 IMI input structure TTL to CMOS. (Courtesy of International Microcircuits, Inc.)

moderate amounts of capacitive currents. On the 2000-cell IMI chip, for example, there are 141 buffer NFETs of roughly 500 Ω on resistance and about 56 buffer PFETs of roughly the same impedance. By using various parallel combinations of buffer and array transistors, relatively low "on" resistances can be generated.

A typical CMOS to TTL driver circuit and its layout on the I/O grid shown in Chapter 3 are shown in Figs. 9.41 and 9.42, respectively. The number of buffer FETs that must be paralleled is determined from consideration of the "on" resistance required.

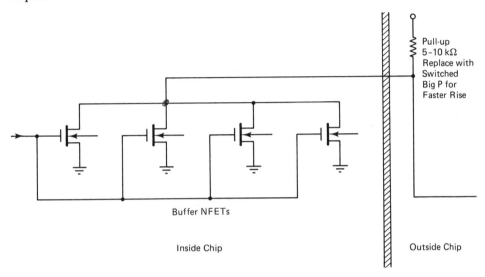

Figure 9.41 High-sink, low-source current output buffer on CMOS chip. Drive each buffer FET with at least two paralleled inverters.

9.15 "ON" RESISTANCE CALCULATIONS

The "on" resistance (R_{on}) required can be calculated as the lesser of the values calculated for the static and dynamic cases. The numbers of PFETs and NFETs which must be paralleled to give the source (low-to-high) and sink (high-to-low) currents are found by dividing the total "on" resistance by the "on" resistance of a given FET. For example, suppose that the required source resistance is 100 Ω and that the "on" resistance of each buffer PFET is 400 Ω. Then four PFETs would have to be paralleled to satisfy this condition. (For a discussion of paralleling, see Section 9.18.)

1. *Static*:

$$\text{PFET } R_{on} = \frac{V_{oh}}{\text{conduction current to be sourced}}$$

$$\text{NFET } R_{on} = \frac{V_{ol}}{\text{conduction current to be sunk}}$$

where V_{oh} and V_{ol} are the output high and low voltages, respectively.

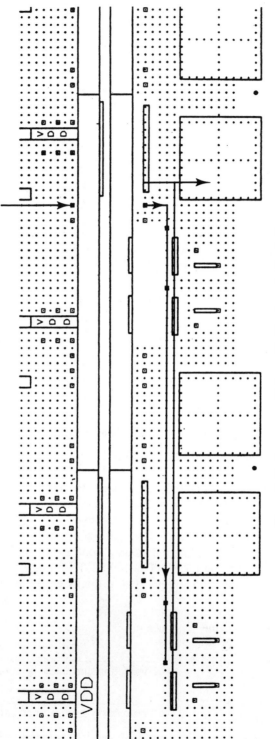

Figure 9.42 Layout on grid sheet of buffer shown in Fig. 9.41.

2. *Dynamic*: This calculation can be considerably more involved than that for the static case for several reasons. First, the resistance of a given FET changes dramatically and nonlinearly as the FET turns on. Second, at the same time the resistance is changing, the drain-to-source and gate-to-source voltages are also changing. Third, in the case of CMOS, the current through the NFET is being affected by the current through its associated PFET, which is undergoing the opposite transition to that of the NFET. Fourth, the voltage across the load capacitance will change nonlinearly (if left to itself) according to the well-known exponential relationships.

To avoid the need to turn to Maxwell's equations or more likely to a simulation program, such as SPICE or TRACAP, every time in order to calculate the "on" resistance and hence the number of buffer stages that must be paralleled, the following method is sometimes useful. The writer uses the word "sometimes" so that users will consider the four factors mentioned above before blindly using the method stated below.

Step 1: Calculate the average value of the displacement current I as

$$I = C \frac{dV}{dT}$$

where C is the total load and output structure capacitance and dV is the switching voltage swing. In the case of TTL, for example, this swing is from 0.45 V to 2.4 V or a dV of approximately 2.0 V. dT is the desired time for the transistion. The shorter the time, the greater the amount of current and hence the lower the resistance of the FET(s) involved.

Step 2: Calculate the "on" resistance of each FET from

$$R_{on} = \textit{one-half the upper switching voltage}\,/I$$

(sometimes the user in his or her judgment may decide to use the average between the upper and lower switching voltages).

Comments on the Method: The only reason that the method described above provides a reasonable answer is that the different nonlinearities tend to balance each other out. For example, unless an exponential value of current versus voltage is used, the value obtained for I is an average between the lower and midpoints of the switching voltage. Because one-half of the upper switching voltage limit is used, the "on" resistance so calculated will be higher than it would be if the average between one-half the upper and lower switching voltages were used. (For high values of the upper switching voltage, the designer may wish to use the average rather than the upper voltage, especially if fairly conservative results are desired.) However, this larger value of "on" resistance is somewhat tempered by the fact that the turn-on action of the FET involved has a "head start" over the charging or discharging of the load capacitor. Hence (in the case of a high-to-low transistion on the output, for example) the *entire* voltage across the output of the FET (which might be 4 V or more in the case of TTL, for example) appears across what is already a low value of resistance because of the aforementioned head start. This means that the displacement current so supplied will in reality be higher *initially* than that calculated even though it will *eventually* be lower than calculated as the output voltage drops (and as it becomes limited by the increasing resistance of the FET to which it is paired). Graphs of PCB capacitance were given in Chapter .

An example will illustrate the point. The following values will be used: $C = 15$ pF, $dV = 2$V, and $dT = 5$ ns (strictly speaking, these are really "deltas," not differentials). The reason for using the former in the formulation is to show the student what can be done if more rigor is desired. The reason for using the latter is to show what normally *is* done.)

Then $I = 6$ mA and $R_{on} = 200$ Ω. This means that two buffer NFETs with 400 Ω "on" resistance each must be paralleled if the 5 ns is for a high-to-low transition. Similarly, two buffer PFETs must be paralleled if the 5 ns applies to a low-to-high transition. If the calculation of "on" resistance from the static case were less, then more PFETs or NFETs would have to be paralleled as shown in the earlier example.

9.16 TRISTATE BUFFERS

A very common output circuit is the Tristate (Tristate is a registered trademark of National Semiconductor) buffer. This is used to enable many circuits to be placed on the same bus. The method used is to make all source and sink impedances of all devices not being used at a given time extremely large. This prevents the bus from being either charged or discharged by the devices that are not in active use at the time. In CMOS this is accomplished by turning off *both* the buffer PFET and the buffer NFET of output signals not in use.

The reader will recall that in many gate array implementations, the input gate to an array PFET is also connected directly to an array NFET because in the vast majority of cases the two types are used together in digital logic. The buffer FETs are not so connected (i.e., the gates of the PFET and NFET are separate). A major reason for doing this is to permit the use of Tristate buffers. Examples of Tristate buffers implemented in CMOS are given in Figs. 9.43(a) and (b). The truth tables are also shown. The reader should be able to confirm the truth tables.

Enable	In	Out
L	H	L
L	L	H
H	X	HI-Z

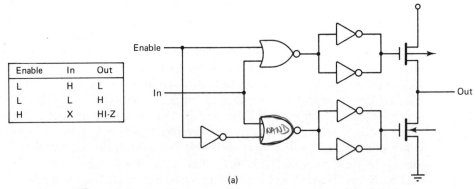

(a)

Figure 9.43 (a) Tristate output buffer configuration; (b) layout; (c) layouts of various other buffer combinations. Student exercise: one of the NOR gates should be a NAND gate. Which one is it?

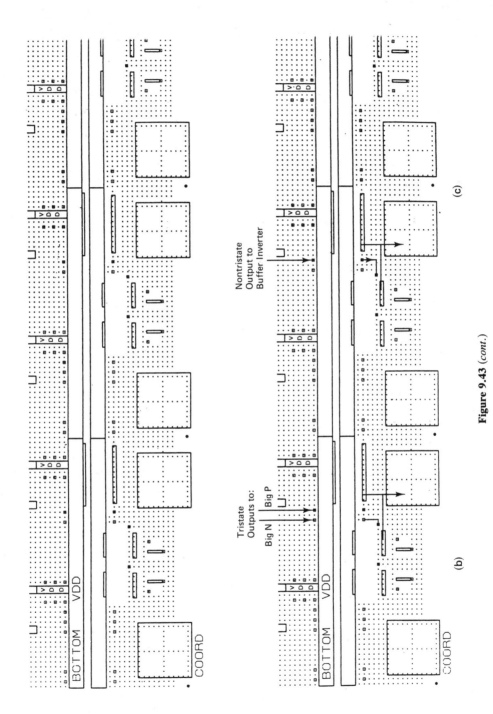

Figure 9.43 (cont.)

9.17 DRIVING THE OUTPUT DEVICES

Having now discussed means of implementing various output devices, attention must be given to driving the buffers in the output devices. Each buffer FET has a W/L ratio many times larger than that of the array FETs. A value of 12 is typical. This means that the capacitive load of the buffer is roughly 12 times that of the array device. If an array inverter drives an inverter made of a buffer PFET and buffer NFET, the total capacitance seen by the array inverter will be 24 times that of either array FET. However, the fan-out seen by the array inverter will be 12, *not* 24. The reason is that fan-out of CMOS array structures is couched in terms of driving the gates of both an array PFET and an array NFET connected together, *not* just one of the devices alone. This often causes a lot of confusion. To drive buffers, array cells are generally paralleled (see Section 9.18).

9.17.1 General Guildelines for Driving Buffers

Normally, one tries to avoid using NAND and NOR gates to drive buffers because of the large multiplier created by the series resistance strings of such devices. This in turn causes large asymmetries in the rise and fall times. More important, the signal is slowed much more by the larger value of resistance in the *RC* time constant. Inverters should be used where possible.

Another common "trap" that first-time designers sometimes fall into is to forget to leave the last inversion stage for the output buffer.

9.18 PARALLELING OF STRUCTURES

To decrease the fan-out seen by a given device, output structures can be paralleled. For example, if a single inverter driving the fan-out 12 buffer discussed in Section 9.17 is replaced by a "triple paralleled inverter" (three inverters in parallel), the fan-out seen by each inverter will be three instead of 12. The propagation time of *that stage* will accordingly be less. However, the fan-out of that which is driving the triple paralleled inverter will have *increased* by two loads (not by a factor of 2). This may force that which is driving the triple parallel inverter stage to also have to be paralleled. The result is that paralleling "ripples back." The gate array designer must keep the output structures in mind even when he or she is considerably far away from them.

Paralleling is also used internally quite extensively with the same caveats. Paralleling of an element is often denoted by the symbol "Px" inside the device symbol, where x is the number of times paralleling is done. Sometimes the number alone is used inside where its usage is unambiguous.

It is not necessary that a given device see an integer number of unit loads. This, too, sometimes causes problems for first-time designers who believe that they are constrained in this fashion. One is dealing with a load structure that is primarily capacitive, and it is the value of this capacitance that is of concern.

Examples of paralleling and the associated layouts are given in Figures 9.44 to 9.47. The latter shows paralleling of a three-input NAND gate.

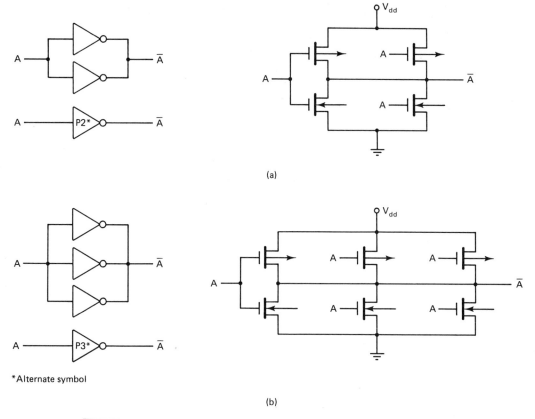

*Alternate symbol

Figure 9.44 Paralleling transistors for greater drive: (a) double paralleled (designated P2); (b) triple paralleled (designated P3).

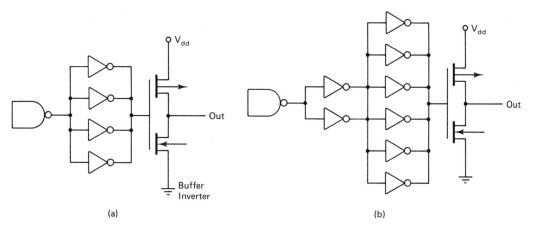

Figure 9.45 Driving output buffers: (a) inverting; (b) noninverting. (Note Third-generation gate arrays have drivers for the buffers built into the macros. Check with vendor.) Each buffer inverter pair is 12 unit loads. Drive each buffer FET with at least two paralleled inverters. The buffer shown in part (b) is 2 to 5 ns slower than the buffer shown in part (a).

Sec. 9.18   Paralleling of Structures

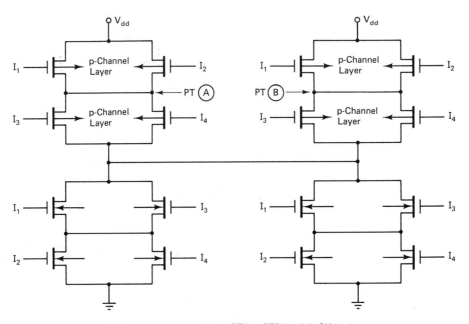

It is not necessary to connect PTA to PTB but it is OK to do so ONLY IF I_1 and I_2 are on the same "p-channel layer" (in this case the topmost one) in BOTH of the paralleled AOIs

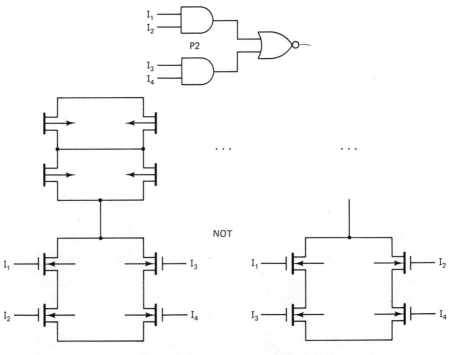

Figure 9.46 Paralleling AOIs.

Layout of Parallel AOI

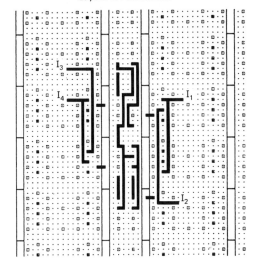

Figure 9.46 (*cont.*)

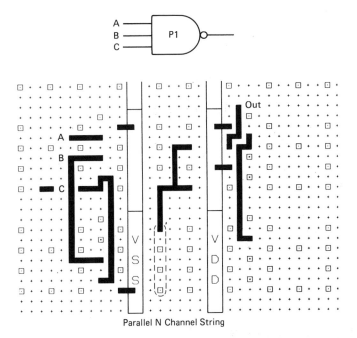

Figure 9.47 Paralleling series strings. The NFET string of the NAND gate is paralleled to decrease the resistance involved in the high-to-low transition. Note how paralleling the PFETs is obtained essentially "free." Usually not done for fewer than 5 inputs. P1 means paralleled 1 series string, and is local terminology only.

Sec. 9.18   Paralleling of Structures

9.18.1 Paralleling AOIs

AOIs pose special precautions when being paralleled. (One should try to avoid paralleling all except inverters where possible, but sometimes this is unavoidable.) The precaution is that the layers of PFETs must correspond exactly in all parallelings. (An analogous statement holds for the NFETs in OAIs.)

Consider the paralleled AOI shown in Fig. 9.46. If I1 and I2 were on different PFET layers in the paralleled version, then when the input to just one AND gate was low (in the case of just two AND gates), a low-impedance path to V_{dd} would be established; and the circuit would try to pull high.

A second precaution which is almost self-evident is that the signals to the NFETs must be in the same strings in the two parallelings. (They do not have to be in the same order going up from V_{ss}, however.) This, too, is shown in Fig. 9.46. (The circuit will not perform the NAND function if this is not the case.)

9.18.2 Paralleling Gates Other Than Inverters

Avoid this when possible. It obviously takes more transistors. A subtlety is that the ripple-back effects of the paralleling are magnified by the number of inputs of the gate being paralleled. For example, a triple paralleled four-input NAND gate will increase the fan-outs of three more signals (by a fan-out of two extra) than would an inverter.

On occasion, it is difficult to avoid. An example of paralleling a NAND gate is shown in Fig. 9.47. It is true that the NFET string alone could be paralleled. This would minimize the skew between the rise and fall times of this particular gate array. However, note that paralleling of the PFETs comes essentially "free" in terms of space and loading to the driving signal once the NFETs have been paralleled.

9.18.3 Other Benefits and Problems with Paralleling

A subtle benefit and also a subtle problem of paralleling is that circuits degrade "gracefully" if one or more of the paralleled structures fails or more likely has an error in it. Such failings will not show up on a functional test and are often hard to find from a dynamic test. Because of process variations (which, as noted elsewhere, can give propagation delay variations of more than 40%) and the fact that the speeds inside the chip have tolerances on them, it is very hard to isolate to such a failed element provided that one of the paralleled structures is intact and correct.

The same property may enable the device to continue to function in the field should it fail there. However, as noted elsewhere, most such failures are due to interconnects to the outside world provided that the circuits have been burned in to weed out infant mortality failures.

9.19 SCHMITT TRIGGERS

It has been mentioned previously that CMOS uses power only when it switches. If the switching times are excessively long, both PFET and NFET in a CMOS pair will be on; and considerably more power will be generated. A well-known way to sharpen rise and fall times as well as to provide hysteresis is via the use of the Schmitt trigger. Examples of this are shown in Figs. 9.48 and 9.49. These are based on modifying switching voltages slightly as shown in Fig. 9.50. By changing the design of the individual transistors, the results can be reversed.

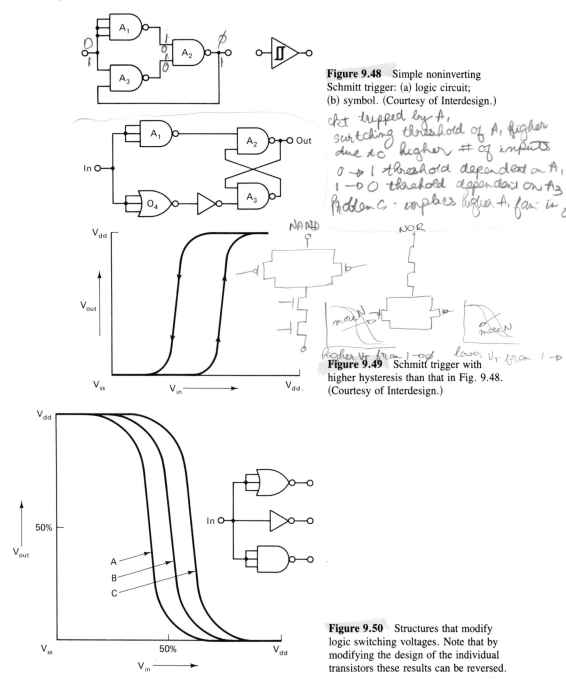

Figure 9.48 Simple noninverting Schmitt trigger: (a) logic circuit; (b) symbol. (Courtesy of Interdesign.)

Figure 9.49 Schmitt trigger with higher hysteresis than that in Fig. 9.48. (Courtesy of Interdesign.)

Figure 9.50 Structures that modify logic switching voltages. Note that by modifying the design of the individual transistors these results can be reversed.

9.19.1 One-Shots (Monostable Multivibrators

A common means of obtaining incremental delays using standard (off the shelf) parts is via the use of one-shots. These devices can also be implemented in gate arrays using Schmitt triggers and external capacitors. An example is shown in Fig. 9.51.

Because of the large resistance of CMOS coupled with the manufacturing tolerance of this resistance and the hard-to-predict parasitic capacitances of the layout itself, one-shots in CMOS are very imprecise and in general should be avoided. On gate arrays, other structures are much more predictable with regard to actual function, although the delays are not that much more precise. These are discussed below.

9.20 USE OF INVERTERS FOR DELAY AND OTHER PURPOSES

9.20.1 Use of Inverters for Delay

Inverters provide a convenient method of delay either alone or in combination with logic gates. *It is cautioned that the delay is not that precise.* The response is, however, generally more predictable than that obtained from one-shots.

Each inverter in 5-μm CMOS, for example, provides a delay of roughly 3 ns for a fan-out of 1. Each two-input cell can provide two such inverters wired "end to end." (A three-input cell can also provide two such inverters, but one of the two can be double paralleled.) Because one, as a rule of thumb, does not use more than about 80% of a given array, there are generally many unused cells available.

9.20.2 Use of Inverters and Logic Gates for Pulse Generation

Inverters used in conjunction with logic gates can provide pulse generators of varying degrees of precision. Such circuits are often useful when imprecise pulse generation needs to be done locally (i.e., in the immediate vicinity of a given portion of the chip). Examples using common NAND and NOR gates are given in Fig. 9.52. Examples using EX-NORs are given in Fig. 9.53(a) and (b). It is left as an exercise to the reader to synthesize the FET diagrams and the layouts from the preceding material. Recall that a two input NAND or NOR gate can be built on the same three-input cell as an independent inverter.

Figure 9.53(a) shows that a pulse is generated from the NAND gate when the input goes from low to high. It is left as an exercise to the reader to explore the effects of varying the number of inverters and to determine what happens in the logic in the lower part of Fig. 9.53(a).

In Fig. 9.53(b), a pulse is generated from the NAND gate when the input goes from high to low. The dotted pulses in the figures mean that the pulse may or may not be there depending upon the relative timing. For this reason, the structures in

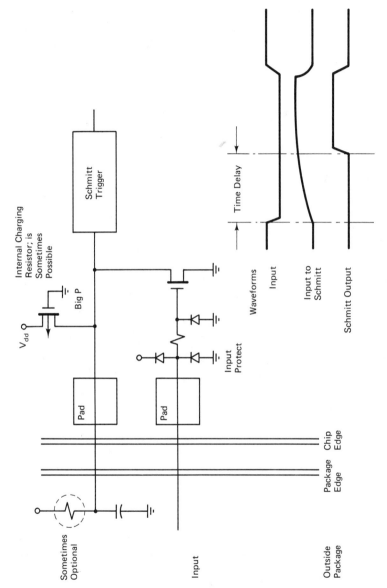

Figure 9.51 Monostable multivibrator (one-shot). One-shots are not recommended for design.

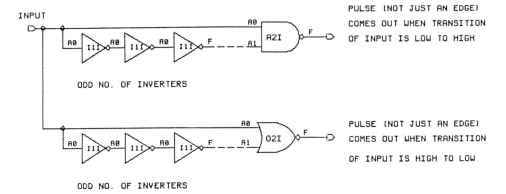

Figure 9.52 Use of inverters to obtain pulses.

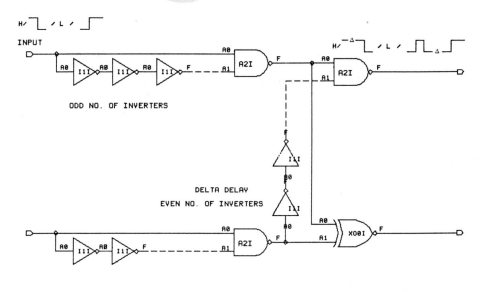

(a)

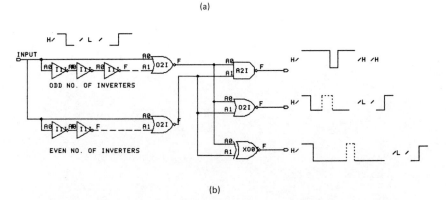

(b)

Figure 9.53a,b Use of EX-NORs to generate pulses. Student exercise: What would the results be if an EX-OR were substituted for the EX-NOR? Slashes denote switching intervals. Dotted lines denote variable number of elements. Deltas denote differential time.

Figures 9.52 and 9.53 have limited value when pulse lengths are very short or precision is required. In all cases, interconnect wiring should be kept as short as possible.

9.21 ROUTING SUBTLETIES

The use of macros has been discussed previously. In this chapter, methods of designing the macros themselves were given. Algorithms for auto P&R (placement and routing) of macros were discussed in Chapter 8. The purpose of this section is simply to point out a few nuances of routing on single-layer metal grids of CMOS gate arrays.

The IMI grid has 21 channels in each group, consisting of one array cell column and its adjoining poly-block column which can be used for vertical routing. See the explanation of the grid in Chapter 3. Of these, about 11 are encumbered in some way, meaning that they cannot be used under all circumstances. The IMI grid is used because it is representative of other gate array grids, as noted previously.

Figure 9.54 is the IMI grid structure with the vertical routing channels labeled for discussion. Each of the dots represents a point in a valid routing channel.

Channels 16, 17, 19, and 20 are the primary "global" routing (i.e., routing among macros which are widely separated) channels. Channels 5, 10, and to a lesser extent channels 7 and 8 are similarly used when they are not being used for local wiring within the macro.

For a NAND or NOR gate, there is only one connection (from PFETs to NFETs) which blocks the use of channels 7 and 8 for global wiring. This connection could be eliminated and the channel freed up if the connections to the NFETs and PFETs are taken from drains on the outside of the power buses. The poly D underpass in the cell is used to route from one side to the other under the power and ground buses.

Because the output is being loaded by a poly D which otherwise would not be there, the designer must be careful to include the one extra poly D in the calculation of propagation delay. For short runs, there is little or no problem.

Channels 4, 6, 9, and 11 (which contain the rows of source/drains) are used only for "dog legs," which are short jogs around the source/drains or for connections to (in the case of TGs) and from the source/drains. Channel 18 is generally used for short runs to connect among polys that are close to one another. Otherwise, doglegging has to be done around the contact points.

Use of channels 3 and 12 will block access to and from the source/drains. Often, either 3 or 12 but not both can be used, that is, the output comes off a source/drain of one side but not the other. This enables routing to be done on the side it does not come off.

Short routes can still be done in channels 3 and 11 even on the side off which the output comes whenever there is a string of FETs on that side. This string occurs, for example, when there is a multi-input NAND or NOR gate or several paralleled inverters, AOIs, and so on. The eight-input AOI shown in Chapter 1 shows that for even a relatively uncomplex structure, *both* channels 3 and 12 are available and one line has been routed in channel 12 in Fig. 1.8(b).

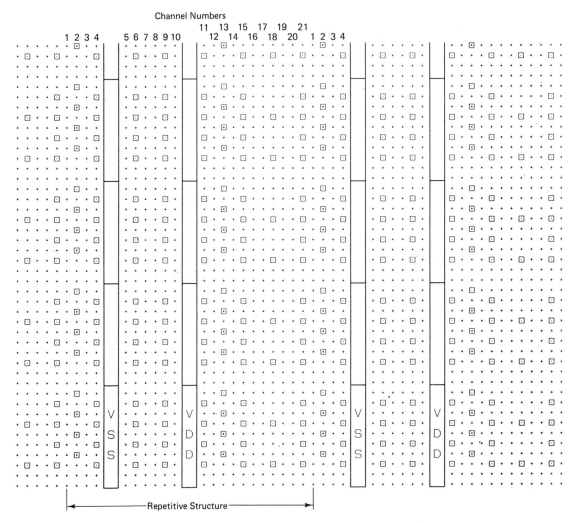

Figure 9.54 IMI grid structure with the channels labeled for discussion.

Channels 2 and 13 contain the contacts to the gates and the poly D. These channels are used either to parallel the gates or as short jogs in conjunction with channels 1 and 3 or 12 and 14. Use of channels 1 and 14 for long runs would deny access to the gates. Because the cells are double entry, as noted earlier, it is possible to use either 2 or 13 but not both for long runs when gates on both sides of a cell are not being used. Short runs are possible especially when adjacent channels are available.

Signals that are routed inside the power buses must be brought outside to effect connection for cells of the IMI type whose gates do not have contacts in the middle. If the signals are going to/from connections to TG I/Os, there is, of ocurse, no problem because the contacts can be made to the source/drains inside the cell. (As noted earlier, the disadvantage of contacts in the middle of gates is that the contacts must be routed around when not being used.) To get the signals outside the power buses, they can be routed to the end of the column, where there is only a short poly (poly B) with an available contact.

A second method which is sometimes used is to use a source/drain contact to bring the signal outside the buses. If this is done, the gates must be appropriately biased by tying them to the power buses. See also the comments in Section 9.7.

Although it is difficult to generalize, non-dog-leg paths are probably slightly preferable to paths made up of many dog legs. A reason is that a dog leg may cover a number of steps (different heights of the die due to the layers of materials that make it up) in strange fashions (like along the side of a step, for example). Runs that stay in the same direction may similarly do this, but there is probably a little less chance for discontinuities with the non-dog-leg path.

9.22 INPUT PROTECTION NETWORKS

Any MOS circuit needs protection from static voltages. This is especially true in the thinner (150 to 250 Å) gate oxides being used in today's MOS. A person walking on a rug can build up a voltage of 10,000 V for example.

The input protection circuits offered by most MOS vendors consist of diode clamps of both high and low signals along with a resistance. The exact details differ, and some of them, especially those on the newer 1Mbit and 4Mbit DRAM (dynamic RAM) chips, are highly proprietary.

A typical protection circuit is that shown in Fig. 9.55(a), (b), and (c). Details of operation were given in Chapter 2. This section briefly describes the macro layout given in Fig. 9.55(b). The reader will note that there is a diode clamp to both high and low voltages formed from the source/drain of the PFETs and NFETs to the n wafer and the p tub, respectively. The reader should recall that the gate extends to both the PFET and the NFET for this particular family of gate arrays. This permits the relatively short connection from V_{dd} to the source to force a high on the gate of the NFET. The corresponding connection to V_{ss} keeps the PFET on. The reader should be able to recognize the structure as a TG. The input comes in on a paralleled

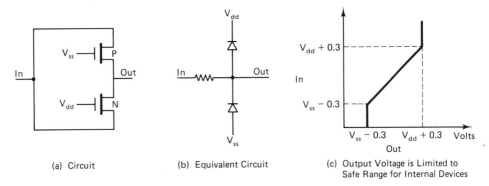

Figure 9.55 (a) Input protect circuit; (a) FET diagram, (b) schematic, (c) voltage response, (d) layout (Courtesy of International Microcircuits, Inc.)

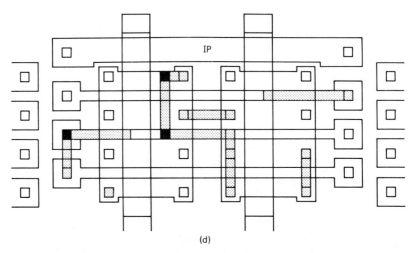

(d)

Figure 9.55 (*cont.*)

source/drain and leaves on a second paralleled source/drain. The reader should readily be able to decipher the remainder of the layout.

9.23 SUMMARY

Previous chapters dealt with gate array design using macros. This chapter deals with design of the macros themselves in the CMOS technology. It is also possible to design portions of gate array chips using techniques delineated in this chapter. Because CMOS is the dominant gate array technology and because there is often more opportunity to design with transistors in this technology, effort is taken to acquaint the reader with the various structures of CMOS. These include transmission gates and floatable drivers as well as structures which are not peculiar to CMOS but which are widely used in CMOS (e.g., the various types of AOIs). Interface circuits to TTL are described.

Chapter 10

Design of Bipolar Macros

10.0 INTRODUCTION*

Gate arrays are built in three major bipolar technologies. These are TTL and its permutations, IIL and its permutations, and ECL and its permutations. Important permutations of TTL are FAST, AS (advanced Schottky logic), ALS (advanced low-power Schottky logic), STTL, and LSTTL (the older junction-isolated version of low-power Schottky TTL and Schottky TTL, respectively).

Important permutations of IIL are ISL (integrated Schottky logic) and STL (Schottky transistor logic, not to be confused with Schottky transistor-transistor logic, STTL). The latter two differ from IIL in that the collectors in ISL/STL are emitters in IIL, and vice versa.

Interestingly, some of the GaAs circuits being designed at TI and other places use ISL/STL structures made from what is known as heterostucture bipolar transistors (HBTs) [1]. Therefore, what is described in this chapter has implications for GaAs as well as silicon technology. The important permutation of ECL is CML (current-mode logic).

Although gate arrays are available in all of the technologies noted above (but not in all permutations of those technologies), this chapter focuses on ISL/STL and ECL. These are representative of the technology groups of which they are part. Because TTL and its permutations are by far the largest family of standard parts, most designers are intimately familiar with them. Therefore, they are not treated. More-

*See Section 9.0 for general flow of the process.

over, there are only a few TTL gate arrays on the market. The same is true of ISL/STL. However, far less information is available on the latter.

The purpose of this chapter is to illustrate some of the ways in which common logic functions can be implemented in the two technologies selected. The reader is referred to the appropriate vendors for more information.

Most bipolar gate arrays have been two-layer metal from the beginning. This contrasts quite sharply with CMOS gate arrays, which went through effectively two generations before two-layer metal was introduced. (The two generations were metal gate CMOS and the early silicon gate CMOS arrays.)

The reason that most bipolar arrays are and always have been two-layer metal is that the voltage swings are considerably lower than those of CMOS and the currents used are much higher. For example, the voltage swing of ISL/STL is in the vicinity of 200 mV, as contrasted to the greater-than-2-V swing of 5-μm CMOS. Therefore, the need to minimize voltage drops in the signal and ground lines is much greater for the bipolar technologies than it is for CMOS; and the processing problems and greater complexity associated with the two (or more) metallization layers were faced much earlier in the development cycles of bipolar arrays.

10.1 ISL AND STL

ISL and STL are treated as one technology by the author because they are very similar in structure and in the way logic is designed using them.

Schematic diagrams of the ISL and STL transistors are shown in Figs. 10.1 and 10.2. The reader will note several similarities between the two technologies. First, each transistor on the array constitutes an inverter stage with multiple outputs taken off the collector. Second, each of the outputs is isolated from the others by means of Schottky diodes. Third, there is a resistor R_B connected to the base that supplies both "injection current" to the base and also acts as a pull-up for the input signal.

Increasing the injection current by using a lower value of resistor increases both the speed and power of the device. The value of the internal pull-up resistor is beyond the control of the user.

Not shown on the schematics is the property that multiple inputs can be wire ANDd to each of the transistors. (In this technology, the use of the terms "transistor" and "gate" in place of the term "cell" is appropriate. Each transistor on the array constitutes a gate in the sense that it is an inverter stage with multiple outputs. Because of the wire ANDing property to be explained in Section 10.1.1, each such inverter is indeed a gate.)

STL differs from ISL primarily in the way in which the collector is clamped to the base of the NPN transistor. In ISL this clamp is via the forward voltage drop of a *pnp* transistor. In STL, a Schottky diode is used between base and collector in a manner similar to that of the Baker clamp of standard LSTTL (see Figs. 10.1 and 10.2). The object in all cases is to keep the *npn* transistor out of saturation.

The benefit of the Schottky base clamp is that it aids in providing higher-speed operation than does the transistor clamp. A potential drawback mentioned by propo-

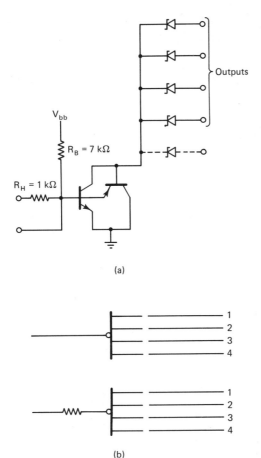

Figure 10.1 Transistor diagram of ISL: (a) schematic; (b) gate symbol. (Courtesy of Signetics.)

nents of ISL is that reportedly [2], [3] it is more difficult to produce with a standard process. The problem is that Schottky diodes with two different "barrier heights" must be used on the same chip. However, TI, Harris, and other manufacturers seem to have mastered it. Unless otherwise noted, the terms "ISL" and "device" will refer to both ISL and to STL.

As noted above, the voltage swings of these devices are very small, typically about 150 to 200 mV. For this reason some rather wide metallization lines are used for the power and ground buses. In Signetics' first-generation version of ISL, for example, the top metallization layer is 12 μm wide, and the first layer is 6 μm wide. (Signals are also routed on these layers.) The voltage swing of STL is the difference between the barrier heights of the clamp diodes and the output diodes.

Both ISL and STL are "collector fan-out" logics as opposed to the "emitter fan-in" logic of TTL and its permutations. Each *npn* transistor has five collectors (in the versions of ISL and STL offered by Signetics and TI, respectively) of which any four can be used. The five are there for flexibility in layout. Metal lines can be routed over the unused collectors, including the fifth one in Signetics' implementations.

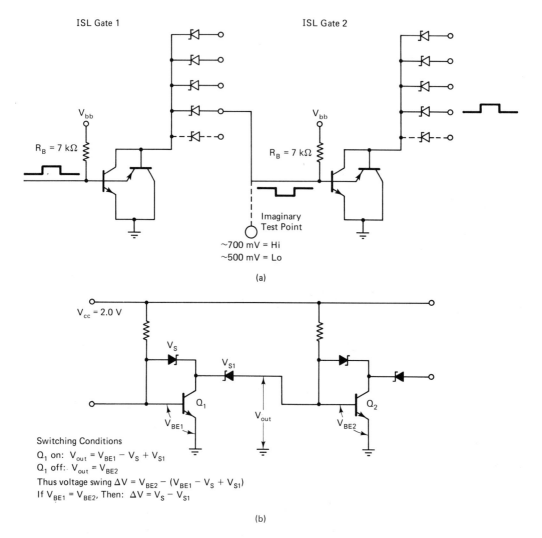

Figure 10.2 Interconnect diagram of (a) ISL; (b) STL. [(a) Courtesy of Signetics; (b) courtesy of Texas Instruments, Inc.]

Each of the five is isolated from the others by a Schottky diode and has a fan-out of 1. Outputs cannot be paralleled to obtain more drive capability as they can, for example, in CMOS. To obtain higher fan-outs, some arrays, such as those in the Signetics 1200 series family, have Schottky buffers built into the middle of the array. These can be used to increase the drive capability of either the internal array transistors or the chip outputs. All five array transistor outputs have essentially the same propagation delays; therefore, none is preferred over another.

Different gate array manufacturers use different symbologies to denote the single-input, multiple-fan-out transistor. TI's symbol is shown in Fig. 10.3(a) and Signetics' symbol is shown in Fig. 10.3(b). Only the output lines actually used as signals are normally shown.

426   Design of Bipolar Macros   Chap. 10

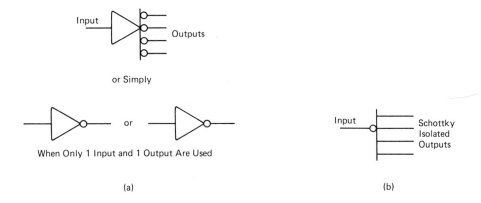

Figure 10.3 (a) STL gate symbol of TI; (b) ISL gate symbol of Signetics.

10.1.1 Wire ANDing

ISL/STL have also been called AND/NAND logic. Unlike CMOS, wire ANDing is fundamental to ISL/STL design. For the benefit of those readers who are not logic designers, wire ANDing is the process of connecting output signals from different gates together to form the AND function. If any one of the outputs connected together is low, the entire grouping is pulled low. This is the property of an AND gate. A major benefit is that the propagation delay of the wire AND is very small, typically a few tenths of a nanosecond. A second major benefit is that no gate is needed, thus saving both active elements and power. The outputs being isolated by Schottky diodes permits this.

The current TI and Signetics STL and ISL gate arrays permit up to 8 and 10 signals, respectively, to be wired together. (These numbers may change as new gate arrays are developed by the respective manufacturers.) The result is then input to the base of a transistor if inversion is to be performed.

10.1.2 Use of Wire ANDing to Obtain Logic Functions

To understand ISL/STL, it is necessary to understand the duality of logic gates. These are given in Appendix B. For example, a multi-input NOR gate is equivalent to an AND gate with the same number of inputs. Each input is inverted, however, to form the same function as the NOR gate.

Any inversion in the signal requires use of an array gate, which in this case is one transistor. Therefore, if a four-input gate is considered, an AND gate would require no transistors (all the signals would simply be wired together), a NAND gate would require one transistor (to invert the result of the wire ANDing), a NOR gate would require four transistors (to invert each of the signals, which then are wire ANDd together), and an OR gate would require five transistors (to invert the output of the NOR gate). These relations are given in Fig. 10.4 (a)–(d).

Figure 10.4(e) illustrates the very high fan-in of ISL. (STL has a fan-in which is typically a little less, but at 8 it is still high.) Figure 10.4(f) shows how a variety of functions can be formed from two outputs.

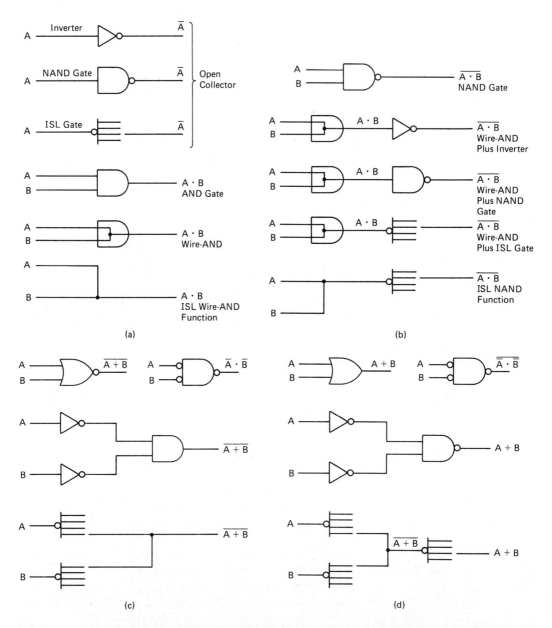

Figure 10.4 Formation of (a) AND gate, (b) NAND gate, (c) NOR gate, and (d) OR gate using ISL/STL, (e) 10-input AND function, (f) wired logic design. [(a) to (e) Courtesy of Signetics; (f) courtesy of Texas Instruments, Inc.]

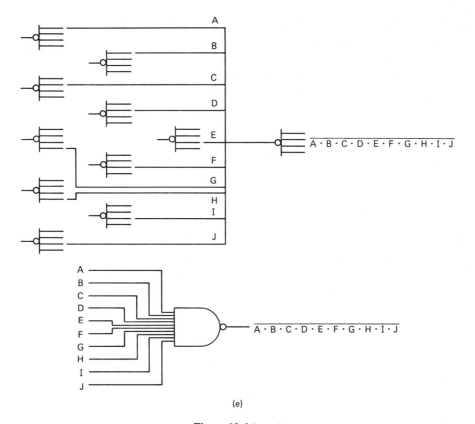

(e)

Figure 10.4 (*cont.*)

10.1.3 Conversion of a Circuit to ISL or STL

Unlike TTL, CMOS, and ECL, there is no family of standard ISL and STL parts. Therefore, to implement a circuit designed in a technology for which there exists a family of standard logic ICs, such as LSTTL or CMOS, it is necessary to convert the schematic using the foregoing principles. Normally, the conversion is not a big problem. It does, however, represent another opportunity for errors to occur.

One way to minimize the errors is to simulate, and hence verify, the circuit designed using ISL or STL macros that will be used in its gate array implementation. The macros may be from the vendor's library or may be user designed. In some cases, macros exist for simulation which do not have layouts associated with them. The user designs the layouts after verifying the logic. This is another reason for learning how to design macros using transistor design.

Obviously, the propagation delays used in the simulations allow for variations among individual layouts. Signetics presently does this with their 1200 series of junction-isolated gate arrays. (The newer, higher-speed oxide-isolated gate arrays of Signetics will have layouts corresponding to the macros used for simulation.)

Alternatively and preferably, the designer can design the circuit from the start using ISL or STL. Examples of conversions are shown in Fig. 10.5.

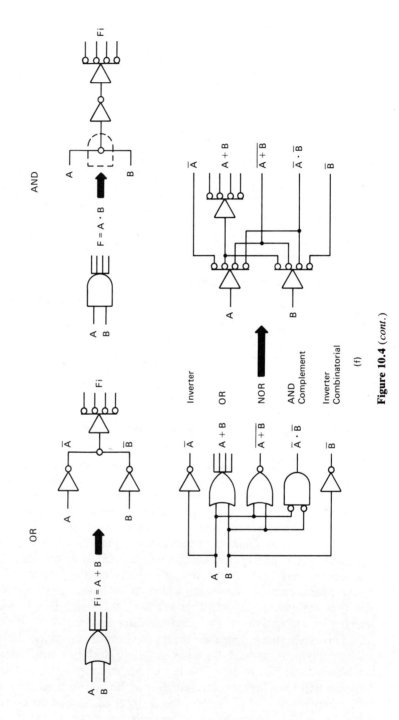

Figure 10.4 (*cont.*)

430

Truth Table

S_2	S_1	S_0	I_0	I_1	I_2	I_3	I_4	I_5	I_6	I_7	$\bar{Y}$
L	L	L	L	X	X	X	X	X	X	X	H
L	L	L	H	X	X	X	X	X	X	X	L
L	L	H	X	L	X	X	X	X	X	X	H
L	L	H	X	H	X	X	X	X	X	X	L
L	H	L	X	X	L	X	X	X	X	X	H
L	H	L	X	X	H	X	X	X	X	X	L
L	H	H	X	X	X	L	X	X	X	X	H
L	H	H	X	X	X	H	X	X	X	X	L
H	L	L	X	X	X	X	L	X	X	X	H
H	L	L	X	X	X	X	H	X	X	X	L
H	L	H	X	X	X	X	X	L	X	X	H
H	L	H	X	X	X	X	X	H	X	X	L
H	H	L	X	X	X	X	X	X	L	X	H
H	H	L	X	X	X	X	X	X	H	X	L
H	H	H	X	X	X	X	X	X	X	L	H
H	H	H	X	X	X	X	X	X	X	H	L

H = HIGH voltage level
L = LOW voltage level
X = Don't care

Functional Description

The "152" is a logical implementation of a single pole, 8-position switch with the switch position controlled by the state of three Select inputs, S_0, S_1, S_2. The logic function provided at the output is:

$Y = I_0 \cdot \bar{S}_0 \cdot \bar{S}_1 \cdot \bar{S}_2 + I_1 \cdot S_0 \cdot \bar{S}_1 \cdot \bar{S}_2 +$
$\quad I_2 \cdot \bar{S}_0 \cdot S_1 \cdot \bar{S}_2 + I_3 \cdot S_0 \cdot S_1 \cdot \bar{S}_2 +$
$\quad I_4 \cdot \bar{S}_0 \cdot \bar{S}_1 \cdot S_2 + I_5 \cdot S_0 \cdot \bar{S}_1 \cdot S_2 +$
$\quad I_6 \cdot \bar{S}_0 \cdot S_1 \cdot S_2 + I_7 \cdot S_0 \cdot S_1 \cdot S_2$

The "152" provides, in one package, the ability to select from eight sources of data or control information.

Logic Diagram

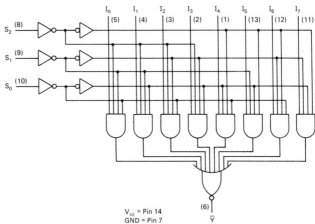

V_{cc} = Pin 14
GND = Pin 7

AC Waveforms

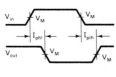

V_M = 1.5 V for 54/74 and 54S/74S
V_m = 1.3 V for 54LS/74LS

DC Characteristics over Operating Temperature Range

	Parameter	Test Conditions		54/74 Min	54/74 Max	54S/74S Min	54S/74S Max	54LS/74LS Min	54LS/74LS Max	Unit
I_{oc}	Output short circuit current	V_{cc} = max V_{out} = 0 V	Mil	−20	−55					mA
			Com	−18	−55					mA
I_{cc}	Supply current	V_{cc} = max			43					mA

AC Characteristics T_A = 25°C

	Parameter	Test Conditions	54/74 C_L = 15 pF R_L = 400 Ω Min	54/74 Max	54S/74S Min	54S/74S Max	54LS/74LS Min	54LS/74LS Max	Unit
t_{plh}	Propagation delay Select to $\bar{Y}$ output	Figure 1			35				ns
t_{phl}					33				ns
t_{plh}	Propagation delay Data to $\bar{Y}$ output	Figure 1			20				ns
t_{phl}					14				ns

(a)

Figure 10.5 Conversion of schematic to AND/NAND logic: (a) unconverted; (b) converted; [(a), (b) Courtesy of Signetics; (c) courtesy of Texas Instruments, Inc.] (c) Another example.

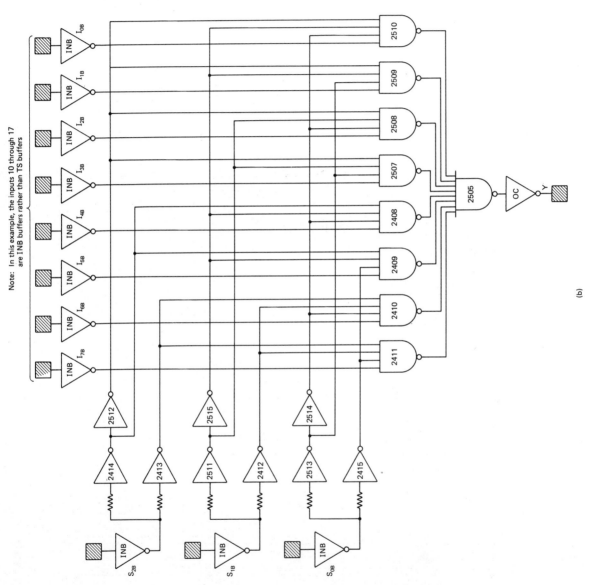

Figure 10.5 (*cont.*)

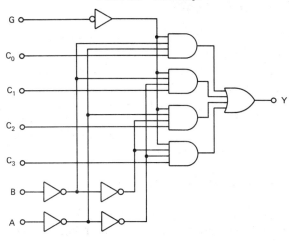

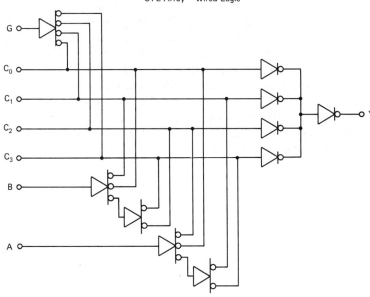

Performance	74S	74LS	STL
Gate Count	10	10	10
Typical Dissipation (mW)	112	15	6
Typical Propagation Delay			
C-Y (ns)	6	13.5	5
G-Y	9	18.5	7.5
A, B-Y	12	22	10

(c)

Figure 10.5 (*cont.*)

10.1.4 Current Hogging

Different gates (transistors) in the array will have different load currents and different values of the ground potential. Suppose that outputs from two or more collectors of two or more transistors are wire ANDd together in such a way that they fan out to two or more transistors.

When one of the gates at the output of the common signal has a ground potential lower than the others, it can hog all of the current when the source signal turns off. The solution to this is to use the resistor input usually provided as an option at the input of each array transistor.

10.1.5 Array Grid Structure

To illustrate the principles discussed above, the Signetics 8A1200 gate array will be used. The grid structure of this was shown in Fig. 3.11 and discussed briefly. This is a two-layer metal gate array, as are most of the bipolar gate arrays. [Signetics no longer offers the ISL array described. However, the information presented applies in general to arrays of other vendors as well as to GaAs ISL and to STL. Raytheon, for example, offers the oxide-isolated ISL arrays designated CGA 50L15 (5040 gates, 150 I/Os) and CGA 35L12 (3584 gates, 124 I/Os). TI uses STL in the VHSIC program and uses I²L (the parent of ISL/STL) in its GaAs (see Chapter 2) arrays. Hence, it is still germane to discuss ISL/STL.]

The structure of an internal portion of this array is as shown in Fig. 10.6. It consists of groups of four transistors. The five possible outputs of each of the four transistors are grouped as shown. Each of the five possible outputs has two possible contact points, labeled L and R, respectively. One or the other but not both may be used. If neither is used, first-layer metal can be routed over the unused output.

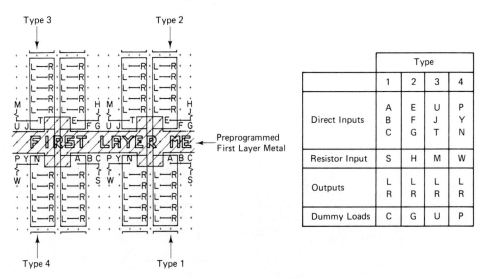

Figure 10.6 Structure of internal portion of Signetics 8A1200 gate array. (Courtesy of Signetics.)

There are four possible inputs to each transistor. Only one may be used. The others are there for routing flexibility, except for the input, which is the resistor input. The latter is used when current hogging may be present.

Each of the small dots represents a potentially allowable routing point. A row or column of such dots represents a routing channel of the array. Such routing channels exist in between the groups of four transistors. Signal lines can be wire ANDd at any point along a line's path.

The groups of four transistors are placed in rows between sets of three first-layer metal power and ground buses. Two of the buses are ground buses; the third (located between the other two) is for power. The result is that first-layer metal is used for horizontal runs and for vertical runs that do not cross the horizontal boundary of power and ground buses. Second-layer metal is used to cross such boundaries and also to cross first-layer signal lines where required. A minor drawback to use of second-layer metal is the extra space required for the "vias" which interconnect first and second metallization layers.

In Raytheon's ISL array, the user can use one transistor to drive the Schottky output diodes of four other transistors. The benefit of doing this is that the Schottky clamp diodes are placed close to their place of actual use.

10.1.6 Examples of Macro Design Using Internal Array Gates

To illustrate the principles involved in laying out ISL/STL macros, the reader will be shown layouts of increasingly more complex circuits using the Signetics grid structure. In each case, the origin of each of the signal lines is not specified but is assumed to be one of the other transistors on the array. This is in line with the method used in presenting the design of CMOS macros. The grid structure, which is from the actual Mylar layout sheet of Signetics, is very "busy" (cluttered) and the lines are somewhat hard to read.

Figure 10.7 gives layouts of several functions designed in ISL. Figure 10.7(a) shows the layout of a three-input NAND gate. The inputs are simply ANDd together and then brought to the input of one transistor. The output can be taken from any four of the five outputs designated L and R (either L or R can be used) which lie directly below the input. See Fig. 10.6 for explanation. Figure 10.7(b) shows the layout of a two-input NOR gate. The input signals coming from the left and right are connected to the inputs of the ISL transistor (circles) and inverted by them. The outputs of the transistors (squares) are wired together to form the AND function of the inverted inputs. The result is the NOR function.

Figure 10.7(c) shows the layout of a 2 × 2 AOI (AND-OR-INVERT) gate. A rather full explanation of the properties of AOIs was given in Chapter 9. It was shown that they are very useful for CMOS. In Fig. 10.7(c) the reader will note that only two transistors are required to build the 2 × 2 AOI. Hence AOIs are also very useful elements for ISL. They are also very useful in GaAs.

In Fig. 10.7(c) two sets of input signals are each wire ANDd together. For clarity, these input signals are assumed to have come from transistor outputs immediately to the left and to the right, although this is not necessary. The rest of the lay-

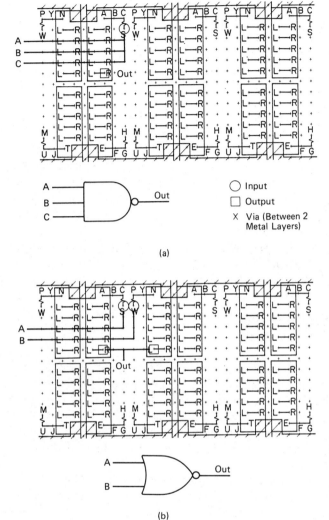

Figure 10.7 Layouts of simple functions in ISL: (a) three-input NAND; (b) two-input NOR; (c) 2 × 2 AOI; (d) degenerate AOI; (e) R-S flip-flop made from NAND gates; (f) EX-OR; (g) eight-input MUX. [(g) Courtesy of Signetics.]

out is similar to that of Fig. 10.7(b) (i.e., the wire ANDs are inverted and the results are themselves wire ANDd together).

Figure 10.7(d) shows the layout of a "degenerate AOI." The were discussed in Chapter 9.

Figure 10.7(e) shows the layout of an R-S flip-flop made from NAND gates. The reader should now be able to trace this circuit. Note that only two transistors are used. As in Fig. 10.7(c), the inputs are assumed to have come from adjacent transistor outputs designated by squares. From the discussion in Section 10.1.2, the reader should also appreciate that a similar flip-flops made from two-input NOR gates would take *twice as many transistors*.

Figure 10.7(f) shows the layout of a two-input EX-OR. The reader will note that only four transistors are used. In this layout, second-layer metal is used

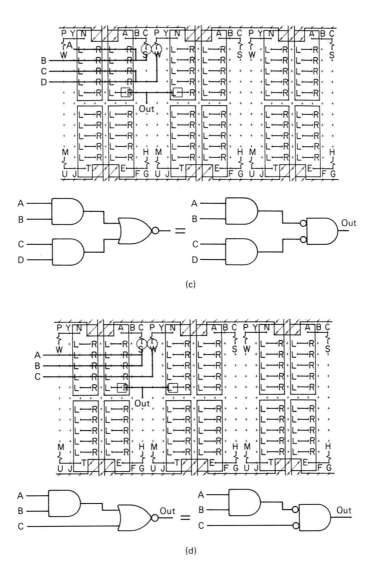

Figure 10.7 (*cont.*)

(horizontal run) and connects to the first-layer metal at the points designated by "x's." The origin of the first-layer metal inputs A and B is assumed to be in other parts of the chip. It is also assumed that signal A splits into the two paths shown elsewhere.

Figure 10.7(g) shows the layout of an 8-to-1 MUX. The logic gate and transistor diagrams are also shown. The heavy black lines represent second-layer metal. The lighter lines are first-layer metal.

Transistor diagrams of other common functions are given in Figs. 10.8(a–d). With the foregoing explanation, the reader should be able to lay out at least some of the simpler elements.

Sec. 10.1   ISL and STL

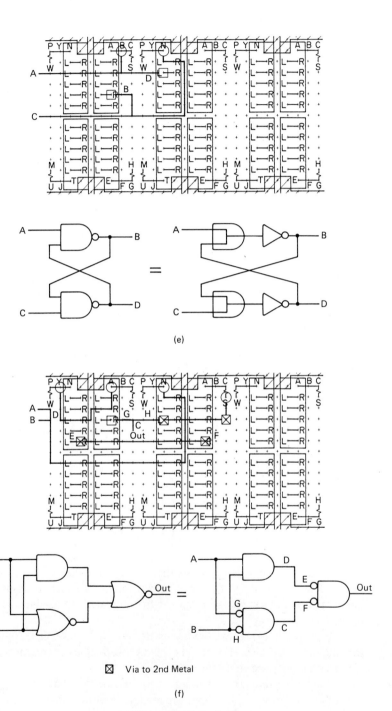

Figure 10.7 (*cont.*)

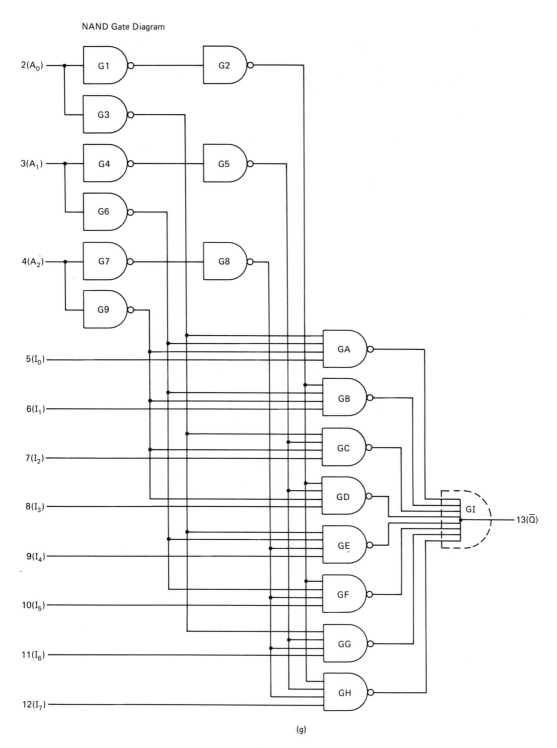

Figure 10.7 (*cont.*)

Sec. 10.1 ISL and STL

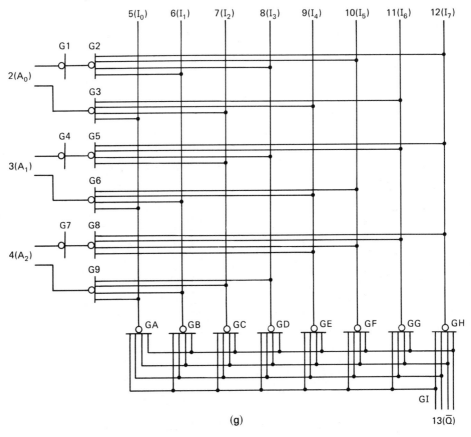

Figure 10.7 (*cont.*)

10.2 ECL

The third major bipolar gate array technology is ECL, emitter-coupled logic. This is the highest-speed silicon technology, offering gate speeds in the subnanosecond range. It is also the highest-powered silicon technology.

10.2.1 Mixed ECL and TTL I/Os on the Same ECL Gate Array Chip

Although most manufacturers offer ECL gate arrays with the capability to use either TTL or ECL I/Os, not all offer the ability to mix such I/Os on a given chip. The AMCC Q700/Q1500/Q3500 series are examples of classes of gate arrays that *do* permit this. Separate ground buses are provided for the TTL and ECL I/Os to prevent TTL voltage spikes from appearing on the ECL ground.

Because CMOS and ECL are on opposite ends of the curves of density, speed,

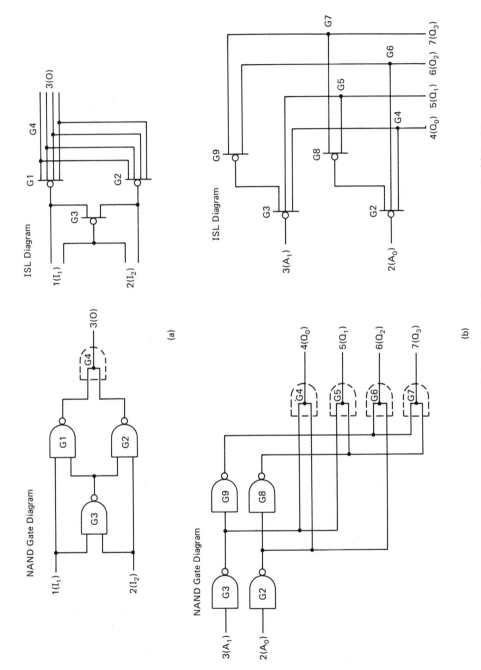

Figure 10.8 Examples of macros designed in ISL: (a) two-input EX-NOR gate; (b) 1-of-4 encoder; (c) edge-triggered D flip-flop; (d) 1-bit full adder. (Courtesy of Signetics.)

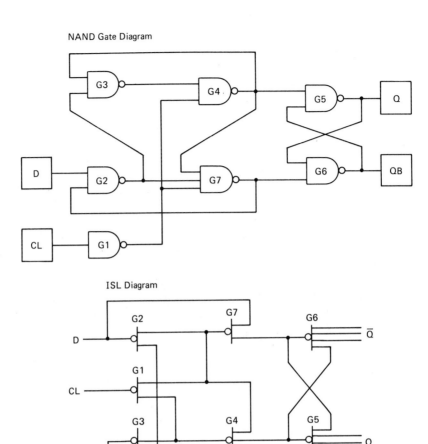

Figure 10.8 (*cont.*)

and power versus technology, it is useful to discuss the two with regard to speed and power.

10.2.2 Speed–Power of ECL Compared to That of CMOS

Multiple levels in one ECL stage. Surprisingly, on a *per gate* basis, the so-called speed–power product of ECL is very close to that of CMOS. The speed–power product in this case is a very misleading indicator, for four reasons. First, in ECL there are *five* different ways of getting multiple levels of gating, with essentially just one level of propagation delay. Three of these are series gating, emitter dotting (wire ORing), and collector dotting (wire ANDing). The fourth is the property that because of the differential nature of the ECL structure (see Section 10.2.5), a signal and its complement are *both* available as the device outputs. Other

NAND Gate Diagram

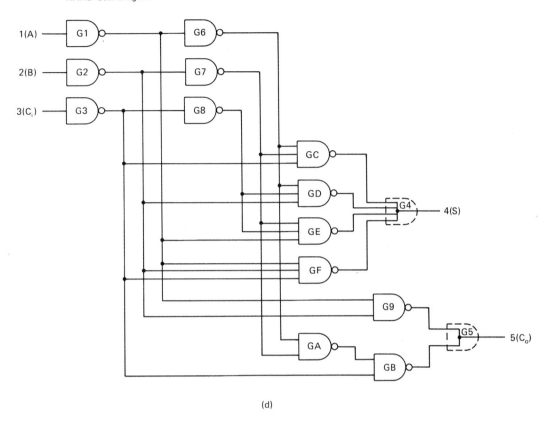

(d)

Figure 10.8 (*cont.*)

technologies require at least one other level of delay (an inverter) to obtain the complement. The fifth is the ability to parallel input transistors to create the OR/NOR function with very little (100 ps, typically) additional delay.

By contrast, in CMOS, there is only one class of structures that enable multiple levels of gating to be combined into fewer levels of delay. This structure is the AOI (AND-OR-INVERT) and its cousin, the OAI (OR-AND-INVERT). (See Chapter 9.)

By the same token, however, (second reason) the actual power dissipated by CMOS is considerably lower than the power dissipation curves would indicate. The reason for this is that CMOS uses power only when switching.

The third reason is that the output impedance of ECL is much lower than that of CMOS. This means that it can drive the capacitances of both the interconnect lines and inputs to the next stage much faster. (The *RC* time constants are lower.) In CMOS, the capacitances of the metal interconnects are almost always negligible compared to the input capacitances of the next stages. In ECL, this is not true.

The fourth reason is that the voltage swings of ECL are lower than those of CMOS. For the ECL 10K family, the voltage swing is on the order of 0.8 to 1.0 V.

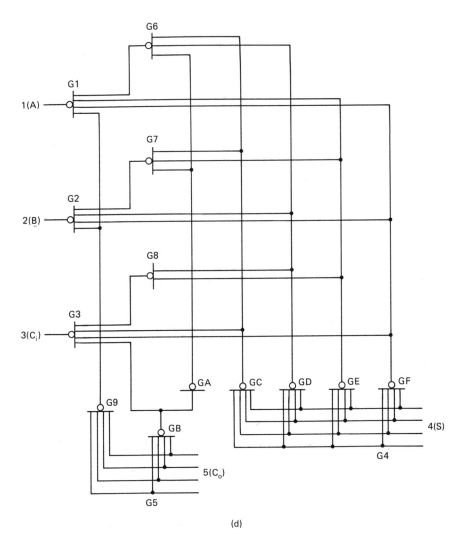

Figure 10.8(d) (*cont.*)

By contrast, the voltage swing of CMOS is normally quoted as being between 30 and 70% of V_{dd} for reliable switching. This means that for $V_{dd} = 5$ V, the voltage swing of CMOS is at least twice that of ECL. (The proposed new 3-V TTL interface standard, if adopted, will essentially eliminate this difference.) If the rate of rise of voltage (volts per nanosecond) were the same for both an ECL input and a CMOS input, the ECL input would reach its switching threshold sooner, all else being equal.

10.2.3 Two Major Families of ECL Parts

Two major ECL families of both standard parts and gate arrays are the so-called "10K" and "100K" series. The former, offered by such companies as Fairchild and Plessey, offers internal gate propagation speeds in the $\frac{3}{4}$-ns range. (Typical numbers are in the range 300 to 400 ps for some processes.) The internal propagation delays of the 10K family are roughly twice that. However, because each macro of the 100K family, as implemented, for example, on Fairchild's gate array, can be programmed during design for one of three different power levels and delay, arrays that employ this feature cannot be classified as simply either 10K or 100K.

Companies offering ECL gate arrays include Motorola (which introduced the family of standard 10K parts) and its second sources National Semiconductor and LSI Logic, Fairchild, Plessey, Signetics, and AMCC.

10.2.4 User-Designed ECL Macros

Unlike other types of logic families, the gate array user has far less chance of designing ECL macros than those of other technologies. The reason is that the major reason for the use of ECL is raw speed. The latter, however, is dependent not only on the amount of power dissipated by an ECL gate and the logic swing but also on the layout. The layout is much more critical for ECL than it is for other technologies, generally because it affects the *raison d"etre* (reason for being) of the ECL gate array. For this reason, the few ECL vendors who allow the user to design the transistor portions of macros often insist on doing the layout as well. (Layout can be critical for the other technologies as well when they are being pushed for speed. However, the forte of the other technologies is not speed.)

The reader should be made aware that the willingness of a gate array vendor to allow a user to design macros (or to do other tasks) is based on economic considerations. The economic considerations include the expectations of future business with that customer. They also include being able to use the customer-designed macro for designs of other customers. The gate array vendor usually also makes a judgment of how much person power will be required to check and characterize the user-designed macro and to interface with the user doing the design.

10.2.5 Basic ECL Structure

The basic structure of ECL is the differential pair as shown in Fig. 10.9. The differential amplifier consisting of Q_0 and Q_1, Q_2, Q_3, and Q_4 is the current steering element that provides the actual logic gating of the circuit. Outputs are taken from the load resistors R_1 and R_2. Although three transistors are shown in parallel with Q_0, only one needs to be there. The three inputs are NORd together at R_1. However, note that the complement (i.e., the OR function) is available at R_2 and that *no additional stage delays are required to obtain the complement*. This is one reason why multiple stages of delay that would exist if the circuit were implemented in CMOS or TTL are obtained in *one* stage of ECL.

The current source is a "stiff" current source which typically consists of a 1-kΩ resistor with 400 μA flowing through it. (The 1 kΩ is much greater than the few to several kilohms "on" resistance of the transistors.)

Just as ISL/STL can be thought of as AND/NAND-type logic, ECL can be thought of perhaps somewhat as OR/NOR-type logic. This is indicated in Fig. 10.9. Multiple signal transistors in parallel with each other form one leg of the differential

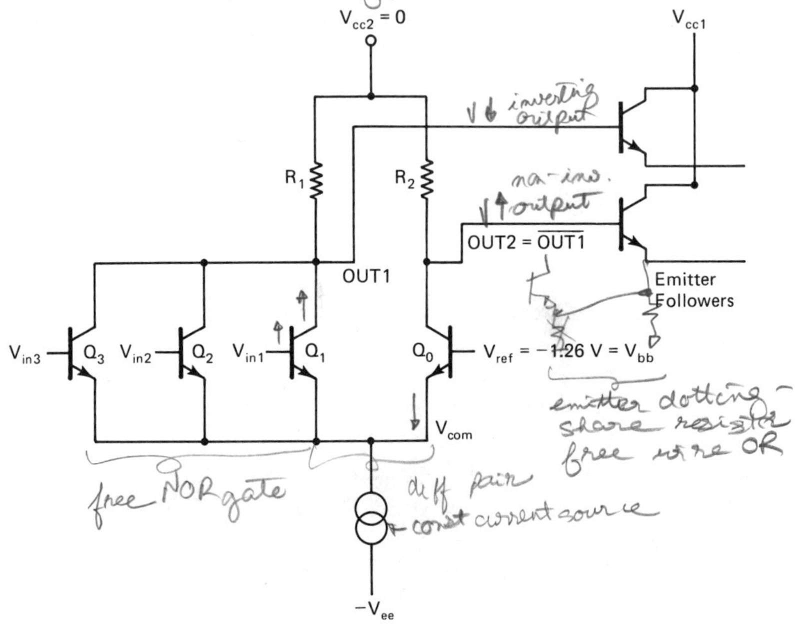

[Figure 10.9 — circuit diagram with handwritten annotations:
"wire OR = weak 0, strong 1"
"wire AND = weak 1, strong 0"
$V_{cc2} = 0$, V_{cc1}, "inverting output", "non-inv. output", OUT2 = $\overline{\text{OUT1}}$, Emitter Followers, OUT1, R_1, R_2, V_{in3}–Q_3, V_{in2}–Q_2, V_{in1}–Q_1, Q_0, $V_{ref} = -1.26\,V = V_{bb}$, V_{com}, "free NOR gate", "diff pair — const current source", "emitter dotting — share resistor — free wire OR", $-V_{ee}$]

- Constant current is steered through either reference XR Q_0 or input XR(S) Q_1, Q_2, Q_3
- Common-emitter voltage V_{com} is clamped one V_{be} drop or ~ 0.8 V below V_{ref} when current flows through Q_0
- If $V_{in} > V_{ref}$, current goes through Q_I not Q_0
- Output (OUT1) and its complement are always available because of the differential nature of the device
- Example $V_I = -1.6$: current is through Q_0 and $V_{com} = -2.06$ V
  $V_I = -0.8$: current is through Q_1 and $V_{com} = -1.6$ (which is < one V_{be} drop below V_{ref})

Figure 10.9 Basic structure of ECL. $I = 1, 2, 3$ (Courtesy of Signetics.)

pair, as described above. Any one of the transistors will produce the change in output state.

The internal voltage- and temperature-compensated bias networks may be added to the basic structure shown. There is a trade-off between adding circuit elements to obtain temperature compensation and the speed of the device.

The output stage consists of emitter followers. When the AND/NAND function is obtained from series gating or collector dotting, the respective emitter followers may be dotted (tied) together to form the wire OR of the product terms. The emitter followers provide level shifting from that of the differential outputs to the voltage levels compatible with the drive requirements of another current switch (a differential "pair," although there could be more than one input transistor, as shown above). The emitter followers also provide isolation to the collectors of the differential pair and a low resistance (on the order of 7 Ω typically) for driving the nodal capacitances of succeeding stages.

In some cases, the emitter followers are placed in the ECL gate array cell so that they can be connected as inputs rather than outputs and thus provide the same level shifting function. Fairchild does this with their GE1000 gate array as does AMCC with their Q700 series. (Fairchild also has emitter followers in the internal gate array cell which can be connected as output emitter followers.)

Figure 10.10 shows the switching characteristics of ECL for the MECL (Motorola ECL) 10K series of devices. The cross-hatched areas indicate the range of specification. Note that the reference voltage to Q_0 is roughly half the voltage swing. Figure 10.11 shows how the transfer curves vary over the temperature range 0 to 75° C.

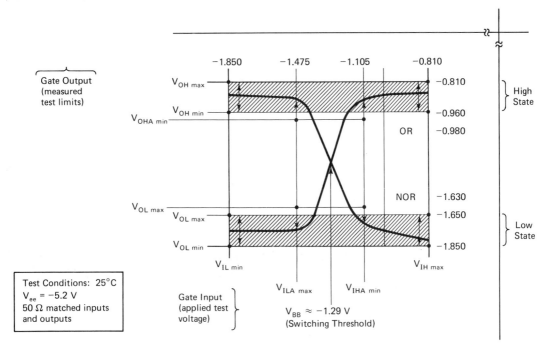

Figure 10.10 Switching characteristics of ECL. (Courtesy of Motorola.)

10.2.6 Delay-Saving Methods in ECL

Three of the methods mentioned above that are used in creating logic functions will now be discussed. The three are series gating and collector and emitter dotting. The other two techniques were discussed above.

Series gating. The OR/NOR function was obtained by paralleling transistors. Series gating is a convenient way to obtain the NAND/AND functions in ECL. Both the AND and the NAND are available for the same reason that the OR/NOR functions are available: that is, because the ECL outputs from R_1 and R_2 are complementary.

In series gating or cascode, in effect, one current source is used to drive two or more sets (often but not always pairs) of transistors which have the property that the

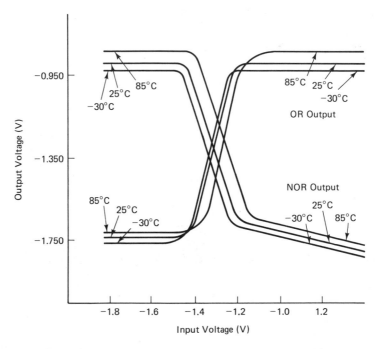

Figure 10.11 Switching characteristics of ECL over temperature. (Courtesy of Motorola.)

collectors of one set are connected to the emitters of the next higher set. Figure 10.12 shows this arrangement. It can be seen how the different product terms are formed. For output A · B to go low, for example, inputs A and B must both be high, as one would expect.

The series gating in Fig. 10.12 is between Q_2/Q_3 and either Q_4/Q_5 or Q_6/Q_7. For each level of series gating, a different reference level is required. This generally limits the number of levels (and hence the number of inputs to a AND/NAND gate) to three.

Collector dotting (wire ANDing). Another method for obtaining the AND/NAND function is the technique of collector dotting. This technique is shown in Fig. 10.13. Connecting the collectors of Q_1 and Q_2 together produces the wire AND as shown.

Emitter dotting (wire ORing). By tying the emitter-follower outputs (not those that are complements of each other) together, the wire OR function can be obtained. Under some logic conditions, tying more than four or five of these outputs together can cause spikes. Nevertheless, the technique is useful when care is exercised. Each such output tied together typically adds about 0.1 ns to the propagation delay.

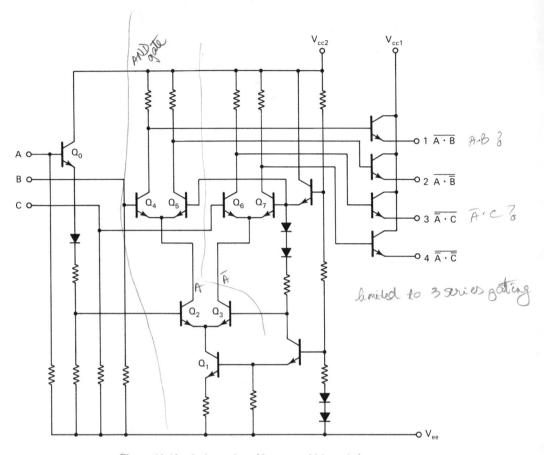

Figure 10.12 Series gating. (Courtesy of Motorola.)

10.2.7 Examples of ECL Gate Array Cells and Common Functions Made From Those Cells

What is analogous to an array cell in CMOS is a grouping of resistors and transistors in ECL. Typical groupings from different manufacturers are shown below.

Fairchild. Figure 10.14 shows the internal structure of the Fairchild GE1000 series array. The reader will note the input emitter-follower groupings q_1, q_2 and q_3, q_4; the four differential pair groupings (G_1, G_2, G_3, and G_4); the emitter followers, q_5, q_6, q_7, and q_8, which would generally be used for outputs; and the reference and load resistors, R_{ref} and R_1, respectively. The three reference resistors allow the macros made from this cell to be power/speed programmed at the mask level.

The transistors in this case are all of one polarity, *npn*, unlike the CMOS case, in which both PFETs and NFETs are present. An example of a macro made with this cell is that of the D flip-flop with enable shown in Fig. 10.15. A second example using the Fairchild GE1000 gate array is shown in Fig. 10.16.

Sec. 10.2 ECL

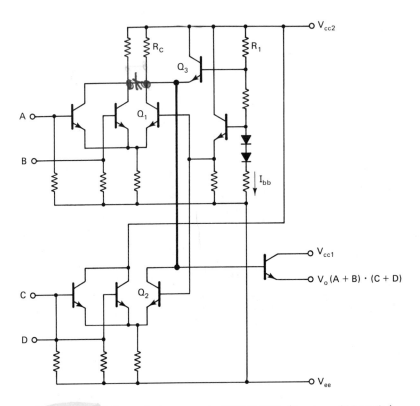

Figure 10.13 Collector dotting to create AND/NAND. (Courtesy of Motorola.)

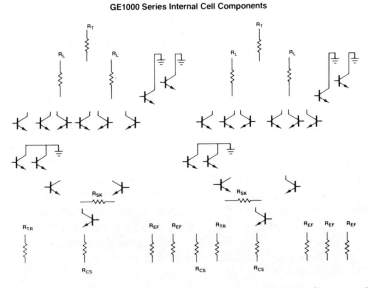

Figure 10.14 Fairchild GE2000 cell showing transistor arrangement. (Courtesy of Fairchild.)

450 Design of Bipolar Macros Chap. 10

D Flip-Flop with Reset

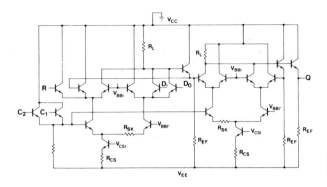

Figure 10.15 D flip-flop macro. (Courtesy of Fairchild.)

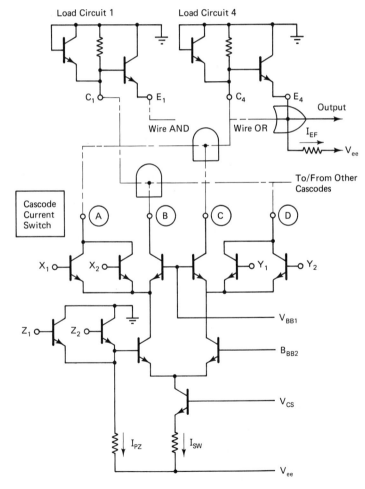

Figure 10.16 Example showing wire OR, wire AND, and series gating. (Courtesy of Fairchild.)

Sec. 10.2 ECL

This is an example showing the wire OR, the wire AND, and series gating. The series gating is between the transistors, with V_{bb2} as a reference voltage and $x_1 + x_2/y_1 + y_2$. (The OR/NOR combinations $x_1 + x_2$ and $y_1 + y_2$ are series gated with the OR/NOR combination of z_1 and z_2. The latter goes through an emitter follower to the base of the transistor, which forms a differential pair with the transistor, whose base input is the reference voltage V_{bb2}. Other functions are as labeled.

Four other macros using the Fairchild GE1000 array are as shown in Fig. 10.17. Figure 10.17(a) is that of a 4-to-1 MUX (multiplexer). In this figure, A_1 and

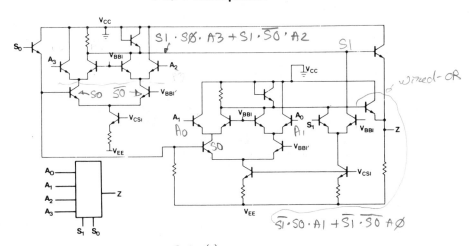

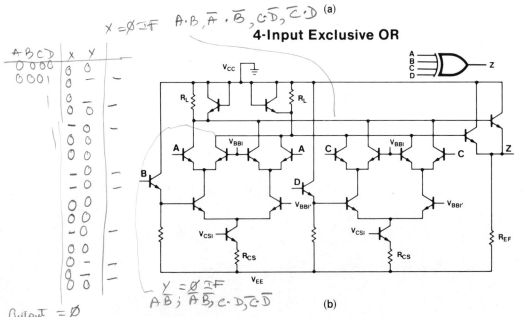

Figure 10.17 Other ECL macros: (a) 4-to-1 multiplexor; (b) 4-input EXOR; (c) adder; (d) 5-input OR/NOR. (Courtesy of Fairchild.)

Full Adder

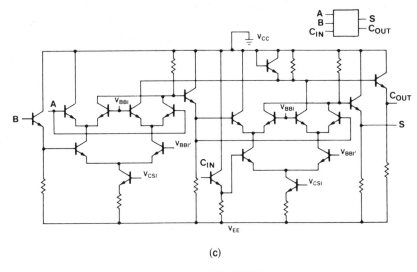

(c)

5-Input OR/NOR

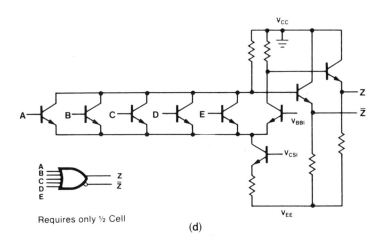

Requires only ½ Cell

(d)

Figure 10.17 (*cont.*)

A_3 are series gated by selection signal S_0 acting through a level-shifting transistor. A_2 and A_4 are series gated by S_0 in the same way in the next block of transistors. Note that there are two reference voltages, V_{bb1} and V_{bb1}'. Signal S_1 is collector dotted with the reference transistor for A_2 and A_4. The other member of the differential pair containing S_1 is similarly collector dotted with the reference transistors of A_1 and A_3. It is left as an exercise to the reader to determine how each of the macros in Fig. 10.17 is formed from the array cell.

AMCC. The AMCC array cell (called a function cell by AMCC to distinguish it from another internal array cell used on some of their ECL gate arrays) is

Sec. 10.2   ECL   453 shown in Fig. 10.18. The reader will note the group of emitter followers, G_1, the unconnected transistors, and the resistors. An example of the use of this cell when configured as a 2-to-1 MUX (multiplexer) with enable is shown in Fig. 10.19. The block-level schematic for the MUX in Fig. 10.19 is shown in Fig. 10.20.

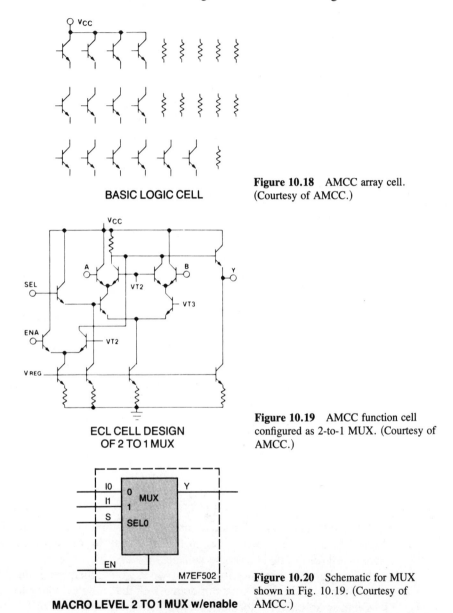

Figure 10.18 AMCC array cell. (Courtesy of AMCC.)

Figure 10.19 AMCC function cell configured as 2-to-1 MUX. (Courtesy of AMCC.)

Figure 10.20 Schematic for MUX shown in Fig. 10.19. (Courtesy of AMCC.)

Motorola array cell structure. The internal cell structure of a Motorola gate array is as shown in Fig. 10.21. It is left as an exercise for the reader to compare it to the structures of the other two vendors.

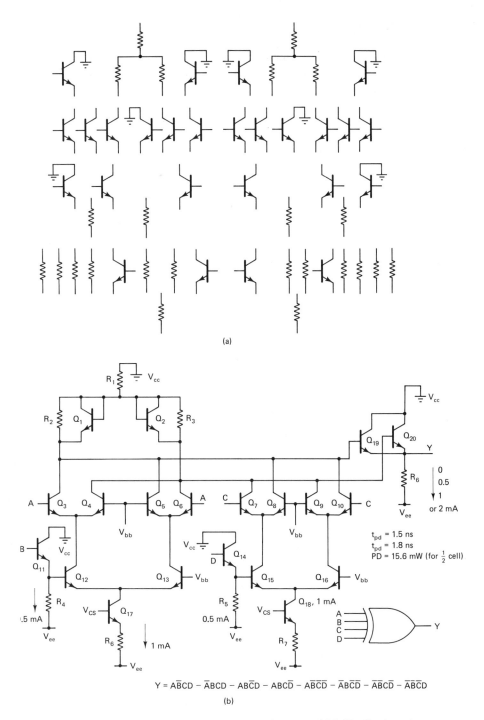

Figure 10.21 Internal cell structure of Motorola arrays: (a) half-cell schematic; (b) four-input exclusive OR gate. Compare to that of Fairchild, Fig. 10.16. (Courtesy of Motorola.)

Sec. 10.2 ECL 455

10.2.8 I/O Structures

I/O structures in ECL take one of two forms. One form is that which interfaces directly to ECL. The other form interfaces to some form of TTL (generally LSTTL). The latter interface also applies to CMOS because CMOS can interface to TTL also (see Chapter 9). The problem with the TTL interface is that level shifting is required. The level shifters are macros on the ECL gate array which add propagation delay to the I/Os. Some ECL arrays from AMCC, Fairchild, and so on, will run entirely from a 5 V and ground power supply. In this case, the voltages are *not* shifted down.

An example of a TTL-to-ECL level shifter is that given in Fig. 10.22. The typical propagation delay added by this translator is 600 ps. Fairchild uses emitter followers on the inputs of internal cells instead of the outputs. When the level shifter feeds directly into a lower-level (in the series gated "tree") current switch, this 600-ps delay is essentially not present. The reason is that the emitter followers which have a comparable time delay are not used. An example of an ECL-to-TTL level shifter is given in Fig. 10.23.

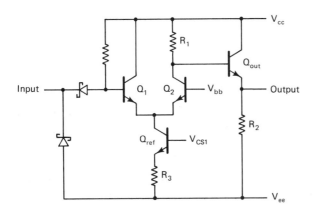

Figure 10.22 Example of TTL-to-ECL input translator. (Courtesy of Fairchild.)

The other class of I/O devices is that required to interface directly with ECL. An example of an ECL input translator is given in Fig. 10.24. This circuit is used when the input goes to the lower level of a series gated switch. The voltage levels for such switches which are lower in the "tree" are less than those above them. Inputs not going to the lower levels do not need this translator. However, it should be noted that the GE1000 of Fairchild uses only two levels of series gating. An array that uses more than two levels will require shifters that shift to the different levels.

Other common cells on ECL gate arrays are ECL receivers, differential receiver drivers, three-state devices, and transceivers. These are sometimes made from cells specially allocated to that purpose and sometimes from array cells.

In addition to the above, most ECL gate arrays have some form of bias generator built into the array. These are necessary to provide the reference voltages for series gating. Normally, these are distributed around the array in some fashion to min-

ECL to TTL Output Interface Circuits

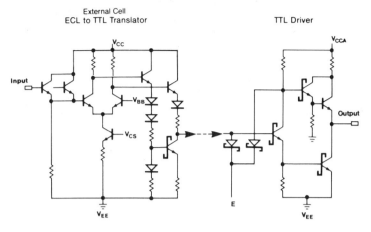

Figure 10.23 Example of ECL-to-TTL level shifter. (Courtesy of Fairchild.)

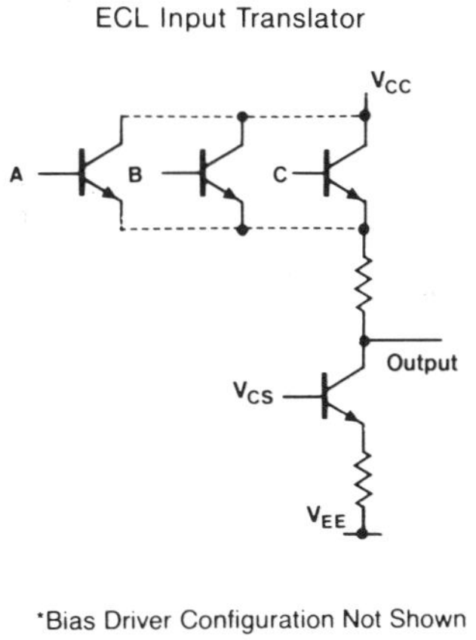

Figure 10.24 Example of ECL-input level translator. (Courtesy of Fairchild.)

imize the effects of voltage drops in the lines. (However, because these bias generators provide drive currents to the *bases* of transistors, the currents, and hence the voltage drops, are quite small for reasonable lengths of interconnect.)

10.2.9 Equalizing Power on the Chip

Because ECL is such a high-powered technology, some manufacturers require the designer to equalize the power dissipated in each quadrant to within ±10% of that dissipated in the other quadrants. This means that the designer must partition the circuit layout to accommodate this requirement. The user performs a manual calculation to ensure that this requirement is met. Some auto P&R programs will accommodate this.

It should be noted that the smaller propagation delays due to lower voltage swings and output impedances of ECL have to be compared against propagation delays of the gates themselves, which are lower than those of CMOS. Therefore, the *percentage* decrease in propagation times due to these effects being less than those of CMOS is not as great as it would be if the propagation delays of CMOS and ECL for a multigate structure were equal. The term *multigate structure* implies the combination of a number of logic functions into essentially one gate delay in ECL.

As noted in Section 2.7.4, some ECL arrays use overhead power even if there is *no* functionality of the chip. This aids in equalizing power on the quadrants (although it is deleterious otherwise and is a design imperfection).

10.3 SUMMARY

The dominant bipolar technologies for gate array use is ECL. Also used are ISL/STL. The makeup of macros in both technologies is described in this chapter. The utility of the wired functions in ISL/STL is shown. The five different methods of obtaining multiple-stage delays in one level of ECL logic are described.

REFERENCES

1. Firstenberg, A. and S. Roosild, "GaAs ICs for new defense systems offer speed and radiation hardness benefits," *Microwave Journal,* Mar. 1985, pp. 145–63.
2. Lohstroh, Jan, "Integrated Schottky Logic," Ph. D. Thesis, Eindhoven University of Technology, Nov. 13, 1981.
3. *ISL Gate Array Design Manual,* Sunnyvale, CA: Signetics, Dec. 1980, Appendix A.

Chapter 11

Processing, Packaging, and Testing

11.0 INTRODUCTION*

Processing, packaging, and testing have been discussed in other chapters where it was appropriate. The purpose of this chapter is not to repeat the previously given material but rather to tie together these three important topics. The three topics are related in that they differ very little when used with the gate array methodology from when they are used to produce standard parts. The largest difference is in the processing. Processing of a gate array wafer is a subset of the processing done on a standard off-the-shelf part.

It is appropriate to include material of this nature because many user organizations develop their own facilities to do some of the final processing of gate array wafers. Moreover, such knowledge is useful, if not essential, when talking with gate array vendors about deliveries and schedules.

The starting point for this chapter are the files containing the interconnects. These will typically encompass designs for two layers of metal, although both single layer and multiple layers of metal are used by some vendors.

Chapter 4 included an overview of the processing steps. These steps can be briefly summarized as follows.

- Make mask(s) that represent the metal interconnects. Each mask represents a level of interconnect. (There are also insulator [between metals] and passivation masks.)

*See Chapter 2 references.

- Use masks to define patterns on a wafer (i.e., finish processing the wafer).
- Test the wafer; identify a good die.
- Saw the wafer to separate the die.
- Package the good die and test the packaged units.

These steps will now be discussed in detail.

11.1 MAKING THE MASK [1], [2], [3]

The first step in making a mask is to make a PG (pattern generator) tape from files containing interconnect data. This task is normally done on an IGS and was discussed briefly in an earlier chapter. The interconnect lines are converted to rectangles, and the resultant rectangles are converted to elements that the pattern generator can use.

The PG is normally used first to make *overlays* or *blow-backs*. These are typically 100× (times) enlargements on Mylar of the actual masks. The overlay is compared to a 100× plot of the original design. If the two match, a *reticle* is made by the same data base on the same PG that produced the overlay. A second indicator of whether the data on the overlay (and later the reticle) is complete is the rectangle count produced by the PG software of the IGS. This can be compared to a similar number generated by the PG. The rectangle count for a typical 2000-cell (three inputs per cell) array is on the order of 95,000 rectangles.

The reticle is a replica of the design on a glass plate. Typically, it is 10× for non DSW (see Section 11.2.2) and 1×, 4×, or 5× for DSW techniques. The reticle is stepped and repeated (in a machine appropriately called a *stepper*) to form the master mask. The master mask (1×) is then copied to form a *submaster,* which is then copied to form one or more *working plates*. When only low quantities of parts are to be produced (as is typical of gate arrays), the working plates and the submasters may be one and the same.

Two major types of masks are used. The less expensive are emulsions on glass. More expensive but more typically used for feature sizes of under 5μm are those made of chrome on AR (antireflecting) glass. Most expensive of all (up to $2000) are the 1× masks used with DSW machines. Because a 1× reticle is stepped across the wafer, a given defect on the recticle is patterned full size on the wafer instead of being shrunk by the 4× or 5× magnification factor when the larger DSW masks are used. Hence much more care must be used in their preparation.

Masks used for precision high-volume work sometimes employ *pellicles*. These are glass coverings of the mask which are placed out of the focal plane of the aligner.

In converting the data files on the IGS to a PG tape, attention must be paid to the type of PR (photoresist) being used. The slight swelling or contraction due to the PR must be accommodated by scaling the conversion appropriately.

Positive PR will shrink when developed. Therefore, the lines on the mask must be made larger. For example, to get 5-μm lines with a positive PR, the drawn sizes must be 6 or even 7 µm. (The actual value also depends on the thickness and viscosity of the PR, development time, and other factors.)

Reticles vary in cost from a few to several hundred dollars each, depending on size and mask vendor. Overlays are typically $35 to $100 each. Chrome plates are typically $30 each, with minimum quantities of two or more. Emulsion plates (masks) are about $6 each. Turnaround times can vary from a few working days to (horrors) several weeks. Two weeks is typical. Most gate array vendors have reserved slots at the mask house being used. Therefore, their turnaround is likely to be faster than that of a user.

Figure 11.1 is a picture of a mask. The resultant wafer was shown in Chapter 1.

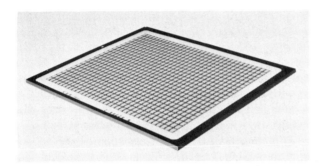

Figure 11.1 Mask.

11.1.1 Runout

It is very important to use the same mask-making machines that produced the original masks of the gate array vendor which were used to fabricate the underlying layers of the gate array wafer. The reason is that each mask machine has a tolerance buildup in its own peculiar amount and direction as it steps across the plate to form the reticle or the master mask. This is called *runout*. Masks made on different machines will have different runouts, which in turn will cause yield to decrease. (Some elements and some dies will be in alignment and other parts will not.) A typical *golden oldie* pattern generator is the D.W. Mann 3000. DSW machines are made by GCA-Mann Ultratech, Optimetrix, Canon, and others. Masks for some of these machines are made using electron beam (E-beam) lithography.

11.1.2 Mask Repair

Because of the high cost of some masks, efforts have been directed toward mask repair. Companies such as Micron Corp. and Ion Beam Technology both in Boston have introduced focused-ion beam machines for this purpose. Some of the techniques were developed at Bell Laboratories. Typically excess material on the masks is knocked off by bombarding the material with ions. Other machines used for mask repair use lasers to thermally destroy excess material. [4]

11.2 PROCESSING

Processing a gate array wafer to customize it involves using masks to define metal layers on a wafer. The first metal layer is defined by the patterning and etching process outlined in Section 11.2.2. An insulating layer is placed between each metal layer and patterned with the openings for the contacts (called "vias") between metal layers. A second metal layer, if used, is then put down, and the process is repeated.

The facilities and equipment described in Section 11.2.1 are those that might be found at a typical user organization or a small gate array vendor. Although 24 to 60 wafers per hour of one design could be processed in such a facility, the main purpose of the facility is to enable small quantities of parts of many designs to be put rapidly in the hands of designers.

11.2.1 Clean Rooms

Processing is done in *class X clean rooms*. X is the number of particles per cubic foot which are greater in diameter than 0.3 μm. Typical values for clean rooms are class 100 and class 10. Some vendors, such as AMCC, run even less. (AMCC runs typically class 2 in most of their area.) As line spacings and feature sizes shrink, more rooms will have to be at the lower figures to obtain satisfactory yields.

Figure 11.2 is a picture of the inside of a typical clean room. Everyone in the room is gowned from head to foot in nonparticulate emitting clothing called "bunny suits." (If the "recipe" for processing is ever lost, the suits are sometimes called "dumb bunny suits.")

Figure 11.2 Inside a clean room. (Courtesy of Sanders Associates, Inc.)

A floor plan of a typical gate array processing facility is shown in Fig. 11.3. The main fabrication areas are at a slight positive pressure differential with respect to the gowning areas of typically 0.1 in. of water. The gowning area is at a similar positive pressure differential with respect to the outside of the facility. These measures prevent outside dust from entering.

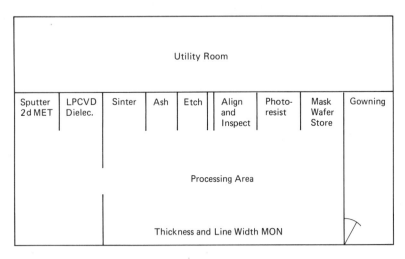

Figure 11.3 Floor plan of typical clean room for doing final processing of gate array wafers.

The air inside the clean rooms is tightly controlled with respect to both temperature and humidity. This is necessary to obtain repeatable results. Chemicals such as PR (photoresist) have properties that are strongly dependent on temperature and humidity. For this reason, anything that does not require a clean-room environment is put in a separate room called a *utility room* or "chase." The utility room is often larger than all of the fabrication areas combined. It contains, for example, a closed deionized water system; protective cabinets for all the gases used; vacuum pumps; the air-conditioning apparatus; and manifolds for vacuum, water, liquid nitrogen (used in blowoff guns and bubblers), and so on.

Everything in the clean room is aimed at keeping the particulate count as low as possible. Special papers that do not emit particles are used. Personnel are prohibited from using cosmetics. Finishes of all materials are sealed where necessary. Lights inside the clean room are yellow because photoresists have minimal sensitivities to that wavelength.

In Fig. 11.2 and subsequent figures, the reader will note the openings in the plexiglass which separates one work area from another. This enables the wafers to be passed from one workstation to the next without having to get into the area occupied by personnel.

At least two types of filters are used to filter the air in clean rooms. The first is often called an "elephant" filter. Its purpose is to catch any large (relatively) particles that might damage the finer mesh filters. The latter are called high-efficiency particulate air (HEPA) filters.

11.2.2 The Process Itself

The process itself can be described succinctly as follows.

1. Take a gate array wafer from stock (it is coated with metal, but metal does not at this point form a good ohmic contact with the silicon).
2. Coat it with PR. This is done in a *spinner*. A drop of PR about the size of a quarter is placed on the wafer mounted on a vacuum chuck in the spinner. The wafer is then rotated at a rate determined by the viscosity, thickness, temperature, and so on, of the PR to cause the PR to be evenly distributed across the wafer by means of centrifugal force. Figure 11.4 shows PR being manually applied. (Automatic dispensers are also common.) A filter-tipped syringe is used. The PR is then dried.

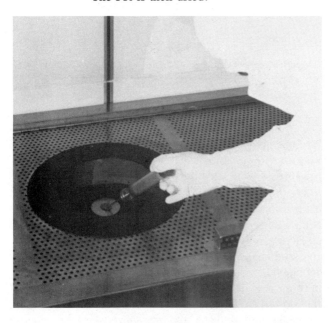

Figure 11.4 Photoresist being manually applied to wafer mounted on spinner. (Courtesy of Sanders Associates, Inc.)

Both positive and negative PRs are used. A positive PR expands slightly when developed, whereas a negative PR does just the opposite. The effect also depends on development time. Examples of positive resists are the Shipley AZ1350, 1400, 1450, 1470, and 2400 series; the Kodak Micro resist 809; the MIT Superfine IC 528; and the Hunt Waycoat HPR 204 and 206. Properties of these resists are given in [5].

3. Align the mask to the wafer and expose the pattern of interconnects of that layer. Several techniques are used here. The process can be performed using

either a projection or a contact "aligner." In the latter case, the mask is in contact with the wafer. This permits better resolution by essentially eliminating the *penumbra* (image spreading due to refraction and diffraction). The disadvantage of the contact aligner is that the mask needs frequent cleaning.

Aligners are mounted on vibration-isolated tables. Some DSW machines, such as the MANN 4800, have their own environmental chambers.

Ultraviolet light at typically 2600 or 3000 Å (wavelength) is passed through the mask to form the image of the pattern on all dies simultaneously if DSW machines are not used. The wafer has previously been covered with PR.

Figure 11.5 is a picture of a contact aligner. The technician is holding the mask. The wafer is placed on a vacuum chuck and maneuvered under the mask using gross and fine controls.

In DSW (direct step on wafer) machines, the reticle itself is imaged on the wafer, and the image is stepped and repeated across the wafer. DSW machines are made by GCA-Mann (model 6300A), Ultratech, Perkin-Elmer, Optimetrix, and Canon, among others. These techniques avoid the degradations that occur in going through successive stages of lithography (i.e., from reticle to master to submaster to working plates). Another benefit of doing this is that warpages in the wafer [which can be as much as 5 to 10 μm (Ghandi p. 536)] can be accommodated.

DSW machines are useful down to about $1\frac{1}{4}$-μm feature sizes. They are "projection" aligners. The reticle that is stepped may contain more than one replication of the die.

A third technique is the use of electron-beam (E beam) machines to pattern the wafer directly or to make the mask. Often, they are used to make the reticle for the DSW machines.

A fourth technique not yet used for gate arrays is x-ray lithography. The very short wavelengths of both E beam and x-ray essentially eliminate the penumbra in production systems and thus permit finer resolution.

These techniques are very expensive. A typical E-beam machine is twice as expensive as a DSW, which in turn is two to four times as expensive (very roughly) as either a PG or a step and repeat. The PRs used for E beam, such as PMMA (polymethyl methacrylate), are much more expensive than those used for optical lithography. However, the increased yields that accrue from the use of such machines coupled with their growth potential has made it attractive to use them. IBM uses the inherent flexibility of the E beam to expose two to four different gate array designs on one wafer.

4. Develop the PR (remove it from areas where metal is not to be protected). This is typically done in a spinner (not the same one used to deposit the PR to begin with).

5. Etch (i.e., remove) metal to define metal lines for that metal layer. This is done either in a planar plasma etcher (dry processing) or by using wet chemicals. A typical formula is (by volume) 16 parts phosphoric acid to 1 part acetic acid to 2 parts water and 1 part nitric acid. This etches aluminum at about 2000

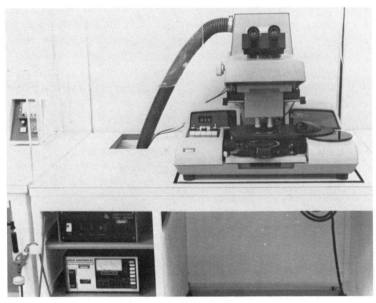

(a)

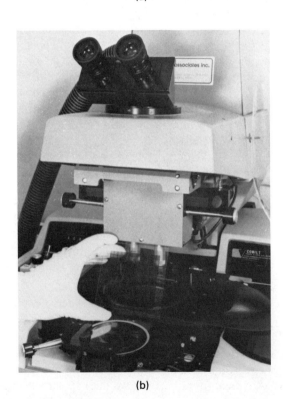

(b)

Figure 11.5 (a) Contact aligner. Control for maneuvering wafer under mask is the dome-shaped object to the right of the mask. (b) Mask before mounting in aligner. (Courtesy of Sanders Associates, Inc.)

Å/min. Therefore, a wafer with 1-μm (10,000 Å)-thick aluminum lines will take about 5 minutes to etch. This emphasizes the point made in an earlier chapter that the bulk of processing time is setup and queue time, not actual "action" time.

Figure 11.6 is a picture of a typical wet etch station. The right side is used for metal etch. The left side is used for etching silicon dioxide used as the insulating layer between metal layers. Barriers prevent interaction of the chemicals used. Precision timers with annunciator horns are overhead. The resistivity of the incoming and outgoing deionized water (used to rinse off the chemicals) is monitored on a meter. Rinsing is done until the final rinse water is close to the 18 MΩ-cm resistivity of the incoming water.

Liftoff versus etching. It is probably safe to say that most patterning processes used in production today of both gate arrays and standard parts use either wet (chemical) or dry (plasma) etching. An alternative process that has been used for many years in the laboratory is that of *lift-off*. In this process, photoresist is used to cover areas where metal is *not* wanted. The wafer is then covered with metal. The remaining photoresist is then expanded by application of a solvent, such as acetone. The swelling of the photoresist breaks the metal where metal covers the steps of the photoresist. The photoresist is then lifted off, carrying the undesired metal with it.

An advantage of the lift-off technique is that it takes advantage of what is normally a very deleterious problem—poor step coverage. For this reason, as aspect ratios of metal lines become greater (in terms of height to width), lift-off may become used more often in production.

Etching metal using lasers. Lasarray Corp. in Irvine, California, offers CMOS gate arrays whose metal can be etched using a laser system which they also build. The claimed advantage is that two hours is required to etch a single layer of metal. Another advantage is that up to four designs can be put on a single wafer. The system *eliminates the need for a mask* because the laser can be run directly off the PG tape. This is obviously a great advantage. (Expensive E-beam machines can also do the same thing.) Current designs use an 8-μm metal pitch.[6]

6. Remove PR which has protected the metal up to this point in areas in which the metal lines are to exist on the chip. This is typically done in an *asher* which is a plasma etcher using welder's-grade oxygen. Figure 11.7 is a picture of a typical unit.
7. Sinter (make good ohmic contact of metal to silicon). To permit removal of the metal, it has not been strongly bonded to the silicon up to this point. Sintering accomplishes such bonding. Typically, the wafer is placed in a precisely controlled furnace for 10 minutes at 425°C (or a lower temperature for a longer period). Figure 11.8(a) shows such a furnace. In this case, it is a diffusion tube used because of the precision involved (even though no impurities will be introduced).

Steps 6 and 7 are common for both single- and double-layer metal systems.

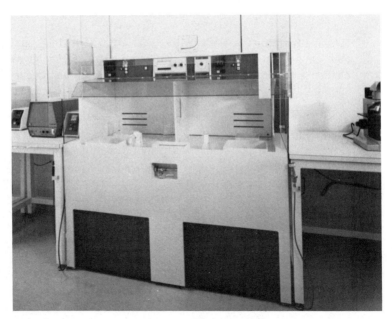

(a)

(b)

Figure 11.6 (a) Typical wet etch station. Vacuum wand for picking up wafer and dry nitrogen blow-off gun are indicated by the vertical hoses on the left and right, respectively. These are used at all stations. (b) Wafer in holder over tank. (Courtesy of Sanders Associates, Inc.)

Figure 11.7 Asher used to remove photoresist. (Courtesy of Sanders Associates, Inc.)

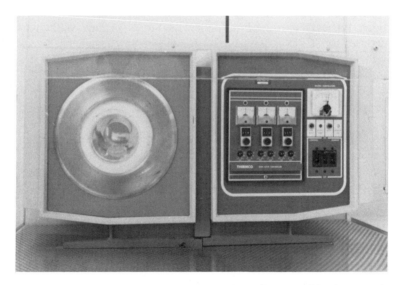

Figure 11.8 Furnace used to sinter metal to silicon. (Courtesy of Sanders Associates, Inc.)

8. Cover with an insulator [glass (SiO_2)] if a second layer of metal is to be used. If only one layer of metal is used, cover with a protective coating [polyimide or silicon nitride (SiN_3)].

   Both the SiO_2 insulator and the SiN_3 protective coating can be deposited in a LPCVD (low-pressure chemical vapor deposition) machine, such as the Pacific Western Coyote. Polyimide can be spun on using a spinner.

9. Pattern and etch glass or protective coating using the masking and exposure

steps described above. Glass is etched using either buffered or unbuffered hydrofluoric acid. The former is six parts ammonium fluoride to one part hydrofluoric acid. This etches non-CVD-deposited glass at about 1200 Å/min. Hydrofluoric acid by itself (no ammonium fluoride) diluted 10:1 with water will etch CVD oxide at about 1800 Å/min. Therefore, it takes only a few minutes to etch the glass.

10. Put down a second layer of metal if used, coat it with PR, and repeat the masking, exposure, and etching steps. If a second layer of metal is not used, send the wafer to a wafer test (see Section 11.3).

Metal can be deposited using either evaporation or sputtering techniques. Figure 11.8(b) is a picture of a sputterer.

The wafer is now ready for a wafer test. This will identify dies on the wafer which appear to be functionally good.

11.3 WAFER TEST [7], [8]

As with packaging, testing is similar to that of "jelly-bean" ICs with a few exceptions. First, the user generally writes the test procedure. While writing such test procedures for ICs is a new activity for most logic designers, it really does not differ radically from writing such procedures for the printed circuit boards that the gate array replaces. One very important difference (and the first that usually comes to the logic designer's mind) is that the inside of a gate array chip cannot be probed with a scope probe the way that a PCB can.

There *are* ways in which the gate array die can be probed internally with microprobers, such as that made by Wentworth. The lines, however, should be spaced several line widths apart or 7 to 15 μm at least. However, the reader should not count on being able to probe any gate array die internally. Normally it is not done; it generally destroys the line, it is very time consuming, and it is strongly *not* recommended. It is mentioned only because it is an option when all else fails. The same prober can, with the same caveats, *cut* such lines if necessary.

A second difference between the jelly-bean IC and the gate array IC is that the user can define the signal paths that will be brought out to the bonding pads. The user must note: *Not every signal that is brought to a bonding pad need be bonded.* The usefulness of bonding pads used for test points will be discussed below.

11.3.1 Wafer Probing

In lieu of probing the gate array die internally, test points can be brought out to the unused bonding pads on the periphery of the die. For example, the IMI 2000-cell gate array die has 116 bonding pads around its periphery. If this die is put in a 40-pin package, there are almost 80 points not used for bonding to which test points can be brought. Care must be taken that the signal lines so brought out do not load the circuit being tested. Otherwise, false readings will result, or worse yet, the circuit will not perform as well as it otherwise would.

Tom Mortimer of Sanders has suggested that special-purpose test pads could also be made internally on the array by leaving metal in otherwise empty areas of the gate array die. This would avoid the need for what are sometimes long signal runs on the die itself. The pads so built would only be used for development of special functions, not for routine testing. Buffers to drive the wafer probes to be connected to the pads could be made from either internal array cells or from buffer cells around the periphery. Because a minimal pad (test only) size is about 1–2 mils, only a few locations on the chip would be suitable.

Probe cards. The die on a wafer are normally probed using a *probe card*. This is a PCB with traces that converge on the center of the PCB to connect to many "fingers" (probe tips) that physically make contact with the die. The fingers are bent so that they stick out below the probe card. This permits the wafer (which is on a movable vacuum chuck) to be brought up under the probe card in such a way that the fingers contact each of the bonding pads to be tested on one die of the wafer. The pattern of the fingers corresponds to that of the bonding pads of the die being tested. Figure 11.9 is a picture of a probe card for the IMI 1960-cell array. The probe card must be connected to external test circuitry. This can be as simple or as complex as the user desires.

Although one probe card design will work with all designs done on one member of a gate array family, separate probe cards are sometimes bought for specific circuit designs even though the pad layout is the same. Consider two circuits done on the IMI 2000-cell gate array. One of them may utilize, for example, 75% of the

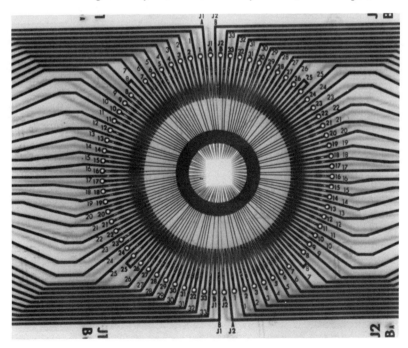

Figure 11.9 Pictures of increasing magnification of probe card. (Courtesy of Sanders Associates, Inc.)

116 bonding pads for either signal connections or for test. The other may utilize a far smaller number. Because high-pad-count probe cards are expensive (around $1000) and because the cost of probe cards is directly dependent on the number of probes, a wise decision may be to purchase a relatively inexpensive probe card for the second circuit. Doing this would save wear and tear on the more expensive card. (Bear in mind that the nature of the gate array business at many user companies which have their own facilities is that many designs with relatively small numbers of parts for each design are done. This implies a lot of setup and tear down of equipment, including probe cards.)

Figure 11.10 is a picture of a wafer probe station. The wafer, mounted on a vacuum chuck, is brought up underneath the probe card. One of the dies is aligned to the fingers of the probe card using axis controls. The size of the die is set via switches on the prober. After external (to the prober) circuitry determines whether the given die is good or bad, the chuck brings the next die on the wafer up under the set of fingers and repeats the test. If a given die is found to be bad, the prober squirts ink on the die. After the test of the entire wafer is finished, the ink marks distinguish the good die from the bad die. The good dies are then separated from the bad by sawing (or less likely these days, scribing) the wafer along the *streets*. The streets are the spacings among the die which have been left specifically for the purpose of die separation. Typically, the wafer is sawed roughly two-thirds of the way through. (This would be a depth of 10 mils on a 14-mil die.) It is then gently rolled in both perpendicular directions typically between two pieces of soft material to effect the separation. The good dies are then packaged.

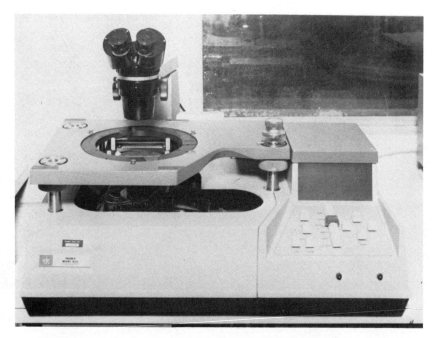

Figure 11.10 Wafer probe station. Switches for setting die size and other parameters are beneath lid above buttons. (Courtesy of Sanders Associates, Inc.)

11.4 PACKAGING

Packaging of a gate array die is no different from packaging any other die. The reason is that gate array die are made via the same processes and sometimes on the same wafer fabrication lines as standard parts. The LSI Logic 5000 series of HS CMOS gate arrays, for example, was made on the same production line of Toshiba that is used to produce the Toshiba 64K DRAM until LSI had its own U.S. facilities built.

The dies on each wafer are separated after probing using a wafer saw using a diamond-tipped blade. This is similar to a radial arm saw except that dimensional tolerancing in all three dimensions is, of course, much more precise.

Figure 11.11 is a picture of such a saw. Each *kerf* or *street* (spacing between adjacent die on the wafer) is sawn about two-thirds of the way through. The wafer is then gently broken. It may also be sawn all the way through using proper backing material.

Figure 11.11 Wafer saw. (Courtesy of Sanders Associates, Inc.)

The number of gross die per wafer is given in Fig. 11.12 and Table 11.1. Most silicon wafers are currently in the 4-in.(100-mm) and 5-in. (125-mm) sizes with a few in the 6-in size. A few manufacturers are experimenting with 8-in. sizes.

Yield as a function of chip size is given in Seed's model shown in Fig. 11.13. Strictly speaking, this model was developed for pure custom and standard dies in which virtually all of the die area is occupied by active and passive elements used in the circuit. As noted earlier, in a gate array die typically only 80% of the active elements are actually used. For this reason, the yield on a gate array die, all else (number of wafer starts per week, same process, same die size, etc.) being equal, the gate array die would be expected to have a somewhat larger yield. The reason is that a given defect has a somewhat lower probability of being in an area which is used.

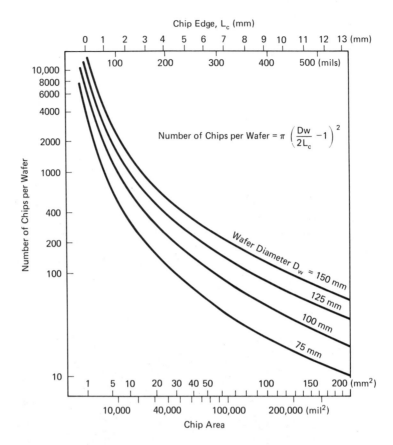

Figure 11.12 Number of chips per wafer. (From Denis J. McGreivy and Kenneth A. Pickar, "VLSI technologies through the 80's and beyond"; registered copyright 1982 by IEEE.)

Die can be packaged in any standard or nonstandard package provided that there is *at least* 20 mils (preferably much more) clearance between all sides of the die and the cavity package sides. A second requirement is that the pin-outs of the chip are configured so as to prevent bonding wires from crossing each other inside the package. The latter problem will occur if the number of chip pads to be bonded on a given side exceeds the number of cavity bonds on that side.

Thermal considerations are the same for gate array packages as for packages containing standard "jelly-bean" parts. Junction temperatures, for example, of bipolar devices should not exceed 175°C (preferably much lower) in military applications and 150°C for commercial applications (unless otherwise specified by the gate array vendor).

Figure 11.14 is a picture of a group of standard packages. Except for the chip carrier (lower right) and chip carrier socket, all the rest are DIPs (dual-in-line packages). Two of the newer package types are chip carriers and pin grid arrays (PGAs).

TABLE 11.1 NUMBER OF CHIPS IN A SILICON WAFER

Chip edge (mm)	Chip area (mm²)	Number of chips per wafer			
		Wafer diameter			
		75 (mm)	100 (mm)	125 (mm)	150 (mm)
1.0	1.0	4,185	7,543	11,882	17,203
2.0	4.0	990	1,809	2,875	4,585
3.0	9.0	415	771	1,236	1,809
4.0	16.0	220	415	672	990
5.0	25.0	133	254	415	616
6.0	36.0	87	169	279	415
7.0	49.0	60	118	198	296
8.0	64.0	43	87	146	220
9.0	81.0	32	65	111	169
10.0	100.0	24	50	87	133
11.0	121.0	18	39	69	106
12.0	144.0	14	32	56	87
13.0	169.0	11	25	46	71
14.0	196.0	9	21	38	60
15.0	225.0	7	7	32	50

Conversion factor: ∼ 40 mils/mm
Source: Denis J. McGreivy and Kenneth A. Pickar, "VLSI technologies through the 80's and beyond," *IEEE Catalog No. EH0192-5,* © 1982 IEEE.

11.4.1 Chip Carriers

Chip carriers are becoming increasingly popular among the semiconductor manufacturers. One reason is the increasing number of package pins that must be accommodated. A second reason is that a lot of today's logic circuitry is built around programmable devices, such as microprocessors, and as such is bus oriented. An important requirement is that all paths of a given bus track each other with respect to propagation delay. The chip carrier is a much better vehicle for satisfying both of these goals than is the DIP (dual-in-line package), with which most logic designers are familiar. The ratio of the shortest to longest path inside the chip carrier is far superior to that of the DIP.

The chip carrier also costs considerably less for larger packages because there is far less ceramic in the single-layer (but see below) ceramic versions. The space taken up by the chip carrier is also much less than that of the DIP. In Fig. 11.14, note that the cavity size of the 64-pin chip carrier is the same as that of the 64-pin DIP that it replaces. Because most of the heat dissipated by a package is dissipated by the leads and the cavity, the thermal conductivity of a chip carrier should be as good as that of an equal-sized DIP of the same material. Studies have shown that this is indeed the case.

Moreover, as packages acquire a larger number of pins, the long, rectangular package simply becomes untenable for the reasons noted above. The largest DIP has

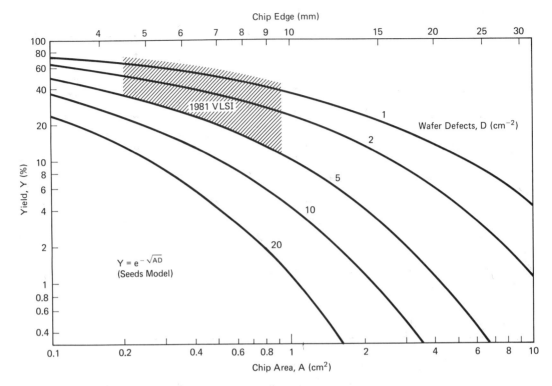

Figure 11.13 Yield versus chip size and defect density. (From Denis J. McGreivy and Kenneth A. Pickar, "VLSI technologies through the 80's and beyond"; registered copyright 1982 by IEEE.)

Figure 11.14 Group of standard IC packages. Hybrid-type packages can also be used but sometimes require a substrate. (Courtesy of Signetics.)

64 pins. By contrast, the largest PGA has 256 and is square, not rectangular. In these applications the packages become expensive partly because of the use of multilayer substrates within the package to accommodate the larger number of pins. This bears out the statement made in Chapter 1 that no longer is die size the principal de-

termining element of IC cost. Rather, costs of packaging and testing play important roles. A smaller die with the same pin-out as a larger one may require the same size package just to get the same number of pins. The cost of this package may be the dominating cost factor.

Two problems have inhibited the use of the chip carrier. One is that the coefficient of expansion of the chip carrier material is considerably different from that of the standard G-10 PCB material. Studies at Wright-Patterson Air Force Base have shown that when ceramic chip carriers are mounted directly on PCBs and thermally cycled over the military temperature range, -55 to $+125°C$, microcracking of the bonds occurs in a significant number of cases. To alleviate this problem, various manufacturers have come up with a number of schemes. The most common method is a socket which is mounted to the PCB and which has a similar coefficient of expansion or is flexible enough to avoid microcracking. An example of this type of socket is given in Fig. 11.14. Clips made by Berg are sometimes used on the sides of chip carriers to help avoid this problem.

The second problem that has inhibited the use of the chip carrier is the lack of suitable equipment to put them on PCBs (in commercial applications, where the temperature cycling is not as extreme as that of the military systems). Although equipment is available in the industry to do this, most manufacturers have such a huge investment in equipment for use with the common DIP package that there is a reluctance to convert. Common wave soldering machines cannot be used with such packages. The reason is that the chip carrier does not have pins which stick down through holes in a PCB. Rather a solder preform is used under the device. The surface tension of the melted solder pulls the chip carrier into precise alignment.

11.4.2 Pin Grid Arrays

Pin grid arrays help overcome many of the thermal cracking problems of chip carriers. They are mounted in such a way that they stand off from the PCB. This gives the needed flexibility. Moreover, they are offered in packages with larger numbers of pin-outs than are current chip carriers. Figure 11.15 shows the PGA in the lower left, with chip carriers of different sizes on the right.

A third advantage of the PGA over the chip carrier is the 100-mil (thousandths of an inch) spacing of the leads. This permits easier routing of traces on the corresponding PCB. (By contrast, the lead spacing of the chip carriers is typically 40 or 50 mils, with a few at lower values.)

The PGA does have a few problems of its own. First, the inner rows of pins cannot be readily probed. Second, it is very difficult to unsolder the device from the PCB. For this reason, a number of manufacturers, such as Raychem, have developed special zero-insertion-force sockets for the PGAs. (Raychem makes use of liquid nitrogen for removal of the device. This is far easier and more practical than it might at first seem and is infinitely easier than unsoldering it.). If used, the size, cost, and benefits of the socket must, of course, be factored into the decision to use the PGA.

Figure 11.15 Pin grid array (PGA) package (lower left). Chip carriers are on the right. (Courtesy of LSI Logic, Inc.)

11.4.3 High-Wattage Circuits

Some technologies, notably ECL, require packages capable of dissipating large amounts of power and still maintaining acceptable junction temperatures. Most of the 1000-gate ECL gate arrays with 80% utilization will use 3 to 4 W of power. A number of packages have been designed for this purpose. Motorola, for example, uses an inverted chip carrier with the heat sink mounted on what is now the top. One thousand linear feet per minute of air is typically blown across this device [Fig. 11.16(a)].

Other schemes involve using thermal bridges to make contact with the package. The bridges lead to a "cold plate" across which air is blown or which is cooled in some other fashion. Minute air gaps between the bridge and the package can cause significant problems because air is such a good insulator. For nonmilitary applications, these air gaps can be replaced with thermal grease to improve the thermal conductivity. A module used by IBM with high-power circuits is shown in Fig. 11.16(b).

11.4.4 Die Bonding

Two kinds of bonding are used for ICs. Eutectic bonding, although more expensive, gives better thermal contact with the package. Epoxy bonding is less expensive, but thermal contact is poorer. A problem in epoxy bonding is making sure that there are no minute "voids" (air bubbles) in the epoxy. These bubbles act as good thermal insulators, and hence are deleterious. Epoxies should be prepared in vacuums to avoid this problem.

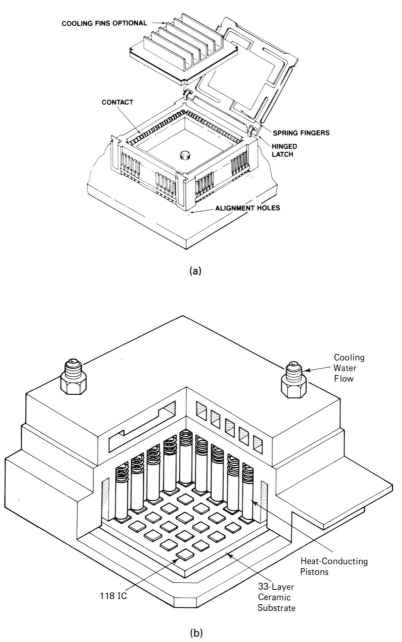

Figure 11.16 High-power package consisting of chip carrier with fins and mounting socket. [(a) Courtesy of Motorola; (b) courtesy of ICE.]

Sec. 11.4   Packaging   479

11.5 TEST CIRCUITS ON THE GATE ARRAY DIE AND WAFER

Various means are used by the gate array manufacturers to verify gate array circuit performance. The purpose of the test structures is twofold. First, it is to verify performance of the wafer lots according to some prearranged sampling scheme. A typical scheme might be to "burn" (process test structures) on 10% of a given wafer lot and then accept or reject the lot based on that sample. Normally, this involves sintering the metal to the silicon. Once this is done, it is not readily removable (i.e., the wafer cannot then later be patterned with the connections of the gate array design). At least one major manufacturer, however, has found that satisfactory (for test purposes) contacts can obtained without sintering. This permits test structures to be exposed and tested and the wafer returned to stock.

The second purpose is to find the problem if something is not working correctly (and possibly to assist in answering the question of whether the responsibility for the problem lies with the wafer manufacturer, the gate array manufacturer, or the designer). The point to be emphasized is that the test structures are normally *not* tested in the process of making routine tests on the die.

A common method is to have a small test circuit which is placed on every die. These small test circuits take one of two forms. The first is a permanent part of the background layers of the gate array die topology. This type of test structure typically consists of test pads attached to one replication of each of the types and sizes of transistors used in the array. The second form is that of a macro which is placed on every die. This macro might be, for example, a D flip-flop with ancillary gating and in the case of CMOS, embedded clock and clock "bar" (i.e., clock inverse) generator circuitry. Typically, these macros require only three or four bonding pads. History built up on these structures can give a very rapid feeling for the wafer quality.

Another method is to use test die that are in the middle of the four quadrants (looking at the wafer with the center of the quadrant system in the center of the wafer) of the wafer plus one from a die in the center of the wafer. This is in recognition that sometimes all or most of the good dies come from one section of the wafer. There are various reasons why the good die sometimes come from only one section of the wafer. A major contributor can be the runout (tolerance buildup across the wafer mask) error of mask-making machines mentioned above. Unlike "jelly-bean" ICs, however, a long period of time may elapse between making the masks for the underlying layers of a gate array and making the masks for the metallization layers. This is particularly true of those users who have their own "back-end" facilities and use and stockpile wafers from a number of gate array vendors. Slight changes in tolerance buildup may occur over this period of time on the machines that originally produced the masks. This is generally a second-order effect, however.

11.6 TESTER INTERFACE [9], [10]

Having built the chip, the question is now how to test it. This generally refers to interfacing the test patterns developed during simulation and fault analysis to a large VLSI tester. Such testers are made by Fairchild (Sentry 21 and 50 series and the earlier VII and VIII), Accutest (the 7900), Terradyne, Tektronix (the 3270), Genrad (GR16 and GR18), and Takeda Riken (the T3340).

Test pattern development was covered in Chapter 7. Logic simulators provide tools for test pattern development. Briefly, the reader will recall that there are really two distinct ways in which a logic simulator such as Tegas5 can enable test pattern development. The first way involves using the input signals used to simulate the circuit as inputs to the tester. This method involves the CHANGE, TESTPATT, and FILEINPUT commands. It will also be recalled that Tegas5 has a primitive element called CLOCK which can be used very efficiently (in terms of memory storage requirements) to generate repetitive signals. (The duty factor of the clock is not restricted to 50%, and the clock does not have to have equal rise and fall times.) The second way involves using automatic test vector generators built into the simulator. (The latter signals could, of course, have been used to simulate the circuit also.)

11.6.1 Tester Interface Programs

Both types of signals need to be converted to formats usable by testers. Tegas, for example, has a nonsupported interface program which configures the test vectors for popular testers. Key elements in a tester interface program are directions to the tester about pin connections to the chip, cycle time of the test patterns, times in the cycle at which the output signals can be strobed, and behavior of the input signals. The latter two features are specified using "time sets."

11.6.2 Time Sets

Input and output signals are grouped into separate time sets. These are time periods equal to the cycle time in which a given input signal can either make a change or not (i.e., it must have one of two modes of behavior). The types of signals allowed are RZ, RO, NRZ (see Fig. 11.17), and NRO, which stand for return to zero, return to one, nonreturn to zero, and nonreturn to one, respectively.

An output time set tells the tester when in the clock cycle the output must be strobed (i.e., when the output has settled to permit sampling of it by the tester).

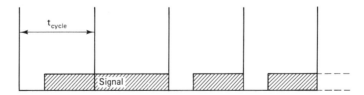

Figure 11.17 Diagram of a time set.

11.6.3 High-Speed Testers

Current major testers operate at up to about 40 to 50 Mhz as noted earlier. This clock frequency is an order of magnitude less than the clock frequencies of some ECL circuits and as much as 60 times less than the clock frequencies of some GaAs circuits.

Tektronix has introduced a very high speed logic analyzer, the DAS 9100 with 91HS8 modules, which is capable of sampling at a 2-GHz rate. Up to 32 channels can be analyzed simultaneously. [11]

11.7 SUMMARY

It is useful for the gate array designer to know something about how gate arrays are processed, including how masks are made. Descriptions of these activities are given in this chapter. Methods of testing the gate array once it is processed are described. Packaging techniques are also given. Test circuits on the gate array die and wafer are mentioned.

REFERENCES

1. "X-ray lithography bids for VLSI dominance," *Electronics,* Dec. 2, 1985, pp. 45–49.
2. Singer, P., "Merchant mask-making in a state of change," *Semiconductor International,* Mar. 1986, pp. 44–49.
3. Iscoff, R., "Photomask and materials review," *Semiconductor International,* Mar. 1986, pp. 82–86.
4. Brown, C., "Ion beam alters submicron circuits," *EE Times,* July 1, 1985, p. 24.
5. Elliot, D., Integrated Circuit Fabrication Technology, New York: McGraw-Hill, 1982.
6. "Shorter wait for custom gate arrays," *Electronic Design,* Oct. 31, 1985, p. 34.
7. "Electron-beam testing of VLSI chips gets practical," *Electronics,* Mar. 24, 1986, pp. 51–54.
8. Lukaszek, W., et.al., "CMOS test chip design for process problem debugging and yield prediction experiments," *Solid State Technology,* Mar. 1986, pp. 87–93.
9. Singer, P., "Test software development," *Semiconductor International,* Sept. 1985, pp. 76–81.
10. McCluskey, E., "Testing semi-custom logic," *Semiconductor International,* Sept. 1985, pp. 118–23.
11. "Logic analysis now possible at unprecedented speed," *Electronic Products,* Sept. 16, 1985, pp. 17–18.

Appendix A

Representative Gate Array Vendors and Products* [1, 2, 3, 4]

Please note: Information in this table is believed to be correct, but no liability is assumed for errors or omissions. Because new products are constantly being developed, the reader should contact the appropriate vendor for current information. For a list of representative GaAs (gallium arsenide) gate arrays, see Table 2.3.

Vendor	Technology[a]	Major CAD tools[b]
CMOS (Silicon gate, oxide isolated unless noted)		
IMI International Microcircuits, Inc. (408)727-2280	21; 1.5 μm; 10,000 gates fall '86	M ws
LSI Logic (408)263-9494	21; 2 μm; 50,000 gates	LDSIII, V, VI (ih) M ws
National (408)721-4532	21; 2 μm; 4800 gates	Logcap; HI-LO; M ws; D ws
GE/Intersil (919)549-3167	21; 2 μm; 13,000 gates	Teg; ih
GM/Hughes	21; 2μm; 41,000 gates	M ws; ih (LSI Logic 2nd source)

[a]nl is n-layer metal; kμm is drawn gate length (feature size) of dimension k; mg is size of largest array in terms of m two-input equivalent gates (a three-input cell would count as $1\frac{1}{2}$ two-input gates).
[b]ws, workstation; M, Mentor; D, Daisy; V, Valid; ih, there is an in-house CAD system.
Simulators: Teg is Tegas5; HI-LO (GenRad/Cirrus); Logcap (Phoenix Data Systems);, Merlyn is auto P&R of VR Information Systems/Tektronix.

Vendor	Technology[a]	Major CAD tools[b]
CMOS (Silicon gate, oxide isolated unless noted)		
AMI (408)246-0330	21; 2 μm; 9700 gates	Mentor;
CDI California Devices, Inc. (408)295-3700	21; 2 μm; 13,000 gates	M ws; D ws
CDC (617)890-4600	21; 2 μm; 6000 gates	Fab'd by National and Fairchild
Universal S.C. (408)279-2830	11; 3 μm; 2400 gates	IBM PC
Fairchild (617)872-4900	21; 2 μm; 6000 gates	M ws; D ws; ih
Motorola (512)928-6000	21; 2 μm; 5000 gates	Logcap; Merlyn (has Bipolar CMOS gate array)
Signetics (408)739-7700	11; 3 μm; 1728 gates	Teg
Harris (305)724-7000	21; 2 μm; 2000 gates	Teg
Mitel (613)592-5630	21; 3 μm	Invented the basic process
Other CMOS		
RCA	21; 3 μm; 6000 gates	Second source to LSI Logic
Interdesign/Ferranti	880 two- and three-input cells; metal gate CMOS; 7 μm	
STC		
Plessey–San Diego		
NEC		
Holt		
Fujitsu		
MCE		
GI		
TI		
GTE		
Siliconix		
ECL (Emitter-Coupled Logic) (See Chapter 1.)		
Motorola (second source is National) (602)244-7100	21; 2800 gates	ih
AMCC (second wafer source is Signetics) (619)450-9333	21; 3500 gates	Teg; ih; M, D ws
Fairchild (408)942-2672	21; 2500 gates	Silvar Lisco; ih; M ws
AMD (408)732-2400	21; 5000 gates	ih
Siemens	9000 gate; 100-ps gate delay (typ.); 20 mW/gate (typ.); 256 I/O pins	

Vendor	Technology[a]	Major CAD tools[b]
	ISL (Integrated Schottky Logic)	
Raytheon (415)965-9211	5000 gates	
Harris (305)724-7000	3000 gates	
	STL (Schottky Transistor Logic)	
TI		
Harris		
	TTL Permutations	
Fairchild	FAST array	
TI (second source to Fujitsu)		
Fujitsu		

REFERENCES

1. VLSI Systems Design Staff, "Survey of custom and semicustom ICs," *VLSI Systems Design,* Dec. 1985, pp. 54–90.
2. Cole, B., "How gate arrays are keeping ahead," *Electronics,* Sept. 23, 1985, pp. 48–52.
3. Gabay, J., "Gate arrays boast highest bipolar densities, subnanosecond speed," *Electronic Products,* Jan. 2, 1986, pp. 25–26.
4. Mullin, M., "High density CMOS gate arrays," *Electronic Design,* Mar. 18, 1986, pp. 86–98.

Appendix B

Major Workstation Vendors[a]

Vendor/phone	Fast simulator
Daisy Systems Corp. (408)773-9111; local: (617)890-2666	Megalogician
Mentor (503)626-7000; local: (617)863-5776	1/4S Zycad Module
Valid (415)940-4000; local: (617)863-5333	Realfast
DAE Systems (408)745-1440; local: (617)879-5575	Zycad
Metheus (503)640-8000	
Avera (408)438-1401	
VIA (617)667-8574	

To Compare Numbers

An *event* is a change in the output of a primitive (Zycad's definition). There are typically $2\frac{1}{2}$ to 3 evaluations per event on the average, but numbers will vary.

Example: The valid realfast processor is rated at 500,000 evaluations per second. This is equivalent to 200,000 events/second. Capacities must also be compared. (At the time of this writing, no other information on Realfast was available.)

Zycad product designator	Primitives	Speed (events/second)
1/4 S Module	16K	0.5^6
LE-1002	64K	10^6 (is one Zycad "S" module)
LE-1004	128K	2×10^6
LE-1032	1048K	$\geq 10 \times 10^6$

[a] Alternative to workstations: Package from, say Silvac Lisco, which runs on VAX (TM) 11/780 and provides multiuser environment.

Appendix C

Simulator CPU Times

In Chapter 7, mathematical formulations for the CPU times required to perform various design for testability tasks were given. In this appendix, typical CPU times to run the various subprograms of Tegas are given.

The CPU is a VAX/11-780 running under VMS 3.6. The gate array circuit is a 1500-gate ECL circuit which has about 93.8% utilization of its cells and I/Os. There are 42 inputs and 30 outputs to/from the chip exclusive of power and ground pads and about 988 signal nets, including those which go to/from the I/Os. (A net is a composite signal path of a macro, including all fan-outs.)

The number of input vectors is about 581.

Because many factors determine the CPU time required for a given operation, the numbers below should be taken only as crude "ballpark" approximations to other combinations of the variables. In particular, in this circuit, none of the modules are nested. Nesting in its various forms and depths (see Chapter 6) would, in general, increase the CPU times required. (However, the design times and risks would, in general, both be decreased by use of nesting.

- *Preprocessor run:* 3 min, 4 sec
- *COPTR:* 3 min, 35 sec
- *Mode 2 simulations:* 1 min, 33 sec to 5 min, 5 sec depending on whether the activity and other reporting commands were used
- *Master Fault File:* 4 min, 32 sec for 3455 total faults partitioned into 1860 equivalent fault classes
- *DFS (Detectable Fault Simulation):* about 3 min, 45 sec
- *Mode 4 fault analysis:* 1 hour, 29 min
- *Mode 5 fault analysis:* 3 hours, 59 min

Formulations of CPU times required are given in Chapter 7 for testability analysis.

Appendix D

Trademarks

The reader should consider all of the trademarks below to be registered by the company listed. The author is not responsible for errors or omissions. Readers in doubt, should check with the appropriate company. Some designations of being registered are given in the text in the first usage of the term.

Product name	Trademark of:
VAX	Digital Equipment Corp.
HI-LO	Cirrus Computers
HI-LO II, III	GenRad
Tegas	GE/Calma
COPTR	GE/Calma
Tristate	National Semiconductor
Lasar	Terradyne
Megalogian	Daisy Systems Corp.
Gatemaster	Daisy Systems Corp.
Realfast	Valid Logic Systems
Fastchip	Valid Logic Systems
Fast	Fairchild
LDS-I, II, III	LSI Logic, Inc.
Quickchip	AMCC
Telenet	GTE
Sentry	Fairchild Test Systems
ISOCMOS	Mitel
Chronoflex	DuPont
Xerox	Xerox
LA120, VT100, VT200	DEC
Macrocell Array	Motorola
Universal Gate Counter	Universal Semiconductor
GDSI, II	GE/Calma
LSSD	IBM Corp.
Helix	Silvar-Lisco
Gards	Silvar-Lisco

Supplemental Reading

NEW GATE ARRAYS, COMPARISONS, AND SO ON: CHAPTER 1

S. FARIELLO, "Gate array boasts of standard cell density," *Integrated Circuits,* Apr. 1985, pp. 33–36.

T. TAKAGAKI ET AL., "A new standard cell approach combined with a gate array design system," Custom IC Conference, May 21, 1985.

"Technology update: custom/semicustom IC's," *Electronic Engineering Times,* May 13, 1985.

R. HAINES, "Chip microcontrollers combine an assortment of functions," *Electronic Design,* Feb. 7, 1985, pp. 87–96; predicts mix of functions on chip.

S. BAKER, "ECL gate array has 1k memory," *Electronic Engineering Times,* Jan. 30, 1984, p. 44.

"IC design: on the eve of a revolution," *Engineering Manager,* June 1985, pp. 20–26; interview with Steve Hohnson.

J. KAMDAR, chairman, "Semi-custom logic today and tomorrow," *Session 8, Electro' 85,* Apr. 24, 1985.

M. TAKECHI, K. IKUZAKI, T. ITOH, M. FUJITA, M. ASANO, A. MASAKI, and T. MATSUNAGA, "A CMOS 12K gate array with flexible 10Kb memory," IEEE International Solid State Circuits Conference, Feb. 24, 1984, paper 17.1; Hitachi's work.

M. HILD, "Converting PALs to CMOS gate arrays," *VLSI Design,* Mar. 1984, pp. 58–62.

W. BURKARD, "Distributors and the application specific IC market: the key issues," *VLSI Design,* Mar. 1984, pp. 68–74; discusses doing gate arrays at distributor's facilities.

W. CURTIS, "Silicon compilation speeds design of complex chips," *Computer Design,* Mar. 1985, pp. 105–10.

P. CROLLA, "An advanced analog array for rapid implementation of semicustom IC designs," Custom IC Conference, May 21, 1985.

T. SANO, IEEE International Solid State Circuits Conference, 1983.

J. WERNER, "Custom/semicustom VLSI design in Japan," *VLSI Design,* July/Aug. 1983, pp. 30–39. 20,000-gate gate array; gate array with memory; 10,000-gate gate isolation gate array with pictures.

J. WERNER, "Custom IC design in Europe, Part 1: The United Kingdom," *VLSI Design,* Jan. 1984, pp. 28–35; "Part 2: Germany, France, and the Netherlands," Feb. 1984, pp. 65–68.

S. BAKER, "Flurry of second-source pacts stirs gate-array market" *Electronic Engineering Times,* Jan. 30, 1984, pp. 43, 45.

D. SCHULTZ and N. PRECKSHOT, "The 10,000 gate hierarchichal logic array: a system designer's tool," *VLSI Design,* Oct. 1983, pp. 71–77.

W. WOLF, "Converting gate array designs to standard cells," *Computer Design,* Mar. 1985, pp. 125–28.

A. GOLDBERGER, chairman, "New programmable logic yields instant custom ICs," Session 11, Electro' 85, Apr. 24, 1985.

"CMOS arrays with 20,000 gates have 1.5 ns delay," *Electronics,* Dec. 15, 1982, pp. 75–76.

C. FEY ET AL. "Selecting cost effective LSI design methodologies," Custom IC Conference, May 21, 1985.

W. TWADDELL, "Standard-cell design, inroads spur specialized gate array growth," *EDN,* Nov. 24, 1983, pp. 49–58.

R. BERESFORD, "Advances in customization free VLSI system designers," *Electronics,* Feb. 10, 1983, pp. 134–45.

"Special report on advanced digital ICs," *Computer Design,* Feb. 1983, pp. 137–167; contains the prediction that "by the end of the decade, over half of all ICs used will be customized designs."

M. BURIC ET AL., "Silicon compilation environments," Custom IC Conference, May 21, 1985.

R. BERESFORD, "Comparing gate arrays and standard cell ICs," *VLSI Design,* Dec. 1983, pp. 30–36.

J. HIVELY, "The case for standard cell semicustom ICs," *Electronic Products,* Feb. 7, 1983, pp. 67–69; defines semicustom to include standard cells (see ICECAP report below).

M. EKLUND, "Semicustom: high density business," ICECAP Report, Aug. 28, 1981; Integrated Circuit Engineering Corp.; Scottsdale, Ariz; defines semicustom to exclude standard cells.

R. MASUMOTO, "Configurable on chip RAM incorporated into high speed logic array, "Custom IC Conference, May 21, 1985.

B. KILSON, D. LAWS, and W. MILLER, "Programmable logic chip rivals gate arrays in flexibility," *Electronic Design,* Dec. 6, 1983, pp. 95–102.

"Marriage isn't for everyone," *Electronic Business,* Mar. 1, 1985, p.10.

P. HAM and D. NEWMAN, "Digital cmos cell library adopts analog circuits," *Electronic Design,* Dec. 6, 1983, pp. 107–14.

W. HOFFMAN, R. BECHADE, C. ERDELYL, and M. CONCANNON, "A comparison of mixed gate array and custom IC design methods," IEEE International Solid State Circuits Conference, Feb. 22, 1984, paper 1.2.; IBM work.

R. WALKER ET AL., "Structured arrays—a new ASIC concept provides the best of gate arrays and standard cells," Custom IC Conference, May 21, 1985.

"Two part array couples memory and logic and adds testing," *Computer Design,* Jan. 1984, p. 218.

S. POLLOCK and T. MROZ, "Analog cells specify worst case performance," *Electronic Design,* Dec. 6, 1983, pp. 119–27.

R. BERESFORD and L. FRANK, "Analog design in digital MOS circuitry," *VLSI Design.* Feb. 1948, pp. 70–76.

"Gate array needs fewer gates for RAM," *Electronics,* Jan. 10, 1983, pp. 86–87; Hitachi CMOS.

R. KNIPE and W. PESCHKE, "Beyond standard cells, *VLSI Design,* May/June 1983, pp. 78–80.

J. CONWAY, R. MASUMOTO, and F. YAKELY, "Boosting on chip RAM lets logic array shoulder a world of new tasks," *Electronic Design,* Feb. 21, 1985, pp. 221–27.

"TI-Fujitsu sign gate array pact," *EE Times,* Jan. 13, 1984, p. 1; mentions Tidal language license.

S. SHIVA and P. KLON, "The VHSIC hardware description language," *VLSI Design,* June 1985, pp. 86–106.

J. CROWLEY, "Standard language covers all aspects of design automation," *Electronic Products,* Oct. 24, 1983, pp. 107–11; on Tidal language.

"Standard proposed for chip design software," *CAE,* Jan./Feb. 1983, pp.10–11; refers to TI's Tidal language.

C. RHODES and T. HERNDON, "Cache-memory functions surface on VLSI chip," *Electronic Design,* Feb. 18, 1982, pp. 159–62.

H. BANKS and K. WIEGNER, "A matter of life and death," *Forbes,* Feb. 28, 1983, pp. 35–38; predicts that by 1990, half of all semiconductor design will be custom.

S. SZIROM, "Custom/semicustom IC industry business update," *VLSI Design,* Jan./Feb. 1983, pp. 30–34; predicts CMOS and bipolar semicustom IC markets will reach $400M and $450M, respectively, in 1986.

"Strategic implications of gate arrays," Mackintosh Consultants, Inc., San Jose, Calif., Jan. 1983.

"Gate market to exceed $6 billion by 1991," *EDN,* Sept. 15, 1982, p. 238; mentions Strategic, Inc. report.

"Semicustom-IC market to reach $2 billion by 1989," *EDN,* May 12, 1983, p. 304; results of Cahners Publishing Research Dept. survey.

ANDERSON/BOGERT STAFF, "Weigh all make-vs.-buy factors for custom/semicustom ICs," *EDN,* Mar. 3, 1982, pp. 127–32.

O. AGRAWAL and J. BECK, "PLDs as semicustom substitutes," *VLSI Design,* June, 1985, pp. 30–46.

"Semicustom market: what role will distributors play?" *Electronic Engineering Times,* Nov. 21, 1983, pp, 15–16.

"Impact of custom circuit alternatives on future product costs," Feb. 1983, Strategic Inc; San Jose, Calif.

"JEDEC's Recommended standards for gate arrays," *VLSI Design,* Jan. 1984, p.50.

A. RAPPAPORT, "CMOS function arrays use silicon efficiently," *EDN,* May 26, 1983, pp. 118–19.

M. BAKER and V. COLI, "Semicustom programmable logic IC's," *Integrated Circuits,* Apr. 1985, pp. 26–32.

D. SUTTON, "A user experience with gate arrays," *Electronic Products,* Oct. 25, 1982, pp. 65–68.

D. SAXBY ET AL, "Erasable PLD program translates logic directly into chips," *EDN,* Feb. 7, 1985, pp. 187–95.

TECHNOLOGIES: CHAPTER 2

W. IVERSEN, "Speed record claimed for GaAs transistor," *Electronics Week,* May 13, 1985, pp. 19–20.

T. C. LEE and S. JAMISON, "GaAs in digital electronics," *Scientific Honeyweller,* Apr. 1985, pp. 1–15.

F. WANLASS and V. ARAT, "Fastest yet CMOS family makes its bid for more Schottky territory," *Electronic Design,* Feb. 7, 1985, pp. 103–12.

R. BERESFORD, "Evaluating gate array technologies; Part 1: A framework," *VLSI Design,* Jan. 1984, pp. 48–52.

R. BERESFORD, "Evaluating gate array technologies; Part 2: Evaluating results," *VLSI Design,* Feb. 1984, pp. 34–41.

"New buried nitride process from Germany," *Semiconductor International,* May 1983, p. 19.

R. GALLAGHER, "Silicon on insulator attains high yields by boundary control," *Electronics,* May 5, 1983, pp. 85–86.

R. GODIN, "IEDM ideas foreshadow future circuit directions," *Electronics,* Nov. 17, 1983, pp. 135–40: discusses vertically stacked CMOS.

D. SNYDER, chairman, "The commercial world of high speed digital GaAs," session 20, Electro' 85, Apr. 25, 1985.

J. GIBBONS, "SOI—A candidate for vlsi?," *VLSI Design,* Mar./Apr. 1982, pp. 54–55: stacked CMOS gate using laser recrystallized silicon.

R. CHWANG and K. YU, "C-HMOS—A N-well bulk CMOS technology for VLSI," *VLSI Design,* Fourth Quarter 1981, pp. 42–47.

"Rad hard CML gate array features ECL and TTL compatible ports," *EDN,* Apr. 11, 1985, p.184.

Y. IKAWA, N. TOYODA, M. MOCHIZUKI, T. TERADA, K. KANAZAWA, M. HIROSE, T. MIZOGUCHI, and A. HOJO, "A 1k GaAs gate array," IEEE International Solid State Circuits Conference, Feb. 22, 1984, paper 3.1.

H. YUAN, W. MCLEVIGE, H. SHIH, and A. HEARN, IEEE International Solid State Circuits Conference, Feb. 22, 1984 paper 3.2.

"Process shrinks bipolar LSI," *Electronics,* Jan. 27, 1983, pp. 48–49.

P. SOLOMON, "A comparison of semiconductor devices for high speed logic," Proceedings of the IEEE, May 1982, pp. 189–215; invited paper with good insights.

V. GANI, W. KLARA, and C. VICARY, "TTL catapults mainframe to the brink of VLSI," *Electronics,* July 14, 1982, pp. 163–68.

I. HOGLUND, "Silicon-on-sapphire reconsidered," *Integrated Circuits,* Apr. 1985, pp. 12–17.

D. FULKERSON, *Structured VLSI design on IIL gate arrays," VLSI Design,* Mar./Apr. 1982, pp. 32–44; mentions having 4000 to 9200 equivalent gates on one gate array: has macros for PLA, A/D, D/A, RAM, and ROM as well as random digital logic functions; has five programmable layers.

"CMOS seen as dominant technology of the '80s," *Electronic Business,* Mar. 1982, pp. 28–29.

R. BERESFORD, "Technology update—1982" *Electronics,* Oct. 20, 1982, pp. 116–34 has SOI technique.

J. GOSCH, "Buried nitride CMOS may double device density," *Electronics,* Jan. 27, 1983, pp. 70–71.

D. BUHANAN, "CML scraps emitter follower for ECL speed, lower power," *Electronics,* Nov. 3, 1982, pp. 93–96.

D. YANEY and C. PEARCE, "The use of thin epitaxial layers for MOS VLSI," *IEDM,* 1981, pp. 236–39.

P. BURGGRAAF, "The forces behind epitaxial silicon trends," *Semiconductor International,* Oct. 1983, pp. 44–51.

M. HAMMOND, "Silicon epitaxy: A 1983 perspective," *Semiconductor International,* Oct. 1983, pp. 58–63.

T. MOROYAN, "CMOS leads to true computer portability," *Computer Design,* Nov. 1983, pp. 95–105.

"Improved epitaxial deposition," *Semiconductor International,* Dec. 1983, pp. 86–87.

G. MOHAN RAO, "MOS Takes a cue from the bipolar world," *VLSI Design,* Nov./Dec. 1982, pp. 52–53; deals with epi for MOS.

A. LONDON, "Two layer metal CMOS vs. two layer poly CMOS," *VLSI Design,* Mar./Apr. 1983, pp. 62–63.

M. GULETT, "Scaling down CMOS design rules," *VLSI Design,* Sept. 1983, pp. 24–29.

J. PFIESTER, J. SHOTT, and J. MEINDL, "Performance limits of NMOS and CMOS," IEEE International Solid State Circuits Conference, Feb. 23, 1984, paper 11.4.

J. HUANG, "Bipolar technology potential for VLSI," *VLSI Design,* July/Aug. 1983, pp. 64–67.

F. GOODENOUGH, "Dielectric isolation arrives for gate arrays," *Electronic Design,* Dec. 6, 1983, pp. 43–44.

R. CHAO, "CMOS logic families diversify design options," *Electronic Design,* Dec. 6, 1983, pp. 203–6.

V. LEE, C. DELL'OCA, P. YIN, and D. MATTISON, "CMOS arrays top Schottky TTL speeds with gate lengths of 2μ," *Electronics,* Mar. 24, 1983, pp. 137–39.

D. CARTER and D. GUISE, "Effects of interconnections on submicron chip performance," *VLSI Design,* Jan. 1984, pp. 63–68.

W. BENZIG, "Shrinking VLSI dimensions demand new interconnection materials," *Electronics,* Aug. 25, 1982, pp. 116–19.

A. DANSKY, "Bipolar circuit design for a 5000 circuit VLSI gate array," *IBM J. Res. Develop.,* May 1981, pp. 116–25; gate array configured as a microprocessor; has histogram of fan-outs.

M. GULETT, "Scaling down CMOS design rules," *VLSI Design,* Sept. 1983, pp. 24–29.

P. HART, T. T. HOF, and F. KLAASSEN, "Device down scaling and expected circuit performance," *IEEE Trans. Electron Devices,* Apr. 1979, pp. 421–29.

J. HUANG, "Bipolar technology potential for VLSI," *VLSI Design,* July/Aug. 1983, pp. 64–67.

R. Percival and M. Fitchett, "Designing a laser personalized gate array," *VLSI Design,* Feb. 1984, pp. 54–61.

K. Raghunathan, "Inside the stacked gate EPROM cell," *Electronic Design,* Aug. 19, 1982, p. 103, shows the stepped source/drain.

J. Robertson, "Say Rockwell, Honeywell get $25M DARPA IC pact," *Electronic News,* Jan. 30, 1984, p. 4; mentions funding of GaAs gate arrays by DARPA.

R. Allen, "Future military systems are drafting GaAs devices," *Electronic Design,* Aug. 4, 1983, pp. 101–18.

L. Lowe, "High speed logic race grows," *Electronics,* Sept. 22, 1982, pp. 110–13; GaAs.

Richard Eden and Arpad Barna, "Is GaAs suitable for VLSI?" *VLSI Design,* Jan./Feb. 1982, pp. 40–43.

K. Chiu, "SWAMI: A zero encroachment local oxidation process," *Hewlett-Packard J.,* Aug. 1982, pp. 31–34.

F. Gaensslen, "MOS devices and ICs at liquid nitrogen temperatures," Proceedings of IEEE International Conference on Circuits and Computers, Oct. 1–3, 1980, pp. 450–52; IEEE Catalog No. 80CH1511-5.

D. Grundy and J. Bruchez, "Switching to bipolar technology for the coming 100,000 gate array," *Electronics,* July 14, 1983, pp. 137–41.

R. Pitts, "ISL girds for battle with CMOS," *Electronics,* Nov. 30, 1982, pp. 110–12; Signetics oxide-isolated ISL.

J. Chen, W. Chin, T. Jen, and J. Hutt, "A high density bipolar logic masterslice for small systems," *IBM J. Res. Development;* May 1981, pp. 142–51.

"Overcoming the limitations to submicron CMOS," *Electronic Design,* Nov. 25, 1982, p. 48.

W. Iversen, "3000 gate array has 600 ps delays," *Electronics,* Feb. 10, 1983, pp. 175–76.

A. Silburt and Richard Foss, "VLSI changes the rules for coping with substrates," *Electronics,* Nov. 17, 1982, pp. 155–59.

L. Wakeman, "Closing in on CMOS latchup," *Integrated Circuits,* Apr. 1985, pp. 38–44.

OVERVIEW OF GATE ARRAY DESIGN: CHAPTER 4

E. Freeman, "Layout verifiers let you complete design of semicustom ICs on a single workstation," *EDN,* Apr. 4, 1985, pp. 79–90.

P. Swartz, "Incremental design rule checking," Custom IC Conference, May 21, 1985.

G. Hromadko and J. Werner, "User perceptions of CAD/CAE systems," *VLSI Design,* Feb. 1984, pp. 18–23.

T. Watanabe, M. Endo, and N. Miyahara, "A new automatic logic interconnection verification system for VLSI design," *IEEE Trans. Computer Aided Design of Integrated Circuits and Systems,* Apr. 1983, pp. 70–81.

"CAD's interface between design and mask making," *Semiconductor International,* Jan. 1984, pp. 128–30.

K. Ueda et al. "A top down layout design system for custom VLSI," Custom IC Conference, May 21, 1985.

C. Ebeling and O. Zajicek, "Validating VLSI circuit layout by wirelist comparison;" IEEE International Conference on Computer Aided Design, Sept. 15, 1983.

E. Kuuttila et al., "Integrated circuit design productivity," Custom IC Conference, May 21, 1985.

R. Rozeboom, "Capable software tools ease semi-custom IC design" *EDN*, May 12, 1983, pp. 185–93.

T. Raymond, "LSI/VLSI design automation," *Computer,* July 1981, pp. 89–101; old but is good tutorial.

H. Daseking, R. Gardner, and P. Weil, "VISTA; A VLSI CAD system," *IEEE Trans. Computer Aided Design of Integrated Circuits and Systems,* Jan. 1982, pp. 36–52.

Frederick Hinchliffe II, "A distributed control system for acquired design automation tools," IEEE International Conference on Computer Aided Design, Sept. 14, 1983.

R. Beresford, "Integrated software tackles VLSI design," *Electronics,* June 16, 1983, pp. 145–46; Phoenix Data Systems products.

Michael Payne, "An integrated VLSI design system," *VLSI Design,* Jan./Feb. 1982, pp. 46–50.

J. Darringer, W. Joyner, C. Berman, and L. Trevillyan, "Logic synthesis through local transformation," *IBM J. Res. Development,* July 1981, pp. 272–80.

V. Shaw and T. O' Connell, "Systematic design approach for high density CMOS gta," IEEE International Conference on Computer Aided Design, Sept. 15, 1983.

A. Feller, R. Noto, D. Smith, B. Wagner, and R. Putatunda, "CAD system coordinates complete semicustom chip design," *Electronics,* June 16, 1983, pp. 116–19; RCA's CAD-DAS system.

S. Garcia and K. Sriaram, "A Survey of IC CAD tools for design, layout, and testing, *VLSI Design,* Sept./Oct. 1982, pp. 68–73.

A. Rappaport, "Logic analysis system ties instrumentation to CAE," *EDN,* Nov. 24, 1983, pp. 73–74.

S. Newell, A. deGeus, and R. Rohrer, "Design automation for ICs," *Science,* Apr. 29, 1983, pp. 465–71.

"Integrated circuit design—hand crafting to automatic," ICECAP Report; ICE Corp.; No. 3, 1983.

"Logic simulation and analysis from the same database," *CAE,* Jan./Feb. 1984 pp. 14–15; Mentor workstation incorporates logic analyzer and enables tests made to be automatically compared to simulation data.

S. Taylor, "Verifying IC layouts," *VLSI Design,* Jan. 1984, pp. 18–25.

FEASIBILITY, CIRCUIT ANALYSIS, AND PARTITIONING: CHAPTER 5

S. Huss, "An efficient approach to *RC* path signal delay calculation for CMOS semicustom ICs," Custom IC Conference, May 21, 1985.

H. Dicken, "Calculating the manufacturing costs of gate arrays," *VLSI Design,* Dec. 1983, pp. 51–55.

"Gate array manufacturing cost/yield," *VLSI Design,* Sept./Oct. 1982, p. 10; mentions Strategic, Inc. study.

"Special techniques for controlling the costs of going custom," *Integrated Circuits,* May/June 1983, pp. 27–31.

S. Cravens, "The semicustom system design process," Custom IC Conference, May 21, 1985.

J. RUBINSTEIN, P. PENFIELD, Jr., and M. HOROWITZ, "Signal delay in *RC* tree networks," *IEEE Trans. Computer Aided Design,* July 1983, pp. 202–11.

T. LIN and C. MEAD, "Signal delay in general *RC* networks with application to timing simulation of digital ICs," 1984 Conference on Advanced Research in VLSI, MIT, Jan. 24, 1984.

D. CARTER and D. GUISE, "Effects of interconnections on submicron chip performance," *VLSI Design* Jan. 1984, pp. 63–68.

A. SINHA, J. COOPER, Jr., and H. LEVENSTEIN, "Speed limitations due to interconnect time constants in VLSI ICs," *IEEE Electron Devices Letters,* Apr. 1982.

Design Manual HC Series, *CMOS Gate Arrays,* California Devices, Inc., 1982, p. 151; temperature variation of conduction factor.

C. LEE ET AL., "Understand system partitioning to optimize custom IC use," *EDN,* June 28, 1984, pp. 255–68.

M. STANSBERRY, "Detailed analysis helps you reduce gate arrays' power consumption," *EDN,* Apr. 4, 1985, pp. 229–36.

SIMULATORS: CHAPTER 6

C. TURCEHETTI ET AL; "A one dimensional approach to the analysis of short channel MOSFETs," Custom IC Conference, May 21, 1985.

J. WERNER and R. BERESFORD, "A system designer's guide to simulators," *VLSI Design,* Feb. 1984, pp. 27–31.

R. SALEH, J. KLECKNER, and A. R. NEWTON, "Iterated timing analysis in SPLICE1," IEEE International Conference on Computer Aided Design, Sept. 14, 1983.

J. GRUNDMANN, "Event driven MOS timing simulator," IEEE International Conference on Computer Aided Design, Sept. 14, 1983.

A. RAPPAPORT, "Capable digital circuit simulators promise breadboard obsolescence," *EDN,* Mar. 17, 1983, pp. 105–26.

G. ROBSON, "Logic design using behavioral models," *VLSI Design,* Jan. 1984, pp. 36–44.

A. RAPPAPORT, "Digital logic simulator software supports behavioral models," *EDN,* May 12, 1983, pp. 95–98; HI-LO 2 simulator.

D. COELHO, "Behavioral simulation of LSI and VLSI circuits," *VLSI Design* Feb. 1984, pp. 42–51.

R. HITCHCOCK, G. SMITH, and D. CHENG, "Timing analysis of computer hardware," *IBM J. Res. Development,* Jan. 1982, pp. 100–105.

W. READ, "A hardware modeling system for MOS LSI circuits, IEEE International Conference on Computer Aided Design, Sept. 15, 1983.

R. Bartel, "Analog simulator interacts with circuit designers," *Electronic Design,* Sept. 16, 1983, pp. 135–40; describes SPICE used in workstation (Mentor).

Y. HOLLANDER, "Using an RTL simulator to simplify VLSI design," *VLSI Design,* Sept. 1983, pp. 60–65.

G. NOGUEZ, "An interactive design checking system based on behaviour simulation," IEEE International Conference on Computer Aided Design, Sept. 13, 1983.

G. BARROS, "A circuit simulation tutorial," *VLSI Design,* June 1985, pp. 110–21.

L. SMITH and R. REZAC, "A simulation engine in the design environment," *VLSI Design,* Dec. 1984, pp. 74–80.

E. Prendergast et al., "A highly automated IC modeling system: MECCA," Custom IC Conference, May 21, 1985.

DESIGN FOR TESTABILITY: CHAPTER 7

E. Freeman, "IC and pc board fault simulation systems predict the consequences of design flaws," *EDN,* Jan. 24, 1985, pp. 77–84.

M. Graf, "Testing—A major concern for VLSI," *Solid State Technology,* Jan. 1984, pp. 101–8.

E. Archambeau, "Test pattern generation for small gate arrays," Wescon '83, paper 33/4; mentions the ARCOP testability analysis program.

T. Powell, "Software gauges the testability of computer designed ICs," *Electronic Design,* Nov. 24, 1983, pp. 149–54; describes the COP testability analysis program of TI.

R. G. Bennetts, *Design of Testable Logic Circuits,* Reading, Mass., Addison-Wesley, 1984.

R. G. Bennetts, "Practical guidelines for designing testable custom/semicustom ICs," *VLSI Design,* Apr. 1984, pp. 64–74.

T. Williams, "Design for testability: What's the motivation?" *VLSI Design,* Oct. 1983, pp. 21–23.

K. Gutfreund, "Integrating the approaches to structured design for testability," *VLSI Design,* Oct. 1983, pp. 34–42.

R. Hess, M. Ausec, R. Biedronski, and R. Rector, "Automated test program generation for semicustom devices," *VLSI Design,* Oct. 1983, pp. 51–61.

R. Chandramouli, "Designing VLSI chips for testability," *Electronics Test,* Nov. 1982, pp. 50–60.

R. Hess, "Testability analysis: An alternative to structured design for testability," *VLSI Design,* Mar./Apr. 1982, pp. 22–29; UTMC work using Sandia Scoap; similar to COPTR.

T. Williams and K. Parker, "Design for testability—A survey," *Proc. IEEE,* Jan. 1983, pp. 98–112.

J. Clark and M. Neighbors, "A method of computer aided test generation," *Solid State Technology,* Nov. 1982, pp. 112–15. D. Resnick, "Testability and maintainability with a new 6K gate array," *VLSI Design,* Mar./Apr. 1983, pp. 34–38.

Y. El-zio, "Classifying, testing, and eliminating VLSI MOS failures," *VLSI Design,* Sept. 1983, pp. 30–35.

T. Mangir, "Design for testability: an integrated approach to VLSI testing," IEEE International Conference on Computer Aided Design, Sept. 13, 1983.

WORKSTATIONS AND OTHER CAD TOOLS: CHAPTER 8

J. Werner and J. Kaye, "Applications libraries for CAE/CAD workstations," *VLSI Design,* Oct. 1983, pp. 62–65.

C. Thames, "Engineering workstations: The dash to the desktop," *Electronic Business,* Sept. 1983, pp. 91–101; good overview.

A. Weiss, "CAD systems: Meeting the challenge of VLSI design," *Semiconductor International,* Mar. 1983, pp. 60–67.

S. OHR, "CAE work station permits rapid analysis and display," *Electronic Design,* Sept. 15, 1983, p. 195; Avera using HP-9000.

J. WERNER, "Sorting out the CAE workstations," *VLSI Design,* Mar./Apr. 1983, pp. 46–55.

G. ROBSON, "Benchmarking the workstations," *VLSI Design,* Mar./Apr. 1983, pp. 58–63.

A. RAPPAPORT, "Evolving CAE workstations furnish increasing sophistication," *EDN,* Apr. 14, 1983, pp. 33–44.

J. WRIGHT, "CAE unifies logic and gate array design," *Computer Design,* Nov. 1983, pp. 145–56.

J. HUBER, "CAE station closes design loop with logic measurements," *Electronic Design,* Oct. 27, 1983, pp. 127–32.

A. RAPPAPORT, "Computer aided engineering tools," *EDN,* Sept. 16, 1983, pp. 106–36.

J. KROUSE, "Selecting a graphic input device for CAD/CAM," *Machine Design,* Oct. 6, 1983, pp. 75–81.

J. WERNER, G. ROBSON, and R. HARRIS, "Comparing the CAE systems in action," *VLSI Design,* Nov. 1983, pp. 24–35.

S. SAPIRO and D. LAUGHLIN, "Work station's software covers entire design cycle," *Electronic Design,* Mar. 3, 1983, pp. 145–50.

J. ANDERSON, "Interactive design of electronic logic," *CAE,* Mar./Apr., 1983, pp. 39–47.

"Super simulation for workstations," *CAE,* Jan./Feb. 1984, p. 20; Daisy Megalogician; 100,000 evaluations per second; 1 million gate simulation capability; compares to Zycad.

S. SAPIRO and R. J. SMITH II, *Handbook of Design Automation,* CAE Systems, Inc., Sunnyvale, Calif., 1984.

G. ROBSON, "Automatic placement and routing of gate arrays," *VLSI Design,* Apr. 1984, pp. 35–45.

VLSI Design Staff, "A review of CAD systems for IC layout," *VLSI Design,* Mar. 1984, pp. 20–27.

G. ROBSON, "Selecting an IC layout CAD system," *VLSI Design,* Mar. 1984, pp. 28–33.

S. TAYLOR, "Symbolic layout," *VLSI Design,* Mar. 1984, pp. 34–43.

H. LAW, "Gate matrix layout: A practical, stylized approach to symbolic layout," *VLSI Design,* Sept. 1983, pp. 49–59.

K. POCEK, "Training system engineers in IC design: Two contrasting approaches," *VLSI Design,* Dec. 1983, pp. 38–46.

J. WERNER, "Recent progress in CAD systems for IC layout," *VLSI Design,* May/June 1983, pp. 48–59.

B. TING and B. TIEN, "Routing techniques for gtas," *IEEE Trans. Computer Aided Design,* Oct. 1983, pp. 301–12.

W. DONATH, "Wire length distribution for placements of computer logic," *IBM J. Res. Development,* May 1982, pp. 272–80.

J. KOCH III, W. MIKHAIL, and W. HELLER, "Influence on LSI package wireability of via availablity and wiring track accessibility," *IBM J. Res. Development,* May 1982, pp. 328–41.

M. SADOWSKA and T. TARNG, "Single layer routing for VLSI: Analysis and algorithms," *IEEE Trans. Computer Aided Design,* Oct. 1983, pp. 246–59.

S. TSUKIYAMA, I. HARADA, M. FUKUI, and I. SHIRAKAWA, "A new global router for gta LSI;" *IEEE Trans. Computer Aided Design,* Oct. 1983, pp. 313–21.

P. Cook, S. Schuster, J. Parrish, V. DiLonardo, and D. Freedman, "1 μm MOSFET VLSI technology, Part III: Logic circuit design methodology and applications," *IEEE J. Solid State Circuits,* Apr. 1979, vol. sc-14, no. 2, pp. 355–68; although old, it has one of the few explanations of the Weinberger layout technique; also discusses Rent's rule.

P. Cohen, "Layout considerations in predicting VLSI performance," *VLSI DESIGN,* Jan./Feb. 1983, pp. 64–65.

T. Watanabe, M. Endo, and N. Miyahara, "A new automatic logic interconnection verification system for VLSI design," *IEEE Trans. Computer Aided Design,* Apr. 1983, pp. 70–82.

Texas Instruments, Inc., *Semiconductor Engineering J.,* Design Automation Special Issue, Summer 1981, vol. 2; no.2.

J. Finkel, "Cutting the cost of hardcopy," *CAE,* Jan./Feb. 1984, pp. 34–44.

Special Issue on VLSI; *IEEE J. Solid State Circuits,* Apr. 1979, old but is one of the few readily available documents that contains Weinberger layout technique.

"Popular IC simulator runs on IBM PC," *Electronics,* Jan. 26. 1984, p. 151.

"Personal CAD's tools run on the IBM PC," *VLSI Design,* Feb. 1984, pp. 82–83.

"Kontron adds gate array design tools to microcomputer development system," *VLSI Design,* Feb. 1984, p. 83.

P. Schleider and A. Rappaport, "Users' attitudes about CAE," *VLSI Design,* June 1985, pp. 74–85.

PACKAGING: CHAPTER 11

"Special report: High density IC packaging," *Electronic Products,* Oct. 24, 1983, pp. 74–101.

J. Farrell III, "Package alternatives for an advanced microprocessor," *Electronic Products,* Oct. 24, 1983, pp. 74–78; discusses the problem of smaller packages with more I/Os; GTAS also face the same problem.

F. Rydwansky, "Chip carriers and pin grid arrays for high density ICs," *Electronic Products,* Oct. 24, 1983, pp. 89–92.

H. Waltersdorf, "A ZIF connector for pin grid arrays," *Electronic Products,* Oct. 24, 1983, pp. 99–102.

L. Mahalingam, J. Andrews, and J. Drye, "Thermal studies of pin array packages," *Semiconductor International,* Aug. 1983, pp. 76–84.

J. Balde, "VLSI packages: Pin grids or chip carriers?" *Circuits Manufacturing,* Sept. 1983, pp. 70–76.

F. Dance, "Interconnecting chip carriers: A review," *Circuits Manufacturing,* May 1983, pp. 56–70.

B. Winchell and E. Winkler, "Packaging to contain the VLSI explosion," *Computer Design,* Sept, 1982, pp. 133–36.

S. Garvey and W. Little, "Micromini refrigerators: An all-in-one hybrid package deal," *Circuits Manufacturing,* Mar. 1983, pp. 45–46.

E. Blackburn, "VLSI packaging reliability," *Solid State Technology,* Jan. 1984, pp. 113–16.

A. Mones and R. Spielberger, "Interconnecting and packaging VLSI chips," *Solid State Technology,* Jan. 1984, pp. 119–22.

G. Fehr, "Logic array packaging," Application Note A33; LSI Logic, Inc., June 1982.

C. McIver, "Flip TAB, copper thick films create the micropackage," *Electronics,* Nov. 3, 1982, pp. 96–99; discusses a water-cooled package for CML used in computer mainframe.

R. Simons, "Thermal management of electronic packages," *Solid State Technology,* Oct. 1983, pp. 131–37; discusses the IBM thermal conduction module.

J. Balde and D. Brown; "Alternatives in VLSI packaging," *VLSI Design,* Dec. 1983, pp. 23–29.

R. Iscoff, "VHSIC/VLSI packaging update," *Semiconductor International,* S. Parris and J. Nelson, "Practical considerations in VLSI packaging," *VLSI Design,* Nov./Dec. 1982, pp. 44–48.

N. Mokhof, "IBM packs high density circuits in and lets them breathe," *Computer Design,* Jan. 1984, pp. 49–52.

TESTING: CHAPTER 11

J. Turino and R. Chapman, "Trends in LSI/VLSI testing," *Semiconductor Internationl,* Jan. 1984, pp. 70–75.

P. Singer, "VLSI test systems: facing the challenge of tomorrow," *Semiconductor International,* Jan. 1984, pp. 82–88.

Rob Walker and Dan King, "Testing logic arrays, Application Note A32A, LSI Logic Corp., Oct. 1982.

S. Ohr, "VLSI/LSI testers: Speed is primary," *Electronic Design,* Feb. 1983, pp. 71–86.

C. Chrones, "Software—A critical dimension in testing LSI/VLSI chips," *Semiconductor International,* Mar. 1983, pp. 72–77.

L. Lawson, "Develop VLSI test software quickly," *Computer Design,* Nov. 1983, pp. 133–41.

"Automatic test equipment," *Electronic Business,* Oct. 1983; pp. 104–40; four articles on VLSI testers of the major manufacturers, including technical and cost problems; good reading for any organization contemplating buying a large VLSI tester.

PROCESSING: CHAPTER 11

A. Weiss, "E-beam lithography: a story of dual identities," *Semiconductor International,* Feb. 1984, pp. 54–60.

"Resistless process etches VLSI patterns," *Electronics,* Jan. 12, 1984, pp. 88–89; Toshiba.

P. Burggraaf, "Advances keep optical aligners ahead of industry needs," *Semiconductor International,* Feb. 1984, pp. 88–95.

J. Maher, Jr., "Selective etching of SiO2 films using RIE," *Semiconductor International,* May 1983, pp. 110–14.

J. Lyman, "Lithography steps ahead to meet the VLSI challenge," *Electronics,* July 14, 1983, pp. 121–28.

Special Issue on VLSI; *IEEE J. Solid State Circuits,* Apr. 1979, vol. sc-14, no. 2; although old, it contains some papers whose information is still valid.

P. BURGGRAAF, "1X mask and reticle technology," *Semiconductor International,* Mar. 1983, pp. 40–47.

HEWLETT-PACKARD J.; Special Issue on IC Process Technology, Aug. 1982.

W. ARNOLD, "A matter of inches: wafer sizes going up," *Electronic Business,* May 1, 1983, p. 172; discusses going to 6 inches and beyond.

P. GARGINI, "Tungsten barrier eliminates VLSI circuit shorts," *Industrial Research and Development,* Mar. 1983, p. 141–47.

J. LYMAN, "Scaling the barriers to VLSI's fine lines," *Electronics,* June 19, 1980; has good comparisons among methods even though it is old.

"HP gives peek at X-ray aligner," *Semiconductor International,* May 1983, pp. 17–18.

"256K DRAMS indicators of leading edge lithography," *Semiconductor International,* May 1983, p. 24.

P. BURGGRAAF, "Wafer and mask imaging systems," *Semiconductor International,* Nov. 1983, pp. 83–89.

D. WANG, "Reactive ion etching of aluminum and its alloys," *Microelectronic Manufacturing and Testing,* Jan. 1984, pp. 19–20.

P. BURGGRAAF, "Diffusion, ion implant, and annealing," *Semiconductor International,* Nov. 1983, pp. 100–5.

A. WEISS, "Etching systems," *Semiconductor International,* Nov. 1983, pp. 114–18.

S. HARRELL, "X-ray lithography for VLSI production," *Microelectronic Manufacturing and Testing,* Jan. 1984, pp. 62–64.

R. ISCOFF, "Die separation, bonding, and encapsulating," *Semiconductor International,* Nov. 1983, pp. 124–27.

J. WILEY and D. ZAKAIB, "Five pitfalls to avoid when obtaining optical photomasks," *VLSI Design,* Nov./Dec. 1982, pp. 28–35.

E. MORGAN and R. SMITH, "Interfacing with E-beam mask suppliers," *VLSI Design,* Nov./Dec. 1982, pp. 36–43.

G. O'CLOCK, L. ERICKSON, and T. MATTORD, "MBE: Precise processing for better EHF devices;" *Microwaves,* Aug. 1981, pp. 101–5.

P. SINGER, "Inspection, measuring, and testing equipment," *Semiconductor International,* Nov. 1983, pp. 135–43.

D. ELLIOT, *Integrated Circuit Fabrication Technology,* New York, McGraw-Hill, 1982.

D. PRAMANIK and A. SAXENA, "VLSI metalization using aluminum and its alloys," *Solid State Technology,* part 1, Jan. 1983, pp. 127–33; part 2, Mar. 1983, pp. 131–38.

DENIS J. MCGREIVY and KENNETH A. PICKAR, "VLSI Technologies through the 80s and Beyond, IEEE Computer Society, Silver Spring, Md., 1982.

L. FRIED ET AL., "A VLSI bipolar metallization design with three level wiring and area array solder connections," *IBM J. Res. Development,* May 1982, pp. 362–71.

H. ROTTMANN, "Metrology in mask manufacturing," *IBM J. Res. Development,* Sept. 1982, pp. 553–60.

J. MARTIN, "Complex ICs drive silicon processing techniques," *Military Electronics/Countermeasures,* Mar. 1983, pp. 56–60.

H. WALKER, "Yield simulation for VLSI ICs," IEEE International Conference on Computer Aided Design, Sept. 15, 1983.

Index

A

Accelerators, simulation, 347–53
Air bridges (*see* GaAs)
Air lines, routing, 337–39
Aligner, 464–65
AOI, AND-OR-INVERT, 11
   asymmetrical, 379
   degenerate, 379
   in complex structures, 379–87
Asher, 469
ASIC, Application Specific IC, 8
Auto P&R (*see* Placement and Routing)

B

Bad machine, 283
BFL, 108
Bipolar (*see also* ECL; ISL/STL; Macros):
   collector resistance, 46
   structures, 43
Bird's beaking, 101

Blowbacks, 460
Bonding pads, 8

C

CAD, computer aided design (*see also* individual topics), 14–15, 220–357
  checking programs, 237
  fault analysis, 280–89
  schematic capture, 237–42
  simulators:
    circuit, Spice, 224–36
    logic, (table) 223, 222–95
    Tegas, 242–68
      encoding, running, 249–67
      features, 243–45
      interfaces, 245–46
      libraries, 246–47
      stimulii, 265–67
  testability analysis (table), 271; 277–80
CAM, Content Addressable Memory, 12

Cascode, 451
Cavity, package, 2
CDI, collector diffused isolation, 42, 99
Cells, 11, 16
CGA, 11
Chip, 4
　carrier (*see* Packages)
Clock distribution, 211–12
CMOS (*see also* Comparisons; MOS; FET), 72–87
　battery operated, useful feature, 85
　and bipolar combined, 41–42
　CMOS/SOS, 73, 75–76
　HS, 9, 73–87
　input protection, 84–85
　interfacing to TTL, 85
　joint gate (vertically stacked), 105–6
　latchup, 75–77, 142–43
　LOCMOS, 91
　metal gate, 73
　propagation delays, 77–81, 197–203
　　temperature variations, 81–82, (table), 198
　　voltage variations, 82
　　wafer processing effects, 82–83
　sharing of source/drains, 68–70, 125–27
　silicon gate, 15, 72–87
　SOCMOS, 91
Comparisons:
　ECL vs. CMOS, 442–44
　　speed-power, 93–94
　　differences in designing, 95–97
　drive capabilities, 95
　GaAs vs. silicon, 107
　Gate array vs.:
　　PAL, 16–17
　　standard cell, 18–20
　　full custom, 21
　　high W/Ls, 70–72
　junction isolated vs. oxide isolated:
　　bipolar, 98–99
　　CMOS, 100–101

logic vs. circuit simulators, 222–23
power dissipation (*see also* Feasibility):
　CMOS vs. bipolar, 94–95
　technology, 91
Configuration control, design, 172–73
Contacts:
　points, 8
　Schottky vs. ohmic, 47
Conversion, technology, 173–79
　to ISL/STL, 429–34
Correct by construction, 322–24
Cost and time drivers, 216–18
CPU times (*see also* Fault analysis), Appendix C
Current hogging, ISL, 434
Cursor movement devices, 322

D

Database, hierarchical, 324
DCFL, 108
Depletion mode, 52–53
Designing remotely, 344–47
Die, 2–3
Digitizing, 304–11
DIP, 2
Doping, 111
DSW, 460–61, 465

E

E-beam, 465
ECL (*see also* Comparisons), 43
　gate arrays:
　　power vs. speed, 50–51
Enhancement mode, 52–53
Epitaxy:
　LPE, 46
　LPCVD, 46
　MBE, 46
　MOVPE, 46
　region, 45–46
　VLE, 46

Equivalent gates, number, 16
Etching, 465–70
 using lasers, 467
 vs. liftoff, 467
Exercises:
 macro design, 369, 370, 373, 384–85, 386, 389, 394, 396, 402
  solutions, 362–86, 395
 routing, 328–34
  solutions, 331, 334

F

Fanout, extraction from netlist, 245
FAST, 43
Fault analysis simulation (*see also* Bad machine; CAD; Stuck-at), 280–96
 CPU times required, 291–92
Faults, undetectable, 288–89
Feasibility, 181–203
 array size, 187–89
  calculator, 188
  table, 189
 assessing, factors (table), 185
 partitioning, 152–53, 207–11
 pinouts, 191–93
  system, Rent's rule, 213–16
 propagation delays (*see also* CMOS), 197–203
  analysis, 200–203
  dependence on transition type, 200–201
 power dissipation (*see also* Comparisons), 194–96
 technology:
  parameters (table), 186
  selection based on logic function, 187
FET, field effect transistor (*see also* MOS; CMOS), 51–86
 cross-section, 7, 54, 61
 scaling, 59–60
 sizes, W/L, 56–58
 self-alignment, 60–64
 symbols, 53–54
 terminal connections, 52
 V-I curve, 55
Full custom, 21
Function cell, 16
Fuse programmable, 17

G

GaAs, 39
 air bridges, 110
 differences between silicon (*see* Comparisons: GaAs vs. silicon)
 gate arrays, table, 106
 gates of, 109
 logic forms, 187
 on silicon, 110–11
 structures, 107–10
  BFL, 108
  DCFL, 108
  SDFL, 108
Gate, three meanings, 16
Gate array, (*see also* Comparisons; CMOS, ECL, GaAs, ISL/STL):
 analog, 13
 CMOS (*see* CMOS)
 die, 8
 ECL (*see also* ECL), 12–13
 examples of use, 33–34
 families, 3, 9, 193–94
 freedoms and restrictions, 203–6
 generations, 15
 high voltage, 14
 memory in, 12–13
  non-volatile, 13
 microprocessors on, 13
 mixed analog and digital, 13
 permutations, 16–19
 representative, 12–14, Appendix A
 sea of gates, CMOS, 146–48
 standards, 36
 structured, LSI logic, 13
 technologies, Chapter 2

VHSIC, 35
  why not use, 32–33
  why use, 23–32
Gate length (*see also* Polysilicon):
  drawn, 56–58
  effective, 56–58
Gate matrix method, layout (*see* Layout)
Generations (*see* Gate array)
Grid systems, 118–48

H

Hardcopy, 353–56
HSCMOS, 9

I

IC, integrated circuit, 4
  mask programmable, example, 5
IGS, 298–311
Interconnects, 87–91
  distribution, 89–91
Interdigitated, 70–72
Ion implant, 46
I/Os, 2
ISL/STL, 47–49
  temperature and bias effects:
    ISL, 199
    STL, 50
Isolation methods:
  CDI, 99
  dielectric, 101–2
  gate 143–46
  junction vs. oxide (*see* Comparisons)
  silicon on insulator, 104–5
  trench 106

K

Kerf, 473

L

Latchup (*see* CMOS)
Layers, FETs, 377
Layout (*see also* Placement and routing):
  CMOS macro, 363–422
  gate matrix method, 340–43
  symbolic, 340–44
  Weinberger image, 339–40
Liftoff, (*see* Etching)
Logic cell array, Xilinx, 17
Logic structures:
  MOS, simple, 67–70

M

Macro, 11, 137–38, 153–54
  design with, 158–62
  design of:
    CMOS (*see also* Comparisons; CMOS), 358–422
    AOI, OAI, 373–89
    counter circuits, 387
    EX-OR, EX-NOR, 382–83, 393–95
    flip-flops, latches, TGs, 389–98
    floatable drivers, 399–400
    input structures, 400–403, 421–22
    MUX, 386, 397–98
    NAND, NOR, INV, 364–73
    output structures, 403–10
  ECL (*see also* Comparisons; ECL), 440–58
    ADDER, 453
    EX-OR, 452, 455
    Input/output structures, 456–57
    MUX, 452
    NAND/AND, 448–50
    OR/NOR, 453
  ISL/STL (*see also* ISL/STL; ECL), 423–44
    adder, 441–43
    AOI, 436–37
    NAND, NOR, 435–36
    D flip-flop, 441–42

ECL (*cont.*)
   EX-OR, EX-NOR, 436–38, 441
   MUX, 436–39
   1 of 4 encoders, 441
   R-S flip-flop, 436–40
   wire ANDing, 427–29
   overview, 358–61
   fixed vs. variable size, 160–61
   library, size, 160
   pastie, 161–62
   vs. design with transistors, 162–65
Macrocells, 16
Mask, 5-7, 459–62, 464–67
   checking, 165–66
   design, 149–51, 155–66
   DSW, 460–62
   floor plan, 156–58
   repair, 461–62
Master fault file, MFF, 280–82
Masterslice, 8
Mistakes, finding, correcting, 178–79
Mobility, 66, 113–14
   effective mass dependence, 114
   PFET greater than NFET, 114
MOS (*see also* FET; CMOS), 51–86
   PMOS, NMOS, 51–56
   PFET, NFET, 51–56
   PCH, NCH, 51–56
Multiwindowing, 316–22

N

NAND, NOR (*see* Macro design)
NCH (*see* MOS, CMOS)
Nesting, 245
   non-uniform, 248
NRE, non-recurring engineering, 19

O

OAI (*see* AOI)
Overlays, 460

P

Packages, 473–79
   chip carriers, 475–77
      thermal cracking, 477
   DIPs, 2, 476
   high power, 478–79
   PGA, pin grid array, 477–78
PAL, 17–18
Paralleling, macro design, 410–14
Partitioning (*see also* Feasibility), 152–53, 207–11
Pattern, 5
PCB, 12
PCH (*see* MOS, CMOS)
PG, pattern generator, 460–61
PGA (*see* Packages)
Physical modeling, simulation, 327
Pin grid array (*see* Packages)
Placement and routing, 327–39
   algorithms, 335–39
PMMA, 465
Polycell array, 18
Polycide, 89
Polyimide, 469
Polysilicon, 60–64, 75, 87–88, 105, 200, 202
Power and ground buses, 10
Power dissipation (*see* Comparisons; Feasibility)
PR, photoresist, 460–61, 464–67, 469
Probe card, 470–72
Propagation delays (*see* Feasibility)

R

RAM, Random Address Memory, 12
Refractory metals, gates, 89
Rent's rule, 213–15, 336–37
Reticle, 460–61
Retrofitting parts, 206–7
Ring oscillators, 102–4
Routing (*see also* Placement and Routing), 335–39, 419–21
Runout, 461

S

Scaling (*see also* CMOS), 59–60
Schmitt triggers, 414–15
SDFL, 108
Seed's model, yield, 473–74, 476
Self-alignment, 60–64
Selloff, 171
Semicustom, 8
Series gating, 447–48
Serpentine, 70–72
Share-a-chip, 203–5
Siemen, 64
Silicon compilers, 21–22
Silicon gate (*see* Polysilicon)
Silicon on insulator, SOI, 39
Simulation:
    vs. breadboarding, 154–55
Simulators (*see* CAD)
SiO2, glass, 10
Spinner, 464–65
Standard cells, 8, 18–20
    benefits, 20
    conversion of gate array to, 20
    die size vs. yield, 19–20
Stepper (*see* Mask, DSW)
Strings, FETs, 377
Stuck-at-model, 280
    adequacy of, 295
Symbolic layout (*see* layout)
    SWAMI technique, 101

T

Technologies, overview, 39–43
Temperature (*see* CMOS propagation delays; ISL/STL)
Test:
    die, wafer, 4
Testability (*see also* CAD), 273–80
Tester interface, 480–82
Testers, list, 480–82
Test vector, 283
Times, task (*see* Cost and Time Drivers)
Time sets, 481
Transconductance, 64–66

U

ULA, 11
Undetectable faults, 288–90

V

Vdd, Vss buses, 10
Vendor, interaction with, 168–71

W

Wafer, 3–6
    saw, dice, 472–73
    test, 470–72
Weinberger image technique (*see* Layout)
W/L ratio (*see also* FET sizes)
    obtaining high values: serpentine vs. interdigitated, 70–72
Wire ANDing:
    ECL, 448–50
    ISL/STL, 427–29
Wire ORing, ECL, 448
Workstations (*see also* CAD; Simulation), 311–27
    PCs used as, 325–27

X

X-ray lithography (*see* E-beam)
XRs, transistors, 53, 54